Antimicrobial Biomaterials: Recent Progress

Antimicrobial Biomaterials: Recent Progress

Editor

Helena Felgueiras

MDPI

Basel • Beijing • Wuhan • Barcelona • Belgrade • Novi Sad • Cluj • Manchester

Editor
Helena Felgueiras
Department of Textile
Engineering
University of Minho
Guimarães
Portugal

Editorial Office
MDPI AG
Grosspeteranlage 5
4052 Basel, Switzerland

This is a reprint of articles from the Special Issue published online in the open access journal *International Journal of Molecular Sciences* (ISSN 1422-0067) (available at: www.mdpi.com/journal/ijms/special_issues/DGFS7Y6GC1).

For citation purposes, cite each article independently as indicated on the article page online and as indicated below:

Lastname, A.A.; Lastname, B.B. Article Title. *Journal Name* **Year**, *Volume Number*, Page Range.

ISBN 978-3-7258-1660-6 (Hbk)
ISBN 978-3-7258-1659-0 (PDF)
doi.org/10.3390/books978-3-7258-1659-0

Contents

International Journal of
Molecular Sciences

MDPI

Editorial

Special Issue "Antimicrobial Biomaterials: Recent Progress"

Helena P. Felgueiras

Centre for Textile Science and Technology (2C2T), University of Minho, Campus de Azurém, 4800-058 Guimarães, Portugal; helena.felgueiras@2c2t.uminho.pt

Biomaterials have demonstrated their ability to serve as effective drug delivery platforms, enabling targeted and localized administration of therapeutic agents. The development of biomaterial-based drug delivery systems has been particularly beneficial in the management of various medical conditions [1,2]. In the field of cancer treatment, biomaterial-based nanoparticles and liposomes can precisely deliver chemotherapeutic agents to cancerous cells, thus reducing the impact on healthy tissues and minimizing side effects [3]. In the treatment of cardiovascular diseases, biomaterial-coated drug-eluting stents have been shown to release pharmaceutical agents locally, reducing the risk of restenosis [4]. Moreover, biomaterials have played a crucial role in overcoming the blood–brain barrier, allowing for neurodegenerative diseases like Alzheimer's and Parkinson's to be treated by target-directed drug delivery vehicles [5]. Biomaterial scaffolds in wound healing have also demonstrated the ability to facilitate cell growth while fighting infections [6,7].

The versatile nature of biomaterials continues to drive advancements in drug delivery, paving the way for more effective and patient-centered healthcare solutions [8,9]. Innovations in nanotechnology and gene therapy, for instance, have further enhanced personalized drug delivery strategies based on individual genetic profiles and disease characteristics [10,11]. Real-time monitoring and feedback systems can be engineered to dynamically adjust drug release rates, ensuring that patients receive precise doses when needed [12].

Biomaterials play a crucial role in the development of implantable devices and drug delivery platforms, significantly impacting patients' quality of life. The versatility of biomaterials allows for the engineering of a diverse array of devices and scaffolding systems, each tailored to specific patient demands. However, the long-term use of these implantable devices can be threatened by the adhesion and proliferation of microorganisms, leading to the formation of biofilms or fibrotic responses that can become highly cytotoxic and endanger the patients' quality of life [13]. Pathogenic microorganisms may cause local infection and, consequently, implant failure; further, they can hinder the delivery of therapeutic molecules by specialized delivery platforms, rendering them ineffective. Many alternatives have been proposed over the years to prevent such events, including the use of antiseptics and antibiotics or the physical modification of the biomaterial's surface, with the incorporation of bioactive agents of interest. From specialized polymers and functional groups to silver and, more recently, antimicrobial peptides and natural extracts, different functionalization/modification techniques have been employed in this fight against pathogenic invaders [8,14]. In fact, the interest in the loading of natural and synthetic antimicrobial agents into different inorganic, lipid, and polymeric-based nanomaterials has been reignited [15]. Not only that, but patients with specific infections, such as ventriculostomy-induced infections, have benefitted from the impregnation and decoration of biomaterials with antimicrobial agents [16]. Silver–porous aluminum oxide nanocomposites, for instance can be synthesized using microwave irradiation. The microwave treatment helps incorporate the silver nanoparticles into the porous alumina matrix, creating stable heterojunctions with improved antiseptic and bactericidal properties [17]. Additionally, antibiotics combined with nanomaterials have been shown to

Citation: Felgueiras, H.P. Special Issue "Antimicrobial Biomaterials: Recent Progress". *Int. J. Mol. Sci.* **2024**, *25*, 7153. https://doi.org/10.3390/ijms25137153

Received: 24 June 2024
Accepted: 28 June 2024
Published: 28 June 2024

surpass the problems with multi-drug resistance, turning obsolete antibiotic-based therapies into new and relevant approaches for fighting infections [18]. In another strategy, a low-temperature atmospheric argon plasma brush combined with antibiotic-like agents was used for oral bacterial deactivation [19]. More recently, antimicrobial peptide-based hydrogels were explored as alternative agents, as they are increasingly examined for their potent antimicrobial properties and practical applications, such as in wound healing [20].

The emergence of drug-resistant tuberculosis is a significant global health issue, with *Mycobacterium tuberculosis* being responsible for drug-resistant cases. Effective detection of this bacterium is a great challenge. Researchers have engineered super-selective primer-based real-time PCR assays, through which efficient diagnoses of hetero-resistant tuberculosis in clinical specimens can be accomplished, while contributing to the understanding of the bacterium's drug resistance mechanisms [21]. The challenge to discover new antimycobacterial agents against this bacterium has forced researchers to explore new classes of antimicrobial compounds, including repurposed drugs like gatifloxacin, moxifloxacin, and various natural compounds with anti-tuberculosis properties [22]. Aside from this, other infections have been the target of research for new antimicrobial formulations, some of which are engineered from natural resources, such as quaternized soy protein isolates, which have shown significant antibacterial properties and superior water solubility [23]. Biosynthesized silver chloride nanoparticles from the leaf extract of *Carduus crispus* have also been examined for their antibacterial efficacy, showing great potency against Gram-positive and Gram-negative bacteria [24].

In the food industry, the preservation of fruits and vegetables to prevent propagation of infections is a great challenge. However, some advances are being made, namely the application of compounds based on chitosan, phosphoric acid plus micronutrients, and sweet orange essential oil to coat the aliments and reduce decay. These coatings increase the antioxidant and flavonoid content at harvest and reduce the use of chemical–synthetic fungicides used for preservation [25]. Chitosan beads immobilized with cellobiose dehydrogenase have also been proposed for packaging in the food industry to improve antioxidant, antimicrobial, and cytotoxic properties [26]. Likewise, chitosan can be used for nanoencapsulation of different antimicrobial agents to improve the biological activity of packaging materials against various pathogens [27].

Funding: This work was funded by the European Regional Development Fund through the Operational Competitiveness Program and the Foundation for Science and Technology of Portugal (FCT) under the projects UID/CTM/00264/2020 of Centre for Textile Science and Technology (2C2T) on its components base (https://doi.org/10.54499/UIDB/00264/2020) and programmatic (https://doi.org/10.54499/UIDP/00264/2020). H.P.F. also acknowledges FCT for the auxiliary researcher contract 2021.02720.CEEIND (https://doi.org/10.54499/DL57/2016/CP1377/CT0098).

Conflicts of Interest: The author declares no conflicts of interest.

References

1. Felgueiras, H.P. Emerging antimicrobial and immunomodulatory fiber-based scaffolding systems for treating diabetic foot ulcers. *Pharmaceutics* **2023**, *15*, 258. [CrossRef]
2. Su, L.; Li, Y.; Liu, Y.; An, Y.; Shi, L. Recent advances and future prospects on adaptive biomaterials for antimicrobial applications. *Macromol. Biosci.* **2019**, *19*, 1900289. [CrossRef]
3. Pawar, V.; Maske, P.; Khan, A.; Ghosh, A.; Keshari, R.; Bhatt, M.; Srivastava, R. Responsive Nanostructure for Targeted Drug Delivery. *J. Nanotheranostics* **2023**, *4*, 55–85. [CrossRef]
4. Udriște, A.S.; Burdușel, A.C.; Niculescu, A.G.; Rădulescu, M.; Grumezescu, A.M. Coatings for Cardiovascular Stents-An Up-to-Date Review. *Int. J. Mol. Sci.* **2024**, *25*, 1078. [CrossRef]
5. Trucillo, P. Biomaterials for Drug Delivery and Human Applications. *Materials* **2024**, *17*, 456. [CrossRef]
6. Teixeira, M.A.; Antunes, J.C.; Seabra, C.L.; Tohidi, S.D.; Reis, S.; Amorim, M.T.P.; Felgueiras, H.P. Tiger 17 and pexiganan as antimicrobial and hemostatic boosters of cellulose acetate-containing poly (vinyl alcohol) electrospun mats for potential wound care purposes. *Int. J. Biol. Macromol.* **2022**, *209*, 1526–1541. [CrossRef] [PubMed]
7. Liang, Y.; Liang, Y.; Zhang, H.; Guo, B. Antibacterial biomaterials for skin wound dressing. *Asian J. Pharm. Sci.* **2022**, *17*, 353–384. [CrossRef]

8. Felgueiras, H.P.; Homem, N.C.; Teixeira, M.; Ribeiro, A.; Teixeira, M.O.; Antunes, J.; Amorim, M. Biodegradable wet-spun fibers modified with antimicrobial agents for potential applications in biomedical engineering. *J. Phys. Conf. Ser.* **2021**, *1765*, 012007. [CrossRef]
9. Pei, J.; Yan, Y.; Palanisamy, C.P.; Jayaraman, S.; Natarajan, P.M.; Umapathy, V.R.; Gopathy, S.; Roy, J.R.; Sadagopan, J.C.; Thalamati, D. Materials-based drug delivery approaches: Recent advances and future perspectives. *Green Process. Synth.* **2024**, *13*, 20230094. [CrossRef]
10. Mirza, Z.; Karim, S. Nanoparticles-based drug delivery and gene therapy for breast cancer: Recent advancements and future challenges. *Semin. Cancer Biol.* **2021**, *69*, 226–237. [CrossRef]
11. De Matteis, L.; Martín-Rapún, R.; de la Fuente, J.M. Nanotechnology in personalized medicine: A promising tool for alzheimer's disease treatment. *Curr. Med. Chem.* **2018**, *25*, 4602–4615. [CrossRef] [PubMed]
12. Awwad, S.; Ibeanu, N.; Liu, T.; Velentza-Almpani, A.; Chouhan, N.; Vlatakis, S.; Khaw, P.T.; Brocchini, S.; Bouremel, Y. Real-Time Monitoring Platform for Ocular Drug Delivery. *Pharmaceutics* **2023**, *15*, 1444. [CrossRef] [PubMed]
13. Khatoon, Z.; McTiernan, C.D.; Suuronen, E.J.; Mah, T.F.; Alarcon, E.I. Bacterial biofilm formation on implantable devices and approaches to its treatment and prevention. *Heliyon* **2018**, *4*, e01067. [CrossRef] [PubMed]
14. Tavares, T.D.; Antunes, J.C.; Padrão, J.; Ribeiro, A.I.; Zille, A.; Amorim, M.T.P.; Ferreira, F.; Felgueiras, H.P. Activity of specialized biomolecules against gram-positive and gram-negative bacteria. *Antibiotics* **2020**, *9*, 314. [CrossRef] [PubMed]
15. Mercan, D.-A.; Niculescu, A.-G.; Grumezescu, A.M. Nanoparticles for Antimicrobial Agents Delivery—An Up-to-Date Review. *Int. J. Mol. Sci.* **2022**, *23*, 13862. [CrossRef] [PubMed]
16. Diop, S.; Roujansky, A.; Kallel, H.; Mounier, R. Prevention of Ventriculostomy Related Infection: Effectiveness of Impregnated Biomaterial. *Int. J. Mol. Sci.* **2023**, *24*, 4819. [CrossRef] [PubMed]
17. Bahadur, N.M.; Furusawa, T.; Sato, M.; Kurayama, F.; Siddiquey, I.A.; Suzuki, N. Fast and facile synthesis of silica coated silver nanoparticles by microwave irradiation. *J. Colloid Interface Sci.* **2011**, *355*, 312–320. [CrossRef] [PubMed]
18. Adeniji, O.O.; Nontongana, N.; Okoh, J.C.; Okoh, A.I. The Potential of Antibiotics and Nanomaterial Combinations as Therapeutic Strategies in the Management of Multidrug-Resistant Infections: A Review. *Int. J. Mol. Sci.* **2022**, *23*, 15038. [CrossRef]
19. Malcangi, G.; Patano, A.; Trilli, I.; Piras, F.; Ciocia, A.M.; Inchingolo, A.D.; Mancini, A.; Hazballa, D.; Di Venere, D.; Inchingolo, F.; et al. Therapeutic and Adverse Effects of Lasers in Dentistry: A Systematic Review. *Photonics* **2023**, *10*, 650. [CrossRef]
20. Copling, A.; Akantibila, M.; Kumaresan, R.; Fleischer, G.; Cortes, D.; Tripathi, R.S.; Carabetta, V.J.; Vega, S.L. Recent Advances in Antimicrobial Peptide Hydrogels. *Int. J. Mol. Sci.* **2023**, *24*, 7563. [CrossRef]
21. Günther, G.; Ruswa, N.; Keller, P.M. Drug-resistant tuberculosis: Advances in diagnosis and management. *Curr. Opin. Pulm. Med.* **2022**, *28*, 211–217. [CrossRef] [PubMed]
22. Shehzad, A.; Rehman, G.; Ul-Islam, M.; Khattak, W.A.; Lee, Y.S. Challenges in the development of drugs for the treatment of tuberculosis. *Braz. J. Infect. Dis.* **2013**, *17*, 74–81. [CrossRef] [PubMed]
23. Rani, S.; Kumar, R. A review on material and antimicrobial properties of soy protein isolate film. *J. Polym. Environ.* **2019**, *27*, 1613–1628. [CrossRef]
24. Urnukhsaikhan, E.; Bold, B.-E.; Gunbileg, A.; Sukhbaatar, N.; Mishig-Ochir, T. Antibacterial activity and characteristics of silver nanoparticles biosynthesized from Carduus crispus. *Sci. Rep.* **2021**, *11*, 21047. [CrossRef] [PubMed]
25. Duda-Chodak, A.; Tarko, T.; Petka-Poniatowska, K. Antimicrobial Compounds in Food Packaging. *Int. J. Mol. Sci.* **2023**, *24*, 2457. [CrossRef] [PubMed]
26. Malhotra, B.; Keshwani, A.; Kharkwal, H. Antimicrobial food packaging: Potential and pitfalls. *Front. Microbiol.* **2015**, *6*, 611. [CrossRef]
27. Flórez, M.; Guerra-Rodríguez, E.; Cazón, P.; Vázquez, M. Chitosan for food packaging: Recent advances in active and intelligent films. *Food Hydrocoll.* **2022**, *124*, 107328. [CrossRef]

International Journal of *Molecular Sciences*

MDPI

Article

Synthesis and Characterization of a Sustained Nitric Oxide-Releasing Orthodontic Elastomeric Chain for Antimicrobial Action

Alec McDonald [1], Carly Warden [1], Jinlian Tan [2], Kellianne M. Piell [3], Jill M. Steinbach-Rankins [4], Nandakumar Janakiraman [5], David A. Scott [2], Marsha P. Cole [3,*] and Sudha Gudhimella [1,*]

1 Department of Orthodontics, University of Louisville School of Dentistry, Louisville, KY 40202, USA; alec.mcdonald@louisville.edu (A.M.); cewarden92@gmail.com (C.W.)
2 Department of Oral Immunology and Infectious Diseases, University of Louisville, Louisville, KY 40202, USA; jinlian.tan@louisville.edu (J.T.); david.scott@louisville.edu (D.A.S.)
3 Department of Biochemistry and Molecular Genetics, Louisville, KY 40202, USA; kellianne.piell@louisville.edu
4 Department of Pharmacology and Toxicology, University of Louisville School of Medicine, Louisville, KY 40202, USA; jmstei01@louisville.edu
5 Georgia School of Orthodontics, Atlanta, GA 30350, USA; njanakiraman@gaorthodontics.org
* Correspondence: marcie.cole@louisville.edu (M.P.C.); s0gudh01@louisville.edu (S.G.)

Abstract: The acidic byproducts of bacteria in plaque around orthodontic brackets contribute to white spot lesion (WSL) formation. Nitric oxide (NO) has antibacterial properties, hindering biofilm formation and inhibiting the growth of oral microbes. Materials that mimic NO release could prevent oral bacteria-related pathologies. This study aims to integrate S-nitroso-acetylpenicillamine (SNAP), a promising NO donor, into orthodontic elastomeric ligatures, apply an additional polymer coating, and evaluate the NO-release kinetics and antimicrobial activity against *Streptococus mutans*. SNAP was added to clear elastomeric chains (8 loops, 23 mm long) at three concentrations (50, 75, 100 mg/mL, and a control). Chains were then coated, via electrospinning, with additional polymer (Elastollan®) to aid in extending the NO release. NO flux was measured daily for 30 days. Samples with 75 mg/mL SNAP + Elastollan® were tested against *S. mutans* for inhibition of biofilm formation on and around the chain. SNAP was successfully integrated into ligatures at each concentration. Only the 75 mg/mL SNAP chains maintained their elasticity. After polymer coating, samples exhibited a significant burst of NO on the first day, exceeding the machine's reading capacity, which gradually decreased over 29 days. Ligatures also inhibited *S. mutans* growth and biofilm formation. Future research will assess their mechanical properties and cytotoxicity. This study presents a novel strategy to address white spot lesion (WSL) formation and bacterial-related pathologies by utilizing nitric oxide-releasing materials. Manufactured chains with antimicrobial properties provide a promising solution for orthodontic challenges, showing significant potential for academic-industrial collaboration and commercial viability.

Keywords: orthodontic brackets; S-Nitroso-N-Acetylpenicillamine (SNAP); cytotoxicity; orthodontic ligatures; dental plaque; antibacterial chains; elastomeric chainswhite spot lesions; nitric oxide; SNAP

Citation: McDonald, A.; Warden, C.; Tan, J.; Piell, K.M.; Steinbach-Rankins, J.M.; Janakiraman, N.; Scott, D.A.; Cole, M.P.; Gudhimella, S. Synthesis and Characterization of a Sustained Nitric Oxide-Releasing Orthodontic Elastomeric Chain for Antimicrobial Action. *Int. J. Mol. Sci.* **2024**, *25*, 6982. https://doi.org/10.3390/ijms25136982

Academic Editor: Helena Felgueiras

Received: 29 April 2024
Revised: 10 June 2024
Accepted: 11 June 2024
Published: 26 June 2024

1. Introduction

1.1. Oral Hygiene Barrier and Cariogenic Plaque

Obstructed oral hygiene is a consequence of orthodontic treatment with fixed appliances [1]. Therefore, treatment with fixed appliances may be a risk factor for plaque accumulation and associated pathologies [2–4]. Appliances, such as brackets, wires, coils, and ligatures, hinder access to tooth surfaces and make mechanical cleaning more complex. Additionally, the effectiveness of physiological self-cleansing by the tongue, cheeks, and saliva is reduced [5]. Moreover, after orthodontic treatment, a shift in the bacterial flora is often observed, resulting in higher concentrations of acidogenic bacteria, such as *S. mutans* and *Lactobacilli* [6,7]. The cariogenic potential of these microorganisms is strengthened by their ability to live and proliferate on hard, non-shedding tooth surfaces [8]. Increasing bacterial loads, coupled with sugar intake, cause plaque pH to drop as bacteria ferment dietary carbohydrates, such as sucrose, into acids, contributing to the reduction [8]. A reduced pH below the critical value (5.0–5.5) may lead to the demineralization of hydroxyapatite crystals in the tooth enamel, which eventually results in dental caries and white spot lesions (WSL) [9].

Enamel demineralization is one of the most common consequences of biofilm formation among patients receiving fixed orthodontic treatment, affecting 50% to 70% of patients [1,10,11]. According to a longitudinal clinical study conducted by Zachrisson et al., it was found that all 75 patients who underwent fixed orthodontic treatment exhibited higher mean values for visible plaque, inflammation, and gingival recession compared to their pre-treatment values [2]. These findings are consistent with a study by Boke et al., which also demonstrated significant increases in plaque buildup, gingival inflammation, and recession of gingiva after treatment [12]. Poor oral hygiene and increased plaque have been recognized as contributing factors to prolonged treatment time and compromised treatment outcomes, according to a retrospective study by Beckwith et al. [13]. The study collected data from 140 subjects who completed treatment with fixed appliances in five orthodontic offices. Progress notes with entries indicating less than "good" oral hygiene were linked to an additional two-thirds of a month in estimated treatment time [13]. Therefore, the introduction of antibacterial orthodontic appliances and accessories may help improve oral hygiene, reduce plaque accumulation, and shorten treatment duration.

1.2. White Spot Lesions

Enamel decalcification can become apparent around the fourth week of orthodontic treatment in cases of poor oral hygiene [14]. These decalcifications often manifest as white spot lesions (WSL) appearing as small chalky lines along the bracket periphery [15]. According to a meta-analysis by Venkatachalapathy et al., WSL occurrence in fixed orthodontic treatment is common, with an incidence rate of 45.8% and a prevalence rate of 68.4% [16]. White spot lesions (WSL) represent the first clinical evidence of demineralization and incipient lesions, which can potentially further progress into dentin as caries [17]. As such, it is crucial for orthodontists and dentists to address these lesions and their contributing factors before they become more destructive. Managing WSLs requires a multifactorial approach that includes preventing demineralization, reducing biofilm proliferation, and remineralizing lesions, as well as esthetic-focused approaches such as thinning, micro-abrasion, resin infiltration, and whitening [18].

Implementing effective oral hygiene is the foremost approach to addressing WSLs [19]. The use of adjunctive therapies to supplement insufficient plaque control has been studied, but their protocols and efficacy remain ambiguous [20,21]. Common to orthodontic practices, a one-time application of fluoride varnish prior to treatment did not provide any additional WSL prevention compared to good oral hygiene [19–21]. However, increasing the frequency of application has shown some success [22]. A Cochrane review found that applying fluoride varnish every six weeks during orthodontic treatment can aid in remineralization and prevent further demineralization [22]. In pursuit of sustained fluoride deposition, researchers have explored the use of fluoride-releasing bonding systems at the base of orthodontic brackets [23,24]. Several studies have shown that such systems effectively reduce enamel demineralization. Furthermore, it has been noted that fluoride-releasing adhesives do not compromise the bond strength of orthodontic brackets. However, further research is necessary to determine the long-term effects and cost-effectiveness of these adhesives.

Studies involving orthodontic elastomeric ligatures with fluoride-releasing properties have produced varying results. For instance, one study found that the use of fluoridated elastomers was ineffective in reducing streptococcal and/or anaerobic bacterial growth over a clinically relevant period [25]. This randomized, prospective, longitudinal clinical trial included 30 subjects and had a split-mouth crossover design. After a mean treatment period of 40 days, a logistic regression model was used to analyze the presence or absence of streptococcal or anaerobic bacteria in relation to the use of fluoridated elastomers. Results revealed that the use of fluoridated elastomers did not significantly reduce either streptococcal (p = 0.288) or anaerobic bacterial counts (p = 0.230) [25]. In a study by Miura et al., it was demonstrated that the use of fluoridated elastomers was insufficient at preventing enamel decalcification [26]. Additionally, it was suggested the incorporation of fluoride may negatively impact the elastomer's physical properties, resulting in faster deterioration [27–29]. It is generally regarded that the anticaries action of fluoride is related to remineralization, rather than reducing the overall number and concentration of cariogenic bacteria [30,31]. Perhaps a shift in adjunctive therapies and appliances should be directed towards the etiology of the demineralization and the introduction of effective means to reduce biofilm populations, rather than targeting the palliative measures of fluoride at the enamel surfaces.

1.3. Biological Implications of Nitric Oxide

Until the late 1980s, nitric oxide (NO) was primarily viewed as an environmental pollutant, generated by industrial processes, automobile exhausts, and electrical storms, as well as a toxic component of cigarette smoke [32–35]. However, Ignarro and Furchgott's research in the 1980s recognized NO's role in physiology, revealing it to be the endothelium-derived relaxing factor responsible for vasodilation and blood pressure regulation [36–38]. Since then, researchers have uncovered numerous functions of nitric oxide and acknowledge NO's role in preventing platelet activation and adhesion, inhibiting bacterial proliferation and biofilm formation, and facilitating signaling within the immune system [33,39–41]. Nitric oxide's antibacterial properties make it a promising candidate for future medical device development. NO exhibits both nitrosative and oxidative effects, producing a range of reactive species that disrupt cellular functions and structures by interacting with microbial proteins, DNA, and enzymes [33,42]. At extreme concentrations, the cytotoxic effects of nitric oxide may become apparent, largely dependent on its reactivity with various radicals, more specifically superoxide ($O_{2\bullet}^{-}$) in the formation of peroxynitrite ($ONOO^{-}$) [43]. Peroxynitrite is a highly reactive nitrogen species (RNS) that can cause significant damage to cellular components, including lipids, proteins, and DNA [44]. It has been shown that peroxynitrite may cause oxidative stress and tissue damage, as well as dysregulation of cellular signaling and impaired mitochondrial function [45–47]. Yet, NO is generally considered safe at lower concentrations due to controlled dosing and targeted delivery [48]. Furthermore, nitric oxide has demonstrated effective activity against both Gram-positive and Gram-negative bacteria, such as *Staphylococcus aureus*, *Escherichia coli*, and *Pseudomonas aeruginosa* [49].

Therapies leveraging NO's actions could prove beneficial for those with associated health complications and diseases, including high blood pressure, hypertension, endothelial dysfunction, and bacteria-related pathologies [50,51]. Biomedical researchers are interested in developing modified materials for the external delivery of nitric oxide (NO) to specific sites [52]. The nitric oxide donor, S-nitroso-acetylpenicillamine (SNAP), is acclaimed for its ability to integrate into elastomeric materials like silicones and polyurethanes while exhibiting long-term stability and retention following sterilization techniques [53–55] Nitric oxide is released from these materials through processes such as thermal decomposition, metal catalysis, or exposure to light [56]. Recent studies involving the incorporation of SNAP into catheters and various medical devices have produced encouraging results by displaying antimicrobial defense against *S. aureus*, *Staphylococcus epidermidis*, and *Proteus mirabilis* [57,58].

1.4. Previous Findings and Addition of Polymer

In the first part of this research, carried out by Warden et al., orthodontic elastomeric chains (6 loops, 17 mm long) were treated with 75 mg of SNAP using the swell–encapsulation–shrink method [59]. Initially, three solutions with concentrations of 75, 125, and 200 mg/mL were developed, along with controls. However, once washed and dried, only the ligatures soaked in the 75 mg/mL SNAP solution, as well as the controls, maintained their elastic properties. Therefore, the study continued using only the 75 mg/mL SNAP chains. Following successful impregnation of the SNAP, NO flux was measured daily for 2 h/day for 4 days to determine the duration and magnitude of NO release. The samples demonstrated nitric oxide release for three days and showed no release over baseline on the fourth day [59]. The experimental chains in Warden's study also demonstrated a good level of efficacy in defending against the oral pathogen *S. mutans* [59]. Biofilm formation on surfaces adjacent to the ligatures and planktonic growth in solution were both significantly decreased with experimental chains compared to controls ($p < 0.05$) [59]. While the chains demonstrated bactericidal effects and nitric oxide release, it is important to note that most common orthodontic practices replace patients' elastomeric ligatures every 4–6 weeks [60]. Therefore, for this application to become more relevant in clinical settings, the NO-release period should be further extended. Orthodontic elastomeric chains demonstrating NO release for several weeks to a month may be more appropriate. Sustained nitric oxide release can be achieved in two ways: (1) by interfering with the factors initiating the NO release from SNAP and (2) by introducing a barrier to prevent NO from escaping. Since SNAP's nitric oxide-donating capabilities are driven by light, it can be inferred that limiting the amount of light contacting the SNAP-impregnated chain may result in less expulsion of NO [56].

Application of an additional polymer coating to the SNAP-impregnated ligatures may provide a means for sustained release of nitric oxide. Elastollan® (BASF, Ludwigshafen, Germany) is a thermoplastic polyurethane used in various commercial products due to its favorable properties, such as high wear and abrasion resistance, high tensile strength, and resistance to tear propagation, making it a viable option in orthodontic materials [61]. Coating the SNAP-impregnated ligatures with fine polymer fibers through controlled dispensing techniques like electrospinning could enable predictable and consistent amounts of coating. Electrospinning, a voltage-driven fabrication process, utilizes a specific electrohydrodynamic phenomenon to produce small fibers from a polymer solution [62]. The technology has been successfully used to generate fibers from various materials, including natural and synthetic polymers in the form of solution or melt [63].

With the deficient amount of literature regarding the integration of nitric oxide-releasing compounds for antibacterial action in orthodontics, we aimed to expand upon the work of Warden et al. in a combined attempt to fabricate an antibacterial elastomeric chain of clinical relevance. In this study, we hypothesize that incorporating the NO-donor molecule, S-nitroso-acetylpenicillamine, into existing orthodontic elastomeric ligatures and applying an additional polymer coating via electrospinning will result in a sustained release of nitric oxide (NO) over a thirty-day period. Furthermore, we predict that these modified ligatures will exhibit antimicrobial defense effects against *Streptococcus mutans*.

2. Results

2.1. SNAP Impregnation

SNAP was successfully incorporated into existing clear elastomeric ligatures at three concentrations. All experimental and control samples returned to the same dimensions after drying and experimental chains exhibited a light-to-dark gradient of green color from lowest to highest concentration, respectively (Figure 1). The masses of ligatures were measured prior to and following impregnation (Table 1). The Spearman correlation coefficient was calculated to evaluate the strength of the relationship between the concentration of added SNAP and the change in chain mass. The coefficient was found to be 0.88 ($p < 0.001$), indicating a strong positive correlation between the two variables (Figure 2). The controls exhibited no increase in mass. In part 1 of this study, conducted by Warden et al., chains soaked in concentrations of 125 and 200 mg/mL SNAP/THF broke under tensile stress [59]. Ligatures incubated in 75 mg/mL solution as well as control chains maintained their elastic properties; therefore, in this experiment, all remaining procedures were carried out using the 75 mg/mL SNAP/THF samples.

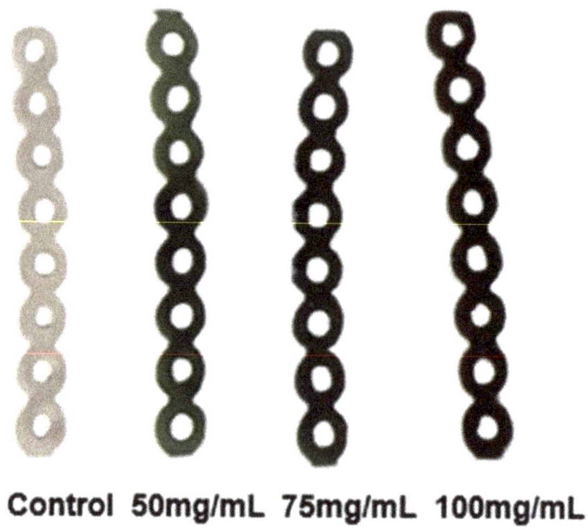

Figure 1. Experimental chains following SNAP impregnation demonstrating a light-to-dark gradient of green color.

Table 1. Average mass of an elastomeric chain (n = 6) before and after the addition of SNAP. Control chains were incubated in THF and dried under hood for 48 h.

	Before (mg)	After (mg)	% Change
Control (n = 6)	22.65 (SD = 0.63)	22.17 (SD = 0.55)	−2.12
50 mg/mL (n = 6)	22.50 (SD = 0.71)	24.94 (SD = 0.53)	+10.84
75 mg/mL (n = 6)	22.72 (SD = 0.77)	26.38 (SD = 0.77)	+16.11
100 mg/mL (n = 6)	22.42 (SD = 0.75)	27.06 (SD = 0.70)	+20.70

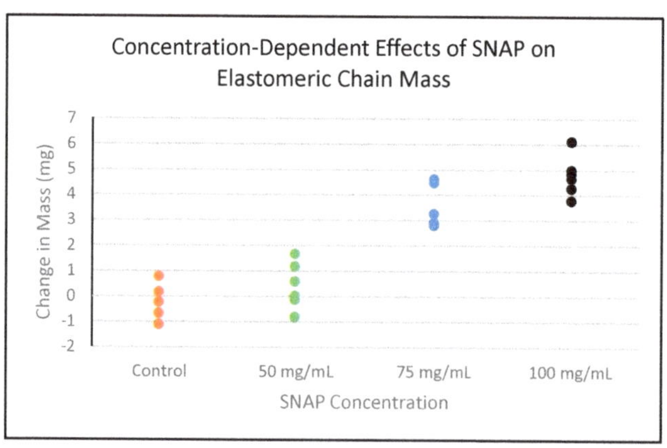

Figure 2. Scatterplot showing the relationship between different concentrations of SNAP and the corresponding changes in chain mass after drying. The Spearman correlation coefficient was found to be 0.88 ($p < 0.001$), indicating a strong positive correlation.

2.2. Polymer Electrospinning

Experimental chains (75 mg/mL SNAP/THF) were successfully coated with the Elastollan® polymer solution (Figure 3). Once dried, the samples exhibited similar stretch compared to their non-coated predecessors and controls.

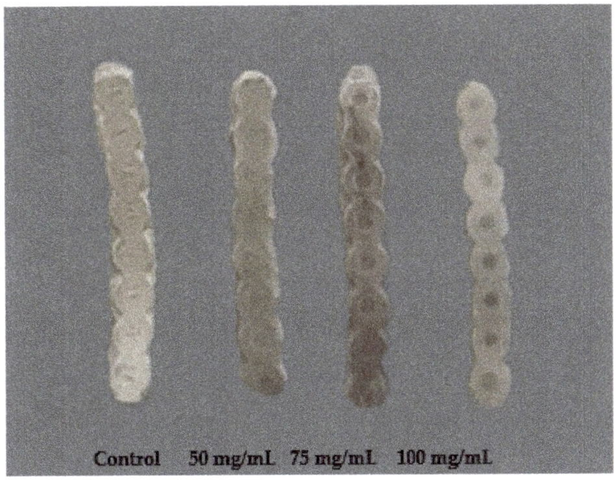

Figure 3. Elastollan® coated experimental chains.

2.3. Nitric Oxide Release Testing

The experimental ligature (75 mg/mL SNAP/THF + Elastollan®) exhibited a large burst of NO release followed by a sharp decrease and then steady-state NO flux at 2 h on each day of testing with the NOA (Figure 4). On the first day of testing, the chain exhibited NO released above the reading capability of the machine and maintained this reading. This flux gradually decreased for twenty-nine days until showing no NO release over baseline on the thirtieth day of testing.

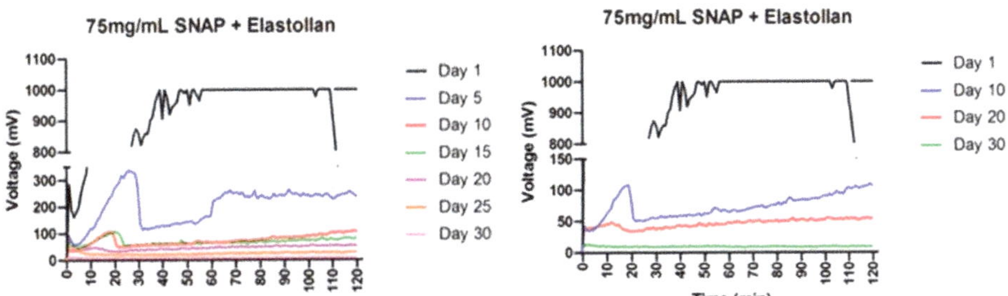

Figure 4. NO-release measurements. Every 5-day reading (**left**), every 10-day reading (**right**).

2.4. Bacterial Testing

2.4.1. Planktonic Growth in Surrounding Solution

Growth curves in solutions containing experimental samples or SNAP-only ligatures vs. control exhibited markedly lower growth by 24 h and inhibition was maintained for 120 h (Figure 5). On average, at 24 h, vials containing experimental chains experienced a 24.39% decrease in planktonic growth compared to control chains. At 120 h, vials containing experimental ligatures experienced a 35.56% decrease compared to controls.

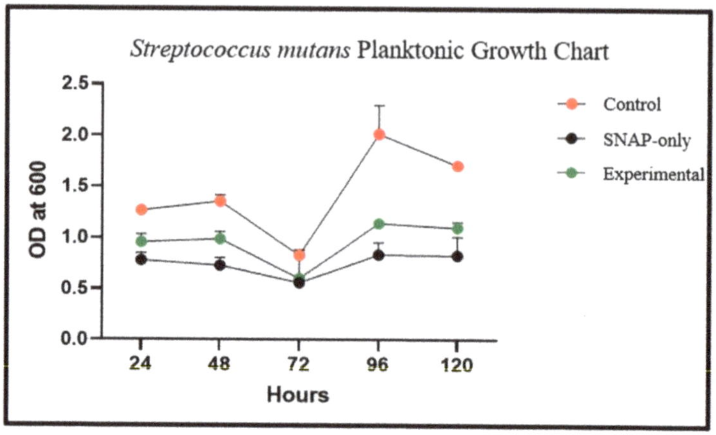

Figure 5. *S. mutans* planktonic growth curves measured at 24, 48, 72, 96, and 120 h. SNAP-only chains were impregnated with 75 mg/mL SNAP/THF. Experimental chains were impregnated with 75 mg/mL SNAP/THF and coated with Elastollan®.

2.4.2. Biofilm Formation on Surrounding Surfaces

After plating 12-well plates with *S. mutans* biofilm and staining with crystal violet, wells containing experimental samples or SNAP-only chains exhibited a notable reduction in biofilm formation compared to controls (Figure 6). Statistical analysis revealed a significant decrease in biofilm formation on the well surface for both experimental and SNAP-only chains at 24, 48, and 72 h ($p < 0.0001$) (Figure 7). Additionally, there was no statistically significant difference in biofilm formation on the wells' surfaces between those containing a coated experimental chain and those containing a SNAP-only chain. This indicates that the additional coating did not interfere with the bactericidal activity.

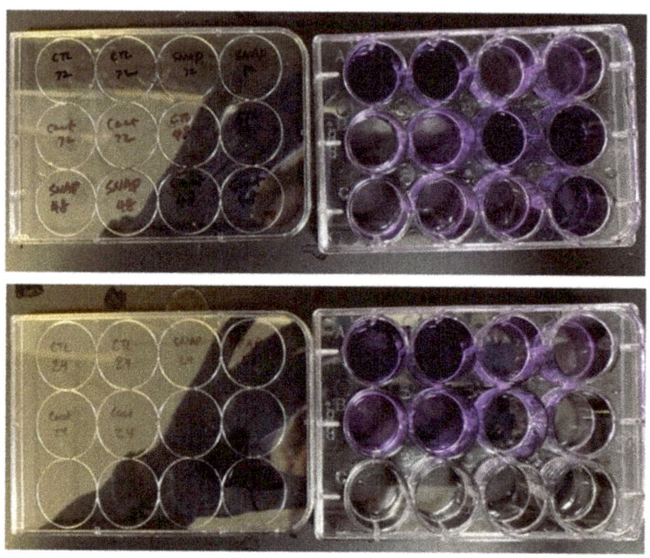

Figure 6. Twelve-well plate with *S. mutans* biofilms and crystal violet staining. All tests were performed in duplicate.

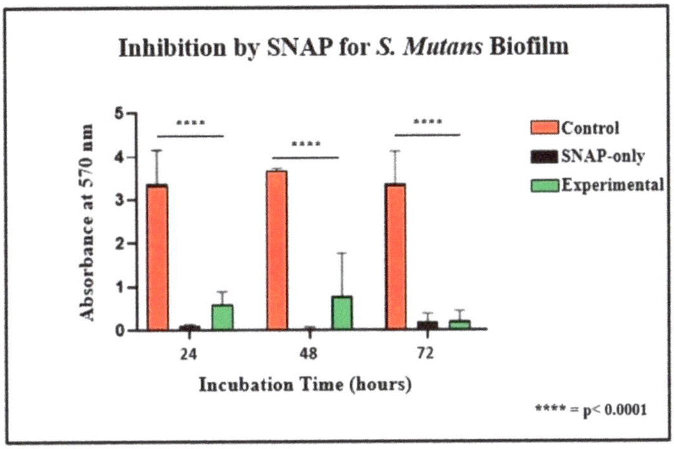

Figure 7. Biofilm formation on surfaces adjacent to the chains at 24, 48, and 72 h.

3. Discussion

Maintaining proper oral hygiene practices with fixed orthodontic appliances can be challenging and, if overlooked, may lead to various detrimental effects on the teeth and supporting structures. The rough and irregular surfaces of the fixed appliances create new areas for bacterial adhesion and proliferation, making brushing and flossing more complex [5,64]. These bacteria can multiply and release harmful metabolic byproducts that can damage the enamel surface and make the teeth more susceptible to irreversible harm. The initial sign of such damage appears as white spot lesions (WSLs), which have a white chalky appearance and are caused by enamel decalcification resulting from the acidic metabolites of fermentable sugars by bacteria. While esthetically unfavorable, WSLs also provide a window for further irreversible tooth damage. If left untreated, the incipient lesion may progress to the underlying dentin and result in dental caries and cavitation. Addressing WSLs promptly can help prevent more severe dental problems from developing [17].

Efforts to increase oral hygiene literacy and adjunctive oral hygiene practices in patients with fixed orthodontic appliances have been implemented but their effectiveness remains uncertain [24,25]. Fluoridated bonding agents and elastomers can serve as corrective measures aimed at remineralizing damaged enamel; however, targeting the etiology of white spot lesions rather than remedial therapies may be more beneficial. As a key contributor responsible for the development of white spot lesions, *Streptococcus mutans* demonstrates susceptibility to the endogenously produced compound, nitric oxide (NO) [33,39]. Nitric oxide displays a variety of physiological functions, including roles in vasodilation and bacterial suppression [39]. Research exploring the biomedical use of nitric oxide's antibacterial properties has been conducted but is deficient in dentistry and orthodontics [54,56–58]. Introducing materials with antibacterial properties of NO offers the potential to suppress bacterial-related pathologies, such as white spot lesions.

S-nitroso-acetylpenicillamine (SNAP) is a nitric oxide donor, responsive to light, that can be incorporated into rubber-like materials, including silicones and polyurethanes, and has effectively demonstrated bacterial suppression in studies [43,54,56,57,65,66]. By incorporating SNAP into orthodontic elastomeric ligatures and subjecting them to ordinary light exposure, significant nitric oxide release can occur, effectively repressing bacterial proliferation. However, the efficacy of these chains depends on several factors, such as the amount of SNAP loaded into the material, SNAP's degradation rate, and the susceptibility of the microbe to the nitric oxide molecule. In this study, experimental orthodontic ligatures were impregnated with 75 mg of SNAP/THF, coated with an additional Elastollan® polymer, and tested for their nitric oxide-releasing capabilities and antibacterial effectiveness. This study builds upon Part 1 of the initial research conducted by Warden et al., which demonstrated the release of nitric oxide from elastomeric ligatures for three consecutive days [59]. The experimental samples in this study were fabricated with the intent of sustained nitric oxide release. Compared to the control chains, the coated experimental samples in this study demonstrated nitric oxide release for twenty-nine days until showing no release over baseline on the thirtieth day. The experimental ligatures utilized in this study exhibited a significant reduction in *Streptococcus mutans* biofilm production on surfaces adjacent to the chains compared to their controls ($p < 0.0001$). Additionally, solutions containing the experimental samples exhibited reduced *S. mutans* planktonic growth. Moreover, these ligatures exhibited no statistically significant differences in biofilm inhibition compared to their non-coated predecessors from Part 1 completed by Warden et al. [59]. These data suggest that the addition of a polymer coating did not limit the antibacterial effectiveness to a sub-functional status. The rise in optical density (OD600) readings at 96 h may be indicative of the regrowth of residual bacteria. Future studies should investigate the effect of subculturing *S. mutans* as time progresses, followed by OD600 measurement.

If the strength and elasticity of these experimental ligatures remain uncompromised, this technology could have the potential to reduce the prevalence of plaque-mediated pathologies associated with fixed orthodontic appliances. By incorporating compounds that release products naturally found in human physiology, such as nitric oxide, greater biocompatibility and patient acceptance may be achieved. A limitation of this study included only characterizing the functions of chains impregnated with one concentration of SNAP. Given more time to conduct this study, samples of varying concentrations should be prepared and tested for their nitric oxide expulsion and antibacterial effectiveness, as well. The experimental sample in this study was chosen to remain consistent with the findings of Warden et al. [59]. In their study, it was noted that elastomeric chains soaked in higher concentrations of 125 and 200 mg/mL SNAP broke under tensile stress; therefore, the remainder of their data was gathered using ligatures composed of 75 mg/mL of SNAP. Also, adjustments to the thickness of the Elastollan® polymer coating could yield different NO and antibacterial effects. However, it is important to maintain a thickness that still renders the elastomeric chain functionable in orthodontics.

Finally, the future goals of this project are aimed at assessing the functional capacity of the ligatures in an orthodontic setting as well as evaluating their biocompatibility in the oral cavity. Future goals also include testing their antibacterial action against other common oral microbes, such as *Porphyromonas gingivalis* and *Streptococcus gordonii*, as well as testing antibacterial action for a longer period, such as thirty days. An antibacterial assessment of longer duration may reinforce the nitric oxide expulsion data revealed in this study. As the experimental chains in this study released quantitative amounts of nitric oxide for twenty-nine days, it is our aim for bacterial growth to be stunted for a period that parallels this duration.

4. Materials and Methods

The following methods for SNAP Impregnation, Nitric Oxide Release Testing, and Bacterial Testing were adapted from Warden et al. [59].

4.1. SNAP Impregnation

S-nitroso-acetylpenicillamine (Pharmablock, Shanghai, China) was incorporated into an existing clear elastomeric ligature (8 loops, 23 mm long; J-120 Closed Clear Energy Chain, Rocky Mountain Orthodontics, Denver, CO, USA) using a swell–encapsulation–shrink method (Figure 8). SNAP was dissolved in 5 mL glass vials containing 4 mL tetrahydrofuran (THF) (Sigma-Aldrich, St. Louis, MO, USA) to produce three solutions at concentrations of 50, 75, and 100 mg/mL, respectively. Control vials contained 4 mL THF without SNAP. Both experimental and control ligatures were immersed in these solutions for 1 h, then removed, washed with PBS buffer (Fisher Scientific, Waltham, MA, USA), and left to dry in darkness for 48 h within a fume hood. Successful impregnation was visually confirmed by observing the characteristic green color of SNAP (Figure 9). The masses of the samples were measured before and after impregnation. The Spearman correlation coefficient was then calculated to assess the strength of the relationship between the concentration of added SNAP and the change in chain mass. Subsequently, elastomeric chains were stored in dark, dry aluminum foil wrappings.

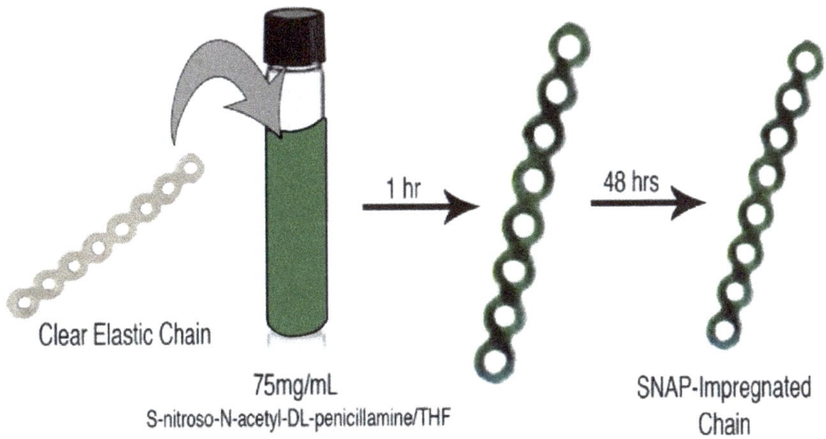

Figure 8. Schematic of swell–encapsulation–shrink method for fabricating chains.

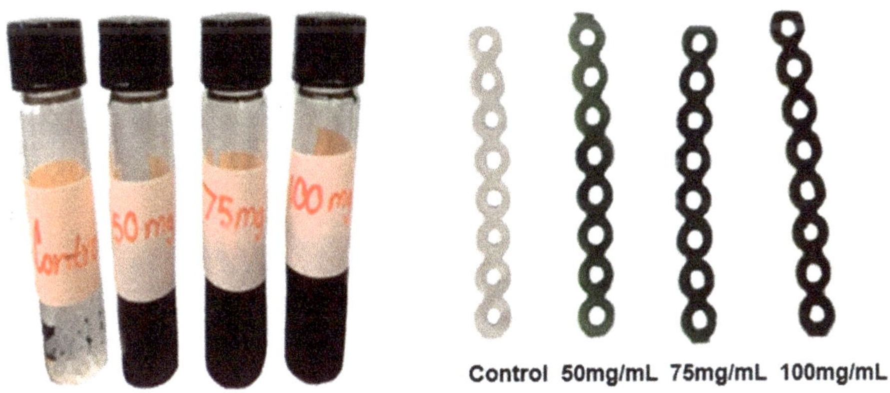

Figure 9. Chains incubating in SNAP + THF solutions at varying concentrations (**left**). SNAP-impregnated chains at varying solutions (**right**).

4.2. Polymer Electrospinning

Via electrospinning, experimental ligatures (75 mg/mL SNAP/THF) were coated with an additional Elastollan® (BASF, Ludwigshafen, Germany) polymer matrix. This concentration was chosen to remain consistent with Warden et al.'s study [59]. Given the opportunity, samples of 50 mg/mL and 100 mg/mL were coated and tested, as well. A polymer solution of Elastollan® was produced by weighing out 10% w/v% polymer (100 mg Elastollan® in 1 mL solvent) dissolved in THF:DMF 1:1 v:v (Figure 10; Sigma-Aldrich, St. Louis, MO, USA). The solution was incubated at 37 °C overnight.

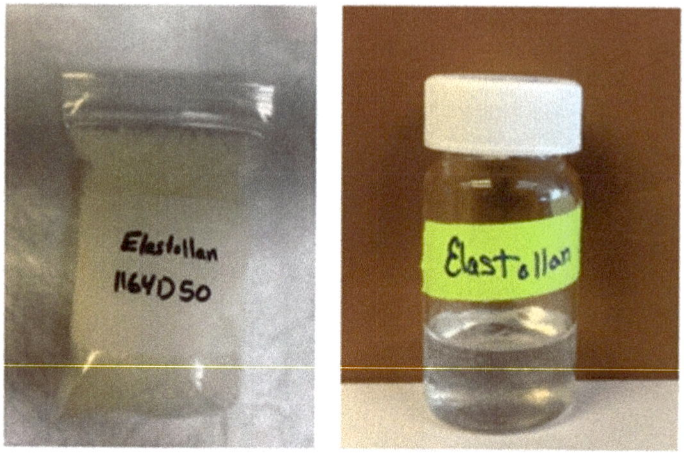

Figure 10. Elastollan® polymer pellets (**left**) and final polymer solution (**right**).

On the day of spinning, 6 mL of polymer solution was transferred to a plastic syringe with needle attached (Becton Dickinson, Franklin Lakes, NJ, USA). The syringe was emptied of all air bubbles and attached firmly to an automatic dispenser (Figure 11). Experimental ligatures were adhered to a mandrel located 15 cm from the syringe needle tip (Figure 12). The flow rate of the dispenser was set to 6 mL/h. and mandrel set to 300 rpm. The electrospinning was completed under an electric field of 20 kV. After 30 min, the electrospinning process was paused, allowing the chains to dry until flipped. They then ran for an additional 30 min coating the other side. The ligatures were dried and removed. All electrospinning was completed in the dark.

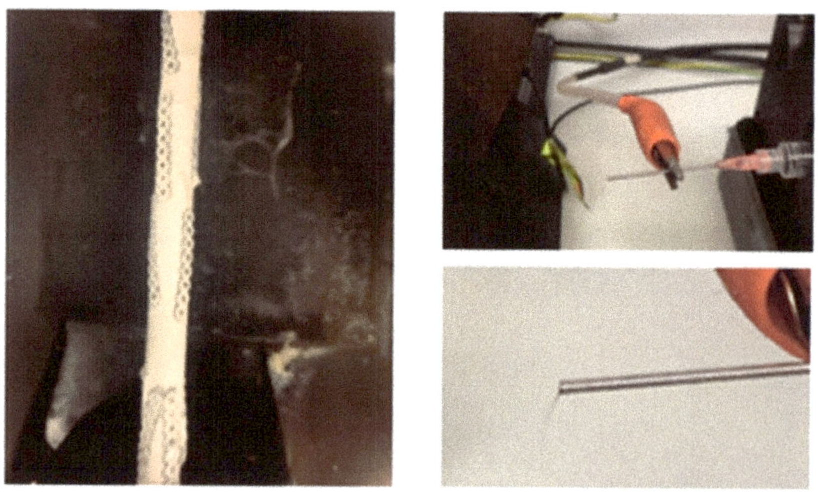

Figure 11. Photographs of the chains on the rotating rod (**left**) and polymer threads dispensing from the syringe (**right**).

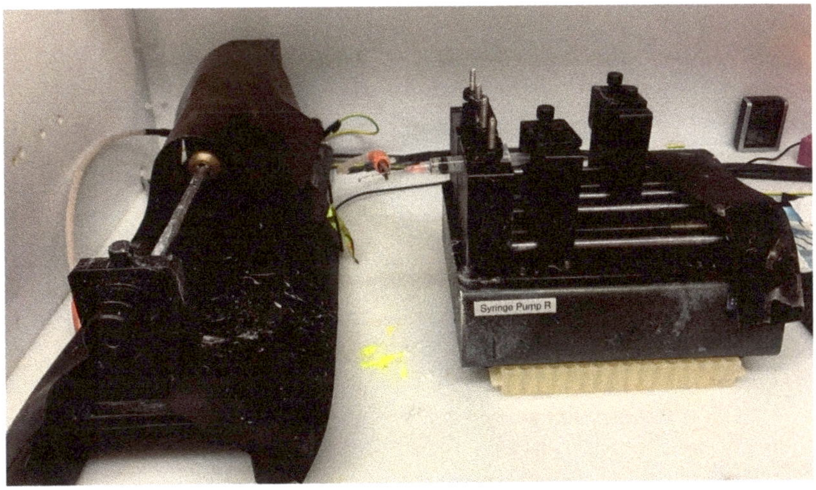

Figure 12. Electrospinning polymer coating to SNAP-impregnated chains. Pictured is the rotating rod housing the chains (**left**) and the syringe dispenser containing the polymer solution (**right**).

4.3. Nitric Oxide Release Testing

Nitric Oxide flux from a single ligature (from the 75 mg/mL SNAP/THF + Elastollan®) was measured daily for 2 h/day for 30 days using a Sievers Chemiluminescence Nitric Oxide Analyzer (GE Analytical Instruments, Boulder, CO, USA) to determine duration and magnitude of NO release (Figure 13). The chain sample was submerged in a glass container with 4 mL of PBS within a water bath set at 37 °C. Nitric oxide released from the sample was then transported into the detection chamber (Figure 13). Readings were conducted until a consistent flux was attained. During periods of non-testing, ligatures were kept in amber glass vials with 1 mL of PBS at 37 °C.

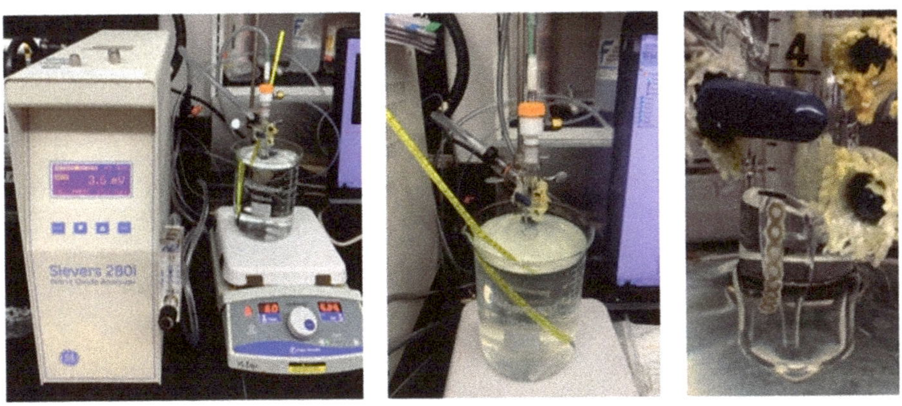

Figure 13. Sievers Chemiluminescence Nitric Oxide Analyzer (**left**), NO-release measurement configuration (**middle**), and chains suspended in PBS (**right**).

4.4. Bacterial Testing

The bactericidal properties of the obtained sample (75 mg/mL SNAP/THF + Elastollan®) were tested against the common oral microbe, *S. mutans*, in two separate microbial studies to determine planktonic growth surrounding the chain and biofilm formation on the ligature itself. Control samples and ligatures impregnated with 75 mg of SNAP only were tested, as well [59].

4.5. Planktonic Growth in Surrounding Solution

A standardized solution of KPSP2 *Streptococcus mutans* (American Type Culture Collection, Manasses, VA, USA) in brain–heart infusion (BHI; Becton Dickinson, Franklin Lakes, NJ, USA) was added to a vial with 4 mL solution of BHI + 2% sucrose (Figure 14). The ratio of inoculant to media was 1:30. Each vial contained either (1) two control chains, (2) two chains impregnated with 75 mg/mL SNAP/THF, or (3) two coated experimental chains (75 mg/mL SNAP/THF + Elastollan®). All tests were performed in duplicate totaling 6 vials. Vials were incubated at 37 °C, aerobically, without shaking, in the dark. Then, 1 mL of solution was removed at 24, 48, 72, 96, and 120 h, and optical density at 600 nm was measured using Eppendorf BioPhotometer Plus (Eppendorf AG, Hamburg, Germany) and recorded.

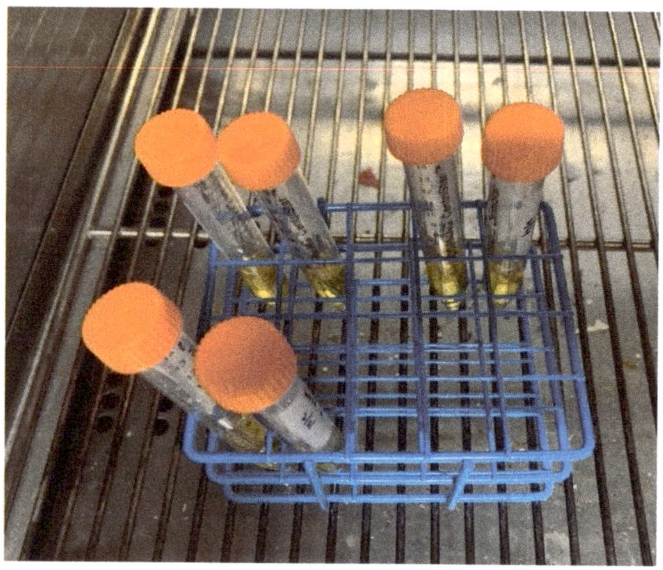

Figure 14. *S. mutans* in 4 mL solution of BHI + 2% sucrose. (**Top left vials**) control chains, (**top right vials**) chains impregnated with 75 mg/mL SNAP/THF, (**bottom left**) experimental chains (75 mg/mL SNAP/THF + Elastollan®).

4.6. Biofilm Formation on Surrounding Surfaces

In a 12-well plate, a standardized solution of *S. mutans* was added to a 1 mL solution of BHI supplemented with 2% sucrose. The ratio of Inoculant to media was 1:30. Experimental, SNAP-only, and control chains (1 chain/well) were suspended in the solution and incubated at 37 °C, aerobically, for periods of 24, 48, and 72 h (Figure 15). All samples were removed, wells were washed with 2 mL PBS, and 500 µL 0.1% crystal violet stain (0.3% stock; Becton Dickinson, Franklin Lakes, NJ, USA) was added to each well. Following incubation for 30 min, wells were washed again with water and allowed to dry. Then, 1 mL 30% acetic acid (MilliporeSigma, Burlington, MA, USA) was added to each well to dissolve the crystal violet, and absorbance was measured at 570 nm. All tests were performed in duplicate.

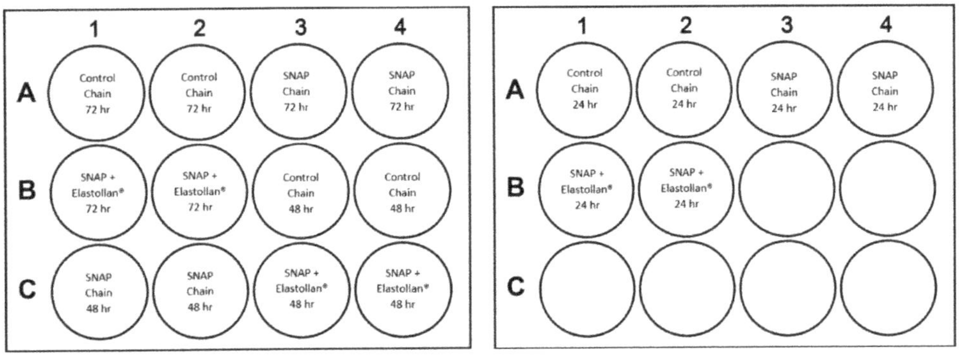

Figure 15. Biofilm on surrounding surfaces 12-well plate schematic.

5. Conclusions

The nitric oxide-releasing molecule, S-nitroso-acetylpenicillamine, can be successfully incorporated into an existing orthodontic elastomeric ligature. Following impregnation, samples exhibited NO release over a twenty-nine-day period after the addition of a polymer coating. Ligatures demonstrating successful NO release inhibited both *S.coccus mutans* planktonic growth and growth on surfaces adjacent to the chain. Manufactured chains integrated with antimicrobial properties present a promising solution for combating white spot lesions and other bacterial-related issues in orthodontics. Furthermore, the advancement of these chains creates opportunities for collaboration between academia and industry, as well as the significant potential for commercial viability.

Author Contributions: Conceptualization, S.G.; Methodology, A.M., C.W., J.T., K.M.P., J.M.S.-R., M.P.C., D.A.S., N.J. and S.G.; Software, A.M., J.T., K.M.P. and M.P.C.; Resources, J.M.S.-R., M.P.C. and D.A.S.; Writing—Original Draft Preparation, A.M.; Writing—A.M.; Supervision, J.M.S.-R., M.P.C., D.A.S. and S.G.; Funding Acquisition, S.G. All authors have read and agreed to the published version of the manuscript.

Funding: Orthodontic Faculty Development Fellowship Award (OFDFA) award from the American Association of Orthodontists Foundation (AAOF) and University of Louisville School of Dentistry Faculty Research Grant.

Institutional Review Board Statement: This study did not require approval by the Institutional Review.

Informed Consent Statement: Not Applicable.

Data Availability Statement: The data that support the findings of this study are available from the corresponding author upon reasonable request.

Conflicts of Interest: The authors declare no conflicts of interest.

References

1. Palomares, N.B.; Celeste, R.K.; Oliveira, B.H.; Miguel, J.A. How does orthodontic treatment affect young adults' oral health-related quality of life? *Am. J. Orthod. Dentofac. Orthop.* **2012**, *141*, 751–758. [CrossRef]
2. Zachrisson, S.; Zachrisson, B.U. Gingival condition associated with orthodontic treatment. *Angle Orthod.* **1972**, *42*, 26–34. [CrossRef] [PubMed]
3. Gwinnett, A.J.; Ceen, R.F. Plaque distribution on bonded brackets: A scanning microscope study. *Am. J. Orthod.* **1979**, *75*, 667–677. [CrossRef]
4. Balenseifen, J.W.; Madonia, J.V. Study of dental plaque in orthodontic patients. *J. Dent. Res.* **1970**, *49*, 320–324. [CrossRef] [PubMed]
5. Forsberg, C.M.; Brattström, V.; Malmberg, E.; Nord, C.E. Ligature wires and elastomeric rings: Two methods of ligation, and their association with microbial colonization of Streptococcus mutans and lactobacilli. *Eur. J. Orthod.* **1991**, *13*, 416–420. [CrossRef] [PubMed]

6. Lundström, F.; Krasse, B. Streptococcus mutans and lactobacilli frequency in orthodontic patients; the effect of chlorhexidine treatments. *Eur. J. Orthod.* **1987**, *9*, 109–116. [CrossRef]
7. Lucchese, A.; Bondemark, L.; Marcolina, M.; Manuelli, M. Changes in oral microbiota due to orthodontic appliances: A systematic review. *J. Oral Microbiol.* **2018**, *10*, 1476645. [CrossRef]
8. Struzycka, I. The oral microbiome in dental caries. *Pol. J. Microbiol.* **2014**, *63*, 127–135. [CrossRef]
9. García-Godoy, F.; Hicks, M.J. Maintaining the integrity of the enamel surface: The role of dental biofilm, saliva and preventive agents in enamel demineralization and remineralization. *J. Am. Dent. Assoc.* **2008**, *139*, 25S–34S. [CrossRef]
10. Skidmore, K.J.; Brook, K.J.; Thomson, W.M.; Harding, W.J. Factors influencing treatment time in orthodontic patients. *Am. J. Orthod. Dentofac. Orthop.* **2006**, *129*, 230–238. [CrossRef]
11. Hadler-Olsen, S.; Sandvik, K.; El-Agroudi, M.A.; Øgaard, B. The incidence of caries and white spot lesionsp in orthodontically treated adolescents with a comprehensive caries prophylactic regimen—A prospective study. *Eur. J. Orthod.* **2012**, *34*, 633–639. [CrossRef]
12. Boke, F.; Gazioglu, C.; Akkaya, S.; Akkaya, M. Relationship between orthodontic treatment and gingival health: A retrospective study. *Eur. J. Dent.* **2014**, *8*, 373–380. [CrossRef]
13. Beckwith, F.R.; Ackerman, R.J.; Cobb, C.M., Jr.; Tira, D.E. An evaluation of factors affecting duration of orthodontic treatment. *Am. J. Orthod. Dentofac. Orthop.* **1999**, *115*, 439–447. [CrossRef] [PubMed]
14. Chapman, J.A.; Roberts, W.E.; Eckert, G.J.; Kula, K.S.; González-Cabezas, C. Risk factors for incidence and severity of white spot lesions during treatment with fixed orthodontic appliances. *Am. J. Orthod. Dentofac. Orthop.* **2010**, *138*, 188–194. [CrossRef] [PubMed]
15. Kleter, G.A. Discoloration of dental carious lesions (a review). *Arch. Oral Biol.* **1998**, *43*, 629–632. [CrossRef]
16. Sundararaj, D.; Venkatachalapathy, S.; Tandon, A.; Pereira, A. Critical evaluation of incidence and prevalence of white spot lesions during fixed orthodontic appliance treatment: A meta-analysis. *J. Int. Soc. Prev. Community Dent.* **2015**, *5*, 433–439. [CrossRef]
17. Zabokova-Bilbilova, E.; Popovska, L.; Kapusevska, B.; Stefanovska, E. White spot lesions: Prevention and management during the orthodontic treatment. *Prilozi* **2014**, *35*, 161–168. [CrossRef] [PubMed]
18. Morrier, J.J. Leucomes et traitement orthodontique. Prévention, traitement [White spot lesions and orthodontic treatment. Prevention and treatment]. *L'Orthod. Fr.* **2014**, *85*, 235–244. [CrossRef]
19. Khoroushi, M.; Kachuie, M. Prevention and Treatment of White Spot Lesions in Orthodontic Patients. *Contemp. Clin. Dent.* **2017**, *8*, 11–19. [CrossRef]
20. Kirschneck, C.; Christl, J.J.; Reicheneder, C.; Proff, P. Efficacy of fluoride varnish for preventing white spot lesions and gingivitis during orthodontic treatment with fixed appliances-a prospective randomized controlled trial. *Clin. Oral Investig.* **2016**, *20*, 2371–2378. [CrossRef]
21. Brêtas, S.M.; Macari, S.; Elias, A.M.; Ito, I.Y.; Matsumoto, M.A. Effect of 0.4% stannous fluoride gel on Streptococci mutans in relation to elastomeric rings and steel ligatures in orthodontic patients. *Am. J. Orthod. Dentofac. Orthop.* **2005**, *127*, 428–433. [CrossRef] [PubMed]
22. Benson, P.E.; Parkin, N.; Dyer, F.; Millett, D.T.; Furness, S.; Germain, P. Fluorides for the prevention of early tooth decay (demineralised white lesions) during fixed brace treatment. *Cochrane Database Syst. Rev.* **2013**, *12*, CD003809. [CrossRef] [PubMed]
23. Menon, S.; Sreedharan, S.; Suma, S.; Balakrishnan, P.; Kumar, M. The effect of a fluoride-releasing orthodontic bonding agent on caries incidence: A randomized clinical trial. *Am. J. Orthod. Dentofac. Orthop.* **2021**, *160*, 417–425. [CrossRef]
24. Kim, Y.J.; Kim, Y.; Cha, J.Y.; Hwang, C.J. Fluoride-releasing orthodontic adhesives: A systematic review and meta-analysis. *J. Dent. Res.* **2020**, *99*, 1109–1117.
25. Benson, P.E.; Douglas, C.W.; Martin, M.V. Fluoridated elastomers: Effect on the microbiology of plaque. *Am. J. Orthod. Dentofac. Orthop.* **2004**, *126*, 325–330. [CrossRef] [PubMed]
26. Miura, K.K.; Ito, I.Y.; Enoki, C.; Elias, A.M.; Matsumoto, M.A. Anticariogenic effect of fluoride-releasing elastomers in orthodontic patients. *Braz. Oral Res.* **2007**, *21*, 228–233. [CrossRef] [PubMed]
27. Banks, P.A.; Chadwick, S.M.; Asher-McDade, C.; Wright, J.L. Fluoride-releasing elastomerics—A prospective controlled clinical trial. *Eur. J. Orthod.* **2000**, *22*, 401–407. [CrossRef] [PubMed]
28. Mattick, C.R.; Mitchell, L.; Chadwick, S.M.; Wright, J. Fluoride-releasing elastomeric modules reduce decalcification: A randomized controlled trial. *J. Orthod.* **2001**, *28*, 217–219. [CrossRef] [PubMed]
29. Miethke, R.R. Comment on determination of fluoride from ligature ties. *Am. J. Orthod. Dentofac. Orthop.* **1997**, *111*, 33A. [CrossRef]
30. Marquis, R.E. Antimicrobial actions of fluoride for oral bacteria. *Can. J. Microbiol.* **1995**, *41*, 955–964. [CrossRef]
31. Bowden, G.H. Effects of fluoride on the microbial ecology of dental plaque. *J. Dent. Res.* **1990**, *69*, 653–683. [CrossRef] [PubMed]
32. Furchgott, R.F. Endothelium-Derived Relaxing Factor: Discovery, Early Studies, and Identifcation as Nitric Oxide (Nobel Lecture). *Angew. Chem. Int. Ed.* **1999**, *38*, 1870–1880. [CrossRef]
33. Carpenter, A.W.; Schoenfisch, M.H. Nitric oxide release: Part II. Therapeutic applications. *Chem. Soc. Rev.* **2012**, *41*, 3742–3752. [CrossRef] [PubMed]
34. Rana, S.V.S. *Environmental Pollution: Health and Toxicology*; Alpha Science Int'l Ltd.: Oxford, UK, 2006; p. 14.
35. Alloway, B.; Ayres, D.C. *Chemical Principles of Environmental Pollution*; CRC Press: Boca Raton, FL, USA, 1997; p. 7.
36. Ignarro, L.J.; Buga, G.M.; Wood, K.S.; Byrns, R.E.; Chaudhuri, G. Endothelium-derived relaxing factor produced and released from artery and vein is nitric oxide. *Proc. Natl. Acad. Sci. USA* **1987**, *84*, 9265–9269. [CrossRef] [PubMed]

37. Hetrick, E.M.; Schoenfisch, M.H. Analytical chemistry of nitric oxide. *Annu. Rev. Anal. Chem.* **2009**, *2*, 409–433. [CrossRef] [PubMed]
38. Palmer, R.M.; Ashton, D.S.; Moncada, S. Vascular endothelial cells synthesize nitric oxide from L-arginine. *Nature* **1988**, *333*, 664–666. [CrossRef] [PubMed]
39. Rizk, M.; Witte, M.B.; Barbul, A. Nitric oxide and wound healing. *World J. Surg.* **2004**, *28*, 301–306. [CrossRef] [PubMed]
40. Taite, L.J.; Yang, P.; Jun, H.W.; West, J.L. Nitric oxide-releasing polyurethane-PEG copolymer containing the YIGSR peptide promotes endothelialization with decreased platelet adhesion. *J. Biomed. Mater. Res. Part B Appl. Biomater.* **2008**, *84*, 108–116. [CrossRef] [PubMed]
41. Bogdan, C. Nitric oxide and the immune response. *Nat. Immunol.* **2001**, *2*, 907–916. [CrossRef]
42. Jones, M.L.; Ganopolsky, J.G.; Labbé, A.; Wahl, C.; Prakash, S. Antimicrobial properties of nitric oxide and its application in antimicrobial formulations and medical devices. *Appl. Microbiol. Biotechnol.* **2010**, *88*, 401–407. [CrossRef]
43. Radi, R. Peroxynitrite, a stealthy biological oxidant. *J. Biol. Chem.* **2013**, *288*, 26464–26472. [CrossRef] [PubMed]
44. Pacher, P.; Beckman, J.S.; Liaudet, L. Nitric oxide and peroxynitrite in health and disease. *Physiol. Rev.* **2007**, *87*, 315–424. [CrossRef] [PubMed]
45. Wang, P.; Zweier, J.L. Measurement of nitric oxide and peroxynitrite generation in the postischemic heart. Evidence for peroxynitrite-mediated reperfusion injury. *J. Biol. Chem.* **1996**, *271*, 29223–29230. [CrossRef] [PubMed]
46. Olufunmilayo, E.O.; Gerke-Duncan, M.B.; Holsinger, R.M.D. Oxidative Stress and Antioxidants in Neurodegenerative Disorders. *Antioxidants* **2023**, *12*, 517. [CrossRef] [PubMed]
47. Szabó, C.; Ischiropoulos, H.; Radi, R. Peroxynitrite: Biochemistry, pathophysiology and development of therapeutics. Nature reviews. *Drug Discov.* **2007**, *6*, 662–680. [CrossRef] [PubMed]
48. Vong, L.B.; Nagasaki, Y. Nitric Oxide Nano-Delivery Systems for Cancer Therapeutics: Advances and Challenges. *Antioxidants* **2020**, *9*, 791. [CrossRef]
49. Poh, W.H.; Rice, S.A. Recent Developments in Nitric Oxide Donors and Delivery for Antimicrobial and Anti-Biofilm Applications. *Molecules* **2022**, *27*, 674. [CrossRef] [PubMed]
50. Hanson, S.R.; Hutsell, T.C.; Keefer, L.K.; Mooradian, D.L.; Smith, D.J. Nitric oxide donors: A continuing opportunity in drug design. *Adv. Pharmacol.* **1995**, *34*, 383–398. [CrossRef] [PubMed]
51. Lundberg, J.O.; Gladwin, M.T.; Weitzberg, E. Strategies to increase nitric oxide signalling in cardiovascular disease. Nature reviews. *Drug Discov.* **2015**, *14*, 623–641. [CrossRef]
52. Frost, M.C.; Reynolds, M.M.; Meyerhoff, M.E. Polymers incorporating nitric oxide releasing/generating substances for improved biocompatibility of blood-contacting medical devices. *Biomaterials* **2005**, *26*, 1685–1693. [CrossRef]
53. Wo, Y.; Li, Z.; Brisbois, E.J.; Colletta, A.; Wu, J.; Major, T.C.; Xi, C.; Bartlett, R.H.; Matzger, A.J.; Meyerhoff, M.E. Origin of Long-Term Storage Stability and Nitric Oxide Release Behavior of CarboSil Polymer Doped with S-Nitroso-N-acetyl-D-penicillamine. *ACS Appl. Mater. Interfaces* **2015**, *7*, 22218–22227. [CrossRef] [PubMed]
54. Stokes, G.S.; Potts, J.T.; Lotz, M.; Bartter, F.C. New Agent in the Treatment of Cystinuria: N-acetyl-D-penicillamine. *Br. Med. J.* **1968**, *1*, 284–288. [CrossRef] [PubMed]
55. Brisbois, E.J.; Davis, R.P.; Jones, A.M.; Major, T.C.; Bartlett, R.H.; Meyerhoff, M.E.; Handa, H. Reduction in Thrombosis and Bacterial Adhesion with 7 Day Implantation of S-Nitroso-N-acetylpenicillamine (SNAP)-Doped Elast-eon E2As Catheters in Sheep. *J. Mater. Chem. B* **2015**, *3*, 1639–1645. [CrossRef] [PubMed]
56. Goudie, M.J.; Brisbois, E.J.; Pant, J.; Thompson, A.; Potkay, J.A.; Handa, H. Characterization of an S-nitroso-N-acetylpenicillamine-based nitric oxide releasing polymer from a translational perspective. *Int. J. Polym. Mater.* **2016**, *65*, 769–778. [CrossRef]
57. Vogt, C.; Xing, Q.; He, W.; Li, B.; Frost, M.C.; Zhao, F. Fabrication and characterization of a nitric oxide-releasing nanofibrous gelatin matrix. *Biomacromolecules* **2013**, *14*, 2521–2530. [CrossRef] [PubMed]
58. Colletta, A.; Wu, J.; Wo, Y.; Kappler, M.; Chen, H.; Xi, C.; Meyerhoff, M.E. S-Nitroso-N-acetylpenicillamine (SNAP) Impregnated Silicone Foley Catheters: A Potential Biomaterial/Device To Prevent Catheter-Associated Urinary Tract Infections. *ACS Biomater. Sci. Eng.* **2015**, *1*, 416–424. [CrossRef] [PubMed]
59. Warden, C.; Tan, J.; Piell, K.; Janakiraman, N.; Meyerhoff, M.; Steinbach-Rankins, J.; Cole, M.; Gudhimella, S. A novel, nitric oxide-releasing elastomeric chain for antimicrobial action: Proof of concept. *Mater. Res. Express* **2021**, *8*, 095309. [CrossRef]
60. Andreasen, G.F.; Bishara, S. Comparison of alastik chains with elastics involved with intra-arch molar to molar forces. *Angle Orthod.* **1970**, *40*, 151–158. [CrossRef] [PubMed]
61. Elastollan® (TPU)—The Thermoplastic Polyurethane of BASF (n.d.). Available online: https://plastics-rubber.basf.com/global/en/performance_polymers/products/elastollan.html (accessed on 10 November 2022).
62. Electrospinning. Nanoscience Instruments (21 September 2022). Available online: https://www.nanoscience.com/techniques/electrospin/ (accessed on 31 October 2022).
63. Xue, J.; Wu, T.; Dai, Y.; Xia, Y. Electrospinning and Electrospun Nanofibers: Methods, Materials, and Applications. *Chem. Rev.* **2019**, *119*, 5298–5415. [CrossRef]
64. Ong, E.; McCallum, H.; Griffin, M.P.; Ho, C. Efficiency of self-ligating vs conventionally ligated brackets during initial alignment. *Am. J. Orthod. Dentofac. Orthop.* **2010**, *138*, 138.e1–138.e7. [CrossRef]

65. Choudhary, M.; MacDonald, L.; Meyerhoff, M.E. S-Nitroso-N-acetylpenicillamine (SNAP)-Based Antithrombotic and Antimicrobial Polymers for Biomedical Applications: Nitric Oxide (NO) to the Rescue. *J. Med. Chem.* **2020**, *63*, 10732–10751.
66. Wo, Y.; Xu, L.C.; Li, Z.; Matzger, A.J.; Meyerhoff, M.E.; Siedlecki, C.A. Antimicrobial nitric oxide releasing surfaces based on S-nitroso-N-acetylpenicillamine impregnated polymers combined with submicron-textured surface topography. *Biomater. Sci.* **2017**, *5*, 1265–1278. [CrossRef] [PubMed]

International Journal of
Molecular Sciences

MDPI

Review

Recent Advances in Antimicrobial Peptide Hydrogels

Aryanna Copling [1,†], Maxwell Akantibila [2,†], Raaha Kumaresan [3], Gilbert Fleischer [2], Dennise Cortes [2], Rahul S. Tripathi [2], Valerie J. Carabetta [2,*] and Sebastián L. Vega [3,4,*]

1 Department of Molecular & Cellular Biosciences, Rowan University, Glassboro, NJ 08028, USA; coplin22@students.rowan.edu
2 Department of Biomedical Sciences, Cooper Medical School of Rowan University, Camden, NJ 08103, USA; akanti18@students.rowan.edu (M.A.); fleischeg0@rowan.edu (G.F.); cortesd@rowan.edu (D.C.); tripat45@rowan.edu (R.S.T.)
3 Department of Biomedical Engineering, Rowan University, Glassboro, NJ 08028, USA; kumare76@students.rowan.edu
4 Department of Orthopedic Surgery, Cooper Medical School of Rowan University, Camden, NJ 08103, USA
* Correspondence: carabetta@rowan.edu (V.J.C.); vegas@rowan.edu (S.L.V.); Tel.: +1-856-956-2736 (V.J.C.); +1-856-256-5522 (S.L.V.)
† These authors contributed equally to this work.

Abstract: Advances in the number and type of available biomaterials have improved medical devices such as catheters, stents, pacemakers, prosthetic joints, and orthopedic devices. The introduction of a foreign material into the body comes with a risk of microbial colonization and subsequent infection. Infections of surgically implanted devices often lead to device failure, which leads to increased patient morbidity and mortality. The overuse and improper use of antimicrobials has led to an alarming rise and spread of drug-resistant infections. To overcome the problem of drug-resistant infections, novel antimicrobial biomaterials are increasingly being researched and developed. Hydrogels are a class of 3D biomaterials consisting of a hydrated polymer network with tunable functionality. As hydrogels are customizable, many different antimicrobial agents, such as inorganic molecules, metals, and antibiotics have been incorporated or tethered to them. Due to the increased prevalence of antibiotic resistance, antimicrobial peptides (AMPs) are being increasingly explored as alternative agents. AMP-tethered hydrogels are being increasingly examined for antimicrobial properties and practical applications, such as wound-healing. Here, we provide a recent update, from the last 5 years of innovations and discoveries made in the development of photopolymerizable, self-assembling, and AMP-releasing hydrogels.

Keywords: antibiotic resistance; biofilm; click chemistry; photopolymerization; AMP-releasing; self-assembling; hydrogel; antimicrobial peptide

Citation: Copling, A.; Akantibila, M.; Kumaresan, R.; Fleischer, G.; Cortes, D.; Tripathi, R.S.; Carabetta, V.J.; Vega, S.L. Recent Advances in Antimicrobial Peptide Hydrogels. *Int. J. Mol. Sci.* **2023**, *24*, 7563. https://doi.org/10.3390/ijms24087563

Academic Editor: Helena Felgueiras

Received: 6 April 2023
Revised: 17 April 2023
Accepted: 18 April 2023
Published: 20 April 2023

1. Introduction

1.1. Medical Device Infections Are a Global Concern

Surgically implanted medical device use has greatly increased over the past 50 years, improving the quality of life of millions of people worldwide. Advancements in the types of biomaterials available has expanded and improved medical devices, such as catheters, stents, pacemakers, prosthetic joints, and orthopedic devices (reviewed in [1]). The introduction of a foreign material into the body comes with an inherent risk of microbial colonization and subsequent infection [2,3]. The infection can occur during the actual surgical procedure or post-operatively, with the source of infection frequently being the improper sanitization of surgical equipment, the hands of healthcare workers, or the colonization of the patient [4]. Infections of implanted medical devices are especially challenging to treat, due the presence of bacterial biofilms [5]. Bacteria can exist in a free-living, planktonic state or as part of a multicellular biofilm community. When a free-living bacterium attaches to a surface, it switches to a sessile lifestyle and there is a drastic change in gene expression.

In particular, the genes required to produce extracellular polymeric substance (EPS) are induced. The biofilm matrix is composed of mostly polysaccharides, proteins, and extracellular DNA, which encases the growing colony in a slime layer that protects the bacteria against environmental insults [6–8].

In the natural environment, nearly every bacterial species can assemble into biofilms [9]. From a clinical standpoint, biofilms protect bacteria from host immune defenses and antibiotics, making them difficult to eradicate [10–12]. Clinically, biofilm-based infections of surgically implanted devices often lead to device failure, which then requires treatment with high doses of antibiotics, followed by the removal and replacement of the infected device. This puts the patient in need of additional, potentially risky surgeries and increases the risk of mortality [13]. Current interventions to address device-associated infections are also costly and often ineffective due to the prevalence of antibiotic-resistant strains and the increased risk of the reinfection of the new implant [8,10]. As the biofilm matures, essential nutrients become limited and toxic byproducts accumulate, causing the surface layer of the bacteria to revert to a planktonic lifestyle and travel to other body sites to begin the process again. Thus, bacterial biofilms are also a reservoir for chronic infections [6,8]. Biofilms are associated with up to 80% of chronic bacterial infections [14].

1.2. Hydrogels as Promising Additions to Our Antimicrobial Arsenal

To overcome the problem of chronic infections, antimicrobial materials are increasingly being researched and developed. These materials are typically polymers, ceramics, metals, or composites, which have microbicidal activity against bacteria, fungi, viruses, or a combination of the three. One such material being extensively studied is hydrogel (reviewed in [15]). Hydrogels are soft biomaterials consisting of highly hydrated polymeric networks that have been tailored with various moieties for applications as antibacterial, tissue-regenerating, and drug-releasing materials [16,17]. As hydrogels are customizable, many different antimicrobial agents have been incorporated or tethered to them. For example, inorganic molecules, such as metal ions and metallic oxides, have been explored. Common metal ions include silver, gold, and copper and the most common metal oxides are zinc, titanium, and nickel [18]. It is generally believed that metal-loaded hydrogels kill bacteria by disrupting and destroying the bacterial cell wall. Silver-loaded hydrogels have broad-spectrum activity, as they are efficient at killing both Gram-negative and Gram-positive bacteria [19–21]. While silver ions are promising antimicrobial agents, their cytotoxicity and interactions with serum proteins, such as albumin, must be considered [22]. While other metals, such as cobalt, copper, and gold, have been studied as alternative antimicrobial agents, these metals have not been sufficiently explored [15]. As with silver ions, the toxicity and clearance of these metals is a significant concern. Metal oxide-loaded hydrogels kill bacteria through the generation of large amounts of reactive oxygen species following exposure to light [23]. Metal oxide-loaded hydrogels, the most widely used being zinc oxide (ZnO), are highly effective and non-toxic to human cells at low concentrations [24–26]. While inorganic agent-loaded hydrogels have many promising, desirable properties, such as broad-spectrum activity and temperature stability, they have low biocompatibility, making their use in practice a challenge.

As an alternative to metal oxide-laden hydrogels, antibiotic-tethered hydrogels have also been studied. Tethering antibiotics allows for a lower dosage to be used, which in theory should lessen the risk of antibiotic resistance [27]. The three most studied antibiotics used for this application are ciprofloxacin, gentamicin, and vancomycin. The fluoroquinolone ciprofloxacin has a broad spectrum of activity. Ciprofloxacin-loaded hydrogels have excellent antimicrobial properties, with low cytotoxicity and no hemolytic activity [28–30]. Gentamicin is an aminoglycoside antibiotic with broad spectrum activity but carries a risk of nephrotoxicity. Gentamicin-tethered hydrogels allow for the local administration of the antibiotic, which overcomes the other issue with this drug, namely achieving an acceptably high serum concentration [31–33]. Vancomycin is widely used in clinical practice and often considered a last-line drug for many Gram-positive infections, including *Enterococcus* and

Staphylococcus species. Vancomycin-loaded hydrogels have shown promise in protecting against surgical site infections [34]. However, the spread and emergence of drug-resistant infections has dampened the enthusiasm for antibiotic-tethered hydrogels.

1.3. Antimicrobial Peptides as a Potential Solution to the Antibiotic Resistance Problem

The overuse and misuse of antibiotics has led to the rise and spread of drug-resistant infections. In 2019, 1.27 million deaths were attributed to drug-resistant bacteria worldwide [35]. According to a comprehensive antimicrobial resistance report, it was estimated that this number would increase to 10 million deaths per year worldwide by 2050 if new solutions to combat bacterial drug resistance are not identified and implemented [36]. As antibiotic use has increased steadily in the past 50 years, we are now seeing the emergence of multidrug-resistant strains which are non-susceptible to three or more standard-of-care antibiotics. Recently, the emergence of pan-drug resistant strains, which are resistant to all standard-of-care antibiotics, have been reported [37–39]. Due to the increased prevalence of antibiotic resistance, antimicrobial peptides (AMPs) are being increasingly explored as alternative antimicrobial agents. AMPs are host defense peptides and are found in all domains of life, but the majority studied are eukaryotic in origin [40]. They are a diverse class of short amino acid sequences that are effective as broad-spectrum antibacterial agents and in mitigating bacterial biofilms [41,42]. In general, AMPs are cationic peptides and contain both hydrophobic and hydrophilic components. Most of the identified AMPs assemble into α-helical structures and disrupt the bacterial membrane by pore formation, leading to cell death [40,43]. However, other structures, such as β-sheets, are commonly observed as well [44]. While membrane disruption is a common mechanism of action, AMPs may also target nucleic acid synthesis, enzymatic activities, and cell wall biogenesis [45]. Resistance to AMPs is thought to arise and spread more slowly, which makes them attractive as novel therapeutics [46,47]. AMP-tethered hydrogels are being increasingly examined for antimicrobial properties and practical applications. Here, we provide a review of the last 5 years of innovations and discoveries made in the development of AMP-loaded hydrogels. We begin with a discussion of new developments with photopolymerizing hydrogels. Next, we discuss advancements made with self-assembling and AMP-releasing hydrogels. Finally, we end with a brief discussion of remaining challenges and outlook for AMP-loaded hydrogel technology towards clinical practice.

2. Photopolymerizing AMP Hydrogels

2.1. Recent Updates Using Chitosan and Polyethylene Glycol Backbones

The synthesis of hydrogels is typically performed through photopolymerization using visible or ultraviolet (UV) light in the presence of a photoinitiator [48]. Photopolymerization reactions are rapid and step-growth reactions between thiols and vinyl (ene) groups yield highly specific click reactions that can be used to form hydrogels functionalized with thiolated molecules, such as peptides [49,50]. For instance, Alves and coworkers developed AMP-laden chitosan hydrogels by first conjugating norbornene groups into chitosan (NorChit), followed by clicking NorChit with thiolated Dhvar5 AMPs using a light-mediated step-growth reaction (NorChit-Dhvar5) [51]. The Dhvar5 (LLLFLLKKRKKRKY) AMP was chosen due to previously established antimicrobial properties [52,53]. Dhvar5-NorChit thin films were then prepared by spin coating NorChit and NorChit-Dhvar5 solutions onto gold (Au) substrates, as had been carried out previously [53]. These films demonstrated up to a 35% reduction in *Staphylococcus epidermidis* adhesion and increased the killing of *Pseudomonas aeruginosa* in comparison to unmodified chitosan films (Figure 1A). These films are also biocompatible, as demonstrated by toxicity assays performed on human neonatal dermal fibroblasts. This coating strategy can be used to create antibacterial biomaterials, and excitingly, there were remaining pendant norbornene groups, which could be used to add more Dhvar5 or any thiolated AMP to further increase antimicrobial properties.

In a recent study by De Zoysa et al., battacin lipopeptide hydrogels were synthesized by covalently linking N-terminal cysteine containing lipopeptides onto polyethylene glycol

(PEG) macromers using a thiol-ene click chemistry reaction [54]. Battacin was chosen due to its potent in vitro and in vivo antibacterial activity against *P. aeruginosa* and *Staphylococcus aureus* [55]. The battacin-loaded hydrogels (0–10 wt%) were formed by mixing a tetra-branched thiol crosslinker, a diacrylate spacer, PEG, and a photoinitiator. Gelation was initiated by photopolymerization under UV irradiation. The hydrogels were prepared in either methanol or water, and the ones prepared in methanol had better antibacterial and antibiofilm activity. A minimum peptide content of 0.5 wt% with respect to polymer content was required to successfully inhibit planktonic bacterial growth and disperse the mature biofilms of *P. aeruginosa* and *S. aureus*. The hydrogels displayed excellent cytocompatibility and were not hemolytic. Together, these properties make these hydrogels good candidates for clinical applications, such as wound dressings to combat skin infections.

2.2. Development of Photopolymerizable Hydrogels for Clinical Applications

Towards developing hydrogels for wound dressing applications, Cheng and colleagues created a sprayable hydrogel to alleviate the impact of wound infection by using cerium oxide nanoparticles (CeONs), AMPs, and gelatin methacrylate (GelMA) as the photopolymerizable hydrogel matrix [56]. The hydrogel wound dressing was made from GelMA conjugated with dopamine (GelMA-DOPA) by (N-(3-(dimethylamino)-propyl)-N′-ethylcarbodiimide hydrochloride (EDC) and N-hydroxysuccinimide (NHS) coupling. The GelMA-DOPA hydrogels were formed by addition of the photoinitiator Irgacure 2959, CeONs, and the AMP HHC-36 (KRWWKWWRR) as well as exposure to UV light for 30 s to produce AMP-CeONs, which were loaded to GelMA-DOPA hydrogels (Figure 1B). HHC-36 exhibited nearly complete bactericidal activity against both Gram-negative and Gram-positive pathogenic bacteria [57]. The AMP-loaded GelMA-DOPA hydrogels displayed the complete bacterial killing of *S. aureus*, *S. epidermidis*, *P. aeruginosa*, and *Escherichia coli* over the course of 24 h. In addition, the GelMA-AMP-CeONs combination showed the most promising outcomes of enhancing healing speed and the promotion of newly formed skin in mouse wound models over 14 days, partly by the reactive oxygen species (ROS) scavenging property of the CeONs. Hydrogel-treated wounds also resulted in significant native type I and type III collagen deposition, which plays a critical role in wound-healing [58]. The notable advantages of using the AMP-CeONs-loaded GelMA-DOPA hydrogels as a sprayable wound dressing are the pliability, adhesiveness, antimicrobial activity, and skin-remodeling abilities of these novel materials.

Hydrogels have high potential for use as wound dressings, due to their ability to be tailored to mimic the composition and physiochemical properties of the human extracellular matrix. They tend to be highly biocompatible and biodegradable in vivo [59]. However, hydrogels alone tend to be mechanically weak, exhibit poor adhesion to wet tissues, and have limited antimicrobial activity [60,61]. Annabi et al. developed novel hydrogels for sutureless wound closures that address these shortcomings [62]. The backbone of the hydrogel was a combination of GelMA and methacrylol-substituted recombinant human tropoelastin (MeTro). The AMP Tet 213 (KRWWKWWRRC) was tethered to MeTro/GelMA composite hydrogels using visible light-induced crosslinking. Tet 213 has broad spectrum activity and was previously tethered to titanium coatings to inhibit biofilm formation of *S. aureus* in vitro [63]. Increasing the hydrogel polymer concentration resulted in increased adhesive strength, burst pressure, and wound-sealing capabilities compared to commercially available tissue adhesives. The Tet 213-loaded MeTro/GelMA hydrogels inhibited the bacterial colonization of methicillin-resistant *S. aureus* (MRSA) and *E. coli* to a similar extent as ZnO nanoparticles [64]. The Tet 213-loaded MeTro/GelMA hydrogels also supported the growth, spread, and proliferation of 2D surface-seeded and 3D encapsulated fibroblasts in vitro. The hydrogels induced minimal inflammatory responses and were shown to efficiently biodegrade when implanted subcutaneously in a murine animal model. Together, these properties make MeTro/GelMA-AMP hydrogels excellent alternatives to current sutureless wound closure approaches.

Towards developing novel antimicrobial coatings, Liu and coworkers designed an AMP-loaded hydrogel coating, with antibacterial, antithrombotic, and antifouling properties [65]. The hydrogel was composed of sulfobetaine methacrylate (SBMA) and acrylic acid (AAc), which have good hemocompatibility. The hydrogel (SA) was grafted on the inside surface of a polyvinyl chloride (PVC) tube following UV irradiation for one hour. The AMPs WR (WRWRWR-NH_2) or Bac2A (RLARIVVIRVAR-NH_2) were embedded using EDC/NHS chemistry to form SA-WR and SA-Bac2A hydrogels (Figure 1C). The cationic AMP bactenecin is one of the smallest AMPs and its derivative Bac2A has similar Gram-negative coverage but improved Gram-positive activity [66]. WR is an ultrashort AMP, which, when loaded into nanoparticles, displays excellent antimicrobial properties against *S. aureus* [67]. Few platelets aggregated on the surface and the hemolysis ratios of both hydrogels were less than 5%, revealing good hemocompatibility. The SA-Bac2 and SA-WR hydrogels completely prevented *E. coli* and *S. aureus* from growing on the hydrogel surface. After 4 weeks of storage at 37 °C, SA-Bac2 still had antimicrobial activity, but SA-WR lost some activity. This demonstrates the remarkable bioactive stability of the AMPs embedded in the SA-Bac2 hydrogels. Moreover, the embedded AMPs did not influence the antifouling or antithrombotic properties of the hydrogel coating. Excitingly, SA-Bac2 and SA-WR effectively killed *S. aureus* in a rat model with a catheter-induced infection. The SA-Bac2 hydrogel has the potential to be used in clinical practice as a coating to prevent bacterial infections and thrombosis formation for implanted medical devices that will contact blood.

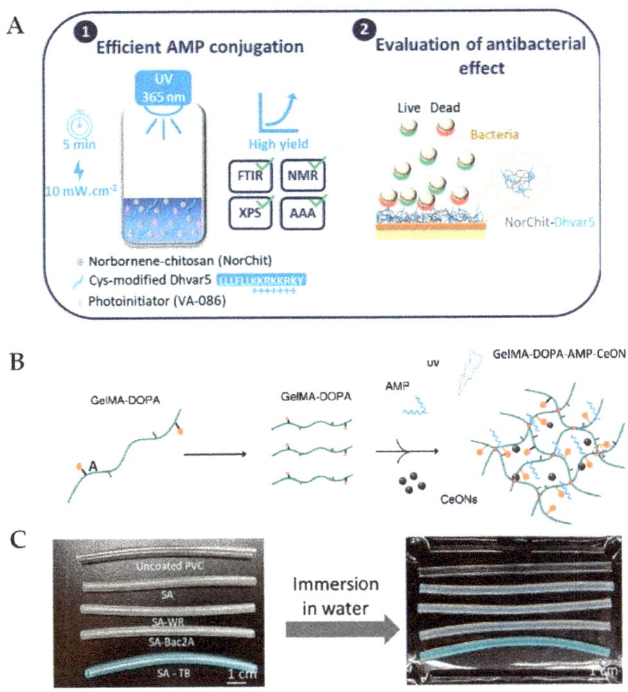

Figure 1. New photopolymerizable, antimicrobial hydrogel formulations created using UV light. (**A**) Norbornene-chitosan (NorChit) was exposed to the photoinitiator VA-086 and UV light (365 nm) at 10 mW/cm^2 for 5 min to initiate the conjugation of cysteine-modified Dhvar5 (NorChit-Dhvar5). The polymers were confirmed and analyzed by Fourier transform infrared spectroscopy (FTIR), proton nuclear magnetic resonance (NMR), X-ray photoelectron spectroscopy (XPS), and amino acid

analysis (AAA). Next, the antimicrobial properties of the NorChit-Dhvar5 films were analyzed and showed a reduction of total adhered and enhanced killing of *S. epidermidis* and *P. aeruginosa*. Reprinted with permission from Ref. [51]. 2022, American Chemical Society. (**B**) GelMA-DOPA polymers were mixed with antimicrobial peptide (AMP) and cerium oxide nanoparticles (CeONs) and exposed to UV light to initiate hydrogel formation. A sprayable antimicrobial and reactive oxygen species (ROS)-scavenging hydrogel was formed. Reprinted with permission from Ref. [56]. 2021, Elsevier. (**C**) Images of various PVC tubes, from top to bottom: uncoated tube, hydrogel (SA) coating, SA-WR, SA-Bac2A, and SA-TB (SA hydrogel coating stained with toluidine blue to demonstrate the uniform and complete coating of the surface). Reprinted with permission from Ref. [65]. 2021, American Chemical Society.

Dental implants are highly prone to infection and Xie and coworkers developed a fine-tuned AMP adhesive system with improved mechanical properties and antimicrobial activity for application in the field of dentistry [68]. The AMP derivative GH12-M2 (GLL-WHLLHHLLH_GSGGG_K) was conjugated to methacrylate (MA) monomers by stirring overnight at room temperature. The GH12-M2 peptide was selected for its activity against oral *Streptococci* [69,70]. The MA-AMP monomers were light-cured for 40 s using visible light. The MA-AMP conjugates had two different C-terminal spacers, GGG and SSSGGG. It was shown that the GGG spacer had superior antimicrobial activity against *Streptococcus mutans*, which may reflect conformational optimization for the improved activity of the AMP. These AMP-loaded hydrogels offer an attractive solution for dental composite restoration failure. Moreover, it was demonstrated that engineering and optimizing the spacer sequence provides the chance to fine-tune the activity of the peptide in the system. The optimization of the peptide linker sequence could be expanded to additional hydrogel applications, such as wound healing and infection prevention on medical devices.

3. Self-Assembling AMP Hydrogels

3.1. Thermosensitive Hydrogels

The main advantage of self-assembling hydrogels over photopolymerizable hydrogels is that they do not require light to form, thereby enabling their use for delivering therapeutics in a minimally invasive manner or in situations where light is not readily available. Hydrogel self-assembly can be triggered by ionic interactions, pH, or temperature [71]. Thermosensitive hydrogels can either transition from liquid to gel above or below a critical solution temperature [72]. For instance, gelatin is liquid at 37 °C and gels below 30 °C [73], whereas poly(N-isopropyl acrylamide) (PNIPAM) is liquid at room temperature and gels at physiologic temperatures [74]. With the aim of designing self-forming antimicrobial hydrogels that enhance wound-healing, Feng and coworkers reported the development of PNIPAM hydrogels containing a joint peptide consisting of an AMP RRWRVIVKW and RADA16 (Ac-RADARADARADARADA-NH_2), termed RA-Amps [75]. The AMP used in this study was found to disrupt bacterial cell membranes and inhibit growth [76,77]. RADA-16 is an amphipathic peptide that self-assembles into a nanofibrous hydrogel [78] and RADA-16 was incorporated into their design as a barrier that prevents the overflow of fluid and cells into the wound site. To promote wound-healing, a mechano growth factor E peptide (MGF-E) was also included due to its well-established role in tissue repair [79]. The hydrogels were formed by dissolving hydrogel components separately, followed by mixing and heating solutions to 37 °C, resulting in rapid assembly (~23 s). The antibacterial properties of the hydrogels were confirmed by agar diffusion assays against *E. coli* and *S. aureus* strains treated with increasing concentrations of RA and RA-Amps. A concentration of 74 μM of RA or RA-Amps resulted in bactericidal activity with over 99% efficiency. PNIPAM/RA-Amps were also loaded with MGF-E peptides and using a murine wound model, the authors found increased the epithelialization, angiogenesis, and formation of fibrous collagen in comparison to commercial wound dressings. While the authors demonstrated the antimicrobial properties of RA-Amps in vitro, this was not confirmed in their in vivo wound model.

3.2. Peptide-Based Self-Assembling Hydrogels

Self-assembling hydrogels can also be completely peptide-based, which form 3D fibrous networks via noncovalent crosslinking between peptides through ionic, hydrophobic, hydrogen, or π–π stacking interactions [80,81]. Certain amphipathic AMPs can form supramolecular hydrogels and self-assembly rates can be controlled by tuning the pH. For instance, Azoulay et al. designed self-forming AMP hydrogels containing the Phe–Lys–Phe (FKF) sequence [82]. FKF peptides are amphipathic and cationic, and these hydrogels were formed by dissolving FKF peptides in an acidic buffer. Self-assembly occurred within minutes and these hydrogels were found to have β-sheet structures, with H-bonding and π–π stacking, at a pH between 3.3 and 4.3. To evaluate their antimicrobial properties in vitro, an agar-based quantitative antibacterial efficacy assay was conducted, significantly reducing the presence of *E. coli*, *P. aeruginosa*, *Acinetobacter baumannii*, and *S. epidermidis* and demonstrating a broad spectrum of activity (Figure 2A). The antimicrobial properties were also evaluated in vivo by applying the hydrogel as a dressing to a wound inoculated with *P. aeruginosa* using a rat wound model. While this study reported a bacterial reduction of 50%, a delay in wound-closing was observed, which could be due to the acidic conditions required for the hydrogel to polymerize.

Amphipathic AMPs can also self-form within a physiological pH range, which can address biocompatibility concerns regarding AMP hydrogels that polymerize under acidic conditions. D-amino acids are more resistant to bacterial proteases [83] and some of these, including lysine, leucine, and valine, have antibacterial properties [84–86]. The KLVFFAK peptide (KK-11) contains a self-assembling motif [87], and Guo et al. synthesized this peptide using D-amino acids (KKd-11) to investigate its antimicrobial activity [88]. The KKd-11 hydrogels were formed with a peptide concentration of 10 mg/mL or higher within 10 min at room temperature. The peptide was first dissolved at a low pH and then the solution was made alkaline with NaOH resulting in hydrogel formation at a pH of 7 or higher. KKd-11 hydrogels prevented the biofilm formation of *E. coli* and *S. aureus* strains, and they killed the bacteria within the biofilms. Given their antimicrobial and inhibitory capabilities and their biocompatibility and higher resistance to proteolytic enzymes, the KKd-11 hydrogels have high potential to be used to prevent wound infections and prevent bacterial growth on medical devices.

In another study, Cao and coworkers synthesized a PAF26 AMP hydrogel that polymerizes at physiologic pH [89]. PAF26 is a hexapeptide with the sequence Ac-RKKWFW-NH_2 that permeabilizes the cell wall, leading to eventual microbial killing [90]. PAF26 is amphipathic in nature and this peptide self assembles into a hydrogel triggered by pH. PAF26 was dissolved in water and the pH slowly increased until a pH of 7.5 was reached, which was determined to be the optimal pH for gelation. After holding at this pH for several minutes, a hydrogel was formed, and the resulting hydrogels were found to have β-sheet structures (Figure 2B). The PAF26-containing hydrogels were co-cultured with fungi or bacteria, and their antimicrobial activity was assessed by the determination of the zone of inhibition and by monitoring growth in the presence of the hydrogels. The PAF26-hydrogels inhibited the growth of *Candida albicans*, *S. aureus*, and *E. coli*. The hydrogels yielded a 100% killing efficiency against a variety of different pathogenic species. These PAF26 hydrogels have the potential for use in clinical practice; however, as the cytotoxicity was not determined, further optimizations may be required to minimize the concentration of PAF26 used that still contains excellent antimicrobial properties.

The research discussed thus far has relied on amphiphilic peptides that form β-sheet secondary structures and rupture cell walls [91]. Armed with this information, Adak et al. synthesized a novel antibacterial lipopeptide, containing palmitic acid-conjugated to a peptide with the sequence NAVSIQKKK (PA-NV) [92]. PA-NV was synthesized using solid-state peptide synthesis. The peptide hydrogel was prepared by dissolving 20 mg PA-NV in phosphate-buffered saline to yield a 2 wt% solution at pH 7.4. The solution was gently heated to dissolve the peptides, and after incubation for 30 min at room temperature, a homogenous PA-NV hydrogels self-assembled. Increasing the hydrogel concentration increased the number of three-dimensional cross-links in the nanofiber structure, which significantly facilitated gelation. The antibacterial activity, cytotoxicity, biocompatibility, and stability of the PA-NA hydrogels against enzymatic degradation were examined. PA-NV hydrogel surfaces were effective at preventing the growth of *E. coli* and *S. aureus* on inoculums of 10^5–10^6 cells/dm^2. However, a larger inoculum of 2.0×10^8 cells/dm^2 or higher showed bacterial growth on the hydrogel surface. It was also found that PA-NV hydrogels are biocompatible, stable against enzymatic degradation by proteinase K, and non-cytotoxic against mammalian cells. This novel lipopeptide has both hydrophobic and hydrophilic domains. The cationic charge of PA-NV hydrogel makes it possible to interact with and disrupt the bacterial membrane, making this the most likely mode of action. The amphiphilic nature of the construct can form β-sheets, which aid in gelation, and produces a biocompatible soft material. These properties make the PA-NV hydrogels exciting and promising candidates for the development of such hydrogels for clinical applications. It will be interesting to see how these hydrogels perform in in vivo model systems.

3.3. AMP Self-Assembling Hydrogels with Multi-Functionality

Self-forming AMP hydrogels can also include additional networks to increase their functionality beyond antimicrobial activity. For example, hyaluronic acid (HA) is a component of the skin extracellular matrix (ECM) and has been shown to accelerate wound-healing [93]. In a study by Suo et al., the antimicrobial properties of the AMP KK(SLKL)$_3$KK were studied with the inclusion of HA [94]. Hydrogels were formed by oxidizing HA, which was then dissolved and mixed with an AMP solution at a neutral pH. The antimicrobial properties were confirmed in vitro by co-incubating *S. aureus* and *E. coli* with soluble AMPs and the AMP-HA complexes at different concentrations. While a higher concentration of AMP-HA compared to AMP alone was required for complete killing, AMP-HA still had excellent antimicrobial activity. Further, the AMP-HA hydrogels greatly reduced healing times in a mouse wound model and promoted increased collagen deposition, wound epithelialization, and angiogenesis (Figure 2C). This hydrogel is promising for the possible development of an antimicrobial hydrogel to promote wound-healing in clinical practice.

Another example was reported by Yang and coworkers, who created a set of RADA16 hydrogels assembled with the AMP Tet 213 [95]. The RADA16-AMP hydrogels self-assembled with a 10 mg/mL RADA-AMP solution within 30 min at room temperature. The antimicrobial activity of RADA-AMP hydrogels compared to RADA-alone hydrogels against *S. aureus* was evaluated. It was found that the RADA-AMP hydrogels were able to continuously release AMPs over 28 days and had reduced *S. aureus* growth to a greater extent. After 28 days, AMPs remained in the hydrogel. In addition, the antimicrobial activity increased as the AMP concentration increased. An in vivo experiment was conducted using rabbits as a model system. A small quantity of bone marrow was extracted from the tibia and inoculated with 1.5×10^8 bacteria/mL to establish an osteomyelitis model. After 4 weeks, RADA and RADA-AMP gels were injected into the same site and, 6 or 10 weeks later, were analyzed by micro-CT. The inflammation caused by the bacteria was limited to the injection site. Moreover, the combination of AMP with RADA16 resulted in new bone formation; the rabbits had more mature bone and greater bone regeneration compared to the controls and RADA-containing hydrogels. Given their antimicrobial efficacy, ability to promote osteogenesis, sustained release of AMPs, and biocompatibility, the RADA16-AMP hydrogels have a high potential to be used in patients to aid in the treatment of bone infections and the promotion of osteogenesis. Future work should concentrate on optimizing the concentration of Tet 213 for potency, the determination of the mechanism of action of RADA-AMP on microbial killing and osteogenesis, and possibly substituting other AMPs with superior antimicrobial or healing properties for other applications.

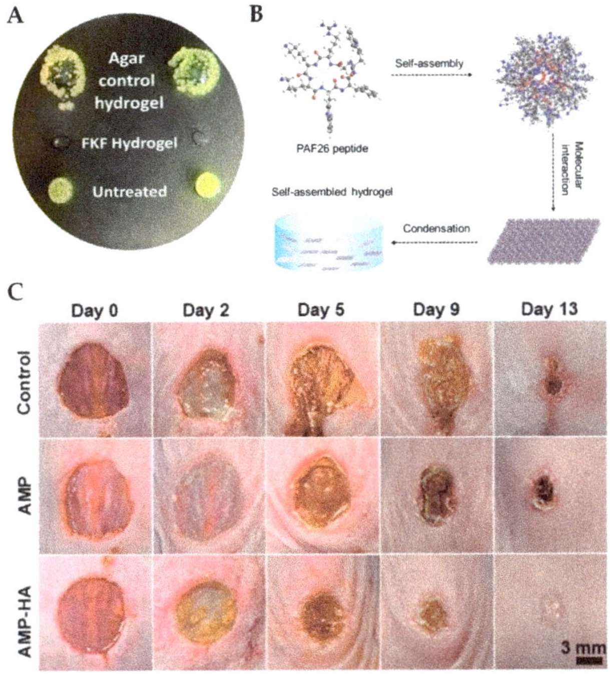

Figure 2. Self-polymerizing hydrogels are effective antimicrobial agents that can promote wound-healing. (**A**) *P. aeruginosa* spotted on an agar plate shows complete inhibition by the FKF hydrogel (center) compared to agar control hydrogels (top) and the untreated positive control (bottom). Reprinted with permission from Ref. [82]. Copyright 2021, Elsevier. (**B**) The proposed self-assembly mechanism of the PAF26 peptide. When the pH of the solution reached 7.5, the molecular stacking interactions lead to the formation of β-sheet structures that self-assemble into micelles. The micelles then are linked together into a chain-like structure, which, following condensation, leads to the formation of a hydrogel. Reprinted with permission from Ref. [89] Copyright 2019, The Royal Society of Chemistry. (**C**) In vivo wound-closure studies from *S. aureus*-infected mice wounds. Wounds were evaluated on days 0, 2, 5, 9, and 13 post-infection. The control group was untreated, while the AMP group was treated with soluble AMP, and the AMP-HA cohort was treated with AMP-loaded hydrogels. Reprinted with permission from Ref. [94]. Copyright 2021, American Chemical Society.

Zhou et al. studied the utility of the AMP, Jelleine-1 (J1), in combination with adenosine diphosphate (ADP). J1 is a natural, amphiphilic AMP that was isolated from the royal jelly of the honeybee, with a sequence of PFKLSLHL-NH_2 [96]. ADP was found to be an important substance for platelet aggregation and coagulation, which was the rational for adding ADP to this hydrogel [97,98]. The J1-ADP hydrogel is formed by dissolving the J1 peptide into dimethyl sulfoxide to make a 100 mM stock solution [99]. An equal volume of ADP is added and stirred to allow for self-assembly into hydrogels. The antimicrobial, hemostatic, and anti-adhesion properties of this J1-ADP hydrogel were assessed. The J1-ADP hydrogel had good antimicrobial activity against some common bacterial and fungal pathogens, including *S. aureus*, *E. coli*, and *C. albicans*. It also had efficacy against MRSA strains. Postoperative peritoneal adhesion is a serious complication of abdominal surgery, which leads to severe abdominal pain, intestinal obstruction, organ disfunction, and additional complications [100–102]. The presence of the J1-ADP hydrogel in an in vivo mouse wound model demonstrated decreased intrabdominal adhesions and more efficient hemostasis, as determined by blood coagulation, platelet activation, and platelet adhesion assays. The injectable quality of these hydrogels makes them suitable for surgical contexts.

Additionally, the anti-adhesive and hemostatic properties of this hydrogel also make it useful for application in large areas that are prone to adhesions. However, this study did not include an analysis of microbial burden with and without the J1-ADP hydrogel in vivo, which could demonstrate an additional infection prevention benefit. Nonetheless, self-forming hydrogels are promising materials for use in a wide variety of clinical applications.

4. AMP-Releasing Hydrogels

With the extended use of nanomaterials to deliver therapeutic agents to the target site, additional control over the release of the payload becomes possible. Any spatiotemporal control of drug release could improve efficacy, limit off-target effects, and minimize cytotoxicity. This technology has been applied in drug delivery systems for specific types of cancer cells [103,104]. Now, this same controlled-release technology is being developed for use with AMP delivery to sites of infection. There are three mechanisms of release that are being currently explored: temperature, pH, and enzymatic [105].

4.1. Temperature Controlled Release of AMPs

Rezaei and coworkers selected chitosan, which is a natural chitin-derived polysaccharide, for study because this material has many desirable properties, such as appropriate biodegradability, low toxicity, and antibacterial activity [106,107]. Thermo-responsive chitosan (TCTS) hydrogels were loaded with varying concentrations, from 0 to 16 μg/mL of the AMP piscidin-1 [108]. Piscidin-1 is a 22-residue, cationic peptide that was originally identified in the aquaculture striped bass. It has demonstrated potent antimicrobial activity against different bacterial and fungal species but is cytotoxic towards mammalian cells [109,110]. Due to the associated cytotoxicity, loading them into a hydrogel with controlled release is highly beneficial. The TCTS hydrogels were formed by dissolving 2% chitosan (*w*/*v*) in acetic acid and stirring for 24 h. To optimize the gelation time, the TCTS hydrogel was cross-linked at 4 °C with various concentrations of β-glycerophosphate disodium salt pentahydrate. It was found that the TCTS-40% β-glycerophosphate hydrogels had the shortest gelation time of 15 min at 37 °C. The formed TCTS hydrogels were loaded with 4, 8, and 16 μg/mL of piscidin-1. The antibacterial activity against susceptible and drug-resistant *A. baumannii* and the biocompatibility and release behavior of the AMP-TCTS hydrogels were determined. At the highest concentration of AMP tested (16 μg/mL), the hydrogels showed excellent bactericidal activity against the drug-resistant *A. baumannii* strains, while all concentrations were effective against the standard strain. On day 1, 40% of the piscidin-1 was released from the TCTS hydrogels and a controlled release of up to 7 days was observed. At day 7, 90% of the AMP had been released. Importantly, the AMP-TCTS hydrogels were not cytotoxic against human fibroblasts. The AMP-TCTS hydrogel represents a potential new, controlled delivery system of AMPs to the site of wound infections for the treatment of infections with drug-resistant pathogens. As piscidin-1 was previously reported to have hemolytic activity [111], it is important to determine this property of the AMP-TCTS hydrogels. In addition, in vivo animal wound models should also be analyzed to determine the antimicrobial and wound-healing properties of this system.

An extension of the temperature-controlled release of antimicrobial cargo is photothermal release. Moorcroft et al. developed a novel, nanoparticle-loaded hydrogel for the light-activated release and photothermal enhancement of AMPs [112]. To assemble the hydrogels, liposomes containing the AMP IK8 (IRIKIRIKCO NH_2) were first prepared. IK8 is a β-sheet forming AMP with a potent broad-spectrum activity [113], which is unfortunately rapidly degraded by proteases [114]. Liposomal encapsulation protected and preserved the antimicrobial activities of IK8. To form hydrogels, a solution of four-arm PEG maleimide was mixed with IK8-encapsulated liposomes and lipid-coated gold nanorods (AuNRs). This solution was mixed thoroughly with PEG dithiol to form the hydrogels. The PEG hydrogel was chosen because gelation occurs through a Michael-type reaction, which allows for tunable gelation rates and mechanical properties, and because of its high permeability by small hydrophilic molecules, it should not interfere with AMP release kinetics [115–117]. The fabricated hydrogels were irradiated with near infrared light (860 nm) at 1.8–2.8 W/cm^2 for 10 min, which enabled the control of the saturation temperature between 50 and 65 °C. Upon heating to 55 °C, about 65% of encapsulated IK8 and AuNRs are released from a 5% hydrogel, causing significant bacterial killing against *P. aeruginosa* and *S. aureus*, with no cytotoxicity. In this study, hydrogels containing AuNRs alone, IK8 alone, and their combination were analyzed, and it was found the presence of AuNRs exhibited better photothermal activity and enhanced the bactericidal activity of IK8 against bacterial biofilms. Based on these findings, this hydrogel has excellent potential for the development of advanced wound dressings that can provide light-triggered and highly effective antimicrobial therapy. However, in vivo studies with animal models and the further optimization of the hydrogel formulation for long-term stability and biocompatibility are required before any practical clinical application can be developed.

4.2. Release of AMPs in Response to pH

The second AMP release mechanism occurs when the pH of the environment changes, causing the release of antimicrobial cargo. Wu and coworkers hypothesized that a pH-sensitive hydrogel could release antimicrobial agents in response to the weakly acidic wound environment in a controlled manner, leading to faster wound-healing and reduced bacterial load [118]. The regions surrounding bacterial infections are typically mildly acidic due to the secretion of lactic acid by the invading bacteria. A pH-sensitive hydrogel was formed by combining oxidized dextran (ODEX) and a peptide called DP7 (VQWRIRVAVIRK). DP7 is a cationic AMP with broad antimicrobial activity that displays synergistic effects with various antibiotics [119,120]. ODEX cross-linked hydrogels are made by mixing a 6% DP7 solution with a 10% dextran-aldehyde in a 1:1 ratio and allowing time for gelation (DP7-ODEX hydrogels). The DP7-ODEX hydrogel forms via pH-sensitive Schiff base reversible covalent bonds, in which DP7 serves as a linker and forms a porous network with ODEX. The Schiff base linkage is broken down in the presence of protons in the environment, as in an acidic environment (Figure 3A). For synergy studies, the 3rd-generation cephalosporin ceftazidime was premixed with DP7 and then combined with the ODEX solution to form CAZ-DP7-ODEX hydrogels. The antimicrobial properties of the hydrogels were analyzed both in vitro and in vivo. Three different pathogenic bacteria were used, *P. aeruginosa*, *S. aureus*, and *E. coli*. The DP7-ODEX hydrogels had superior killing efficacy against all three species compared to control hydrogels. Moreover, when combined with ceftazidime, CAZ-DP7-ODEX hydrogels displayed an additive or synergistic effect against a collection of 27 multidrug-resistant *P. aeruginosa* isolates. Furthermore, the use of these hydrogels in both normal and diabetic mouse models showed faster wound-healing rates and the more complete regeneration of dermis and epidermis skin layers. In addition, the hydrogels were well-tolerated and did not cause any adverse effects in the mice. These results strongly suggest that pH-sensitive hydrogels made from dextran and pH-sensitive peptides have the potential to be an effective treatment for multidrug-resistant bacterial-infected wounds. While promising so far, this study only used a small sample size of mice, which limits the generalizability of the findings. The long-term safety and efficacy of the pH-sensitive hydrogels are worth being further investigated.

Wei et al. used a pH-sensitive hydrogel to prepare a system for the treatment of chronically infected wounds which regulates both the inflammatory response, through the AMPs and enhances collagen deposition and angiogenesis to promote wound-healing, through the addition of platelet-rich plasma (PRP) [121]. Like Zhang et al., the pH-sensitive hydrogel was formed based on Schiff base chemistry using ODEX. The hydrogels were prepared by mixing equal volumes of 100 mg/mL ODEX with 25 mg/mL AMP-modified HA (HA-AMP). The AMP used in this study was cecropin (SWLSKTAKKLFKKIPKKIPKKRFPRPRPWPRPNMI-NH_2), which has a broad antimicrobial spectrum of activity [122,123]. The mechanism of action of cecropin is the disruption of membrane integrity by changing the conformation of the cell membrane, resulting in the leakage of cell contents [124]. To prepare a PRP-loaded hydrogel, PRP was added to a 3.125% solution of HA-AMP prior to mixing with an equal volume of ODEX, to create ODEX/HA-AMP/PRP hydrogels. Both the ODEX/HA-AMP and ODEX/HA-AMP/PRP hydrogels displayed good antimicrobial activity against *E. coli*, *S. aureus*, and *P. aeruginosa* without any cytotoxicity. Spectroscopy was used to monitor AMP release over a span of 14 days. The ODEX/HA-AMP hydrogel released 20% of the AMP on the first day and afterwards had a continuous, sustained released, with 10% AMP remaining on day 14. Next, the release kinetics of the growth factors PDGF, TGF-β, and EGF from the hydrogel were measured using ELISA. The release of each growth factor showed a linear cumulative release for the first 7 days, after which the release rate slowed down. This suggests that the hydrogel provides a stable supply of growth factors during the wound-healing period. Finally, using a diabetic mouse model, ODEX/HA-ADH/PRP hydrogels promoted the skin-healing of infected wounds but were not completely effective in inhibiting inflammation. All things considered, the ODEX/HA-AMP/PRP hydrogels were effective at accelerating wound-healing, lowering the bacterial burden, and balancing inflammatory cell infiltration. Therefore, these hydrogels are promising for the treatment of severely infected wounds, including problematic diabetic foot ulcers.

4.3. Enzymatic Release of AMPs

The third mechanism of AMP release is the enzymatic cleavage of AMPs that are tethered to a hydrogel surface. These enzymes may be host or bacterial in nature. Obuobi and coworkers synthesized AMP-loaded DNA hydrogels based on the high binding affinity between cationic AMPs and polyanionic DNA nanostructures [125]. The AMP L12 is a synthetic peptide (LKKL) with strong broad-spectrum activity against both susceptible and multidrug-resistant bacteria [126]. Hydrogels were formed by mixing a linker and Y-scaffold nanostructure solutions, heating and gradually cooling to 25 °C over two hours. Various concentrations of L12, from 125 to 250 μM, were then gently mixed with the nanostructures to ensure uniform drug dispersion and prevent precipitation. The resulting L12-loaded DNA hydrogels were biocompatible and had significant activity against methicillin-resistant and susceptible *S. aureus* and *E. coli*. As the release of AMPs are triggered by the enzymatic secretion from the bacteria, the DNA hydrogels were tested in the presence and absence of DNase I. The L12 had a half-life of 3 h when subjected to 10 U/mL of DNase I, with a complete release by 12 h, but the half-life could be reduced to 30 min if exposed to more DNase I. These results show that the release of loaded AMPs can be controlled in response to DNA degradation. Ex vivo antimicrobial studies using porcine skin infected with *S. aureus* showed a 4-log reduction in bacterial numbers within 4 h for groups treated with L12-loaded DNA hydrogels. These results demonstrate the potential for treating cutaneous *S. aureus* infections. As an application of these DNA hydrogels is wound-healing, their ability to modulate inflammation was examined by monitoring TNF-α, a pro-inflammatory cytokine released after macrophage activation. In the presence of lipopolysaccharide (LPS), a component of the Gram-negative bacterium cell wall and powerful immunogen, macrophages significantly increased TNF-α release. However, LPS-stimulated cells co-treated with the DNA hydrogels resulted in significant reduction in TNF- α production, thereby demonstrating anti-inflammatory properties and potential to accelerate wound-healing. To confirm the in vitro findings, DNA hydrogels were applied on mice excision wounds and healing properties evaluated for ten days. Wound healing was evaluated by measuring changes in the residual wound area and re-epithelialization rates. There were faster wound-healing rates in the DNA-treated hydrogels than control hydrogels. In addition, the formation of granulation tissues was significantly greater in hydrogel-treated mice than the controls, indicating that these hydrogels could be used for deep or chronic wounds. These DNA hydrogels have high potential, but there may need to be additional optimizations required to increase the stability of the hydrogels, as after 24 h without enzyme, more than 50% of the AMP was already released.

Patients with diabetic wounds are frequently exposed to bacterial infections, which causes the damaged skin tissue to heal more slowly due to hyperglycemia [127]. Jeong and colleagues developed an injectable, controlled AMP-releasing hydrogel for the treatment of infected diabetic wounds with high stability and less toxicity [128]. The hydrogels were prepared by modifying HA with cyclodextrin (CD) to synthesize HA-CD and adamantane (Ad) to form Ad-HA through amidation and esterification, respectively. Equal amounts of HA-CD and Ad-HA were mixed for 15 s to form hydrogels. The KR12 AMP (KRIVQRIKDFLR-NH_2), the shortest alpha helical derivative of the human cathelicidin LL-37 AMP with antimicrobial activity [129,130], was conjugated to Ad-HA hydrogels through a cyclic peptide linker comprising a reactive oxygen species (ROS) and matrix metalloprotease (MMP) cleavable sequence to form Ad-HA-AMP hydrogels. The infection-responsive cyclic linker helps in the controlled release of the AMP only in the presence of both MMP and ROS during infection, which minimizes AMP toxicity (Figure 3B). The Ad-HA-AMP hydrogels displayed enhanced antibacterial activity against *E. coli* and *S. aureus* and increased serum stability compared to KR12 alone. Strikingly, the soluble KR12 lost its antibacterial activity in 72 h, while the hydrogel-embedded KR12 maintained its activity. KR12 is short and is easily degraded by proteolytic enzymes in the body. In an in vitro cytocompatibility assay on fibroblast cells, Ad-HA-AMP hydrogel did not release AMP in the absence of ROS and MMP, thereby inducing cell proliferation and minimizing cytotoxicity. Moreover, in an in vivo diabetic wound-healing mouse model, the AMP-containing hydrogels showed enhanced and accelerated wound-healing activity in infected mice. In contrast, the controls delayed wound-healing and thus did not cure the bacterial infections. The controlled release of the AMP from these hydrogels minimizes side effects, treats infection, and displays efficient wound-healing. These supramolecular hydrogels have a high potential to be developed into new wound coatings that would be particularly useful for the treatment of chronic wounds, especially in diabetic patients.

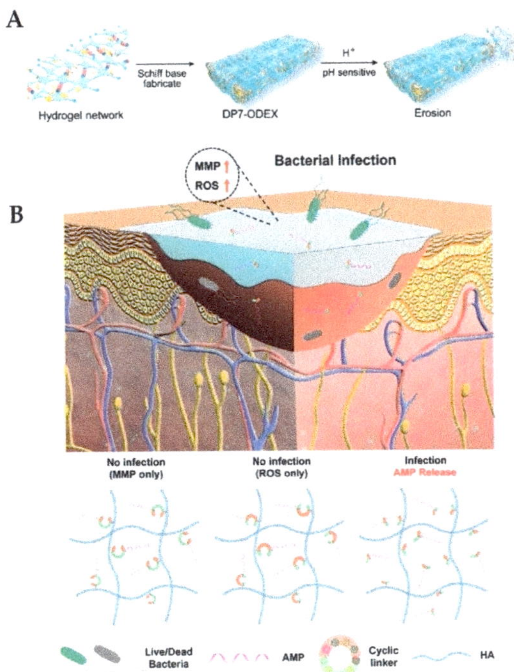

Figure 3. Release mechanisms for AMP-loaded hydrogels. (**A**) Structure of DP7-ODEX hydrogel. The hydrogel network is fabricated by Schiff base chemistry, forming DP7-ODEX. When exposed to low pH, the hydrogel erodes, releasing antimicrobial contents. Reprinted with permission from Ref. [118].

Copyright 2022, Elsevier. (**B**) Schematic representation of the Ad-HA hydrogel conjugated to KR12 (AMP) as an injectable wound coating. The cyclic linker is responsive to both reactive oxygen species (ROS) and matrix metalloproteinases (MMPs) and will only release AMP in the presence of both, during bacterial infection. Reprinted with permission from Ref. [128]. Copyright 2023, American Chemical Society.

4.4. Nanoparticles as Drug Delivery Systems

The use of nanoparticles as drug delivery systems has been increasingly explored in recent years [131]. Among the many nanomaterials available, mesoporous silica nanoparticles (MSNs) are particularly attractive because they are biocompatible and capable of anchoring to multiple different functional groups. In addition, MSNs have a large surface area and a large porous volume, which improve the ability to load therapeutic cargo [132,133]. In a recent study by Ma and coworkers, the objective was to design an MSNs modified with an ovotransferrin-derived AMP that could act as a mechanism to interact with the bacteria and produce antimicrobial activities against them [134]. The amphipathic AMP OVTp12 was derived from egg white ovotransferrin, which displayed excellent antimicrobial activity through the disruption of the bacterial cell membrane [135]. MSNs were synthesized by heating a mixture of hexadecyltrimethyl ammonium bromide and NaOH. Tetraethyl orthosilicate was added to precipitate the MSNs. OVTp12 was coupled to the MSNs using NHS and EDC chemical crosslinking agents, which were mixed in solution for 24 h to form MSNs@OVTp12 particles. In this study, gentamicin was also included and mixed with the MSNs@OVTp12 particles at room temperature for 24 h, forming MSNs@OVTp12@Gen particles. In a weakly acidic environment, the cargo was continuously released over 72 h. The presence of OVTp12 on the surface allowed for effective interaction with *E. coli* and enhanced bacterial killing. MSNs@OVTp12@Gen was found to have effective properties for both attracting and adhering to bacteria and bactericidal activity. In vivo, MSNs@OVTp12@Gen enhanced overall survival and more effectively treated *E. coli* infections in mice than MSNs@Gen or gentamicin treatment alone. In addition, these particles were found to decrease inflammatory cytokine release and were noncytotoxic in vivo. The MSNs@OVTp12@Gen designed and used in this study had the ability to tolerate weakly acidic environments and had a strong, bactericidal effect, which makes them promising for targeted antibiotic delivery to bacterial infection sites. The potential to incorporate MSNs into hydrogel matrices is exciting and should be further explored, as MSNs are bioactive, can improve mechanical properties of the hydrogel, and can release various types of cargo in a controllable manner [136]. Together, an MSN-hydrogel could be designed with superior antimicrobial and healing properties, while limiting the emergence of further resistance.

5. Conclusions, Challenges, and Perspectives

AMP-loaded hydrogels as antibacterial weapons are an attractive alternative solution to traditional antibiotics, which, in the face of the current antimicrobial resistance crisis, is of utmost importance. Exciting new antimicrobial hydrogel coatings are rapidly being investigated and developed. In this review, we highlighted recent work on AMP-loaded hydrogels from the last five years that has advanced the development of photopolymerizable, self-forming, and AMP-releasing hydrogels (Table 1). These studies clearly demonstrate that a wide variety of cationic or amphipathic AMPs retain good antibacterial activity when tethered on or within a hydrogel matrix. Overall, each of the newly studied hydrogels discussed here resulted in decreased cytotoxicity, improved antimicrobial killing, and, in many cases, enhanced wound-healing. It is important to note that the AMP concentrations of many of the studies presented were seemingly arbitrary. High AMP concentrations can be cytotoxic, so future AMP hydrogels could be designed with minimalist AMP concentrations that maximize bactericidal properties while minimizing unwanted cytotoxicity. One promising way to reduce the amount of total AMP is to explore the use of multiple AMPs that have synergistic effects. These interactions can be determined using a simple microbial growth technique called a checkerboard assay, where each AMP is serially diluted horizontally or vertically to examine combinatorial effects [137]. Reducing the amount of AMP required not only further reduces cytotoxic potential but also makes large-scale production more practical. In addition, using a combination of AMPs with different mechanisms of action may provide superior microbicidal activity and further reduce the possibility of the development of downstream AMP resistance.

Table 1. Summary of recently developed AMP hydrogels.

AMP(s)	Antimicrobial Activity	Studied	Unique Features/ Applications	Reference
Dhar5	*S. epidermidis* *P. aeruginosa*	In vitro	Surface coatings	Alves et al. [51]
Battacin	*S. aureus* *P. aeruginosa*	In vitro	Topical antibacterial agents	De Zoysa et al. [54]
HHC-36	MRSA *S. epidermidis* *P. aeruginosa* *E. coli*	In vitro In vivo	Sprayable wound dressing	Cheng et al. [56]
Tet 213	MRSA *E. coli*	In vitro In vivo	Sprayable wound dressing	Annabi et al. [62]
WR and Bac2a	*S. aureus* *E. coli*	In vitro In vivo	Wound dressing coating	Liu et al. [65]
RRWRVIVKW	*S. aureus* *E. coli*	In vitro In vivo	Wound dressing coating	Feng et al. [75]
FKF	*P. aeruginosa* *E. coli* *A. baumannii* *S. epidermidis*	In vitro In vivo	Injectable wound dressing	Azoulay et al. [82]
KK-11 and KKd-11	*S. aureus* *E. coli*	In vitro	Enzymatically stable antimicrobial hydrogels	Guo et al. [88]
PAF26	*S. aureus* *E. coli* *C. albicans*	In vitro	Injectable antimicrobial hydrogels	Cao et al. [89]
NAVSIQKKK	*S. aureus* *E. coli*	In vitro	Biocompatible antimicrobial hydrogels	Adak et al. [92]
KK(SLKL)$_3$KK	*S. aureus* *E. coli*	In vitro In vivo	Injectable wound dressing	Suo et al. [94]
Tet 213	*S. aureus*	In vitro In vivo	Bone-forming antibacterial hydrogels	Yang et al. [95]
Jelleine-1	MRSA/MSSA *E. coli* *C. albicans*	In vitro In vivo	Postoperative, adhesive antibacterial hydrogels	Zhou et al. [99]
Piscidin-1	*A. baumannii*	In vitro	AMP-releasing with temperature	Rezaei et al. [108]
IK8	*S. aureus* *P. aeruginosa*	In vitro	AMP-releasing with light	Moorcroft et al. [112]
DP7	*S. aureus* *E. coli* *P. aeruginosa*	In vitro In vivo	AMP-releasing with pH	Wu et al. [118]

Table 1. *Cont.*

AMP(s)	Antimicrobial Activity	Studied	Unique Features/ Applications	Reference
Cecropin	*S. aureus* *E. coli* *P. aeruginosa*	In vitro In vivo	AMP and PRP hydrogels for wound healing	Wei et al. [121]
L12	MRSA/MSSA *E. coli*	In vitro In vivo	Anti-inflammatory antimicrobial hydrogels	Obuobi et al. [125]
KR12	*S. aureus* *E. coli*	In vitro In vivo	Antimicrobial hydrogels for diabetic wounds	Jeong et al. [128]
OVTp12	*E. coli*	In vitro In vivo	Antibacterial NPs for wound infections	Ma et al. [135]

MRSA: methicillin-resistant *S. aureus*, MSSA: methicillin-susceptible *S. aureus*, NP: nanoparticle.

The potential of these novel AMP-loaded hydrogel formulations to become clinically approved coatings is exciting and likely in the foreseeable future. However, there are multiple considerations and challenges that remain and must be considered before use in clinical practice becomes a reality. What features are the most important when designing a practical antimicrobial coating? The field agrees that potent, broad spectrum antimicrobial activity is important. In addition, little to no hemolytic or cytotoxicity is essential. However, there seems to be some disagreement on whether AMPs would be better suited in a controlled and prolonged release platform or whether they should be covalently tethered. AMP hydrogel self-assembly is guided by hydrophobic forces and π–π stacking or covalent crosslinking using polymers, such as chitosan [138–140]. One notable issue with self-assembling AMP hydrogels is the requirement for high concentrations of AMP, which may introduce off-target cytotoxicity and would increase manufacturing costs. In contrast, AMP hydrogels with a non-AMP backbone can include lower concentrations of AMP while still forming a hydrogel, thereby potentially circumventing limitations seen with self-assembling AMP hydrogels. As hydrogels can be functionalized with multiple bioactive molecules and combined with other nanoparticles, another challenge will be to determine the optimal ratio and compatibility of each component so that they produce synergistic effects.

Without a doubt, it is worthwhile to investigate and optimize different multifunctional hydrogel configurations to develop practical applications for clinical use. It will be important to continue to optimize and broaden the antimicrobial spectrum of activity. For example, a wound coating that features antimicrobial activity against many pathogens, including Gram-negative and -positive bacteria and skin infection-causing fungi, would be ideal. It is already possible to design hydrogels with excellent antimicrobial properties in combination with hemostatic, immunomodulatory, antithrombotic, and healing properties. The maximum number of beneficial factors that can be incorporated into a single hydrogel formulation are not currently known but the possibilities and combinations are limitless. The future may simply involve the production of highly specialized hydrogels tailored to specific medical needs to provide superior therapy. These antimicrobial hydrogels are broadly applicable to many fields, including wound dressings, catheters, biofilm-based medical device infections, dental coatings, and even contact lenses. One thing is certain—the future of AMP-loaded hydrogel technology for the treatment of infections is bright.

Author Contributions: Conceptualization, S.L.V. and V.J.C.; writing—original draft preparation, A.C., R.K., M.A., G.F., D.C., R.S.T., V.J.C. and S.L.V.; writing—review and editing, S.L.V. and V.J.C.; supervision, S.L.V. and V.J.C.; funding acquisition, S.L.V. and V.J.C. All authors have read and agreed to the published version of the manuscript.

Funding: This work was funded by a Biomedical Sciences departmental award from Cooper Medical School of Rowan University to V.J.C. and S.L.V. Research reported in this publication was supported by the National Institute of General Medical Sciences of the National Institutes of Health under Award Number T34GM136492.

Institutional Review Board Statement: Not applicable.

Informed Consent Statement: Not applicable.

Data Availability Statement: Data sharing not applicable.

Acknowledgments: We thank members of the Carabetta and Vega labs for helpful discussions and critical reading of the manuscript.

Conflicts of Interest: The authors declare no conflict of interest.

References

1. Festas, A.; Ramos, A.; Davim, J. Medical devices biomaterials–A review. *Proc. Inst. Mech. Engs Part. L* **2020**, *234*, 218–228. [CrossRef]
2. Pietrocola, G.; Campoccia, D.; Motta, C.; Montanaro, L.; Arciola, C.R.; Speziale, P. Colonization and infection of indwelling medical devices by *Staphylococcus aureus* with an emphasis on orthopedic implants. *Int. J. Mol. Sci.* **2022**, *23*, 5958. [CrossRef]
3. Weinstein, R.A.; Darouiche, R.O. Device-associated infections: A macroproblem that starts with microadherence. *Clin. Infect. Dis.* **2001**, *33*, 1567–1572.
4. Donlan, R.M. Biofilms and device-associated infections. *Emerg. Infect. Dis.* **2001**, *7*, 277–281. [CrossRef]
5. Khatoon, Z.; McTiernan, C.D.; Suuronen, E.J.; Mah, T.F.; Alarcon, E.I. Bacterial biofilm formation on implantable devices and approaches to its treatment and prevention. *Heliyon* **2018**, *4*, e01067. [CrossRef]
6. Bjarnsholt, T. The role of bacterial biofilms in chronic infections. *Apmis* **2013**, *121*, 1–58. [CrossRef]
7. Caldara, M.; Belgiovine, C.; Secchi, E.; Rusconi, R. Environmental, microbiological, and immunological features of bacterial biofilms associated with implanted medical devices. *Clin. Microbiol. Rev.* **2022**, *35*, e0022120. [CrossRef]
8. Gupta, P.; Sarkar, S.; Das, B.; Bhattacharjee, S.; Tribedi, P. Biofilm, pathogenesis and prevention—A journey to break the wall: A review. *Arch. Microbiol.* **2016**, *198*, 1–15. [CrossRef] [PubMed]
9. Flemming, H.-C.; Wingender, J.; Szewzyk, U.; Steinberg, P.; Rice, S.A.; Kjelleberg, S. Biofilms: An emergent form of bacterial life. *Nat. Rev. Microbiol.* **2016**, *14*, 563–575. [CrossRef] [PubMed]
10. Fux, C.A.; Stoodley, P.; Hall-Stoodley, L.; Costerton, J.W. Bacterial biofilms: A diagnostic and therapeutic challenge. *Expert Rev. Anti-Infect. Ther.* **2003**, *1*, 667–683. [CrossRef]
11. Mah, T.F. Biofilm-specific antibiotic resistance. *Future Microbiol.* **2012**, *7*, 1061–1072. [CrossRef] [PubMed]
12. Roilides, E.; Simitsopoulou, M.; Katragkou, A.; Walsh, T.J. How biofilms evade host defenses. *Microbiol. Spectr.* **2015**, *3*. [CrossRef]
13. Josephs-Spaulding, J.; Singh, O.V. Medical device sterilization and reprocessing in the era of multidrug-resistant (MDR) bacteria: Issues and regulatory concepts. *Front. Med. Technol.* **2021**, *2*, 587352. [CrossRef] [PubMed]
14. Jamal, M.; Ahmad, W.; Andleeb, S.; Jalil, F.; Imran, M.; Nawaz, M.A.; Hussain, T.; Ali, M.; Rafiq, M.; Kamil, M.A. Bacterial biofilm and associated infections. *J. Chin. Med. Assoc.* **2018**, *81*, 7–11. [CrossRef] [PubMed]
15. Li, S.; Dong, S.; Xu, W.; Tu, S.; Yan, L.; Zhao, C.; Ding, J.; Chen, X. Antibacterial hydrogels. *Adv. Sci.* **2018**, *5*, 1700527. [CrossRef]
16. Caliari, S.R.; Burdick, J.A. A practical guide to hydrogels for cell culture. *Nat. Methods* **2016**, *13*, 405–414. [CrossRef]
17. Lin, C.C.; Ki, C.S.; Shih, H. Thiol-norbornene photo-click hydrogels for tissue engineering applications. *J. Appl. Polym. Sci.* **2015**, *132*. [CrossRef]
18. Moritz, M.; Geszke-Moritz, M. The newest achievements in synthesis, immobilization and practical applications of antibacterial nanoparticles. *J. Chem. Eng.* **2013**, *228*, 596–613. [CrossRef]
19. Lu, Y.; Mei, Y.; Drechsler, M.; Ballauff, M. Thermosensitive core-shell particles as carriers for ag nanoparticles: Modulating the catalytic activity by a phase transition in networks. *Angew. Chem. Int. Ed. Engl.* **2006**, *45*, 813–816. [CrossRef]
20. Boonkaew, B.; Suwanpreuksa, P.; Cuttle, L.; Barber, P.M.; Supaphol, P. Hydrogels containing silver nanoparticles for burn wounds show antimicrobial activity without cytotoxicity. *J. Appl. Polym. Sci.* **2014**, *131*. [CrossRef]
21. Simon, T.; Wu, C.-S.; Liang, J.-C.; Cheng, C.; Ko, F.-H. Facile synthesis of a biocompatible silver nanoparticle derived tripeptide supramolecular hydrogel for antibacterial wound dressings. *NJC* **2016**, *40*, 2036–2043. [CrossRef]
22. Xu, L.; Li, X.; Takemura, T.; Hanagata, N.; Wu, G.; Chou, L.L. Genotoxicity and molecular response of silver nanoparticle (NP)-based hydrogel. *J. Nanobiotechnol.* **2012**, *10*, 16. [CrossRef] [PubMed]
23. Jones, N.; Ray, B.; Ranjit, K.T.; Manna, A.C. Antibacterial activity of ZnO nanoparticle suspensions on a broad spectrum of microorganisms. *FEMS Microbiol. Lett.* **2008**, *279*, 71–76. [CrossRef] [PubMed]
24. Mohandas, A.; Kumar, P.T.S.; Raja, B.; Lakshmanan, V.K.; Jayakumar, R. Exploration of alginate hydrogel/nano zinc oxide composite bandages for infected wounds. *Int. J. Nanomed.* **2015**, *10* (Suppl. 1), 53–66. [CrossRef]
25. Adams, L.K.; Lyon, D.Y.; Alvarez, P.J. Comparative eco-toxicity of nanoscale TiO_2, SiO_2, and ZnO water suspensions. *Water Res.* **2006**, *40*, 3527–3532. [CrossRef]

26. Schwartz, V.B.; Thétiot, F.; Ritz, S.; Pütz, S.; Choritz, L.; Lappas, A.; Förch, R.; Landfester, K.; Jonas, U. Antibacterial surface coatings from zinc oxide nanoparticles rmbedded in poly(N-isopropylacrylamide) hydrogel surface layers. *Adv. Funct. Mater.* **2012**, *22*, 2376–2386. [CrossRef]
27. Montanari, E.; D'Arrigo, G.; Di Meo, C.; Virga, A.; Coviello, T.; Passariello, C.; Matricardi, P. Chasing bacteria within the cells using levofloxacin-loaded hyaluronic acid nanohydrogels. *Eur. J. Pharm. Biopharm.* **2014**, *87*, 518–523. [CrossRef]
28. Marchesan, S.; Qu, Y.; Waddington, L.J.; Easton, C.D.; Glattauer, V.; Lithgow, T.J.; McLean, K.M.; Forsythe, J.S.; Hartley, P.G. Self-assembly of ciprofloxacin and a tripeptide into an antimicrobial nanostructured hydrogel. *Biomaterials* **2013**, *34*, 3678–3687. [CrossRef]
29. Roy, D.C.; Tomblyn, S.; Isaac, K.M.; Kowalczewski, C.J.; Burmeister, D.M.; Burnett, L.R.; Christy, R.J. Ciprofloxacin-loaded keratin hydrogels reduce infection and support healing in a porcine partial-thickness thermal burn. *Wound Repair Regen.* **2016**, *24*, 657–668. [CrossRef] [PubMed]
30. Ciprofloxacin-loaded keratin hydrogels prevent *Pseudomonas aeruginosa* infection and support healing in a porcine full-thickness excisional wound. *Adv. Wound Care* **2015**, *4*, 457–468. [CrossRef]
31. Posadowska, U.; Brzychczy-Włoch, M.; Drożdż, A.; Krok-Borkowicz, M.; Włodarczyk-Biegun, M.; Dobrzyński, P.; Chrzanowski, W.; Pamuła, E. Injectable hybrid delivery system composed of gellan gum, nanoparticles and gentamicin for the localized treatment of bone infections. *Expert. Opin. Drug. Deliv.* **2016**, *13*, 613–620. [CrossRef]
32. Dorati, R.; De Trizio, A.; Genta, I.; Merelli, A.; Modena, T.; Conti, B. Gentamicin-loaded thermosetting hydrogel and moldable composite scaffold: Formulation study and biologic evaluation. *J. Pharm. Sci.* **2017**, *106*, 1596–1607. [CrossRef]
33. Hu, J.; Yang, L.; Cheng, X.; Li, Y.; Cheng, Y. Aminoglycoside-based biomaterials: From material design to antibacterial and gene delivery applications. *Adv. Funct. Mater.* **2021**, *31*, 2103718. [CrossRef]
34. Gustafson, C.T.; Boakye-Agyeman, F.; Brinkman, C.L.; Reid, J.M.; Patel, R.; Bajzer, Z.; Dadsetan, M.; Yaszemski, M.J. Controlled delivery of vancomycin via charged hydrogels. *PLoS ONE* **2016**, *11*, e0146401. [CrossRef]
35. Antimicrobial Resistance, Global burden of bacterial antimicrobial resistance in 2019: A systematic analysis. *Lancet* **2022**, *399*, 629–655. [CrossRef] [PubMed]
36. O'Neill, J. *Tackling Drug-Resistant Infections Globally: Final. Report and Recommendations*; Government of the United Kingdom: London, UK, 2016.
37. Gashaw, M.; Berhane, M.; Bekele, S.; Kibru, G.; Teshager, L.; Yilma, Y.; Ahmed, Y.; Fentahun, N.; Assefa, H.; Wieser, A.; et al. Emergence of high drug resistant bacterial isolates from patients with health care associated infections at Jimma University medical center: A cross sectional study. *Antimicrob. Resist. Infect. Control.* **2018**, *7*, 138. [CrossRef] [PubMed]
38. Souli, M.; Galani, I.; Giamarellou, H. Emergence of extensively drug-resistant and pandrug-resistant Gram-negative bacilli in Europe. *Eurosurveillance* **2008**, *13*, 19045. [CrossRef]
39. Ghafur, A.; Lakshmi, V.; Kannain, P.; Thirunarayan, M.A. Emergence of pan drug resistance amongst gram negative bacteria! The first case series from India. *J. Microbiol. Infect. Dis.* **2014**, *4*, 86–91. [CrossRef]
40. Kumar, P.; Kizhakkedathu, J.N.; Straus, S.K. Antimicrobial peptides: Diversity, mechanism of action and strategies to improve the activity and biocompatibility in vivo. *Biomolecules* **2018**, *8*, 4. [CrossRef]
41. Batoni, G.; Maisetta, G.; Esin, S. Antimicrobial peptides and their interaction with biofilms of medically relevant bacteria. *Biochim. Biophys. Acta* **2016**, *1858*, 1044–1060. [CrossRef]
42. Zhang, L.J.; Gallo, R.L. Antimicrobial peptides. *Curr. Biol.* **2016**, *26*, R14–R19. [CrossRef]
43. Le, C.F.; Fang, C.M.; Sekaran, S.D. Intracellular targeting mechanisms by antimicrobial peptides. *Antimicrob. Agents Chemother.* **2017**, *61*, e02340-16. [CrossRef]
44. Wang, G. Stuctural studies of antimicrobial peptides provide insight into their mechansims of action. In *Antimicrobial Peptides: Discovery, Design and Novel Therapeutic Strategies*; Wang, G., Ed.; CABI: Boston, MA, USA, 2010; pp. 141–168.
45. Boparai, J.K.; Sharma, P.K. Mini-review on antimicrobial peptides, sources, mechanism and recent applications. *Protein Pept. Lett.* **2020**, *27*, 4–16. [CrossRef] [PubMed]
46. Joo, H.S.; Fu, C.I.; Otto, M. Bacterial strategies of resistance to antimicrobial peptides. *Philos. Trans. R. Soc. Lond. B Biol. Sci.* **2016**, *371*, 20150292. [CrossRef]
47. Assoni, L.; Milani, B.; Carvalho, M.R.; Nepomuceno, L.N.; Waz, N.T.; Guerra, M.E.S.; Converso, T.R.; Darrieux, M. Resistance mechanisms to antimicrobial peptides in Gram-positive bacteria. *Front. Microbiol.* **2020**, *11*, 593215. [CrossRef]
48. Nguyen, K.T.; West, J.L. Photopolymerizable hydrogels for tissue engineering applications. *Biomaterials* **2002**, *23*, 4307–4314. [CrossRef]
49. Fairbanks, B.D.; Schwartz, M.P.; Halevi, A.E.; Nuttelman, C.R.; Bowman, C.N.; Anseth, K.S. A versatile synthetic extracellular matrix mimic via thiol-norbornene photopolymerization. *Adv. Mater.* **2009**, *21*, 5005–5010. [CrossRef] [PubMed]
50. Gramlich, W.M.; Kim, I.L.; Burdick, J.A. Synthesis and orthogonal photopatterning of hyaluronic acid hydrogels with thiol-norbornene chemistry. *Biomaterials* **2013**, *34*, 9803–9811. [CrossRef]
51. Alves, P.M.; Pereira, R.F.; Costa, B.; Tassi, N.; Teixeira, C.; Leiro, V.; Monteiro, C.; Gomes, P.; Costa, F.; Martins, M.C.L. Thiol–norbornene photoclick chemistry for grafting antimicrobial peptides onto chitosan to create antibacterial biomaterials. *ACS Appl. Polym. Mater.* **2022**, *4*, 5012–5026. [CrossRef]

52. Costa, F.M.T.A.; Maia, S.R.; Gomes, P.A.C.; Martins, M.C.L. Dhvar5 antimicrobial peptide (AMP) chemoselective covalent immobilization results on higher antiadherence effect than simple physical adsorption. *Biomaterials* **2015**, *52*, 531–538. [CrossRef] [PubMed]
53. Barbosa, M.; Costa, F.; Monteiro, C.; Duarte, F.; Martins, M.C.L.; Gomes, P. Antimicrobial coatings prepared from Dhvar-5-click-grafted chitosan powders. *Acta Biomater.* **2019**, *84*, 242–256. [CrossRef]
54. De Zoysa, G.H.; Wang, K.; Lu, J.; Hemar, Y.; Sarojini, V. Covalently immobilized battacin lipopeptide gels with activity against bacterial biofilms. *Molecules* **2020**, *25*, 5945. [CrossRef] [PubMed]
55. Qian, C.-D.; Wu, X.-C.; Teng, Y.; Zhao, W.-P.; Li, O.; Fang, S.-G.; Huang, Z.-H.; Gao, H.-C. Battacin (Octapeptin B5), a new cyclic lipopeptide antibiotic from *Paenibacillus tianmuensis* active against multidrug-resistant Gram-negative bacteria. *Antimicrob. Agents Chemother.* **2012**, *56*, 1458–1465. [CrossRef]
56. Cheng, H.; Shi, Z.; Yue, K.; Huang, X.; Xu, Y.; Gao, C.; Yao, Z.; Zhang, Y.S.; Wang, J. Sprayable hydrogel dressing accelerates wound healing with combined reactive oxygen species-scavenging and antibacterial abilities. *Acta Biomater.* **2021**, *124*, 219–232. [CrossRef]
57. Cheng, H.; Yue, K.; Kazemzadeh-Narbat, M.; Liu, Y.; Khalilpour, A.; Li, B.; Zhang, Y.S.; Annabi, N.; Khademhosseini, A. Mussel-inspired multifunctional hydrogel coating for prevention of infections and enhanced osteogenesis. *ACS Appl. Mater. Interfaces* **2017**, *9*, 11428–11439. [CrossRef]
58. Wang, C.; Wang, M.; Xu, T.; Zhang, X.; Lin, C.; Gao, W.; Xu, H.; Lei, B.; Mao, C. Engineering bioactive self-healing antibacterial exosomes hydrogel for promoting chronic diabetic wound healing and complete skin regeneration. *Theranostics* **2019**, *9*, 65–76. [CrossRef] [PubMed]
59. Zhao, L.; Li, X.; Zhao, J.; Ma, S.; Ma, X.; Fan, D.; Zhu, C.; Liu, Y. A novel smart injectable hydrogel prepared by microbial transglutaminase and human-like collagen: Its characterization and biocompatibility. *Mater. Sci. Eng. C* **2016**, *68*, 317–326. [CrossRef]
60. Ghobril, C.; Grinstaff, M.W. The chemistry and engineering of polymeric hydrogel adhesives for wound closure: A tutorial. *Chem. Soc. Rev.* **2015**, *44*, 1820–1835. [CrossRef]
61. Sierra, D.H.; Eberhardt, A.W.; Lemons, J.E. Failure characteristics of multiple-component fibrin-based adhesives. *J. Biomed. Mater. Res.* **2002**, *59*, 1–11. [CrossRef] [PubMed]
62. Annabi, N.; Rana, D.; Shirzaei Sani, E.; Portillo-Lara, R.; Gifford, J.L.; Fares, M.M.; Mithieux, S.M.; Weiss, A.S. Engineering a sprayable and elastic hydrogel adhesive with antimicrobial properties for wound healing. *Biomaterials* **2017**, *139*, 229–243. [CrossRef]
63. Zhao, G.; Zhong, H.; Zhang, M.; Hong, Y. Effects of antimicrobial peptides on *Staphylococcus aureus* growth and biofilm formation in vitro following isolation from implant-associated infections. *Int. J. Clin. Exp. Med.* **2015**, *8*, 1546–1551.
64. Oyarzun-Ampuero, F.; Vidal, A.; Concha, M.; Morales, J.; Orellana, S.; Moreno-Villoslada, I. Nanoparticles for the treatment of wounds. *Curr. Pharm. Des.* **2015**, *21*, 4329–4341. [CrossRef]
65. Liu, K.; Zhang, F.; Wei, Y.; Hu, Q.; Luo, Q.; Chen, C.; Wang, J.; Yang, L.; Luo, R.; Wang, Y. Dressing blood-contacting materials by a stable hydrogel coating with embedded antimicrobial peptides for robust antibacterial and antithrombus properties. *ACS Appl. Mater. Interfaces* **2021**, *13*, 38947–38958. [CrossRef] [PubMed]
66. Wu, M.; Hancock, R.E.W. Improved derivatives of bactenecin, a cyclic dodecameric antimicrobial cationic peptide. *Antimicrob. Agents Chemother.* **1999**, *43*, 1274–1276. [CrossRef] [PubMed]
67. Almaaytah, A.; Mohammed, G.K.; Abualhaijaa, A.; Al-Balas, Q. Development of novel ultrashort antimicrobial peptide nanoparticles with potent antimicrobial and antibiofilm activities against multidrug-resistant bacteria. *Drug Des. Dev. Ther.* **2017**, *11*, 3159–3170. [CrossRef]
68. Xie, S.X.; Song, L.; Yuca, E.; Boone, K.; Sarikaya, R.; VanOosten, S.K.; Misra, A.; Ye, Q.; Spencer, P.; Tamerler, C. Antimicrobial peptide-polymer conjugates for dentistry. *ACS Appl. Polym. Mater.* **2020**, *2*, 1134–1144. [CrossRef]
69. Wang, Y.; Wang, X.; Jiang, W.; Wang, K.; Luo, J.; Li, W.; Zhou, X.; Zhang, L. Antimicrobial peptide GH12 suppresses cariogenic virulence factors of Streptococcus mutans. *J. Oral Microbiol.* **2018**, *10*, 1442089. [CrossRef]
70. Tu, H.; Fan, Y.; Lv, X.; Han, S.; Zhou, X.; Zhang, L. Activity of synthetic antimicrobial peptide GH12 against oral Streptococci. *Caries Res.* **2016**, *50*, 48–61. [CrossRef]
71. Yang, J.-A.; Yeom, J.; Hwang, B.W.; Hoffman, A.S.; Hahn, S.K. In situ-forming injectable hydrogels for regenerative medicine. *Prog. Polym. Sci.* **2014**, *39*, 1973–1986. [CrossRef]
72. Klouda, L.; Mikos, A.G. Thermoresponsive hydrogels in biomedical applications. *Eur. J. Pharm. Biopharm.* **2008**, *68*, 34–45. [CrossRef]
73. Bohidar, H.B.; Jena, S.S. Kinetics of sol–gel transition in thermoreversible gelation of gelatin. *J. Chem. Phys.* **1993**, *98*, 8970–8977. [CrossRef]
74. Haq, M.A.; Su, Y.; Wang, D. Mechanical properties of PNIPAM based hydrogels: A review. *Mater. Sci. Eng. C* **2017**, *70*, 842–855. [CrossRef]
75. Feng, T.; Wu, H.; Ma, W.; Wang, Z.; Wang, C.; Wang, Y.; Wang, S.; Zhang, M.; Hao, L. An injectable thermosensitive hydrogel with a self-assembled peptide coupled with an antimicrobial peptide for enhanced wound healing. *J. Mater. Chem. B* **2022**, *10*, 6143–6157. [CrossRef]

76. Zhang, Y.; Liu, S.; Li, S.; Cheng, Y.; Nie, L.; Wang, G.; Lv, C.; Wei, W.; Cheng, C.; Hou, F.; et al. Novel short antimicrobial peptide isolated from *Xenopus laevis* skin. *J. Pept. Sci.* **2017**, *23*, 403–409. [CrossRef]
77. Zhang, X.; Cheng, Y.; Yang, Y.; Liu, S.; Shi, H.; Lu, C.; Li, S.; Nie, L.; Su, D.; Deng, X.; et al. Polypeptides from the skin of *Rana chensinensis* exert the antioxidant and antiapoptotic activities on HaCaT cells. *Anim. Biotechnol.* **2017**, *28*, 1–10. [CrossRef]
78. Wu, M.; Ye, Z.; Zhu, H.; Zhao, X. Self-assembling peptide nanofibrous hydrogel on immediate hemostasis and accelerative osteosis. *Biomacromolecules* **2015**, *16*, 3112–3118. [CrossRef]
79. Matheny, R.W., Jr.; Nindl, B.C.; Adamo, M.L. Minireview: Mechano-growth factor: A putative product of IGF-I gene expression involved in tissue repair and regeneration. *Endocrinology* **2010**, *151*, 865–875. [CrossRef]
80. Li, J.; Xing, R.; Bai, S.; Yan, X. Recent advances of self-assembling peptide-based hydrogels for biomedical applications. *Soft Matter* **2019**, *15*, 1704–1715. [CrossRef]
81. Qi, G.-B.; Gao, Y.-J.; Wang, L.; Wang, H. Self-assembled peptide-based nanomaterials for biomedical imaging and therapy. *Adv. Mater.* **2018**, *30*, 1703444. [CrossRef]
82. Azoulay, Z.; Aibinder, P.; Gancz, A.; Moran-Gilad, J.; Navon-Venezia, S.; Rapaport, H. Assembly of cationic and amphiphilic β-sheet FKF tripeptide confers antibacterial activity. *Acta Biomater.* **2021**, *125*, 231–241. [CrossRef] [PubMed]
83. Falciani, C.; Lozzi, L.; Pollini, S.; Luca, V.; Carnicelli, V.; Brunetti, J.; Lelli, B.; Bindi, S.; Scali, S.; Di Giulio, A.; et al. Isomerization of an antimicrobial peptide broadens antimicrobial spectrum to Gram-positive bacterial pathogens. *PLoS ONE* **2012**, *7*, e46259. [CrossRef]
84. Fura, J.M.; Sabulski, M.J.; Pires, M.M. D-amino acid mediated recruitment of endogenous antibodies to bacterial surfaces. *ACS Chem. Biol.* **2014**, *9*, 1480–1489. [CrossRef]
85. Leiman, S.A.; May, J.M.; Lebar, M.D.; Kahne, D.; Kolter, R.; Losick, R. D-amino acids indirectly inhibit biofilm formation in *Bacillus subtilis* by interfering with protein synthesis. *J. Bacteriol.* **2013**, *195*, 5391–5395. [CrossRef]
86. Qi, H.; Li, B.; Wang, H.; Cai, Q.; Quan, X.; Cui, Y.; Meng, W. Effects of D-valine on periodontal or peri-implant pathogens: *Porphyromonas gingivalis* biofilm. *J. Periodontol.* **2018**, *89*, 303–314. [CrossRef]
87. Dai, B.; Li, D.; Xi, W.; Luo, F.; Zhang, X.; Zou, M.; Cao, M.; Hu, J.; Wang, W.; Wei, G.; et al. Tunable assembly of amyloid-forming peptides into nanosheets as a retrovirus carrier. *Proc. Natl. Acad. Sci. USA* **2015**, *112*, 2996–3001. [CrossRef]
88. Guo, Z.; Wang, Y.; Tan, T.; Ji, Y.; Hu, J.; Zhang, Y. Antimicrobial D-peptide hydrogels. *ACS Biomater. Sci. Eng.* **2021**, *7*, 1703–1712. [CrossRef]
89. Cao, F.; Mei, L.; Zhu, G.; Song, M.; Zhang, X. An injectable molecular hydrogel assembled by antimicrobial peptide PAF26 for antimicrobial application. *RSC Adv.* **2019**, *9*, 30803–30808. [CrossRef]
90. Muñoz, A.; López-García, B.; Marcos, J.F. Studies on the mode of action of the antifungal hexapeptide PAF26. *Antimicrob. Agents Chemother.* **2006**, *50*, 3847–3855. [CrossRef]
91. Hu, B.; Owh, C.; Chee, P.L.; Leow, W.R.; Liu, X.; Wu, Y.-L.; Guo, P.; Loh, X.J.; Chen, X. Supramolecular hydrogels for antimicrobial therapy. *Chem. Soc. Rev.* **2018**, *47*, 6917–6929. [CrossRef]
92. Adak, A.; Ghosh, S.; Gupta, V.; Ghosh, S. Biocompatible lipopeptide-based antibacterial hydrogel. *Biomacromolecules* **2019**, *20*, 1889–1898. [CrossRef]
93. Price, R.D.; Myers, S.; Leigh, I.M.; Navsaria, H.A. The role of hyaluronic acid in wound healing. *Am. J. Clin. Dermatol.* **2005**, *6*, 393–402. [CrossRef]
94. Suo, H.; Hussain, M.; Wang, H.; Zhou, N.; Tao, J.; Jiang, H.; Zhu, J. Injectable and pH-sensitive hyaluronic acid-based hydrogels with on-demand release of antimicrobial peptides for infected wound healing. *Biomacromolecules* **2021**, *22*, 3049–3059. [CrossRef]
95. Yang, G.; Huang, T.; Wang, Y.; Wang, H.; Li, Y.; Yu, K.; Dong, L. Sustained release of antimicrobial peptide from self-assembling hydrogel enhanced osteogenesis. *J. Biomater. Sci. Polym. Ed.* **2018**, *29*, 1812–1824. [CrossRef]
96. Fontana, R.; Mendes, M.A.; Souza, B.M.d.; Konno, K.; César, L.l.M.M.; Malaspina, O.; Palma, M.S. Jelleines: A family of antimicrobial peptides from the royal jelly of honeybees (*Apis mellifera*). *Peptides* **2004**, *25*, 919–928. [CrossRef]
97. Bandyopadhyay, S.K.; Azharuddin, M.; Dasgupta, A.K.; Ganguli, B.; SenRoy, S.; Patra, H.K.; Deb, S. Probing ADP induced aggregation kinetics during platelet-nanoparticle interactions: Functional dynamics analysis to rationalize safety and benefits. *Front. Bioeng. Biotechnol.* **2019**, *7*, 163. [CrossRef]
98. Hollopeter, G.; Jantzen, H.-M.; Vincent, D.; Li, G.; England, L.; Ramakrishnan, V.; Yang, R.-B.; Nurden, P.; Nurden, A.; Julius, D.; et al. Identification of the platelet ADP receptor targeted by antithrombotic drugs. *Nature* **2001**, *409*, 202–207. [CrossRef]
99. Zhou, J.; Zhang, H.; Fareed, M.S.; He, Y.; Lu, Y.; Yang, C.; Wang, Z.; Su, J.; Wang, P.; Yan, W.; et al. An injectable peptide hydrogel constructed of natural antimicrobial peptide J-1 and ADP shows anti-infection, hemostasis, and anti-adhesion efficacy. *ACS Nano* **2022**, *16*, 7636–7650. [CrossRef]
100. Okabayashi, K.; Ashrafian, H.; Zacharakis, E.; Hasegawa, H.; Kitagawa, Y.; Athanasiou, T.; Darzi, A. Adhesions after abdominal surgery: A systematic review of the incidence, distribution and severity. *Surg. Today* **2014**, *44*, 405–420. [CrossRef]
101. Yang, Y.; Liu, X.; Li, Y.; Wang, Y.; Bao, C.; Chen, Y.; Lin, Q.; Zhu, L. A postoperative anti-adhesion barrier based on photoinduced imine-crosslinking hydrogel with tissue-adhesive ability. *Acta Biomater.* **2017**, *62*, 199–209. [CrossRef]
102. Yu, J.; Wang, K.; Fan, C.; Zhao, X.; Gao, J.; Jing, W.; Zhang, X.; Li, J.; Li, Y.; Yang, J.; et al. An ultrasoft self-fused supramolecular polymer hydrogel for completely preventing postoperative tissue adhesion. *Adv. Mater.* **2021**, *33*, 2008395. [CrossRef]
103. Alejo, T.; Uson, L.; Arruebo, M. Reversible stimuli-responsive nanomaterials with on-off switching ability for biomedical applications. *J. Control. Release* **2019**, *314*, 162–176. [CrossRef]

104. Luo, Z.; Jin, K.; Pang, Q.; Shen, S.; Yan, Z.; Jiang, T.; Zhu, X.; Yu, L.; Pang, Z.; Jiang, X. On-demand drug release from dual-targeting small nanoparticles triggered by high-intensity focused ultrasound enhanced glioblastoma-targeting therapy. *ACS Appl. Mater. Interfaces* **2017**, *9*, 31612–31625. [CrossRef]
105. Wang, C.; Hong, T.; Cui, P.; Wang, J.; Xia, J. Antimicrobial peptides towards clinical application: Delivery and formulation. *Adv. Drug Deliv. Rev.* **2021**, *175*, 113818. [CrossRef]
106. Dang, Q.; Liu, K.; Zhang, Z.; Liu, C.; Liu, X.; Xin, Y.; Cheng, X.; Xu, T.; Cha, D.; Fan, B. Fabrication and evaluation of thermosensitive chitosan/collagen/α, β-glycerophosphate hydrogels for tissue regeneration. *Carbohydr. Polym.* **2017**, *167*, 145–157. [CrossRef]
107. Dai, T.; Tanaka, M.; Huang, Y.-Y.; Hamblin, M.R. Chitosan preparations for wounds and burns: Antimicrobial and wound-healing effects. *Expert Rev. Anti-Infect. Ther.* **2011**, *9*, 857–879. [CrossRef]
108. Rezaei, N.; Hamidabadi, H.G.; Khosravimelal, S.; Zahiri, M.; Ahovan, Z.A.; Bojnordi, M.N.; Eftekhari, B.S.; Hashemi, A.; Ganji, F.; Darabi, S.; et al. Antimicrobial peptides-loaded smart chitosan hydrogel: Release behavior and antibacterial potential against antibiotic resistant clinical isolates. *Int. J. Biol. Macromol.* **2020**, *164*, 855–862. [CrossRef]
109. Lee, E.; Shin, A.; Jeong, K.W.; Jin, B.; Jnawali, H.N.; Shin, S.; Shin, S.Y.; Kim, Y. Role of phenylalanine and valine10 residues in the antimicrobial activity and cytotoxicity of piscidin-1. *PLoS ONE* **2014**, *9*, e114453. [CrossRef]
110. Kumar, A.; Tripathi, A.K.; Kathuria, M.; Shree, S.; Tripathi, J.K.; Purshottam, R.K.; Ramachandran, R.; Mitra, K.; Ghosh, J.K. Single amino acid substitutions at specific positions of the heptad repeat sequence of piscidin-1 yielded novel analogs that show low cytotoxicity and in vitro and in vivo anti-endotoxin activity. *Antimicrob. Agents Chemother.* **2016**, *60*, 3687–3699. [CrossRef]
111. Jiang, Z.; Vasil, A.I.; Vasil, M.L.; Hodges, R.S. "Specificity determinants" improve therapeutic indices of two antimicrobial peptides piscidin 1 and dermaseptin S4 against the Gram-negative pathogens *Acinetobacter baumannii* and *Pseudomonas aeruginosa*. *Pharmaceuticals* **2014**, *7*, 366–391. [CrossRef]
112. Moorcroft, S.C.T.; Roach, L.; Jayne, D.G.; Ong, Z.Y.; Evans, S.D. Nanoparticle-loaded hydrogel for the light-activated release and photothermal enhancement of antimicrobial peptides. *ACS Appl. Mater. Interfaces* **2020**, *12*, 24544–24554. [CrossRef]
113. Ong, Z.Y.; Gao, S.J.; Yang, Y.Y. Short synthetic β-sheet forming peptide amphiphiles as broad spectrum antimicrobials with antibiofilm and endotoxin neutralizing capabilities. *Adv. Funct. Mater.* **2013**, *23*, 3682–3692. [CrossRef]
114. Ong, Z.Y.; Cheng, J.; Huang, Y.; Xu, K.; Ji, Z.; Fan, W.; Yang, Y.Y. Effect of stereochemistry, chain length and sequence pattern on antimicrobial properties of short synthetic β-sheet forming peptide amphiphiles. *Biomaterials* **2014**, *35*, 1315–1325. [CrossRef]
115. Jansen, L.E.; Negrón-Piñeiro, L.J.; Galarza, S.; Peyton, S.R. Control of thiol-maleimide reaction kinetics in PEG hydrogel networks. *Acta Biomater.* **2018**, *70*, 120–128. [CrossRef]
116. Phelps, E.A.; Enemchukwu, N.O.; Fiore, V.F.; Sy, J.C.; Murthy, N.; Sulchek, T.A.; Barker, T.H.; García, A.J. Maleimide cross-linked bioactive PEG hydrogel exhibits improved reaction kinetics and cross-linking for cell encapsulation and in situ delivery. *Adv. Mater.* **2012**, *24*, 64–70. [CrossRef]
117. Lin, C.-C.; Metters, A.T. Hydrogels in controlled release formulations: Network design and mathematical modeling. *Adv. Drug Deliv. Rev.* **2006**, *58*, 1379–1408. [CrossRef]
118. Wu, S.; Yang, Y.; Wang, S.; Dong, C.; Zhang, X.; Zhang, R.; Yang, L. Dextran and peptide-based pH-sensitive hydrogel boosts healing process in multidrug-resistant bacteria-infected wounds. *Carbohydr. Polym.* **2022**, *278*, 118994. [CrossRef]
119. Wu, X.; Li, Z.; Li, X.; Tian, Y.; Fan, Y.; Yu, C.; Zhou, B.; Liu, Y.; Xiang, R.; Yang, L. Synergistic effects of antimicrobial peptide DP7 combined with antibiotics against multidrug-resistant bacteria. *Drug Des. Dev. Ther.* **2017**, *11*, 939–946. [CrossRef]
120. Wu, X.; Wang, Z.; Li, X.; Fan, Y.; He, G.; Wan, Y.; Yu, C.; Tang, J.; Li, M.; Zhang, X.; et al. In vitro and in vivo activities of antimicrobial peptides developed using an amino acid-based activity prediction method. *Antimicrob. Agents Chemother.* **2014**, *58*, 5342–5349. [CrossRef]
121. Wei, S.; Xu, P.; Yao, Z.; Cui, X.; Lei, X.; Li, L.; Dong, Y.; Zhu, W.; Guo, R.; Cheng, B. A composite hydrogel with co-delivery of antimicrobial peptides and platelet-rich plasma to enhance healing of infected wounds in diabetes. *Acta Biomater.* **2021**, *124*, 205–218. [CrossRef]
122. Osusky, M.; Zhou, G.; Osuska, L.; Hancock, R.E.; Kay, W.W.; Misra, S. Transgenic plants expressing cationic peptide chimeras exhibit broad-spectrum resistance to phytopathogens. *Nat. Biotechnol.* **2000**, *18*, 1162–1166. [CrossRef]
123. Spohn, R.; Daruka, L.; Lázár, V.; Martins, A.; Vidovics, F.; Grézal, G.; Méhi, O.; Kintses, B.; Számel, M.; Jangir, P.K.; et al. Integrated evolutionary analysis reveals antimicrobial peptides with limited resistance. *Nat. Commun.* **2019**, *10*, 4538. [CrossRef] [PubMed]
124. Yang, Y.; Wu, D.; Wang, C.; Shan, A.; Bi, C.; Li, Y.; Gan, W. Hybridization with insect cecropin A (1–8) improves the stability and selectivity of naturally occurring peptides. *Int. J. Mol. Sci.* **2020**, *21*, 1470. [CrossRef] [PubMed]
125. Obuobi, S.; Tay, H.K.-L.; Tram, N.D.T.; Selvarajan, V.; Khara, J.S.; Wang, Y.; Ee, P.L.R. Facile and efficient encapsulation of antimicrobial peptides via crosslinked DNA nanostructures and their application in wound therapy. *J. Control. Release* **2019**, *313*, 120–130. [CrossRef] [PubMed]
126. Khara, J.S.; Obuobi, S.; Wang, Y.; Hamilton, M.S.; Robertson, B.D.; Newton, S.M.; Yang, Y.Y.; Langford, P.R.; Ee, P.L.R. Disruption of drug-resistant biofilms using de novo designed short α-helical antimicrobial peptides with idealized facial amphiphilicity. *Acta Biomater.* **2017**, *57*, 103–114. [CrossRef] [PubMed]
127. Bowling, F.L.; Rashid, S.T.; Boulton, A.J.M. Preventing and treating foot complications associated with diabetes mellitus. *Nat. Rev. Endocrinol.* **2015**, *11*, 606–616. [CrossRef] [PubMed]
128. Jeong, S.H.; Cheong, S.; Kim, T.Y.; Choi, H.; Hahn, S.K. Supramolecular hydrogels for precisely controlled antimicrobial peptide delivery for diabetic wound healing. *ACS Appl. Mater. Interfaces* **2023**, *15*, 16471–16481. [CrossRef]

129. Jacob, B.; Park, I.S.; Bang, J.K.; Shin, S.Y. Short KR-12 analogs designed from human cathelicidin LL-37 possessing both antimicrobial and antiendotoxic activities without mammalian cell toxicity. *J. Pept. Sci.* **2013**, *19*, 700–707. [CrossRef]
130. Kamysz, E.; Sikorska, E.; Jaśkiewicz, M.; Bauer, M.; Neubauer, D.; Bartoszewska, S.; Barańska-Rybak, W.; Kamysz, W. Lipidated analogs of the LL-37-derived peptide fragment KR12-structural analysis, surface-active properties and antimicrobial activity. *Int. J. Mol. Sci.* **2020**, *21*, 887. [CrossRef]
131. Jelinkova, P.; Mazumdar, A.; Sur, V.P.; Kociova, S.; Dolezelikova, K.; Jimenez, A.M.J.; Koudelkova, Z.; Mishra, P.K.; Smerkova, K.; Heger, Z.; et al. Nanoparticle-drug conjugates treating bacterial infections. *J. Control. Release* **2019**, *307*, 166–185. [CrossRef]
132. Argyo, C.; Weiss, V.; Bräuchle, C.; Bein, T. Multifunctional mesoporous silica nanoparticles as a universal platform for drug delivery. *Chem. Mater.* **2014**, *26*, 435–451. [CrossRef]
133. Polo, E.; Collado, M.; Pelaz, B.; del Pino, P. Advances toward more efficient targeted delivery of nanoparticles in vivo: Understanding interactions between nanoparticles and cells. *ACS Nano* **2017**, *11*, 2397–2402. [CrossRef]
134. Ma, B.; Chen, Y.; Hu, G.; Zeng, Q.; Lv, X.; Oh, D.H.; Fu, X.; Jin, Y. Ovotransferrin antibacterial peptide coupling mesoporous silica nanoparticle as an effective antibiotic delivery system for treating bacterial infection in vivo. *ACS Biomater. Sci. Eng.* **2022**, *8*, 109–118. [CrossRef]
135. Ma, B.; Guo, Y.; Fu, X.; Jin, Y. Identification and antimicrobial mechanisms of a novel peptide derived from egg white ovotransferrin hydrolysates. *LWT* **2020**, *131*, 109720. [CrossRef]
136. Zengin, A.; Castro, J.P.O.; Habibovic, P.; van Rijt, S.H. Injectable, self-healing mesoporous silica nanocomposite hydrogels with improved mechanical properties. *Nanoscale* **2021**, *13*, 1144–1154. [CrossRef] [PubMed]
137. Howell, M.; Wenc, A.K.; Donaghy, C.M.; Wasche, D.V.; Abissi, I.; Naing, M.D.; Pierce, S.; Angeles-Boza, A.M. Exploring synergy and its role in antimicrobial peptide biology. In *Methods in Enzymology*; Hicks, L.M., Ed.; Academic Press: Cambridge, MA, USA, 2022; Volume 663, pp. 99–130.
138. Cui, H.; Webber, M.J.; Stupp, S.I. Self-assembly of peptide amphiphiles: From molecules to nanostructures to biomaterials. *Pept. Sci. Orig. Res. Biomol.* **2010**, *94*, 1–18. [CrossRef]
139. Tomasini, C.; Castellucci, N. Peptides and peptidomimetics that behave as low molecular weight gelators. *Chem. Soc. Rev.* **2013**, *42*, 156–172. [CrossRef]
140. Dasgupta, A.; Mondal, J.H.; Das, D. Peptide hydrogels. *RSC Adv.* **2013**, *3*, 9117–9149. [CrossRef]

International Journal of
Molecular Sciences

MDPI

Article

Microwave Irradiation vs. Structural, Physicochemical, and Biological Features of Porous Environmentally Active Silver–Silica Nanocomposites

Aleksandra Strach [1], Mateusz Dulski [2,*], Daniel Wasilkowski [3], Oliwia Metryka [1], Anna Nowak [3], Krzysztof Matus [4], Karolina Dudek [5], Patrycja Rawicka [6], Jerzy Kubacki [6], Natalia Waloszczyk [7], Agnieszka Mrozik [3] and Sylwia Golba [2]

1 Doctoral School, University of Silesia, Bankowa 14, 40-032 Katowice, Poland; aleksandra.strach@us.edu.pl (A.S.); oliwia.metryka@us.edu.pl (O.M.)
2 Institute of Materials Engineering, Silesian Center for Education and Interdisciplinary Research, 75 Pulku Piechoty 1A, 41-500 Chorzow, Poland; sylwia.golba@us.edu.pl
3 Institute of Biology, Biotechnology, and Environmental Protection, Faculty of Natural Sciences, University of Silesia, Jagiellońska 28, 40-032 Katowice, Poland; daniel.wasilkowski@us.edu.pl (D.W.); anna.m.nowak@us.edu.pl (A.N.); agnieszka.mrozik@us.edu.pl (A.M.)
4 Materials Research Laboratory, Silesian University of Technology, Konarskiego 18A, 44-100 Gliwice, Poland; krzysztof.matus@polsl.pl
5 Łukasiewicz Research Network, Institute of Ceramics and Building Materials, Cementowa 8, 31-938 Cracow, Poland; karolina.dudek@icimb.lukasiewicz.gov.pl
6 A. Chełkowski Institute of Physics, University of Silesia, 75 Pulku Piechoty 1, 41-500 Chorzow, Poland; patrycja.rawicka@us.edu.pl (P.R.); jerzy.kubacki@us.edu.pl (J.K.)
7 Faculty of Chemistry, Silesian University of Technology, 44-100 Gliwice, Poland; natalia.szulc@polsl.pl
* Correspondence: mateusz.dulski@us.edu.pl

Abstract: Heavy metals and other organic pollutants burden the environment, and their removal or neutralization is still inadequate. The great potential for development in this area includes porous, spherical silica nanostructures with a well-developed active surface and open porosity. In this context, we modified the surface of silica spheres using a microwave field (variable power and exposure time) to increase the metal uptake potential and build stable bioactive Ag_2O/Ag_2CO_3 heterojunctions. The results showed that the power of the microwave field (P = 150 or 700 W) had a more negligible effect on carrier modification than time (t = 60 or 150 s). The surface-activated and silver-loaded silica carrier features like morphology, structure, and chemical composition correlate with microbial and antioxidant enzyme activity. We demonstrated that the increased sphericity of silver nanoparticles enormously increased toxicity against *E. coli*, *B. cereus*, and *S. epidermidis*. Furthermore, such structures negatively affected the antioxidant defense system of *E. coli*, *B. cereus*, and *S. epidermidis* through the induction of oxidative stress, leading to cell death. The most robust effects were found for nanocomposites in which the carrier was treated for an extended period in a microwave field.

Keywords: porous silica spheres; silver nanoparticles; porosity; surface area; microbial activity; enzyme activity

Citation: Strach, A.; Dulski, M.; Wasilkowski, D.; Metryka, O.; Nowak, A.; Matus, K.; Dudek, K.; Rawicka, P.; Kubacki, J.; Waloszczyk, N.; et al. Microwave Irradiation vs. Structural, Physicochemical, and Biological Features of Porous Environmentally Active Silver–Silica Nanocomposites. *Int. J. Mol. Sci.* **2023**, *24*, 6632. https://doi.org/10.3390/ijms24076632

Academic Editor: Helena Felgueiras

Received: 30 December 2022
Revised: 7 March 2023
Accepted: 17 March 2023
Published: 1 April 2023

1. Introduction

Silver, for centuries, has been considered a miracle medicine to cure wounds and protect food from spoilage. The development of silver in nanometric form conceived as nanoparticles (Ag NPs) with a size smaller than 100 nm has set a new direction in microbiologically-active agents [1,2]. The positive biological aspect describes two different mechanisms of action (direct and indirect).

The direct approach explains cell death by changing the permeability of the cell membrane due to the adhesion of Ag NPs to the wall [3]. This mechanism is related to bacterial cell wall thickness [4]. The alternative hypothesis showed that a decrease in silver nanoparticle size leads to apoptosis due to phagocytosis and interactions with the protein structures of bacterial cells [5]. Silver nanoparticles also tend to denature the ribosome, inhibiting protein synthesis and impairing the function of bacterial β-galactosidase [6]. Ag NPs may affect the expression of enzymes involved in cellular oxidation due to the generation of reactive oxygen species (ROS) [7], leading to oxidative stress and DNA damage [6,8].

In turn, the indirect mechanism of action refers to the gradual release of Ag^+ ions, with the degree of ion release depending on the development of the nanoparticle surface. Released silver ions covalently attach to the sulfhydryl groups of bacterial proteins, disrupting membrane transport and inhibiting enzyme activity [9]. The induction of oxidative stress by both silver nanoparticles and ions contributes to the development of antibiotic resistance by increasing the tolerance to oxidative stress. After repeated exposure to Ag NPs, pathogenic bacteria can produce adhesion proteins that cause Ag NP aggregation [10]. For this purpose, there are many developmental directions for new silver-based structures with extended antimicrobial effects that will limit bacterial resistance formation.

Other positive features of silver refer to long-time stability at normal conditions ('slow aging') [11,12] but relatively low at higher temperatures [13,14]. Silver aging and ion-release potential correlate with the particle size, oxidation rate [15–19] as well as synthesis conditions [10], whereby the oxidation is triggered by the contact of silver with an aqueous solution or higher temperature following in the sequence Ag(0) > Ag(I) > Ag(II) [20–22]. Silver oxidation at suitable water solutions and environmental conditions, including accessibility to atmospheric water vapor, promotes the crystallization of Ag_2CO_3 [23–25].

Silver carbonate crystallized in such conditions is considered a semiconductor with antibacterial and photocatalytic features. Similar to metallic or oxidized silver, silver carbonate exhibits features conducive to, e.g., the inactivation of *E. coli* both in limited accessibility of light and with a significant rise of the antimicrobial effect under visible light [26,27]. Similar effects were reported for *S. aureus* and *P. aeruginosa* under ambient light. Moreover, the antimicrobial effect resulted from releasing silver ions due to the dissolution of silver particles in the suspension or after visible light illumination [28,29]. Other applications of silver carbonate are focused on pollution removal, fuel production [30–32], and the photodegradation of rhodamine B (RhB) [33,34], methylene blue [35,36], and ethylparaben [37].

Unfortunately, despite biological-friendly effects, nanostructured silver features negative aspects distinctive for human health, animals, and 'good' microorganisms. The negative mechanism of silver resulted from the uncontrolled release of ions from nanoparticle surfaces and their strong tendency to aggregate. It provides a local increase in ecotoxicity or cytotoxicity, even at low metal concentrations [38]. Silver carbonate exhibits relatively poor stability and is prone to self-photo corrosion [39].

Therefore, stabilization of the silver or silver-based nanostructure is usually realized through core-shelled Ag_2O/Ag_2CO_3 [39], $Ag_2CO_3@Al_2O_3$ [40], $Ag_2S@Ag_2CO_3$ [41], or $Ag_2CO_3@Ti_2O$ [42] composites. Another promising solution is the immobilization of silver within the silicon dioxide (=silica). Silica, in this context, ought to be considered two-folded; the first positive aspect is related to high thermal and chemical stability, large surface area, and low density, while the second corresponds to its porosity [43–46]. Porosity is divided into micro- (<2 nm), meso- (2–50 nm), and macroporosity (>50 nm) [47,48]. Among them, microporous or mesoporous features poor mass transfer capacity and high surface areas that increase the metal-capturing potential. On the contrary, macroporous silica exhibits a high external mass transfer rate due to its interconnected larger pores at a relatively low specific surface area limiting metal capturing potential [49,50]. An increase in the macroporous silica's functionality can be reached by chemical or microwave-assessed modification. Such surface treatment should prevent metal aggregation, improve suitable nanostructures' stability and dispersion, and allow silver ions to be control-released [51,52]. As a result of surface modification, silver or silver-based nanostructures ought to be stabilized as $Ag@SiO_2$, $Ag_2O@SiO_2$, $Ag_2CO_3@SiO_2$, or even $Ag_2O/Ag_2CO_3@SiO_2$.

The paper summarizes a new insight into engineering environmentally applicable silica-based composites. For this purpose, a changeable microwave field and synthesis conditions follow the surface morphology—mainly the surface area, pore volume, and pore size. A structural modification was directed to increase the uptake potential of silica and stimulate stable, bioactive Ag_2O/Ag_2CO_3 heterojunctions. The correctness of the synthesis assumptions was verified by analyzing the silver particle shape, size, and phase composition, as well as the molecular interaction of silver with silica. Morphological, structural, and chemical parameters were correlated with potential antimicrobial effects against reference Gram-positive and Gram-negative bacterial strains. Finally, the activities of three antioxidant enzymes, catalase (CAT), peroxidase (PER), and superoxide dismutase (SOD), were analyzed.

2. Synthesis

The powder silver–silica nanocomposite was fabricated during a chemical reaction in a colloidal suspension. For synthesis, silver nitrate (purity above 99.9999% with trace metal basis; Merck Life Science Sp.z.o.o., Darmstadt, Germany), sodium hydroxide (purity above 99.99%; Avantor Company, Gliwice, Poland), and commercially available spherical amorphous silica powder (S1: d = 15–20 nm, purity: 99.5+%, surface area: 170–200 m^2/g, and true density: 2.4 g/cm^3 [53]; and S2: d = 20–30 nm, purity: 99.5+%, surface area: 180–600 m^2/g, and true density: 2.4 g/cm^3 [54]), were used.

In the first step, three individual flasks contained: 10 g of silica powder (S1 and S2) mixed with 200 mL of distilled water in addition to 2 g of silver nitrate and 2 g of sodium hydroxide dissolved in 20 mL of distilled water, each (Figure 1). Next, the colloidal silica suspension was subjected to microwave irradiation for 60 and 150 s in 150 and 700 W to activate the silica surface. Then, the water solution of sodium hydroxide was added dropwise to the flasks with the microwave-treated silica suspension and continuously stirred on a magnetic stirrer. Afterward, the water solution of the silver nitrate was added by Pasteur pipette drop by drop to the suitable flasks to fabricate the silver–silica colloids. The final products were individually filtered on a polyethylene filter, washed, and dried at room temperature to remove impurities. Finally, eight samples were prepared according to variable synthesis conditions (Table 1).

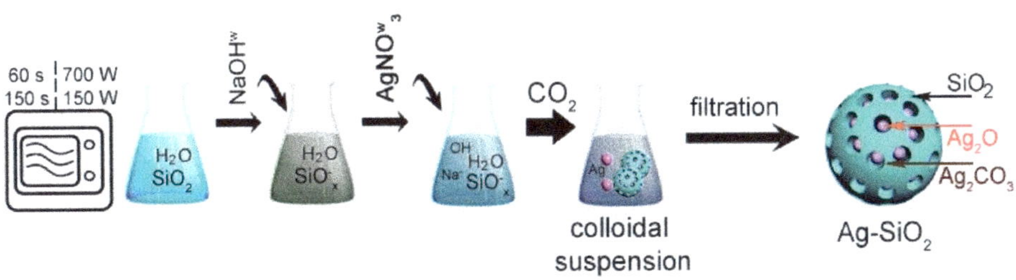

Figure 1. The synthesis of silver–silica spherical nanocomposites.

Table 1. The samples were prepared at variable synthesis conditions.

	P = 150 W		P = 700 W	
	t = 60 s	t = 150 s	t = 60 s	t = 150 s
Ag-S1	AR1.1	AR1.2	AR1.3	AR1.4
Ag-S2	AR1.5	AR1.6	AR1.7	AR1.8

3. Results and Discussion

3.1. Morphology, Surface Parameters, and Particle Analysis

The nanocomposite particle diameters varied from d = 140 - 161 nm (Table 2, Figure 2). A statistically significant difference of d_H occurred depending on the type of silica (S1 or

S2), and the microwave filed exposure parameters (P = 150 or 700 W and t = 60 or 150 s). Thus, Ag-S1 composites (AR1.1, AR1.3) featured higher hydrodynamic particle diameters than those corresponding to Ag-S2 (*** $p < 0.001$).

Table 2. Physicochemical properties of Ag-SiO_2 nanocomposites. d_H: hydrodynamic diameter, PDI: polydispersity index, and z: Zeta potential. The data are summarized with the standard deviation.

Sample Code	d_H (nm)	PDI (-)	z (mV)
AR1.1	146.3 ± 4.8	0.35 ± 0.02	−39.9 ± 0.93
AR1.2	150.9 ± 10.1	0.35 ± 0.03	−37.6 ± 0.86
AR1.3	161.2 ± 3.8	0.33 ± 0.03	−37.5 ± 1.28
AR1.4	154.0 ± 7.5	0.39 ± 0.02	−38.6 ± 1.45
AR1.5	151.4 ± 3.2	0.34 ± 0.03	−38.5 ± 1.99
AR1.6	159.2 ± 10.6	0.37 ± 0.06	−37.2 ± 0.28
AR1.7	140.1 ± 6.1	0.36 ± 0.03	−37.0 ± 1.79
AR1.8	160.8 ± 7.5	0.36 ± 0.05	−38.8 ± 0.65

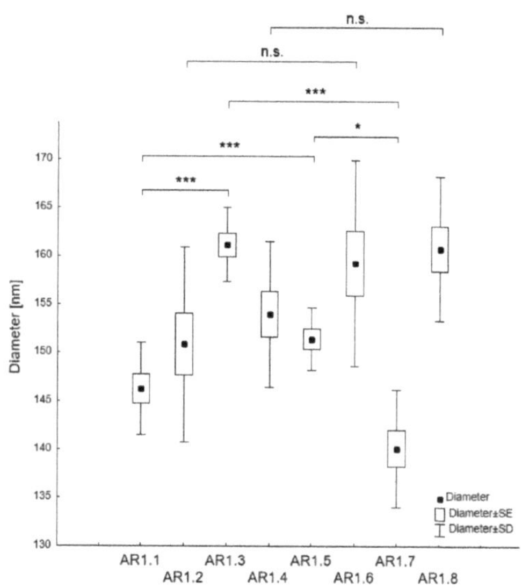

Figure 2. Categorized box-and-whisker plot for particles of Ag-SiO_2 nanocomposites prepared at variable conditions with a marked level of significance (* $p < 0.05$ and *** $p < 0.001$).

Ag-S1 prepared at a microwave field of 700 W showed statistically higher values in the mean hydrodynamic diameter (d_H) than Ag-S1 samples exposed to a microwave field of 150W (* $p < 0.05$). Conversely, Ag-S2 nanocomposites displayed the opposite trend of changes considering the microwave power (*** $p < 0.001$). Samples exposed to microwave fields of different power and at t = 150 s had no statistically significant differences. Higher values for silver-silica nanocomposites than those declared by the producer for reference silicas confirmed the silica surface activation due to microwave irradiation and the synthesis conditions.

Moreover, strongly modified silica particle surfaces through over-melt of the edges led to the formation of aspherical (=rode-like) porously-altered structures. On the other hand, the effect of molten silica edges was enhanced by the catalytic properties of silver that stimulated surface modification. Nevertheless, the PDI lower than 0.4 yielded a monodisperse character of the samples with a low degree of aggregation in suspension (the Zeta potential values changed from −37 to −40 mV).

The synthesis condition yielded similar SSA_{BET} values independent of the silicas' type (S1 or S2) (Table 3). Twice lower values for Ag-S1 and six times lower than for Ag-S2 confirmed a huge role of surface modification and the size of porously-altered structure on the sorption potential of S1 and S2 silicas. In this context, S2 silica seemed more sensitive to microwave irradiation and synthesis conditions than S1 (Table 3). Total pore volume (TPV) and porosity outcomes grouped the samples into two categories with a TPV value lower or higher than 1.0. The first group classified samples prepared at the shorter exposure of silica on microwave irradiation (AR1.1, AR1.3, AR1.5, and AR1.7), while the second group classified samples at a longer time (AR1.2, AR1.4, AR1.6, and AR1.8). The pore size diameter, D_{pore}, changed between 30 and 50 Å. Lower pore size diameters resulted from the differentiated morphology of the silica, including higher internal pore wall roughness. However, the pore size values turned out to be much lower for the fabrication route at a longer time but independent from microwave irradiation power (Table 3).

Table 3. Nitrogen sorption parameters for Ag-SiO_2 prepared at variable microwave power (P = 150 and 700 W) and time (t = 60 and 150 s). SSA_{BET}: specific surface area with standard deviation, V_p: micropore volume, S_P: micropore surface, D_{pore}: pore diameter, TVP: total pore volume, and *: *t*-plot micropore volume. **: *t*-plot micropore area.

Sample Code	SSA_{BET} (m^2/g)	V_p (ml/g) *	S_P (m^2/g) **	D_{pore} (nm)	TPV (cm^3/g)	Porosity (%)
AR1.1	106.45 ± 0.36	0.00326	9.60	47.02	1.2514	75.02
AR1.2	109.66 ± 0.56	0.00622	15.69	35.78	0.9808	70.19
AR1.3	100.70 ± 0.43	0.00459	11.97	43.56	1.1713	73.76
AR1.4	106.41 ± 0.55	0.00594	15.03	32.37	0.8612	67.39
AR1.5	96.04 ± 0.51	0.00561	14.16	51.17	1.2288	74.68
AR1.6	112.90 ± 0.56	0.00613	15.61	34.99	0.9878	70.33
AR1.7	100.33 ± 0.57	0.00465	12.50	46.69	1.0967	72.47
AR1.8	108.04 ± 0.57	0.00607	15.32	35.76	0.9659	69.86

Finally, hierarchical cluster analysis (HCA) considering SSA_{BET} and D_{pore} grouped samples into two or five clusters dependent on the Sneath criterion (Figure S1 in Supplementary Materials). Application of less restrictive criterion (66%) grouped samples among exposition time of silica on microwave irradiation (t = 60 or 150 s) into two clusters (AR1.1, AR1.3, AR1.5, AR1.7) and (AR1.2, AR1.4, AR1.6, AR1.8). According to a more restrictive criterion (33%), samples were divided additionally among irradiation time and power. Hence, three clusters separated nanocomposites synthesized at a shorter time and lower power: AR1.1 (red), AR1.5 (green), and higher power: AR1.3, AR1.7 (blue) from those of more extended time: AR1.2, AR1.6, AR1.8 (violet) and AR1.4 (yellow) (Figure S1 in Supplementary Materials). Samples belonging to violet and yellow clusters featured similar results illustrating a lower impact of power in this route. At the same time, D_{pore} values were much lower than for the fabrication route at a longer time but independent from microwave irradiation power (Table 3, Figure S1 in Supplementary Materials). A similar trend of changes was found for micropore volume and area, with lower values for samples fabricated at shorter times (Table 3).

HCA analysis confirmed two independent routes of synthesis: (i) one with the silica surface activation at the default power and a longer time (t = 150 s) and (ii) the second obtained at a shorter time (t = 60 s) (Figure S1 in Supplementary Materials). The first route ensured a higher surface area with a smaller pore size, while the second one generally favored forming a system with a smaller active surface and a higher pore size (Table 3). In such a route, we additionally distinguished samples with the highest D_{pore} values (AR1.1 and AR1.5) and the lowest (AR1.4 and AR1.8), thus, suggesting that the most optimal parameters for surface modification were achieved at a short time (t = 60 s) and low microwave power (P = 150 W). Furthermore, synthesizing the Ag-Si1 and Ag-S2 nanocomposites was morphologically destructive, especially for S2. However, S1 and S2 silicas still demonstrated potential sorption towards silver capture.

Proof of silver capturing success was provided through microscopic observations performed by SEM and TEM. SEM observations followed local chemical alterations on a macro scale, while TEM was used to observe morphological and chemical changes on a nanoscale level. According to the SEM-EDS results, the silver content was generally lower than 4.0 at.%, while the highest silver content was found for nanocomposites prepared at t = 150 s and P = 700 W. At the same time, nanostructures based on the S1 silica featured slightly higher silver content (Table 4).

Table 4. Average atomic element concentration for silver–silica nanocomposites estimated using SEM-EDS and TEM-EDS with statistical analysis (mean ± SD, n = 3: SEM, n = 5: TEM).

Sample Code	SEM-EDS O (at.%)	SEM-EDS Si (at.%)	SEM-EDS Ag (at.%)	TEM-EDS O (at.%)	TEM-EDS Si (at.%)	TEM-EDS Ag (at.%)
AR1.1	75.0 (4.6)	21.8 (5.0)	3.2 (1.2)	77.2 (13.1)	12.9 (2.9)	9.9 (1.5)
AR1.2	77.5 (3.2)	20.2 (3.0)	2.4 (1.5)	85.1 (4.6)	14.1 (4.5)	0.8 (0.5)
AR1.3	73.9 (4.5)	23.0 (5.2)	3.1 (0.9)	78.0 (4.9)	19.8 (3.7)	2.2 (1.3)
AR1.4	76.2 (1.0)	20.0 (1.4)	3.8 (0.6)	78.4 (1.0)	21.1 (0.8)	0.5 (0.2)
AR1.5	78.7 (0.7)	18.9 (1.2)	2.4 (0.6)	75.9 (8.7)	15.2 (1.6)	8.9 (3.9)
AR1.6	76.7 (1.8)	20.9 (1.4)	2.5 (0.4)	83.3 (1.9)	16.0 (1.8)	0.6 (0.2)
AR1.7	74.3 (1.5)	24.7 (1.1)	0.9 (0.6)	82.8 (2.4)	16.8 (2.0)	0.4 (0.2)
AR1.8	74.2 (1.2)	23.0 (1.8)	2.8 (1.0)	83.9 (2.8)	16.0 (2.8)	0.1 (0.1)

More local TEM-EDS observations revealed higher silver content for samples prepared at t = 60 s and P = 150 W with a low silver concentration for silicas microwave-irradiated for a longer time (Table 4). Furthermore, S1 and S2 silicas, regardless of microwave treatment (power and time), permitted the crystallization of heterogeneously distributed spherical or aspherical silver particles (Figure 3 and Figure S2 in the Supplementary Materials). Silicas treated through high microwave irradiation (AR1.3, AR1.4, AR1.7, and AR1.8) featured the highest number of aspherical particles. In comparison, no aspherical Ag NPs were found for a low microwave field and short time (Table 5). Ag-S1 and Ag-S2 highlighted similar shape-forming tendencies, except for Ag-S2 prepared at P = 700 W and t = 60 s with a higher percentage of aspherical particles (Table 5). A difference in particle shape resulted in the necessity of data limitation and calculation of the average particle size considering only the fully spherical shape. As a result, the particle size distribution histogram was fitted using a lognormal function (logN) with values changing from d = 1.72 to 6.31 nm (Figure 3, Table 5, and Figure S2 in Supplementary Materials).

The diffraction pattern revealed cell parameters a_0 = 4.085(7) Å and space group Fm-3m. Inter-planar d-spacing indicated higher phase differentiation, i.e., the face-cantered cubic silver, the monoclinic AgO, and trigonal Ag_2O (Figure 3). TEM data concluded that longer microwave field exposure time led to the crystallization of Ag NPs with a larger diameter. Furthermore, a lower microwave field (P = 150 W) formed smaller and spherical Ag NPs. In contrast, the P = 700 W enforced the crystallization of larger and more aspherical silver nanoparticles.

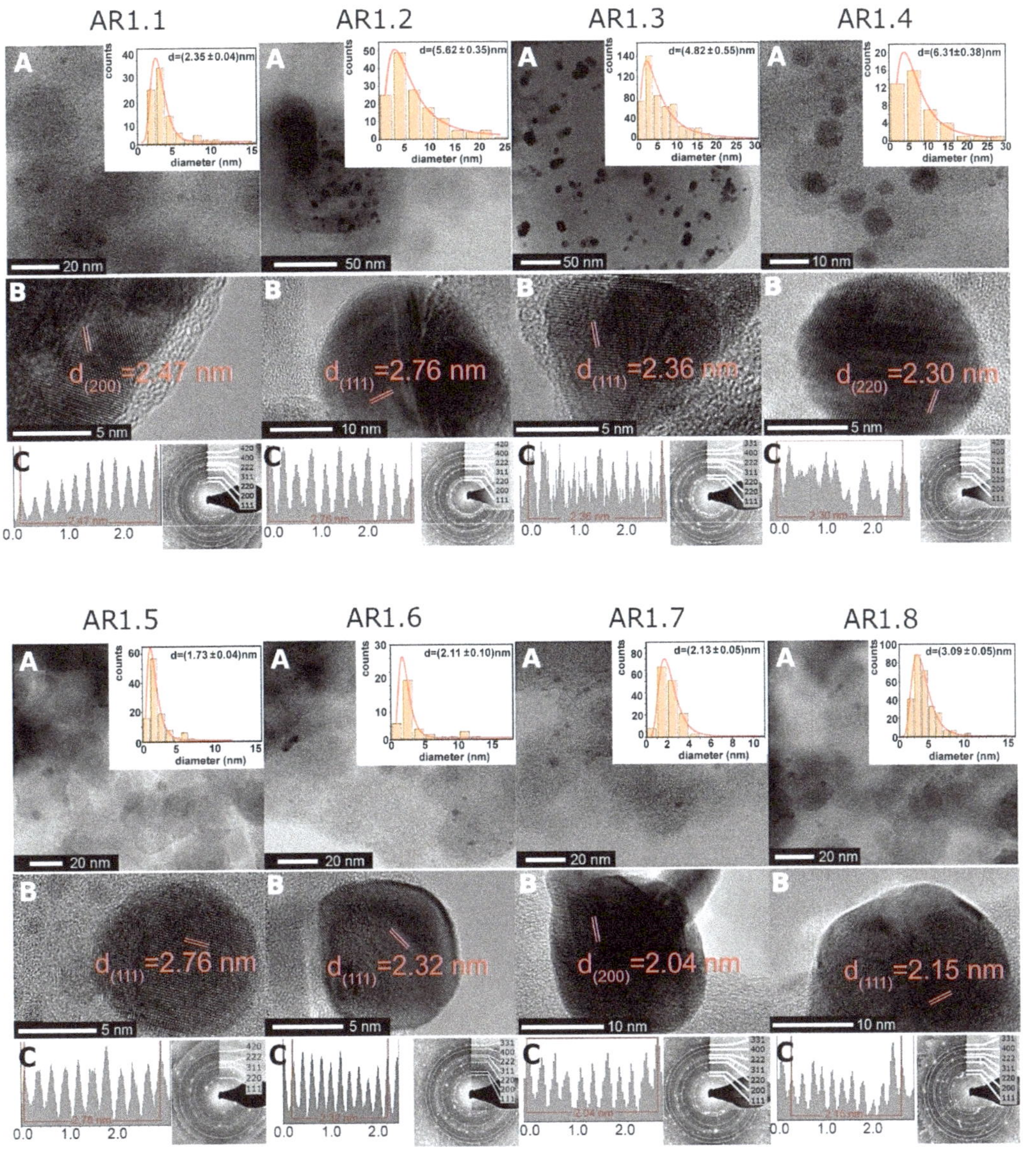

Figure 3. (**A**,**B**) HRTEM microimages for Ag-SiO_2 nanocomposites (AR1.1–AR1.8) with the inter-planar d-spacing plot along the silver core and the particle size distribution of Ag nanoparticles. The data on histograms were fitted with a logN function. (**C**) Inter-planar spacing plot along the silver nanoparticle core and SAED pattern.

Table 5. The average diameter of silver nanoparticles obtained from lognormal function approximation for Ag-SiO_2 prepared at variable microwave power (P = 150 and 700 W) and time (t = 60 and 150 s). Silver nanoparticle diameter summarized with statistical error and the percentage content of spherical Ag NPs.

Sample Code	Ag NP Diameter (nm)	% Spherical NPs
AR1.1	2.35 ± 0.04	100%
AR1.2	5.61 ± 0.35	96%
AR1.3	4.82 ± 0.55	46%
AR1.4	6.31 ± 0.38	67%
AR1.5	1.73 ± 0.04	100%
AR1.6	2.10 ± 0.10	98%
AR1.7	2.13 ± 0.05	59%
AR1.8	3.09 ± 0.05	65%

3.2. Structural Analysis

The X-ray diffraction measurements revealed the co-association of amorphous with a broad hump at 2θ up to 30°, and crystalline phases revealed two polymorphic forms of silver carbonate (Ag_2CO_3) (Figure 4). One crystallized in a hexagonal crystal system ($P3_1c$), while the second had a monoclinic structure ($P2_1/m$).

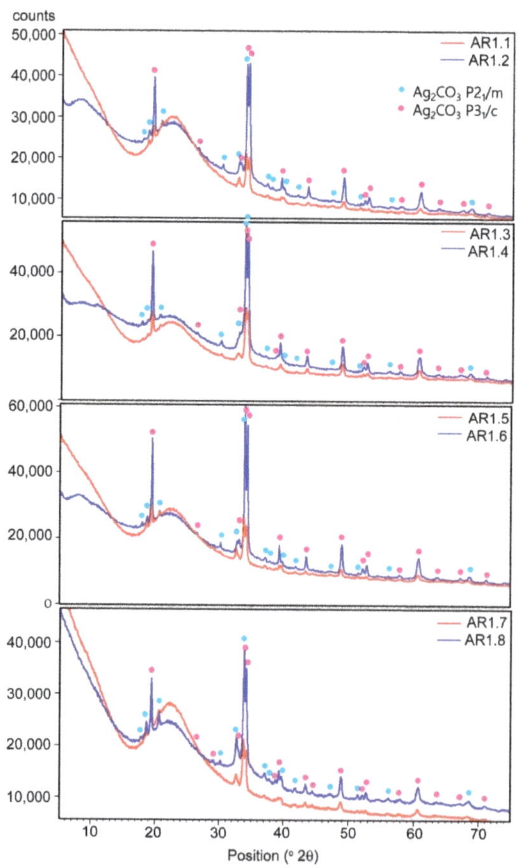

Figure 4. XRD patterns of the Ag-SiO_2 nanocomposites.

The lattice parameters determined from the Rietveld refinement are summarized in Table 6 for individual phases. Under different synthesis conditions, minor shifts of the diffraction lines and differences in the lattice parameters indicate a disturbed structure of silver carbonates due to phase crystallization. Structure disturbances in the crystal lattice resulted from the incorporation of foreign ions or the presence of vacancies. In all samples, the amorphous phase was the predominant one.

Table 6. Lattice parameters of the silver carbonates were determined from the Rietveld refinement.

Phase (Space Group)		Ag_2CO_3 ($P2_1$)			Ag_2CO_3 ($P3_1c$)	
Lattice Parameters	a_0 (Å)	b_0 (Å)	c_0 (Å)	β (°)	a_0 (Å)	c_0 (Å)
AR1.1	3.255(7)	9.547(7)	4.855(1)	92.20(2)	9.189(8)	6.405(4)
AR1.2	3.258(4)	9.561(5)	4.8656(3)	92.20(7)	9.201(2)	6.411(7)
AR1.3	3.251(7)	9.609(1)	4.876(9)	91.98(2)	9.198(8)	6.405(1)
AR1.4	3.263(6)	9.557(8)	4.866(1)	92.22(2)	9.204(1)	6.421(4)
AR1.5	3.255(9)	9.545(9)	4.863(6)	92.25(1)	9.194(7)	6.404(9)
AR1.6	3.258(9)	9.559(1)	4.861(1)	92.13(6)	9.200(1)	6.413(3)
AR1.7	3.255(9)	9.558(4)	4.864(1)	92.15(2)	9.198(4)	6.406(8)
AR1.8	3.260(1)	9.560(9)	4.865(9)	92.13(1)	9.204(5)	6.414(2)

However, considering the diffraction peaks' intensity, the nanocomposites differ in the content of polymorphic silver carbonates. In addition, a more significant proportion of crystalline phases was observed in the samples produced in a short duration of microwave irradiation. The diffraction peaks of silver and silver oxides were not identified due to their low content or nanocrystalline nature.

The XRD results provided incomplete information about the silica structure and mutual interactions between individual nanocomposite materials. As a result, the XRD outcomes were supplemented by Raman spectroscopy. Regardless of the type of silica, the microwave treatment and synthesis conditions provided extra information on the silica network (Figure 5). The introduction of silver by synthesis combined with the morphological change and the formation of porous rod-like structures forced the formation of a disordered silica network with numerous NBOHC, E′, or neutral oxygen center structural defects marked by high-intensity features at 278 and 323 cm^{-1} [55–57].

The series of variable point defects is comparable to Ag-S1 and Ag-S2 composites and independent of the microwave field, including the power and time. Deformation of the silica lattice also revealed the appearance of intermediate-order silica superstructures in the form of bands around 382, 458, and 693 cm^{-1} originating from O-Si-O bending vibrations within n-member rings with n > 4 [58,59] and movement within the $Si_2O_7^-$ unit [60] (Figure 5).

The determination of the degree of depolymerization, structural ordering, and silica potential to capture metal ions determine the number and type of Q^n units (n = 0–4 and denotes the number of bridging oxygen (BO) per tetrahedral SiO_4) (850–1250 cm^{-1}) [61–67]. As a result, the molecular image of Ag-S1 and Ag-S2 nanocomposites, regardless of the synthesis conditions, consists of three Gaussian peaks of similar intensity around 895, 945, and 1005 cm^{-1} associated with Si-O* stretching modes in Q^1, Q^2, and Q^3 units with three, two, and one non-bridging oxygen (NBO), respectively.

Moreover, the Raman spectra of AR1.2, AR1.4, AR1.6, and AR1.8 are characterized by additional bands around 540, 1025, and 1100 cm^{-1}, originating from Si-O* vibrations within the metal-activated Q^3 unit (Figure 5) [63,68,69]. According to the physicochemical and morphological studies, silicas with larger pores featured a capillary effect with tremendous potential to capture more significant amounts of metal, which led to the formation of stronger binding between silver nanoparticles and the interior of silica pore walls (Table 2 and chemical analysis).

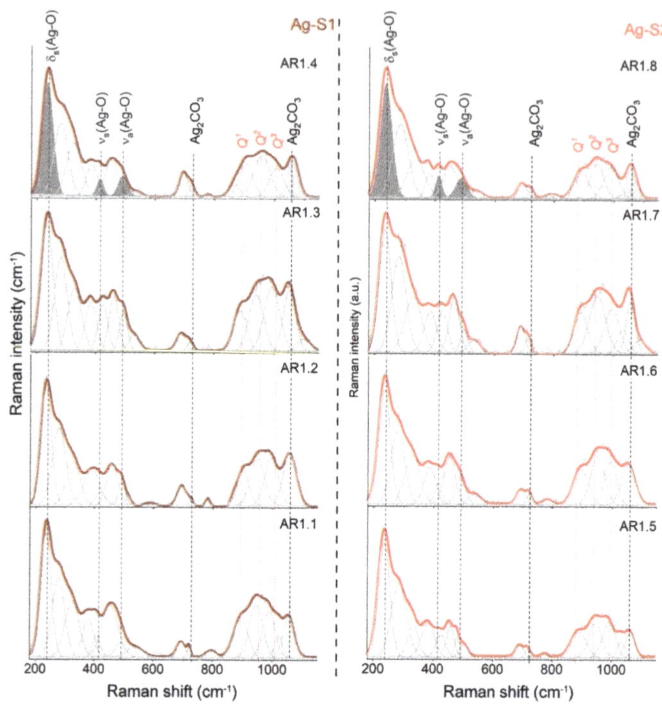

Figure 5. Raman spectra of silver–silica (Ag-SiO_2) nanocomposites summarized in the 180–1150 cm^{-1} region. Spectra were fitted using the Voigt function at a minimum number of components. Q^1, Q^2, and Q^3 represent individual structural units of amorphous silica. Grey-colored bands are referred to as silver (I and III) oxide (dark) and silver carbonate (light). Ag-S1 and Ag-S2 refer to silver-silica nanocomposites that were prepared on S1 silica (d = 15–20 nm, [53]) and S2 silica (d = 20–30 nm, [54]).

A signal of silver oxide appeared after the fitting procedure as three well-separated bands at 233, 417, and 490 cm^{-1} (Figure 5). In contrast, the band arrangement corresponds to the oxide with the crystallographic structure closely related to $Ag^IAg^{III}O_2$ [70,71]. Consequently, the band at 488 cm^{-1} comes from Ag(I) cations linearly coordinated by oxygen in the AgO structure, while at 417 cm^{-1}, the vibration is within the square-planar $Ag^{III}O_4$ unit. The lowest-laying band at 233 cm^{-1} is derived from the symmetric bending mode of the $Ag^{III}O_4$ unit [72,73], while Raman's active asymmetric stretching mode overlaps with the O-Si-O motion.

Some differences in the band positions resulted from the presence of non-bridging silica bonds that enforced the formation of Si-O-Ag stabilizing the nanoparticles within the carrier. The bands at 709 and 1068 cm^{-1} arose from the silver carbonate [74–76]. Two separated CO_3^{2-} stretching vibration-related bands occurred between 1350 and 1650 cm^{-1}, confirming the silver carbonate with an aragonite-type structure [77]. One idea to explain two different forms of silver was the assumption of a core-shell structure formation [78].

Another idea illustrated a crystallization of silver oxide (Ag_2O) and silver carbonate (Ag_2CO_3) simultaneously. In this approach, Ag_2O had to be pulled through capillary action into the pore and stabilized inside the silica carrier through unsaturated bonds. At the same time, silver carbonate crystallized immediately only at surface-bonded silver oxide with great accessibility of atmospheric CO_2. Both hypotheses proved the molecular sorption potential of S1 and S2 (Figure 5).

3.3. Surface Chemical Studies

X-ray photoemission spectroscopy (XPS) was performed to determine the surface atomic concentration (Table 7) and chemical states of silver, oxygen, silicon, and carbon (Figures 6 and 7).

Table 7. Surface atomic concentration values obtained for Ag-SiO_2 derived from XPS.

Sample Code	C 1s	O 1s	Si 2p	Ag 3d
		(at.%)		
AR 1.1	5.31	63.99	30.42	0.28
AR 1.2	3.55	64.28	32.07	0.10
AR 1.3	2.49	64.15	32.67	0.69
AR 1.4	2.41	65.35	32.03	0.20
AR 1.5	2.97	64.79	31.92	0.32
AR 1.6	4.60	62.79	32.44	0.17
AR 1.7	3.42	64.58	31.49	0.51
AR 1.8	2.58	64.32	33.00	0.10

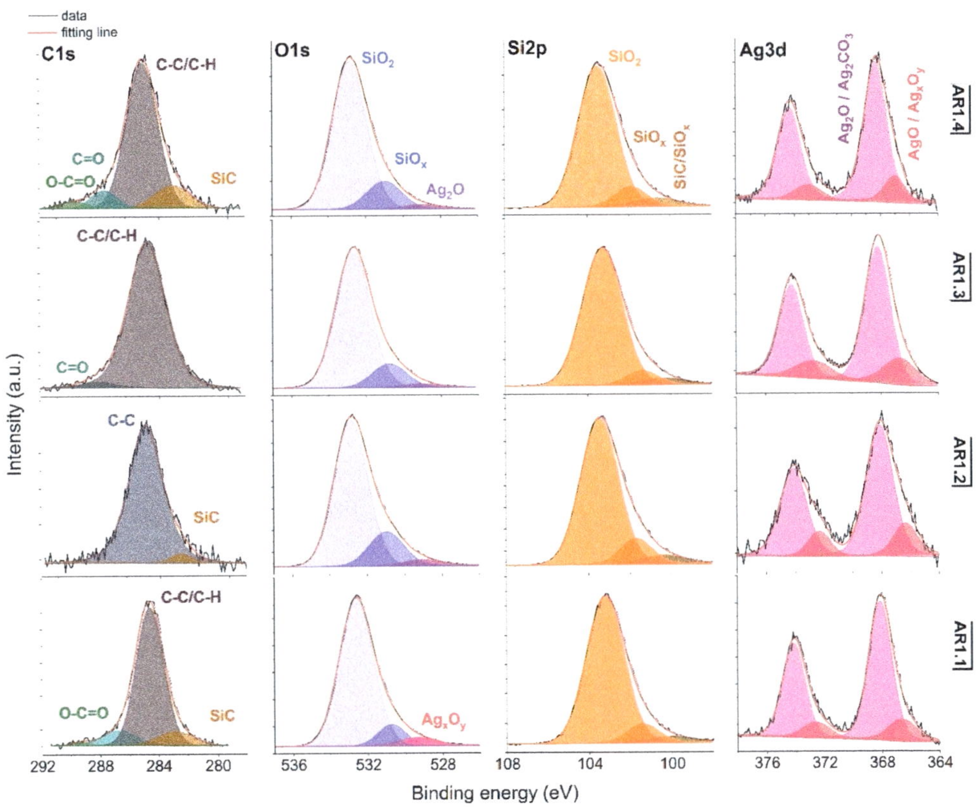

Figure 6. XPS spectra for Ag-S1 with Si 2p, O 1s, Si 2p, and Ag 3d core levels. AR1.1-AR1.4 refer to Ag-S1 silver-silica nanocomposites (S1: d = 15–20 nm, [53]).

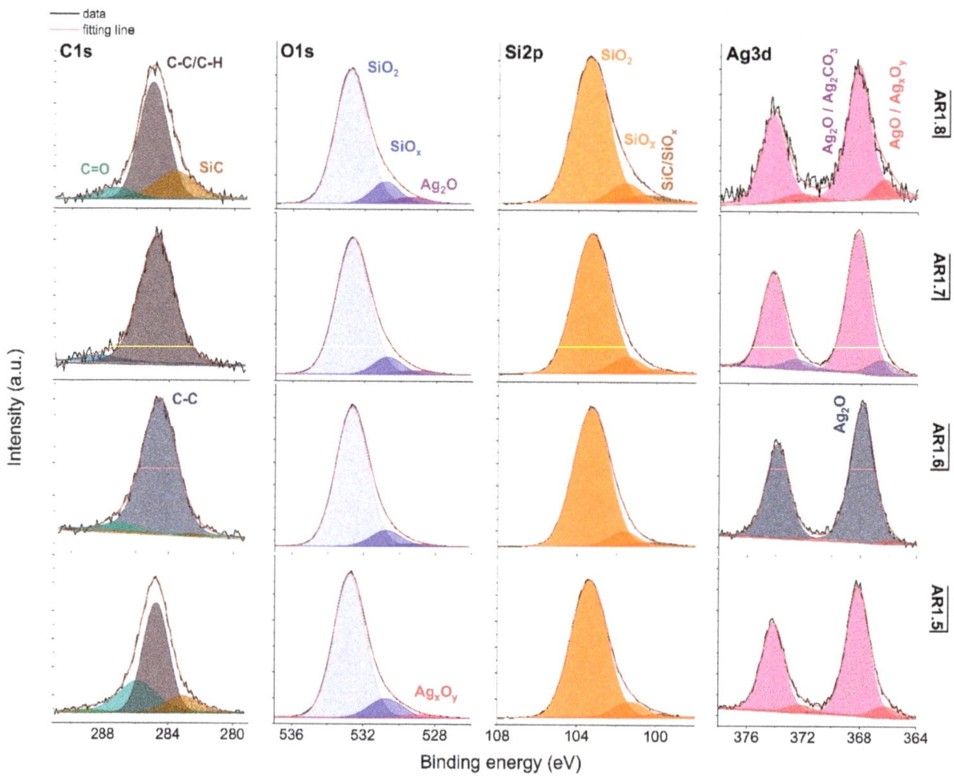

Figure 7. XPS spectra for Ag-S2 with Si 2p, O 1s, Si 2p, and Ag 3d core levels. AR1.5-AR1.8 refer to Ag-S2 silver-silica nanocomposites (S2: d = 20–30 nm, [54]).

Generally, the deconvoluted Ag 3d core lines of Ag-S1 (Figure 6) and Ag-S2 (Figure 7) consist of two or three doublets with binding energies around 368.3, 368.1, 368.2, and 367.6 eV. The maxima of individual peaks point to Ag, AgO, Ag=O, and Ag_2CO_3, wherein unambiguous separation of metallic and silver oxides is difficult because of the close binding energy values [79]. A clue confirming metallic nanoparticles is the signal originating from non-stochiometric silver (Ag_xO_y) with binding energy from 366.1 to 366.9 eV due to the adsorption of different silver oxygen structures [80].

The relative atomic concentration of silver differs from 0.10 up to 0.69 at.%. Nanocomposites exposed longer to microwave irradiation had lower silver capture potential (Table 7). Conversely, the highest uptake potential yielded silver–silica nanocomposites fabricated at P = 700 W and t = 60 s with 0.69% at.% for Ag-S1. From a chemical point of view, the optimal development of the silica surface is stated to be with high power and a short microwave irradiation time.

The chemical states observed during the analysis of the Ag 3d core line correspond to the chemical state observed for the O 1s line of AgO (530.0 eV), Ag_2O (528.8 eV), SiO_2 (532.7–532.9 eV), and non-stoichiometric SiO_x (531.9 eV) (Figures 6 and 7). Interestingly, in practice, the XPS results as a surface-sensitive technique do not provide chemical evidence of silver carbonate, regardless of the more bulk-sensitive XRD. Hence, these results indirectly suggest forming a core-shell system with a silver oxide layered by silver carbonate. The XPS and XRD data proofed assumed heterojunction silver structured composites. Microwave irradiation application caused the absorption of ambient carbon and the sorption or desorption of oxygen by silicon. As a result, the Si 2p core level fitting revealed the line corresponding to SiC, SiO_x, or SiO_{2-x} (Figures 6 and 7). The surface carbon formed C-C/C-H bonds, Si-C, CO^+O_2/Ag (284.8eV), O-C-O (288.3 eV), C-OH (285.8–286.7 eV), C=O (287.3–287.6 eV), and O-C=O (288.9–289.3 eV) (Figures 6 and 7).

4. Biological Studies

4.1. Microbial Activity of Newly Synthesized NCs

The antibacterial activity of all the Ag-SiO_2 nanocomposites was evaluated against three reference bacterial strains: Gram-negative *Escherichia coli* (ATCC® 25922™), and Gram-positive *Bacillus cereus* (ATCC® 11778™), *Staphylococcus epidermidis* (ATCC® 12228™). The model bacteria proved sensitive to all tested NCs—in particular, the silver nanoparticle responses were found to be strain-specific silver nanoparticle responses (Table 8). Generally, *B. cereus* was characterized by the highest sensitivity to NCs, for which the average MBC values ranged from 5 to 10 mg L^{-1}.

Table 8. The MBC and IC_{50} values of synthesized NCs against *E. coli*, *B. cereus*, and *S. epidermidis*.

	MBC (mg L^{-1})	IC_{50} (mg L^{-1})
AR1.1	25/**10**/25	5.35/**1.83**/7.21
AR1.2	**5**/10/10	**2.11**/5.43/5.95
AR1.3	25/**5**/25	**1.28**/**1.22**/7.97
AR1.4	7.5/**1.5**/10	5.32/6.11/**2**
AR1.5	25/**10**/25	13.30/**5.54**/10.69
AR1.6	**2.5**/10/7.5	**0.31**/0.67/1.69
AR1.7	25/**10**/**10**	12.66/**3.75**/4.87
AR1.8	**5**/10/7.5	1.18/7.79/**0.41**

Based on the calculated IC_{50} values, the toxic effect for *B. cereus* decreased in the following order: AR1.6 > AR1.3 > AR1.1 > AR1.7 > AR1.2 > AR1.5 > AR1.4 > AR1.8. The opposite effect was observed for *E. coli* and *S. epidermidis*, which proved to be more resistant to stress caused by NCs. Interestingly, among all studied bacterial strains, AR1.8 and AR1.6 had the highest lethal effect on the *S. epidermidis* cells with IC_{50} = 0.41 and 1.69 mg L^{-1}, respectively.

Comparing the IC_{50} results generated for Gram-negative *E. coli*, the decreasing bactericidal influence of the tested NCs can be ordered as follows: AR1.6 > AR1.8 > AR1.3 > AR1.2 > AR1.4 > AR1.1 > AR1.7 > AR1.5. A unique and different response to the synthesized NCs characterized all bacterial strains. Nevertheless, all studied NCs were highly toxic to *E. coli*, *B. cereus*, and *S. epidermidis* (MBC 2.5–25 mg L^{-1}), with AR1.6 as the most toxic nanocomposite to all three strains. Its high toxicity yielded small-size silver nanoparticles that were likely low-bound to the silica carrier, providing free passage through the outer layers of microbial cells.

Furthermore, the percentage reduction in the spherical shape of Ag NPs can increase the porosity of tested Ag-SiO_2 and further enhance the contact of NCs with the surface of bacterial strains compared to completely spherical Ag NPs. This property of synthesized nanocomposites may also increase their biological activity due to the higher surface-to-volume ratio [81–83].

Previous studies have indicated that inorganic NPs with uneven surfaces and irregular shapes are characterized by more significant activity, which could be attributed to the faster release of metal ions from the surface [81,84,85]. This phenomenon is demonstrated by the collected results, as spherical AR1.1 and AR1.2 proved to be less toxic than the rest of Ag-SiO_2. Considering the d_H, the homoagglomeration of Ag-SiO_2 nanocomposites should be comparable for all silver–silica nanocomposites.

However, the presence of compounds in the culture medium, bacterial secretions, and the bacteria themselves may alter the toxicity of the tested nanocomposites [86,87]. Hence, weaker electrostatic interaction of the agglomerates with the surface of bacterial cells explains the lower lethal effect of AR1.2 compared to other nanostructures.

Similarly, a more toxic effect of Ag NPs was found for triangular nanostructures on *Escherichia coli* and *Pseudomonas aeruginosa* cells regarding spherical nanoparticles [88]. Other studies reported the opposite effect, i.e., spherical Ag NPs had a more significant bactericidal effect on both Gram-negative (*Klebsiella pneumoniae* and *Pseudomonas aeruginosa*) and Gram-positive (*Bacillus subtilis* and *Staphylococcus aureus*) bacteria compared with NCs with a rod shape [89].

Acharya et al. [90] demonstrated that Ag@SiO_2 core-shell nanoparticles (~17 nm) had a more substantial antibacterial effect on *B. subtilis*, *S. aureus*, *Serratia marcescens*, and *K. pneumoniae* than singular Ag NPs (~14 nm) due to controlled and sustainable release of Ag(I). Correspondingly, Gankhuyag et al. [91] and Pal et al. [92] recorded an enhanced bactericidal activity of SiO_2@AgNPs and Ag-NP hybrid silica films against *E. coli* cells, respectively. A similar effect was reported for the core-shell silver–silica nanosystem against highly resistant *Mycobacterium tuberculosis* [93]. The so-called "Trojan horse effect" as a biocidal mechanism of AR-nanocomposites can be assigned to the results of the presented work.

Furthermore, releasing active Ag^+ ions from the SiO_2 carrier as an effective antibacterial agent would explain their highly toxic effects on tested microorganisms [81]. It also suggests that depositing or capping Ag NPs in core-shell systems increases their toxicity, enabling their stabilization and preventing the direct formation of Ag-NP aggregates [94]. These findings also follow Ahmed et al. [95], who observed a significant toxicological impact of Ag NPs obtained through microwave-mediated synthesis against *E. coli*, *P. aeruginosa*, *S. aureus*, and *K. pneumoniae*.

4.2. Antioxidant Enzymes Activity in Response to NC Stress

The activity of the triad antioxidant enzymes was measured to examine the effect of the tested NCs on a catalytic antioxidant cell system, including catalase (CAT) and peroxidase (PER), where breakdown generated O_2 and H_2O from H_2O_2. Furthermore, the third examined enzyme was superoxide dismutase (SOD), catalyzing the dismutation of $O_2^{\cdot -}$ to O_2 and H_2O_2. The obtained findings showed unique catalytic profiles for each bacterial strain and the explicit dependency of the primary antioxidant defense system on the properties of Ag-SiO_2 (Figure 8).

Generally, the treatment of *E. coli* with AR1.1, AR1.3, AR1.5, and AR1.7 had a stimulating effect on CAT activity, especially in the presence of AR1.1 (23.61 U mg^{-1}). Successively, all nanocomposites had a significant enhancing impact on PER activity except AR1.2. For instance, the addition of AR1.7 caused the highest increase (2.3 fold) in PER activity (6.04 U mg^{-1}) compared to the control. Intriguingly, AR1.3 and AR1.7 were the only structures that stimulated SOD activity in *E. coli* cells, while inhibition of its functioning was observed in the presence of remaining Ag-S1 and Ag-S2.

In the case of the *B. cereus* strain, the NC treatment manifested diverse effects on CAT, PER, and SOD activities (Figure 9). Among all tested NCs, only AR1.1 and AR1.5 stimulated CAT and SOD activity, resulting in 13- and 21.5-fold as well as 3- and 5.5-fold increases compared to the control, respectively. In contrast, the conducted research revealed an inhibitory effect of AR1.2, AR1.4, AR1.6, and AR1.8 on CAT and SOD activities, while these nanostructures had the opposite impact on PER functioning.

Treatment of *S. epidermidis* with individual NCs demonstrated their divergent effects on the activity of CAT, PER, and SOD (Figure 10). We found that the presence of AR1.5 and AR1.8 caused an increase in CAT activity. In contrast, AR1.1, AR1.3, and AR1.4 treatments resulted in a decrease in CAT. Compared to the control, the most significant inhibitory effect (10-fold lower) was observed after exposure to AR1.4.

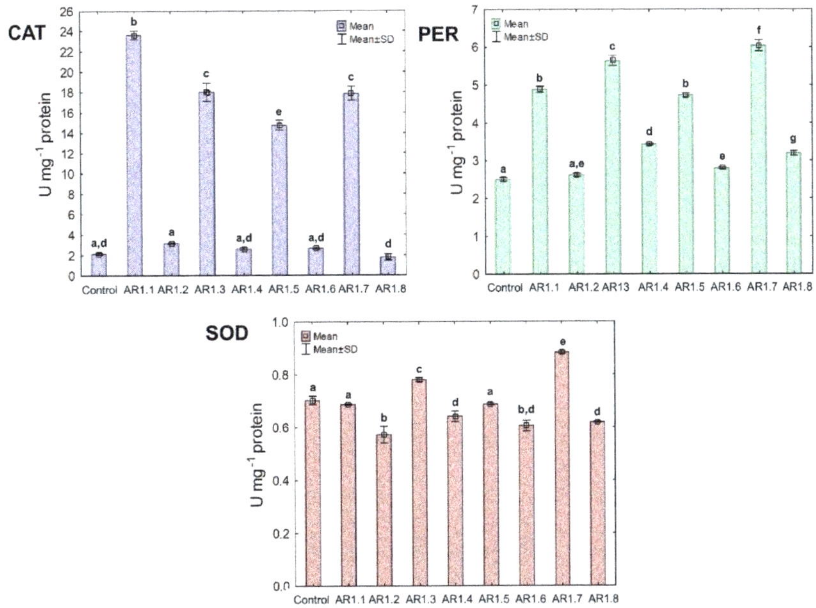

Figure 8. The activities of CAT, PER, and SOD in *E. coli* exposed to NCs at IC_{50} (mean ± SD; n = 3). The same letter(s) indicate no significant statistical differences at $p < 0.05$ among means between NP-treated cells.

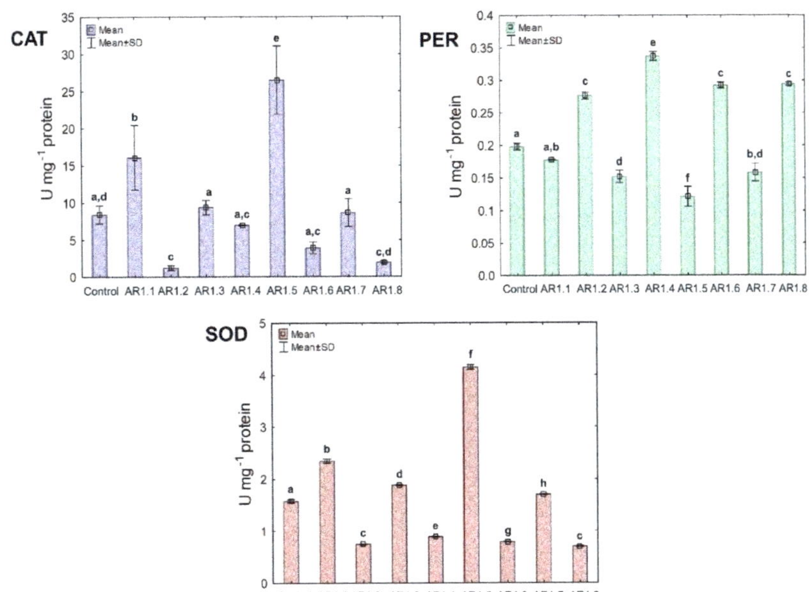

Figure 9. The activities of CAT, PER, and SOD in *B. cereus* exposed to NCs at IC_{50} (mean ± SD; n = 3). The same letter(s) indicate no significant statistical differences at $p < 0.05$ among means between NC-treated cells.

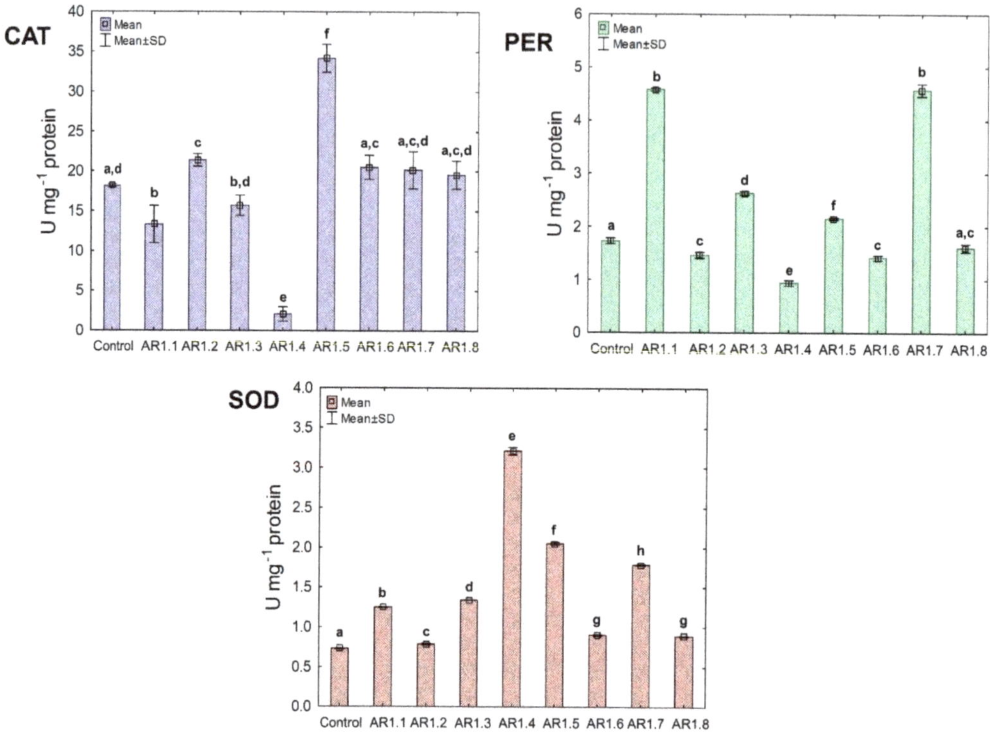

Figure 10. The activities of CAT, PER, and SOD in *S. epidermidis* exposed to NCs at IC_{50} (mean ± SD; n = 3). The same letter(s) indicate no significant statistical differences at $p < 0.05$ among means between NC-treated cells.

In turn, treatment of *S. epidermidis* with AR1.1 and AR1.7 resulted in the highest (3-fold) increase in PER activity, while the addition of AR1.4 caused the most significant decrease (1.6-fold). Interestingly, the application of Ag-SiO_2 increased the SOD activity, particularly AR1.4, which had the most distinguishable effect on its functioning.

The collected data for the catalytic profiles of the three strains indicate that the Ag-SiO_2 nanocomposites altered the antioxidant defense systems of *E. coli*, *B. cereus*, and *S. epidermidis* compared to the control. The observed changes in the CAT, PER, and SOD activity can be ascribed to the induction of oxidative stress in bacterial cells. Previous studies have shown that inorganic Ag NPs can generate reactive oxygen species (ROS), such as hydroxyl radicals ($^{\bullet}OH$), hydrogen peroxide (H_2O_2), or singlet oxygen ($^1O_2$) [81,96]. Despite conducting intensive research into the effect of nanoparticles on the antioxidant system of microorganisms, the current research concerns only selected strains, enzymes, and specific nanostructures.

Therefore, it is crucial to explore this subject further to understand the characteristics of the biological activity of engineered nanostructures. Analysis of the collected results shows that Ag-SiO_2 could generate and increase the level of H_2O_2 in bacterial cells due to the noticeable and significant changes in CAT and PER activities, resulting in stimulation of the antioxidant defense system. Both Ag NPs and the released metal ions can generate ROS but can also interact with protein groups modifying their functioning, consequently causing their inactivation [97–99].

The biochemical data would explain the significant decrease of both CAT and PER, especially in *B. cereus* and *S. epidermidis* cells, after treatment with selected Ag-SiO_2. However,

SOD activity in all three strains was the least affected by the nanocomposites, indicating the active defense systems of *E. coli, B. cereus,* and *S. epidermidis* against $^1O_2$ generation. AR1.3 and AR1.7 in *E. coli* culture resulted in a characteristic pattern of stimulated antioxidant activity of CAT, PER, and SOD.

This phenomenon could be explained by the stabilized and increased induction of ROS previously mediated by irradiation of AR1.3 and AR1.7 with more substantial microwave power. It has been reported that radiosensitization and other chemical phenomena applied in the synthesis of nanomaterials may also affect ROS synthesis in biological systems [100]. Nevertheless, this pattern is not reproducible in other strains, which confirms the dependence of the observed changes on the physicochemical properties of nanocomposites and the unique characteristics of microorganisms. Furthermore, the increased toxic effects of AR1.2, AR1.4, AR1.6, and AR1.8 on SOD and CAT in *B. cereus* cultures may be due to the protein denaturation and elevated concentration of H_2O_2.

Similarly, Yuan et al. [101] observed a considerable inhibition of SOD and CAT in *P. aeruginosa* and *S. aureus* exposed to Ag NPs, which positively correlated with ROS generation and plausible inactivation of enzyme activities. By comparison, Liao et al. [99] recorded decreased CAT, PER, and SOD activity in *P. aeruginosa* cells in the presence of Ag NPs. In different experiments, Jain et al. [102] demonstrated a positive correlation of an elevated ROS generation with increased SOD and PER activities in *E. coli, P. putida, B. cereus,* and *S. aureus* exposed to Ag NPs. Similarly, a considerable increase in the CAT activity of *E. coli* and *B. cereus* was recorded in the presence of SiC/Ag/CE/T nanocomposites [103]. Our collected data showed antioxidant activities are a strain-specific trait and depend on the type of nanomaterial.

4.3. Statistical Data Exploration

Cluster analysis and PCA were performed for the physicochemical and biological parameters to evaluate NC treatment variability and determine the relationships with the whole dataset. The results from PCA analysis explained 82.11%, 79.05%, and 69.66% of the data variability for *E. coli, B. cereus,* and *S. epidermidis*, respectively (Figure 11). Two clusters were distinguished for each strain for the correlation biplot projection along PC1.

Similar compositions of the projections were observed for the tested strains, where AR1.1, AR1.3, AR1.5, and AR1.7 were arranged into one plot, whereas the second cluster contained AR1.2, AR1.4, AR1.6, and AR1.8 (Figure 11A,C,E). The coordination biplot for *E. coli* revealed a strong negative correlation of PER with SSA_{BET} and CAT with V_P and S_P (Figure 11B). In turn, for *B. cereus*, the most distinguishable negative correlation was observed for the silver particle diameter (D_{Ag}), CAT, and SOD (Figure 11D). The obtained PCA projection for *S. epidermidis* demonstrated a strong negative correlation of SOD with the number of particles with a spherical shape (NP_{sf}). At the same time, the other parameters were more scattered (Figure 11F).

The cluster analysis displays a characteristic relationship between the obtained results and NC treatments, providing thematically similar patterns for all tested strains (Figure 12). The dendrogram projections obtained for *E. coli, B. cereus,* and *S. epidermidis* revealed that SSA_{BET}, D_{pore}, and S_P were the most discriminating analyses. At the same time, among the measured enzymes, the most differentiating one was CAT. According to the acquired dendrogram, the most characteristic and discriminating were AR1.5 for the *E. coli* and *B. cereus* strains, while AR1.4 had the most divergent effects on *S. epidermidis*.

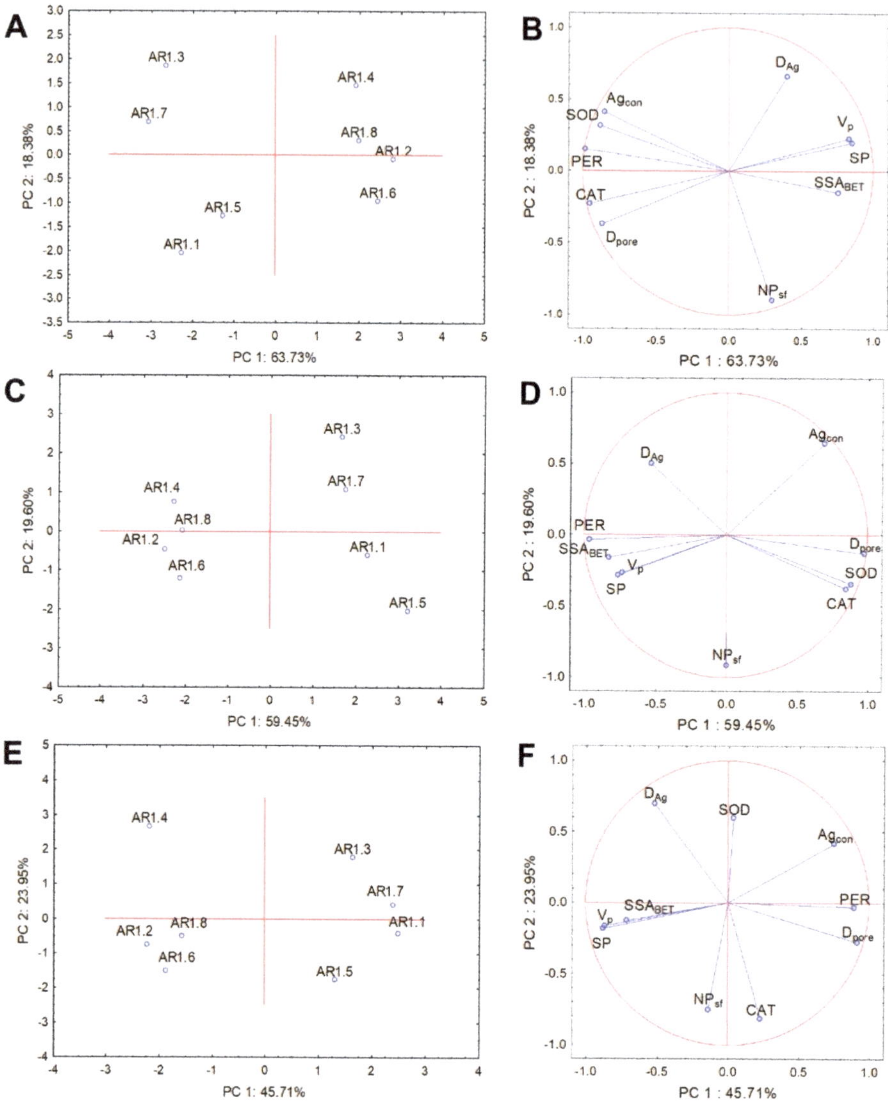

Figure 11. Projection of the individual plots and PCA analysis of CAT, PER, and SOD activity with silver particle diameter (D_{Ag}), number of particles with a spherical shape (NP_{sf}), pore volume (V_P), specific surface area (SSA_{BET}), micropore surface (S_P), pore diameter (D_{pore}), and silver content (Ag_{con}) along PC1 and PC2 for the NP-treated IC_{50} bacterial strains: *E. coli* (**A**,**B**), *B. cereus* (**C**,**D**), and *S. epidermidis* (**E**,**F**).

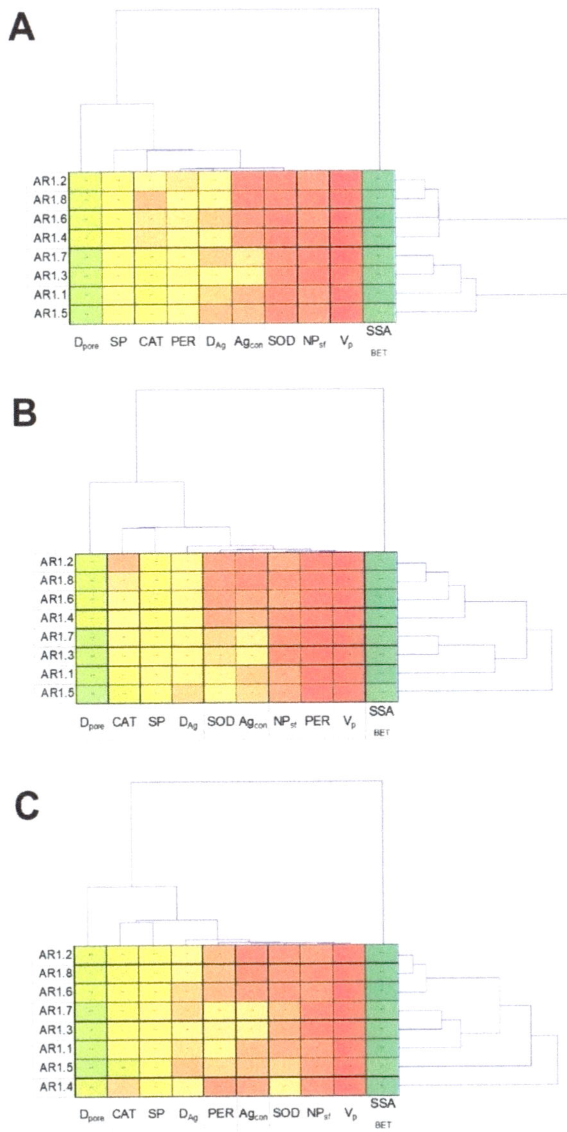

Figure 12. Projection of cluster analysis dendrograms for bacterial strains *E. coli* (**A**), *B. cereus* (**B**), and *S. epidermidis* (**C**) exposed to NCs at IC_{50}.

5. Materials and Methods

5.1. Structural Analysis

An X'PertPro MPD PANalytical X-ray diffractometer (Malvern PANalytical, Almelo, the Netherlands) with a Cu K_α radiation (λ = 1.54 Å) was used for analyzing the crystal structure of Ag-SiO_2. The Rietveld method refined the lattice parameters using HighScore Plus (version 5.1, Malvern PANalytical, Almelo, The Netherlands) software and the ICCD PDF-4+ 2023 database.

The three-dimensional silica network was followed through a WITec confocal Raman microscope CRM alpha 300 R (WITec Wissenschaftliche Instrumente und Technologie GmbH, Ulm, Germany) equipped with an air-cooled solid-state laser (λ = 532 nm). The excitation laser radiation was coupled into a microscope through a polarization-maintaining single-mode optical fiber with a 50 μm diameter. The monochromatic light was focused on the sample by an air Olympus MPLAN (100x/0.90NA) objective. Raman scattered light was passed through a multi-mode fiber (50 μm diameter) into a monochromator with a 600 line/mm grating and a CCD camera.

The spectrometer monochromator was checked before the measurements using a silicon plate (520.7 cm^{-1}). Raman spectra were accumulated over 20 scans with an integration time of 20 s and a resolution of 3 cm^{-1}. The post-processing analysis, such as baseline correction and cosmic ray removal, was performed in the WITecProjectFive Plus (version 5.1.1, WITec Wissenschaftliche Instrumente und Technologie GmbH, Ulm, Germany). The peak fitting analysis was performed in GRAMS (version 9.2, Thermo Fisher Scientific, Waltham, MA, USA) software.

5.2. Morphological Studies and Chemical Composition

TEM micrographs with chemical imagining were collected with a probe Cs-corrected S/TEM Titan 80-300 FEI microscope equipped with an EDAX EDS detector. The images were recorded in STEM mode, using the HAADF (high-angle annular dark field) and bright-field (BF) detectors. At the same time, the local chemical content and chemical mapping were realized using an Energy Dispersive Spectrometer (EDS). A scanning electron microscope equipped with an energy dispersive X-ray spectrometer (SEM-EDS) (Phenom ProX (ThermoFisher, Eindhoven, The Netherlands) at an accelerating voltage of 15 kV was used to estimate the chemical content in the microscale.

The X-ray photoelectron spectroscopy (XPS) measurements were made using a Physical Electronics PHI 5700 spectrometer (Physical Electronics Inc., Chanhassen, MN, USA). Photoelectrons were excited by monochromatized Al Kα radiation from the sample surface. The survey spectrum showed the presence of main core level lines from C, O, Si, and Ag with no evidence of impurities. The high-resolution XPS spectra for all core levels were corrected using the Iterated Shirley algorithm for the background signal. At the same time, the bands were fitted by a composition of Gaussian and Lorentzian lines in Multipack software. The binding energy value was corrected for minor surface charging effects by referring to the C 1s line at 284.5 eV.

5.3. Estimation of Nanoparticles Size

The diameter of the nanoparticles was measured at 25 °C on a Zetasizer Nano ZS (Malvern Panalytical, Grovewood Road, Malvern, UK). The analyzer was equipped with a He-Ne laser wavelength of 633 nm. The particle diameter was measured using dynamic light scattering (DLS), allowing us to obtain the diffusion coefficient (D) and the hydrodynamic diameter using the Stokes–Einstein equation (1). The measurements were performed in polystyrene cuvettes with an optical path of 1 cm at a detection angle of 173°.

Before measurement, all samples were diluted in water and sonicated with an ultrasonic homogenizer (Omni Sonic Ruptor 400, PerkinElmer, Kennesaw, GA, USA). The sonication time was repeated thrice for 5 min (continuous mode; ultrasonic power did not exceed 20%) until a homogeneous suspension, and repeatable results were obtained. Each sample was measured ten times, and the results were averaged. The averaged values of hydrodynamic diameter (d_H) and polydispersity coefficient (PDI) are summarized in Table 2.

$$d_H = \frac{kT}{3\pi\eta D} \tag{1}$$

where d_H: hydrodynamic diameter, k: Boltzmann's constant, T: absolute temperature, η: viscosity of the diluent, and D: diffusion coefficient.

5.4. Determination of Nanoparticle Surface Charge

The Zeta potential was measured on Malvern's Zetasizer Nano ZS (Malvern Panalytical, Grovewood Road, Malvern, UK) at 25 °C and in a U-shaped cuvette (DTS1070). The detection angle was 173°. The electrophoretic mobility (U_E) was determined using laser Doppler velocimetry (LDV), wherein Zeta potential was recalculated using Henry's equation (2). Samples were prepared analogously to the sample preparation procedure for measuring the diameter. Each sample was measured five times, and the average values and standard deviations are shown in Table 2.

$$U_E = \frac{2\varepsilon z f(Ka)}{3\eta} \quad (2)$$

where U_E: electrophoretic mobility, ε: dielectric constant, z: Zeta potential, *f(Ka)*: Henry's function, and η: viscosity.

5.5. Determination of the Surface Area, Pore Volume, and Pore Diameter

The specific surface area (SSA_{BET}), micropore volume (V_P), area (S_P), and pore size (D_{pore}) were determined using a Gemini VII 2390a analyzer (Micromeritics Instruments Corp., Norcross, GA, USA). The measurements were performed at the boiling point of nitrogen (−196 °C), and final data was obtained using the Brunauer–Emmet–Teller (BET) method and *t*-plot analysis. The total pore volume (TPV) was calculated from the gas sorption isotherm at p/p_0 close to saturation pressure (0.995 p/p_0).

The porosity was calculated as the TPV ratio to the sum of the TPV and solid particle volume (Table 3). Samples, before the measurements, were thermal-treated at 250 °C for three hours to remove gases and vapors adsorbed on the surface during the synthesis. This analysis was achieved with a VacPrep 061 degassing system (Micromeritics Instruments Corp., Norcross, GA, USA). Samples were not analyzed immediately after the degassing procedure was kept at 60 °C. The instrument's accuracy was verified for each use by analyzing a carbon black reference material of known surface area (P/N 004-16833-00 from Micromeritics, Norcross, GA, USA).

5.6. Microbial Studies

5.6.1. Bacterial Strains and Toxicological Studies

The assessment of microbial response to silver-silica nanocomposites (NCs) was performed using bacterial strains purchased from the American Type Culture Collection (ATCC) comprising Gram-negative *Escherichia coli* (ATCC® 25922™) and Gram-positive *Bacillus cereus* (ATCC® 11778™) and *Staphylococcus epidermidis* (ATCC® 12228™).

The minimum bactericidal concentration (MBC) and half-maximal inhibitory concentration (IC_{50}) were estimated using the dilution method. For this purpose, standard solutions (200 mg L^{-1}) with individual NCs were suspended in a sterile H_2O_{MP} (ultrapure Millipore water) and sonicated before application using Vibra-Cell™ at 20 kHz for 10–20 min. Primarily, the lysogeny broth (LB) medium (tryptone 10 g L^{-1}, NaCl 10 g L^{-1}, and yeast extract 5 g L^{-1}) was inoculated with a bacterial suspension in 0.85% NaCl-containing cells. The logarithmic growth phase was achieved until optical density OD_{600} = 0.1 (~10^7 CFU mL^{-1}).

Subsequently, NCs were added to the bacterial cultures with final concentrations ranging from 0.1 to 100 mg L^{-1} and were incubated at 37 °C with shaking conditions (140 rpm). The control samples in this research were bacterial cells cultured without NCs. Following 24 h incubation, dilution series of each culture were prepared in 0.85% NaCl, subcultured on LB agar plates, and incubated for 24 h at 37 °C. Afterward, bacterial colonies were counted and expressed as colony-forming units (CFU mL^{-1}). The lowest concentration of an antimicrobial agent, defined as MBC, was established as a 100% mortality rate of bacteria [104,105], whereas IC_{50} values of each NCs were estimated with the AAT Bioquest calculator [106].

5.6.2. Antioxidant Enzymes Response to NC Treatment

The influence of NCs on the antioxidant cell defense system was examined by analyzing the activity of three antioxidant enzymes: catalase (CAT), peroxidase (PER), and superoxide dismutase (SOD) in *E. coli*, *B. cereus*, and *S. epidermidis* cells. For this purpose, bacteria were exposed to NCs at concentrations equal to IC_{50} values for 24 h (at 37 °C and 140 rpm).

Enzymes were isolated from overnight cultures through Hegemans' method [107]. Then, bacterial cultures were centrifuged (4 °C, 5000 rpm, and 20 min), and the obtained pellet was suspended in 50 mM phosphate buffer (pH 7.0). Afterward, the enzymes were released from cells during sonication (six times for t = 15 s with t = 30 s intervals and f = 20 kHz). The supernatant obtained after centrifugation (4 °C, 12 000 rpm, and 20 min) was subjected to the study of microbial antioxidants.

The activity of CAT was assessed spectrophotometrically based on H_2O_2 decomposition—measured as a decrease in the absorbance at λ_{240} in time. The specific enzyme activity was calculated using the molar extinction coefficient (36,000 $dm^3\ mol^{-1}\ cm^{-1}$). The Sigma-Aldrich colorimetric protocol based on the measurement of an absorbance intensification at λ_{420} equivalent to the increase of the purpurogallin product was applied to evaluate the PER activity.

Successively, SOD activity was determined by the reduction of tetrazolium salt and a decrease in the dye's color intensity (λ_{450}) using a commercial reagent kit (cat. 19160; Sigma-Aldrich, St. Louis, MI, USA). The specific activity of SOD was calculated according to Zhang et al. [108]. The samples' total protein concentrations in the isolated protein fractions were quantified using the Coomassie Brilliant Blue G-250 reagent method [109]. Finally, the CAT, PER, and SOD activities were expressed as U mg^{-1} of protein.

5.7. Statistical Analysis

Hydrodynamic particle size diameter was repeated in ten independent measurements for each sample. SEM-EDS and TEM-EDS measurements were done in three and five individual places. The toxicological and antioxidant enzyme response data were repeated three times. Data were statistically treated and presented as mean ± standard deviation values (± SD) each time. The statistical significance between the given Ag-S1 and Ag-S2 NCs and mean hydrodynamic diameter or their effects on enzyme activities were followed using a one-way ANOVA and Tukey's Honest Significant Difference test (HSD). The substantial variations in experimental groups were represented in figures by annotated letters as $p < 0.05$ (*) and $p < 0.001$ (***) levels of significance. All experimental data were subjected to multivariate analysis. Hydrodynamic particle size was graphically presented using box-and-whiskers graphs. The specific surface area was correlated with pore size diameter by applying hierarchical cluster analysis (HCA) at the usage of the furthest neighbor, Euclidean distance, and application of the Sneath criterion (33% and 66%). In turn, the multifaceted analysis covering various aspects of the influence of NCs on the first line of the antioxidant defense system was performed using principal component analysis (PCA). The projection of the topographic map by comparing an entire set of physicochemical and biological variables was used to establish similar groups of objects forming clusters.

All the graphical/statistical processing data were estimated using MS Office 2019 (Microsoft Inc., Redmond, WA, USA), OriginPro2023 (OriginLab Corporation, Northampton, MA, USA), and the Statistica 13.3 software package (TIBCO Software Inc., Palo Alto, CA, USA).

6. Conclusions

The presented outcomes demonstrated the influence of different microwave irradiation conditions on modifying the absorption of silver-based nanostructures captured on silica carriers. Detailed studies showed that microwave treatment modified the silicas morphologically, increased the carrier's shape and open porosity, and lowered the active surface.

These effects became stronger and observable through a shorter exposure to microwave irradiation. However, modifying the surface of silica spheres using a microwave field both at t = 60 and 150 s, as well as P = 150 and 700 W, allowed for silver uptake and its crystallization as stable bioactive Ag_2O/Ag_2CO_3 heterojunctions. Based on the statistical approach, the applied silica surface modification procedures divided the results into two routes in which the Ag-SiO_2 systems were significantly different, physicochemically and structurally. The power of the microwave field proved to have a minor effect on carrier modification compared with the time (t = 60 or 150 s). We showed that increased time and power of microwave irradiation gradually increased the size of the silver particles. Increasing the microwave irradiation time and power enhanced the silver nanoparticle's asphericity. Finally, an extended exposure time to the microwave field corresponded with lower porosity and content of silver captured by the silica.

The surface-activated and silica carriers with Ag_2O/Ag_2CO_3 were used to study antimicrobial activity and the effects on antioxidant enzyme functioning. We found that a greater sphericity of silver nanoparticles resulted in more considerable toxicological impacts against *E. coli*, *B. cereus*, and *S. epidermidis*. Furthermore, such structures affect the antioxidant defense system of *E. coli*, *B. cereus*, and *S. epidermidis* through enzymatic activity, causing the induction of oxidative stress and leading to cell death. The most robust effects were found for nanocomposites in which the silica carrier was treated for an extended period in a microwave field.

Supplementary Materials: The following supporting information can be downloaded at: https://www.mdpi.com/article/10.3390/ijms24076632/s1.

Author Contributions: Conceptualization, M.D.; methodology, A.S., M.D., D.W., O.M., A.N., K.M., P.R., N.W. and S.G.; formal analysis and validation, M.D., D.W., O.M., A.N., K.D. and S.G.; investigation, A.S., M.D., D.W., O.M., A.N., K.D., K.M., J.K., N.W., A.M. and S.G.; writing—original draft preparation, M.D.; writing—review and editing, A.S., M.D., D.W., O.M., A.N. and A.M.; visualization, A.S., M.D., D.W. and O.M.; supervision, M.D.; project administration, M.D.; funding acquisition, M.D. All authors have read and agreed to the published version of the manuscript.

Funding: M.D. is thankful for the financial support from the National Center of Science (NCN) based on 2020/37/B/ST8/03637.

Institutional Review Board Statement: Not applicable.

Informed Consent Statement: Not applicable.

Data Availability Statement: Data are stored in the cloud and stick in the form of a backup.

Conflicts of Interest: The authors declare no conflict of interest. The funders had no role in the study's design; in the collection, analyses, or interpretation of the data; in the writing of the manuscript, or in the decision to publish the results.

References

1. Pareek, V.; Gupta, R.; Panwar, J. Do Physico-Chemical Properties of Silver Nanoparticles Decide Their Interaction with Biological Media and Bactericidal Action? A Review. *Mater. Sci. Eng. C* **2018**, *90*, 739–749. [CrossRef] [PubMed]
2. Chang, T.-H.; Liu, Y.-W.; Lin, Y.-H.; Lin, J.-J.; Huang, J.-W.; Hussain, A.; Chang, P.-F.L. Silver Nanoparticles on Nanoscale Silica Platelets (AgNP/NSP) and Nanoscale Silica Platelets (NSP) Inhibit the Development of Fusarium Oxysporum f. Sp. Niveum. *ACS Appl. Bio Mater.* **2019**, *2*, 4978–4985. [CrossRef] [PubMed]
3. Tăbăran, A.-F.; Matea, C.T.; Mocan, T.; Tăbăran, A.; Mihaiu, M.; Iancu, C.; Mocan, L. Silver Nanoparticles for the Therapy of Tuberculosis. *Int. J. Nanomed.* **2020**, *15*, 2231–2258. [CrossRef]
4. Qing, Y.; Cheng, L.; Li, R.; Liu, G.; Zhang, Y.; Tang, X.; Wang, J.; Liu, H.; Qin, Y. Potential Antibacterial Mechanism of Silver Nanoparticles and the Optimization of Orthopedic Implants by Advanced Modification Technologies. *Int. J. Nanomed.* **2018**, *13*, 3311–3327. [CrossRef]
5. Korzekwa, K.; Kędziora, A.; Stańczykiewicz, B.; Bugla-Płoskońska, G.; Wojnicz, D. Benefits of Usage of Immobilized Silver Nanoparticles as Pseudomonas Aeruginosa Antibiofilm Factors. *Int. J. Mol. Sci.* **2022**, *23*, 284. [CrossRef]
6. Tang, S.; Zheng, J. Antibacterial Activity of Silver Nanoparticles: Structural Effects. *Adv. Healthc. Mater.* **2018**, *7*, 1701503. [CrossRef]
7. He, H.; Meng, S.; Li, H.; Yang, Q.; Xu, Z.; Chen, X.; Sun, Z.; Jiang, B.; Li, C. Nanoplatform Based on GSH-Responsive Mesoporous Silica Nanoparticles for Cancer Therapy and Mitochondrial Targeted Imaging. *Microchim. Acta* **2021**, *188*, 154. [CrossRef]

8. Bapat, R.A.; Chaubal, T.V.; Joshi, C.P.; Bapat, P.R.; Choudhury, H.; Pandey, M.; Gorain, B.; Kesharwani, P. An Overview of Application of Silver Nanoparticles for Biomaterials in Dentistry. *Mater. Sci. Eng. C* **2018**, *91*, 881–898. [CrossRef] [PubMed]
9. Yin, I.X.; Zhang, J.; Zhao, I.S.; Mei, M.L.; Li, Q.; Chu, C.H. The Antibacterial Mechanism of Silver Nanoparticles and Its Application in Dentistry. *Int. J. Nanomed.* **2020**, *15*, 2555–2562. [CrossRef]
10. Lee, S.H.; Jun, B.-H. Silver Nanoparticles: Synthesis and Application for Nanomedicine. *Int. J. Mol. Sci.* **2019**, *20*, 865. [CrossRef]
11. Min, Y.; Akbulut, M.; Kristiansen, K.; Golan, Y.; Israelachvili, J. The Role of Interparticle and External Forces in Nanoparticle Assembly. *Nat. Mater.* **2008**, *7*, 527–538. [CrossRef] [PubMed]
12. Stevenson, A.P.; Blanco Bea, D.; Civit, S.; Antoranz Contera, S.; Iglesias Cerveto, A.; Trigueros, S. Three Strategies to Stabilise Nearly Monodispersed Silver Nanoparticles in Aqueous Solution. *Nanoscale Res. Lett.* **2012**, *7*, 151. [CrossRef]
13. Sotiriou, G.A.; Pratsinis, S.E. Engineering Nanosilver as an Antibacterial, Biosensor and Bioimaging Material. *Curr. Opin. Chem. Eng.* **2011**, *1*, 3–10. [CrossRef]
14. Dulski, M.; Peszke, J.; Włodarczyk, J.; Sułowicz, S.; Piotrowska-Seget, Z.; Dudek, K.; Podwórny, J.; Malarz, K.; Mrozek-Wilczkiewicz, A.; Zubko, M.; et al. Physicochemical and Structural Features of Heat Treated Silver-Silica Nanocomposite and Their Impact on Biological Properties. *Mater. Sci. Eng. C* **2019**, *103*, 109790. [CrossRef] [PubMed]
15. Panáček, A.; Kvítek, L.; Prucek, R.; Kolář, M.; Večeřová, R.; Pizúrová, N.; Sharma, V.K.; Nevěčná, T.; Zbořil, R. Silver Colloid Nanoparticles: Synthesis, Characterization, and Their Antibacterial Activity. *J. Phys. Chem. B* **2006**, *110*, 16248–16253. [CrossRef]
16. Sharma, V.K.; Yngard, R.A.; Lin, Y. Silver Nanoparticles: Green Synthesis and Their Antimicrobial Activities. *Adv. Colloid Interface Sci.* **2009**, *145*, 83–96. [CrossRef]
17. Sotiriou, G.A.; Pratsinis, S.E. Antibacterial Activity of Nanosilver Ions and Particles. *Environ. Sci. Technol.* **2010**, *44*, 5649–5654. [CrossRef]
18. Sotiriou, G.A.; Meyer, A.; Knijnenburg, J.T.N.; Panke, S.; Pratsinis, S.E. Quantifying the Origin of Released Ag+ Ions from Nanosilver. *Langmuir* **2012**, *28*, 15929–15936. [CrossRef]
19. Xiu, Z.; Zhang, Q.; Puppala, H.L.; Colvin, V.L.; Alvarez, P.J.J. Negligible Particle-Specific Antibacterial Activity of Silver Nanoparticles. *Nano Lett.* **2012**, *12*, 4271–4275. [CrossRef]
20. Sotiriou, G.A.; Sannomiya, T.; Teleki, A.; Krumeich, F.; Vörös, J.; Pratsinis, S.E. Non-Toxic Dry-Coated Nanosilver for Plasmonic Biosensors. *Adv. Funct. Mater.* **2010**, *20*, 4250–4257. [CrossRef] [PubMed]
21. Owan, R.; Handy, R. Viewpoint: Formulating the Problems for Environmental Risk Assessment of Nanomaterials. *Environ. Sci. Technol.* **2007**, *41*, 5582–5588. [CrossRef]
22. Stone, V.; Nowack, B.; Baun, A.; van den Brink, N.; von der Kammer, F.; Dusinska, M.; Handy, R.; Hankin, S.; Hassellöv, M.; Joner, E.; et al. Nanomaterials for Environmental Studies: Classification, Reference Material Issues, and Strategies for Physico-Chemical Characterisation. *Sci. Total Environ.* **2010**, *408*, 1745–1754. [CrossRef]
23. Lakbita, O.; Rhouta, B.; Maury, F.; Senocq, F.; Amjoud, M.; Daoudi, L. Influence of the Crystal Structure of Ag2CO3 on the Photocatalytic Activity under Visible Light of Ag2CO3-Palygorskite Nanocomposite Material. *Appl. Surf. Sci.* **2019**, *464*, 205–211. [CrossRef]
24. Mergenbayeva, S.; Atabaev, T.S.; Vakros, J.; Mantzavinos, D.; Poulopoulos, S.G. Photocatalytic Degradation of 4-Tert-Butylphenol Using Solar Light Responsive Ag_2CO_3. *Catalysts* **2022**, *12*, 1523. [CrossRef]
25. Ren, H.-T.; Jia, S.-Y.; Wu, S.-H.; Zhang, T.-H.; Han, X. Phase Transformation Synthesis of Novel Ag2O/Ag2CO3/g-C3N4 Composite with Enhanced Photocatalytic Activity. *Mater. Lett.* **2015**, *142*, 15–18. [CrossRef]
26. Feng, Q.L.; Wu, J.; Chen, G.Q.; Cui, F.Z.; Kim, T.N.; Kim, J.O. A Mechanistic Study of the Antibacterial Effect of Silver Ions on Escherichia Coli and Staphylococcus Aureus. *J. Biomed. Mater. Res.* **2000**, *52*, 662–668. [CrossRef]
27. Yamanaka, M.; Hara, K.; Kudo, J. Bactericidal Actions of a Silver Ion Solution on Escherichia Coli, Studied by Energy-Filtering Transmission Electron Microscopy and Proteomic Analysis. *Appl. Environ. Microbiol.* **2005**, *71*, 7589–7593. [CrossRef]
28. Yao, Z.T.; Chen, T.; Li, H.Y.; Xia, M.S.; Ye, Y.; Zheng, H. Preparation, Characterization, and Antibacterial Activity of Shell Waste Loaded with Silver. *J. Mater. Sci.* **2013**, *48*, 8580–8587. [CrossRef]
29. Xu, C.; Liu, Y.; Huang, B.; Li, H.; Qin, X.; Zhang, X.; Dai, Y. Preparation, Characterization, and Photocatalytic Properties of Silver Carbonate. *Appl. Surf. Sci.* **2011**, *257*, 8732–8736. [CrossRef]
30. Zhang, H.; Chen, G.; Bahnemann, D.W. Photoelectrocatalytic Materials for Environmental Applications. *J. Mater. Chem.* **2009**, *19*, 5089–5121. [CrossRef]
31. Teoh, W.Y.; Scott, J.A.; Amal, R. Progress in Heterogeneous Photocatalysis: From Classical Radical Chemistry to Engineering Nanomaterials and Solar Reactors. *J. Phys. Chem. Lett.* **2012**, *3*, 629–639. [CrossRef] [PubMed]
32. Kamat, P.V.; Bisquert, J. Solar Fuels. Photocatalytic Hydrogen Generation. *J. Phys. Chem. C* **2013**, *117*, 14873–14875. [CrossRef]
33. Guo, S.; Bao, J.; Hu, T.; Zhang, L.; Yang, L.; Peng, J.; Jiang, C. Controllable Synthesis Porous Ag2CO3 Nanorods for Efficient Photocatalysis. *Nanoscale Res. Lett.* **2015**, *10*, 193. [CrossRef]
34. Dai, G.; Yu, J.; Liu, G. A New Approach for Photocorrosion Inhibition of Ag2CO3 Photocatalyst with Highly Visible-Light-Responsive Reactivity. *J. Phys. Chem. C* **2012**, *116*, 15519–15524. [CrossRef]
35. Lončarević, D.; Vukoje, I.; Dostanić, J.; Bjelajac, A.; Đorđević, V.; Dimitrijević, S.; Nedeljković, J.M. Antimicrobial and Photocatalytic Abilities of Ag_2CO_3 Nano-Rods. *ChemistrySelect* **2017**, *2*, 2931–2938. [CrossRef]
36. Zhou, L.; Liang, L.; Talifu, D.; Abulizi, A. Sonochemical Fabrication of Ag2CO3 Nanomaterial and Influencing Factors on Photocatalytic Properties. *IOP Conf. Ser. Mater. Sci. Eng.* **2017**, *167*, 012032. [CrossRef]

37. Petala, A.; Nasiou, A.; Mantzavinos, D.; Frontistis, Z. Photocatalytic Evaluation of Ag2CO3 for Ethylparaben Degradation in Different Water Matrices. *Water* **2020**, *12*, 1180. [CrossRef]
38. Panácek, A.; Kolár, M.; Vecerová, R.; Prucek, R.; Soukupová, J.; Krystof, V.; Hamal, P.; Zboril, R.; Kvítek, L. Antifungal Activity of Silver Nanoparticles against *Candida* Spp. *Biomaterials* **2009**, *30*, 6333–6340. [CrossRef]
39. Yu, C.; Li, G.; Kumar, S.; Yang, K.; Jin, R. Phase Transformation Synthesis of Novel Ag2O/Ag2CO3 Heterostructures with High Visible Light Efficiency in Photocatalytic Degradation of Pollutants. *Adv. Mater.* **2014**, *26*, 892–898. [CrossRef]
40. Buckley, J.J.; Gai, P.L.; Lee, A.F.; Olivi, L.; Wilson, K. Silver Carbonate Nanoparticles Stabilised over Alumina Nanoneedles Exhibiting Potent Antibacterial Properties. *Chem. Commun.* **2008**, *34*, 4013–4015. [CrossRef]
41. Yu, C.; Wei, L.; Zhou, W.; Dionysiou, D.D.; Zhu, L.; Shu, Q.; Liu, H. A Visible-Light-Driven Core-Shell like Ag2S@Ag2CO3 Composite Photocatalyst with High Performance in Pollutants Degradation. *Chemosphere* **2016**, *157*, 250–261. [CrossRef]
42. Yu, C.; Wei, L.; Chen, J.; Xie, Y.; Zhou, W.; Fan, Q. Enhancing the Photocatalytic Performance of Commercial TiO2 Crystals by Coupling with Trace Narrow-Band-Gap Ag_2CO_3. *Ind. Eng. Chem. Res.* **2014**, *53*, 5759–5766. [CrossRef]
43. Rossi, N.R.; de Menezes, B.R.C.; Sampaio, A.D.G.; da Silva, D.M.; Koga-Ito, C.Y.; Thim, G.P.; Paes-Junior, T.J.D.A. Silver-Coated Silica Nanoparticles Modified with MPS: Potential Antimicrobial Biomaterials Applied in Glaze and Soft Reliner. *Polymers* **2022**, *14*, 4306. [CrossRef] [PubMed]
44. Otari, S.V.; Patel, S.K.S.; Kalia, V.C.; Kim, I.-W.; Lee, J.-K. Antimicrobial Activity of Biosynthesized Silver Nanoparticles Decorated Silica Nanoparticles. *Indian J. Microbiol.* **2019**, *59*, 379–382. [CrossRef] [PubMed]
45. Canhaci, S.J.; Perez, R.F.; Borges, L.E.P.; Fraga, M.A. Direct Conversion of Xylose to Furfuryl Alcohol on Single Organic–Inorganic Hybrid Mesoporous Silica-Supported Catalysts. *Appl. Catal. B Environ.* **2017**, *207*, 279–285. [CrossRef]
46. Chen, K.; Fan, W.; Huang, C.; Qiu, X. Enhanced Stability and Catalytic Activity of Bismuth Nanoparticles by Modified with Porous Silica. *J. Phys. Chem. Solids* **2017**, *110*, 9–14. [CrossRef]
47. Jaroniec, M.; Kruk, M.; Olivier, J.P. Standard Nitrogen Adsorption Data for Characterization of Nanoporous Silicas. *Langmuir* **1999**, *15*, 5410–5413. [CrossRef]
48. Holland, B.T.; Blanford, C.F.; Do, T.; Stein, A. Synthesis of Highly Ordered, Three-Dimensional, Macroporous Structures of Amorphous or Crystalline Inorganic Oxides, Phosphates, and Hybrid Composites. *Chem. Mater.* **1999**, *11*, 795–805. [CrossRef]
49. Ai, C.; Gong, G.; Zhao, X.; Liu, P. Macroporous Hollow Silica Microspheres-Supported Palladium Catalyst for Selective Hydrogenation of Nitrile Butadiene Rubber. *J. Taiwan Inst. Chem. Eng.* **2017**, *77*, 250–256. [CrossRef]
50. Li, H.; Hou, B.; Wang, J.; Huang, X.; Chen, C.; Ma, Z.; Cui, J.; Jia, L.; Sun, D.; Li, D. Effect of Hierarchical Meso–Macroporous Structures on the Catalytic Performance of Silica Supported Cobalt Catalysts for Fischer–Tropsch Synthesis. *Catal. Sci. Technol.* **2017**, *7*, 3812–3822. [CrossRef]
51. Jayan, H.; Sun, D.-W.; Pu, H.; Wei, Q. Mesoporous Silica Coated Core-Shell Nanoparticles Substrate for Size-Selective SERS Detection of Chloramphenicol. *Spectrochim. Acta Part A Mol. Biomol. Spectrosc.* **2023**, *284*, 121817. [CrossRef] [PubMed]
52. Ambrogi, V.; Pietrella, D.; Donnadio, A.; Latterini, L.; Di Michele, A.; Luffarelli, I.; Ricci, M. Biocompatible Alginate Silica Supported Silver Nanoparticles Composite Films for Wound Dressing with Antibiofilm Activity. *Mater. Sci. Eng. C* **2020**, *112*, 110863. [CrossRef] [PubMed]
53. Silicon Oxide Nanopowder/SiO2 Nanoparticles (SiO2, 99.5+%, 15-20nm, S-Type, Spherical). Available online: https://www.us-nano.com/inc/sdetail/410 (accessed on 29 December 2022).
54. Silicon Dioxide SiO2 Nanopowder/Nanoparticles (SiO2, 99+%, 20-30 Nm). Available online: https://www.us-nano.com/inc/sdetail/408 (accessed on 29 December 2022).
55. Griscom, D.L. The Electronic Structure of SiO2: A Review of Recent Spectroscopic and Theoretical Advances. *J. Non-Cryst. Solids* **1977**, *24*, 155–234. [CrossRef]
56. Devine, R.A.; Arndt, J. Correlated Defect Creation and Dose-Dependent Radiation Sensitivity in Amorphous SiO_2. *Phys Rev. B Condens. Matter* **1989**, *39*, 5132–5138. [CrossRef]
57. Skuja, L. Optically Active Oxygen-Deficiency-Related Centers in Amorphous Silicon Dioxide. *J. Non-Cryst. Solids* **1998**, *239*, 16–48. [CrossRef]
58. Furukawa, T.; Fox, K.E.; White, W.B. Raman Spectroscopic Investigation of the Structure of Silicate Glasses. III. Raman Intensities and Structural Units in Sodium Silicate Glasses. *J. Chem. Phys.* **1981**, *75*, 3226–3237. [CrossRef]
59. Awazu, K.; Kawazoe, H. Strained Si–O–Si Bonds in Amorphous SiO2 Materials: A Family Member of Active Centers in Radio, Photo, and Chemical Responses. *J. Appl. Phys.* **2003**, *94*, 6243–6262. [CrossRef]
60. Robinet, L.; Coupry, C.; Eremin, K.; Hall, C. The Use of Raman Spectrometry to Predict the Stability of Historic Glasses. *J. Raman Spectrosc.* **2006**, *37*, 789–797. [CrossRef]
61. Bykov, V.N.; Koroleva, O.N.; Osipov, A.A. Structure of Silicate Melts: Raman Spectroscopic Data and Thermodynamic Simulation Results. *Geochem. Int.* **2009**, *47*, 1067. [CrossRef]
62. McMillan, P.F.; Wolf, G.; Poe, B.T. Vibrational Spectroscopy of Silicate Liquids and Glasses. *Chem. Geol.* **1992**, *96*, 351–366. [CrossRef]
63. Matson, D.W.; Sharma, S.K.; Philpotts, J.A. The Structure of High-Silica Alkali-Silicate Glasses. A Raman Spectroscopic Investigation. *J. Non-Cryst. Solids* **1983**, *58*, 323–352. [CrossRef]

64. Parkinson, B.G.; Holland, D.; Smith, M.E.; Larson, C.; Doerr, J.; Affatigato, M.; Feller, S.A.; Howes, A.P.; Scales, C.R. Quantitative Measurement of Q3 Species in Silicate and Borosilicate Glasses Using Raman Spectroscopy. *J. Non-Cryst. Solids* **2008**, *354*, 1936–1942. [CrossRef]
65. Konijnendijk, W.L.; Stevels, J.M. The Structure of Borosilicate Glasses Studied by Raman Scattering. *J. Non-Cryst. Solids* **1976**, *20*, 193–224. [CrossRef]
66. Alessi, A.; Agnello, S.; Buscarino, G.; Gelardi, F.M. Structural Properties of Core and Surface of Silica Nanoparticles Investigated by Raman Spectroscopy. *J. Raman Spectrosc.* **2013**, *44*, 810–816. [CrossRef]
67. McMillan, P. A Raman Spectroscopic Study of Glasses in the System CaO-MgO-SiO_2. *Am. Mineral.* **1984**, *69*, 645–659.
68. Sharma, S.K.; Mammone, J.F.; Nicol, M.F. Raman Investigation of Ring Configurations in Vitreous Silica. *Nature* **1981**, *292*, 140–141. [CrossRef]
69. McMillan, P. Structural Studies of Silicate Glasses and Melts-Applications and Limitations of Raman Spectroscopy. *Am. Mineral.* **1984**, *69*, 622–644.
70. McMillan, J.A. Magnetic Properties and Crystalline Structure of AgO. *J. Inorg. Nucl. Chem.* **1960**, *13*, 28–31. [CrossRef]
71. McMillan, J.A. Higher Oxidation States of Silver. *Chem. Rev.* **1962**, *62*, 65–80. [CrossRef]
72. Bosworth, Y.M.; Clark, R.J.H. Intensity Studies on the Raman-Active Fundamentals of Some Square-Planar (MX4n-) Ions and of the Dichloroiodate(1-) Ion. Calculation of Parallel and Perpendicular Bond Polarizability Derivatives and Bond Anisotropies. *Inorg. Chem.* **1975**, *14*, 170–177. [CrossRef]
73. Waterhouse, G.I.; Bowmaker, G.A.; Metson, J.B. The Thermal Decomposition of Silver (I, III) Oxide: A Combined XRD, FT-IR and Raman Spectroscopic Study. *Phys. Chem. Chem. Phys.* **2001**, *3*, 3838–3845. [CrossRef]
74. Slager, T.L.; Lindgren, B.J.; Mallmann, A.J.; Greenler, R.G. Infrared Spectra of the Oxides and Carbonates of Silver. *J. Phys. Chem.* **1972**, *76*, 940–943. [CrossRef]
75. Morimoto, T.; Aoki, K. Effect of Adsorbed Water on the Reaction of Silver(I) Oxide with Carbon Dioxide. *Langmuir* **1986**, *2*, 525–528. [CrossRef]
76. Martina, I.; Wiesinger, R.; Jembrih-Simbürger, D.; Schreiner, M. Micro-Raman Characterisation of Silver Corrosion Products: Instrumental Set Up and Reference Database. *E-Preserv. Sci.* **2012**, *9*, 1–8.
77. Allen, J.A.; Scaife, P.H. Infrared Spectra in the Ag_2O-CO_2-Ag_2CO_3 System. *Aust. J. Chem.* **1966**, *19*, 715–724. [CrossRef]
78. Shinoda, K.; Hato, M.; Hayashi, T. Physicochemical Properties of Aqueous Solutions of Fluorinated Surfactants. *J. Phys. Chem.* **1972**, *76*, 909–914. [CrossRef]
79. Naumkin, V.; Kraut-Vass, A.; Gaarenstroom, S.W.; Powell, C.P. *NIST X-Ray Photoelectron Spectroscopy Database*; National Institute of Standards and Technology: Gaithersburg, MA, USA, 2000.
80. Mennig, M.; Schmitt, M.; Schmidt, H. Synthesis of Ag-Colloids in Sol-Gel Derived SiO_2-Coatings on Glass. *J. Sol-Gel. Sci. Technol.* **1997**, *8*, 1035–1042. [CrossRef]
81. Slavin, Y.N.; Asnis, J.; Häfeli, U.O.; Bach, H. Metal Nanoparticles: Understanding the Mechanisms behind Antibacterial Activity. *J. Nanobiotechnology* **2017**, *15*, 65. [CrossRef]
82. Khan, I.; Saeed, K.; Khan, I. Nanoparticles: Properties, Applications and Toxicities. *Arab. J. Chem.* **2019**, *12*, 908–931. [CrossRef]
83. Linklater, D.P.; Baulin, V.A.; Le Guével, X.; Fleury, J.-B.; Hanssen, E.; Nguyen, T.H.P.; Juodkazis, S.; Bryant, G.; Crawford, R.J.; Stoodley, P.; et al. Antibacterial Action of Nanoparticles by Lethal Stretching of Bacterial Cell Membranes. *Adv. Mater.* **2020**, *32*, e2005679. [CrossRef] [PubMed]
84. Sajid, M.; Ilyas, M.; Basheer, C.; Tariq, M.; Daud, M.; Baig, N.; Shehzad, F. Impact of Nanoparticles on Human and Environment: Review of Toxicity Factors, Exposures, Control Strategies, and Future Prospects. *Environ. Sci. Pollut. Res.* **2015**, *22*, 4122–4143. [CrossRef]
85. Ahmed, B.; Solanki, B.; Zaidi, A.; Khan, M.S.; Musarrat, J. Bacterial Toxicity of Biomimetic Green Zinc Oxide Nanoantibiotic: Insights into ZnONP Uptake and Nanocolloid–Bacteria Interface. *Toxicol. Res.* **2019**, *8*, 246–261. [CrossRef] [PubMed]
86. Handy, R.D.; von der Kammer, F.; Lead, J.R.; Hassellöv, M.; Owen, R.; Crane, M. The Ecotoxicology and Chemistry of Manufactured Nanoparticles. *Ecotoxicology* **2008**, *17*, 287–314. [CrossRef] [PubMed]
87. Ha, M.K.; Shim, Y.J.; Yoon, T.H. Effects of Agglomeration on in Vitro Dosimetry and Cellular Association of Silver Nanoparticles. *Environ. Sci. Nano* **2018**, *5*, 446–455. [CrossRef]
88. Raza, M.A.; Kanwal, Z.; Rauf, A.; Sabri, A.N.; Riaz, S.; Naseem, S. Size- and Shape-Dependent Antibacterial Studies of Silver Nanoparticles Synthesized by Wet Chemical Routes. *Nanomaterials* **2016**, *6*, 74. [CrossRef]
89. Acharya, D.; Singha, K.M.; Pandey, P.; Mohanta, B.; Rajkumari, J.; Singha, L.P. Shape Dependent Physical Mutilation and Lethal Effects of Silver Nanoparticles on Bacteria. *Sci. Rep.* **2018**, *8*, 201. [CrossRef]
90. Acharya, D.; Mohanta, B.; Pandey, P.; Nasiri, F. Antibacterial Properties of Synthesized Ag and Ag@SiO2 Core–Shell Nanoparticles: A Comparative Study. *Can. J. Phys.* **2018**, *96*, 955–960. [CrossRef]
91. Gankhuyag, S.; Bae, D.S.; Lee, K.; Lee, S. One-Pot Synthesis of SiO_2@Ag Mesoporous Nanoparticle Coating for Inhibition of Escherichia Coli Bacteria on Various Surfaces. *Nanomaterials* **2021**, *11*, 549. [CrossRef]
92. Pal, S.; Nisi, R.; Licciulli, A. Antibacterial Activity of In Situ Generated Silver Nanoparticles in Hybrid Silica Films. *Photochem* **2022**, *2*, 479–488. [CrossRef]

93. Montalvo-Quirós, S.; Gómez-Graña, S.; Vallet-Regí, M.; Prados-Rosales, R.C.; González, B.; Luque-Garcia, J.L. Mesoporous Silica Nanoparticles Containing Silver as Novel Antimycobacterial Agents against Mycobacterium Tuberculosis. *Colloids Surf. B Biointerfaces* **2021**, *197*, 111405. [CrossRef]
94. Malekzadeh, M.; Yeung, K.L.; Halali, M.; Chang, Q. Preparation and Antibacterial Behaviour of Nanostructured Ag@SiO2–Penicillin with Silver Nanoplates. *New J. Chem.* **2019**, *43*, 16612–16620. [CrossRef]
95. Ahmed, F.; AlOmar, S.Y.; Albalawi, F.; Arshi, N.; Dwivedi, S.; Kumar, S.; Shaalan, N.M.; Ahmad, N. Microwave Mediated Fast Synthesis of Silver Nanoparticles and Investigation of Their Antibacterial Activities for Gram-Positive and Gram-Negative Microorganisms. *Crystals* **2021**, *11*, 666. [CrossRef]
96. Canaparo, R.; Foglietta, F.; Limongi, T.; Serpe, L. Biomedical Applications of Reactive Oxygen Species Generation by Metal Nanoparticles. *Materials* **2020**, *14*, 53. [CrossRef]
97. Manke, A.; Wang, L.; Rojanasakul, Y. Mechanisms of Nanoparticle-Induced Oxidative Stress and Toxicity. *BioMed Res. Int.* **2013**, *2013*, 942916. [CrossRef]
98. Raghunath, A.; Perumal, E. Metal Oxide Nanoparticles as Antimicrobial Agents: A Promise for the Future. *Int. J. Antimicrob. Agents* **2017**, *49*, 137–152. [CrossRef]
99. Liao, S.; Zhang, Y.; Pan, X.; Zhu, F.; Jiang, C.; Liu, Q.; Cheng, Z.; Dai, G.; Wu, G.; Wang, L.; et al. Antibacterial Activity and Mechanism of Silver Nanoparticles against Multidrug-Resistant *Pseudomonas Aeruginosa*. *Int. J. Nanomed.* **2019**, *14*, 1469–1487. [CrossRef] [PubMed]
100. Howard, D.; Sebastian, S.; Le, Q.V.-C.; Thierry, B.; Kempson, I. Chemical Mechanisms of Nanoparticle Radiosensitization and Radioprotection: A Review of Structure-Function Relationships Influencing Reactive Oxygen Species. *Int. J. Mol. Sci.* **2020**, *21*, 579. [CrossRef]
101. Yuan, Y.-G.; Peng, Q.-L.; Gurunathan, S. Effects of Silver Nanoparticles on Multiple Drug-Resistant Strains of *Staphylococcus Aureus* and *Pseudomonas Aeruginosa* from Mastitis-Infected Goats: An Alternative Approach for Antimicrobial Therapy. *Int. J. Mol. Sci.* **2017**, *18*, 569. [CrossRef]
102. Jain, N.; Bhargava, A.; Rathi, M.; Dilip, R.V.; Panwar, J. Removal of Protein Capping Enhances the Antibacterial Efficiency of Biosynthesized Silver Nanoparticles. *PLoS ONE* **2015**, *10*, e0134337. [CrossRef] [PubMed]
103. Borkowski, A.; Cłapa, T.; Szala, M.; Gąsiński, A.; Selwet, M. Synthesis of SiC/Ag/Cellulose Nanocomposite and Its Antibacterial Activity by Reactive Oxygen Species Generation. *Nanomaterials* **2016**, *6*, 171. [CrossRef]
104. Wiegand, I.; Hilpert, K.; Hancock, R.E.W. Agar and Broth Dilution Methods to Determine the Minimal Inhibitory Concentration (MIC) of Antimicrobial Substances. *Nat. Protoc.* **2008**, *3*, 163–175. [CrossRef]
105. Bagchi, B.; Kar, S.; Dey, S.K.; Bhandary, S.; Roy, D.; Mukhopadhyay, T.K.; Das, S.; Nandy, P. In Situ Synthesis and Antibacterial Activity of Copper Nanoparticle Loaded Natural Montmorillonite Clay Based on Contact Inhibition and Ion Release. *Colloids Surf. B Biointerfaces* **2013**, *108*, 358–365. [CrossRef]
106. Sebaugh, J.L. Guidelines for Accurate EC50/IC50 Estimation. *Pharmaceut. Statist.* **2011**, *10*, 128–134. [CrossRef]
107. Hegeman, G.D. Synthesis of the Enzymes of the Mandelate Pathway by *Pseudomonas Putida* I. Synthesis of Enzymes by the Wild Type. *J. Bacteriol.* **1966**, *91*, 1140–1154. [CrossRef] [PubMed]
108. Zhang, C.; Bruins, M.E.; Yang, Z.-Q.; Liu, S.-T.; Rao, P.-F. A New Formula to Calculate Activity of Superoxide Dismutase in Indirect Assays. *Anal. Biochem.* **2016**, *503*, 65–67. [CrossRef] [PubMed]
109. Bradford, M.M. A Rapid and Sensitive Method for the Quantitation of Microgram Quantities of Protein Utilizing the Principle of Protein-Dye Binding. *Anal. Biochem.* **1976**, *72*, 248–254. [CrossRef] [PubMed]

International Journal of
Molecular Sciences

MDPI

Review

Prevention of Ventriculostomy Related Infection: Effectiveness of Impregnated Biomaterial

Sylvain Diop [1,*], Ariane Roujansky [2], Hatem Kallel [2,3] and Roman Mounier [4,5]

1 Anesthesiology and Intensive Care Department, Marie Lannelongue Surgical Hospital, 133 Avenue de la Résistance, 92350 Le Plessis Robinson, France
2 Intensive Care Unit, Cayenne General Hospital, Av des Flamboyants, 97306 Cayenne, France
3 Tropical Biome and Immuno-Pathology CNRS UMR-9017, Inserm U 1019, University of French Guiana, 97300 Cayenne, France
4 Department of Neuro-ICU, GHU-Paris, Paris University, 75014 Paris, France
5 INSERM U955, Team 15, Biomedical Research Institute, University Paris-Est-Creteil (UPEC), 94000 Paris, France
* Correspondence: menes.diop@gmail.com; Tel.: +33-(1)-40-94-86-37; Fax: +33-(1)-40-94-85-86

Abstract: External ventricular drain(EVD) exposes the patient to infectious complications which are associated with significant morbidity and economic burden. Biomaterials impregnated with various antimicrobial agents have been developed to decrease the rate of bacterial colonization and subsequent infection. While promising, antibiotics and silver-impregnated EVD showed conflicting clinical results. The aim of the present review is to discuss the challenges associated with the development of antimicrobial EVD catheters and their effectiveness from the bench to the bedside.

Keywords: external ventricular drain; biofilm; bacteria; silver nanoparticles; antibiotics impregnated; biomaterial

Citation: Diop, S.; Roujansky, A.; Kallel, H.; Mounier, R. Prevention of Ventriculostomy Related Infection: Effectiveness of Impregnated Biomaterial. *Int. J. Mol. Sci.* **2023**, *24*, 4819. https://doi.org/10.3390/ijms24054819

Academic Editor: Helena Felgueiras

Received: 14 February 2023
Revised: 24 February 2023
Accepted: 25 February 2023
Published: 2 March 2023

1. Introduction

Bacteria are among the oldest life form on earth [1]. The most ancient proof of their existence goes back to 3.8 billion years ago [1]. Over time they colonized the terrestrial environment and were exposed to extreme conditions [2,3]. Early in their evolution, bacteria acquired the capacity to produce biofilm (the most ancient trace of biofilm goes back around 3.2 and 3.4 billion years ago), a particular lifestyle able to link bacteria together and provide advantageous mechanisms of protection ensuring their survival against hostile environmental conditions [4]. Bacteria play an essential role in human health and development [4,5]. All parts of the human epithelium (skin, gut, lungs, among others) are colonized by a large diversity of bacteria, fungi, or viruses forming microbiota. Most of these bacteria live under biofilm conditions [4].

In the twentieth century major medical advances have been made regarding the development of long-term (prosthesis, pacemaker) or short-term (central venous catheter, EVD) implanted medical devices, revolutionizing the prognosis of many diseases [6]. Concomitantly, it raises the problematic of associated infectious complications. From several decays, bacteria adhesion, and biofilm formation at the surface of biomaterials have been incriminated as the main pathway of infection [7,8]. Consequently, manufacturers proposed new innovating materials which were expected to decrease the rate of bacterial colonization and infection. In this review, we focus on the effectiveness of EVD impregnated with antimicrobials agent (antibiotics or silver) to prevent ventriculostomy-related infection (VRI).

2. Background

EVD are widely used in neurosurgery to control cerebral hypertension mainly related to subarachnoid hemorrhage (SAH) or traumatic brain injury [9]. It consists of a catheter inserted through the skull into the ventricles by a neurosurgeon, allowing the drainage of the cerebrospinal fluid (CSF) and the monitoring and control of the intracranial pressure [9]. It also exposes the patient to VRI leading to higher morbidity and economic burden [10]. The pooled incidence of VRI is 11.4 per 1000 catheters per day and the main risk factors identified are SAH, intraventricular hemorrhage, and CSF leakage at the point of insertion [10]. Duration of EVD catheterization remained a debated risk factor, but in the majority of the studies, VRI occurred around day 10 [11,12]. The most frequent bacteria implicated in VRI are Gram-positive cocci (GPC) belonging to the head skin flora (*Staphylococcus epidermidis*, *Staphylococcus aureus*, *Staphylococcus* spp., *Streptococcus* spp.) [11]. It has been postulated that VRI involves prior colonization of the surface device by bacterial biofilm originating from the skin microbiota [13]. Periprocedural prophylactic and prolonged intravenous antibiotics failed to demonstrate convincing evidence to prevent VRI [11,14]. Consequently, EVD impregnated with an antimicrobial agent has been developed to reduce the risk of bacterial adhesion, biofilm formation, and VRI development.

3. Silver-Impregnated EVD

Currently, numerous medical devices impregnated or coated with silver nanoparticles have been developed such as urinary catheters, central venous catheters (CVC), or EVD. The most studied and worldwide available silver-impregnated EVD is the Silverline^{TM} EVD. The catheter is made in polyurethane recovered with 1% of silver nanoparticles and 1% of insoluble silver salt. According to the manufacturers (Spielberg KG), it allows a continuous release of silver ions with a broad-spectrum activity of up to 32 h [15].

3.1. *Antimicrobial Effect of Silver*

The antimicrobial properties of silver have been known for thousands of years. Ancient Egyptians were familiar with the use of various metals, such as lead or silver, to treat or prevent infectious diseases [16]. Silver has good biocompatibility with mammalian cells and a broad-spectrum antimicrobial activity against both GPC, Gram-negative rods (GNR), and fungi even at low concentrations. The antimicrobial action is mediated by the direct toxicity of silver ions on bacteria through several mechanisms such as the generation of reactive oxygen species, damage of intracellular structure and proteins, alteration of signal transduction pathway, or electron chain transport [16]. Silver ion has a high affinity to peptidoglycan. It seems less effective in GPC because the large peptidoglycan wall could prevent the silver ion from reaching the bacteria cytoplasm [17]. Silver ions also demonstrated an anti-biofilm activity in vitro against both GPC and GNR. The antibacterial activity is very short because, in vivo, the silver ion quickly binds an anion (such as chloride) and precipitates [18]. Thus, to be effective, silver must be released continuously from the biomaterial surface. The antimicrobial effect of silver nanoparticles also depends on several factors such as size, shape, colloidal state, and the concentration of silver ions generated over time [19]. It leads to the development of numerous impregnation methods on biopolymer allowing the release of a small concentration of silver ions over time.

3.2. *Synthesis of Silver-Impregnated Biomaterial and Experimental Results*

Numerous synthesis methods for silver nanoparticles are available such as chemical (i.e., chemical reduction, electrochemical synthesis, pyrolysis method) or physical methods (i.e., arc discharge, laser irradiation). Recently a green synthetic process has also been developed, limiting the use of toxic chemical compounds [18]. The challenge associated with the synthesis of silver nanoparticles-impregnated biomaterial is multi-faceted: It must allow the synthesis of small-size nanoparticles (<100 nm) homogeneously distributed along the catheter and released continuously at a predictable rate over time, without local or systemic toxic effect. Polyurethane is a biopolymer with wide application in the medical field because of its biocompatibility and its advantageous physical properties [20].

Saveleyev et al. experimentally tested the synthesis and effectiveness of silver nanoparticles-impregnated polyurethane catheters [20]. The nanoparticles synthesized had a spherical shape and a size of 10 to 110 nm without changing the nature of the biomaterial. The impregnated material demonstrated bactericidal and bacteriostatic activity against *S. aureus*, GNR (*Pseudomonas aeruginosa*, *Enterobacter aerogenes*, *Klebsiella pneumoniae*, *Proteus mirabilis*, and *Escherichia coli*), and fungi [20]. Another study tested the impregnation of silver nanoparticles on polyurethane CVC grafted with acrylic acids. Scanning electron microscopy (SEM) showed that silver nanoparticles had a mean size of 45 nm, however, the concentration of silver in the biomaterial was very low. The antimicrobial effect was observed against *E coli* and methicillin-resistant *S. aureus* (MRSA) strains [21]. The data from the Manufacturer showed that Silverline^{TM} catheters have antimicrobial activity on GPC, GNR, and *Candida* (with a concentration of 10^7 to 10^8 cfu/mL according to the strain considered) when assessed with the roll culture plate method [15].

Bayston et al. investigated specifically in vitro Silverline^{TM} EVD effectiveness against different strains of *Staphylococci* and *Escherichia coli* at a concentration of 10^4 cfu/mL during in and out flow conditions and found a rapid decrease of the antimicrobial activity over time, presumably due to the large size and the low density of the silver nanoparticles. Indeed, SEM observation showed that the silver nanoparticles had a diameter of 500 nm and were not uniformly disposed onto the catheter. Silverline^{TM} EVD were unable to kill 100% of the bacteria attached during flow conditions [22]. The application of a conditioning film on the catheter, mimicking in vivo conditions did not influence the antimicrobial effect of the EVD [22].

Galiano et al. investigated the concentration of silver by atomic absorption spectroscopy on an artificial CSF fluid crossing continuously (10 mL per hour at 37 °C) polyurethane ventricular shunt (VS) impregnated with silver nanoparticles (Silverline^{TM}) and sampled every 24 h. They found no silver in each sample and expressed concerns about the effective delivery of silver by the catheter over time [23]. Moreover, they found no difference in bacterial growth when *S. aureus* and *E. coli* strains were exposed to Silverline^{TM} and a control catheter [23].

4. Effectiveness of Silver-Impregnated EVD in Clinical Practice

In the Infectious Disease Society of America (IDSA) guidelines regarding the prevention of healthcare-associated ventriculitis and meningitidis, published in 2017, the use of antimicrobial-impregnated EVD is recommended, but Silverline^{TM} EVD are not specifically mentioned, whereas antibiotics-impregnated EVD (AI-EVD) are [10]. Results of the SILVER randomized clinical trial (RCT), including 325 patients, found a significant decrease in EVD infection with Silverline^{TM} compared to unprocessed EVD (12.3 % and 21.4%, $p = 0.043$; respectively). VRI was defined as bacteria identified on Gram stain or isolated by culture in a CSF sample [24]. Another RCT assessing the effectiveness of silver-impregnated lumbar drains compared to unprocessed drains included 48 patients and found a similar rate of infection-related devices in both groups (4.2% and 16.7%, $p = 0.16$; respectively). Infection was defined as a positive CSF culture or at least one sign of meningitidis and (1) increased CSF white blood cell count, proteins level, or decreased glucose level or (2) microorganisms seen on Gram stain or (3) colonization of catheter tip [25]. A large prospective study comparing 146 silvers-impregnated EVD with 188 AI-EVD and 161 unprocessed EVD, found no difference in the incidence of CSF infection [26]. A meta-analysis of one RCT and four cohort studies (two retrospective and two prospective) for a total of 943 patients found that silver-impregnated EVD were associated with a lower risk of infection (RR = 0.60; 95% CI [0.40–0.90]). The authors also reported that there was no difference in mortality whatever the type of catheter used (RR = 1.17; 95% CI [0.76–1.81]) [27]. In a meta-analysis including one RCT and one prospective study, there was no difference in the incidence of VRI between silver-impregnated EVD and unprocessed EVD (OR = 0.33; 95% CI [0.07–1.69]; $p = 0.18$) [28]. Another meta-analysis of six observational studies found no significant benefice of silver EVD on the rate of VRI (OR = 0.71; 95% CI [0.46–1.08]; $p = 0.11$) [29]. Similarly, in a large meta-analysis of 12 studies comparing silver-impregnated and unprocessed CVC: there was no difference in the rate of colonization (OR = 0.907; 95% CI [0.758–1.087]; $p = 0.290$) and in the rate of catheter-related bloodstream infection (CRBSI) (OR = 0.721; 95% CI [0.476–1.094]; $p = 0.124$) [30]. Clinical results of silver compared to unprocessed EVD are resumed in Table 1.

Table 1. Effectiveness of silver impregnated compared to unprocessed EVD in clinical practice.

Comparison between Silver Impregnated and Unprocessed EVD				
Years of Publication/Authors	**Type of Study**	**Catheter**	**Number of Patients**	**Incidence of VRI/Infection Rate According to the Type of EVD**
Keong et al., 2012 [24]	Randomized controlled trial	Silver-impregnated EVD (Silverline™) vs. unprocessed EVD	278 patients (138 vs. 140)	12.3% vs. 21.4% $p = 0.043$
Jakobs et al., 2018 [25]	Randomized controlled trial	Silver-impregnated ELD (Silverline™) vs. unprocessed ELD	48 patients (24 in each group)	4.2% vs. 16.7% $p = 0.16$
Jamjoom et al., 2018 [26]	Prospective cohort	Silver-impregnated EVD (Silverline™) vs. unprocessed EVD	307 patients (146 vs. 161)	13.7% vs. 7.5% $p = 0.09$
Wang et al., 2013 [28]	Meta-analysis (1 RCT, 1 prospective study)	Silver-impregnated EVD (Silverline™) vs. unprocessed EVD	317 patients (157 vs. 160)	OR = 0.33; 95% CI [0.07–1.69]; $p = 0.18$
Konstantelias et al., 2015 [27]	Meta-analysis (1 RCT, 2 prospective, and 2 retrospective studies)	Silver-impregnated EVD (Silverline™) vs. unprocessed EVD	943 patients (491 vs. 452)	RR = 0.60; 95% CI [0.40–0.90]
Atkinson et al., 2016 [29]	Meta-analysis (2 prospective and 4 retrospective studies)	Silver-impregnated EVD (Silverline™) vs. unprocessed EVD	1057 patients (504 vs. 553)	OR = 0.71; 95% CI [0.46–1.08]; $p = 0.11$

ELD: External Lumbar Drain. EVD: External Ventricular Drain. OR: Odds Ratio. RCT: Randomized Controlled Trial. RR: Relative Risk.

Adverse Effect

The toxicity of silver ions is a function of their concentration. Silver toxicity seems to be low in the human body. Chronic exposition or ingestion of silver leads to a deposit in tissue and organs which usually are not life-threatening [31]. In the case of EVD, silver is directly delivered into the cerebral parenchyma and ventricles. An animal study showed a cerebral inflammatory response when the brain was exposed to a silver clip [32]. An experimental study assessing the potential toxic effect of ventricular shunt impregnated with silver nanoparticles showed reassuring results. The concentration of silver ions, in a fluid sampled after crossing the catheter, was negligible, thus, limiting the potentiality of adverse effects [23]. Another experimental study on silicone catheters impregnated with silver nanoparticles inserted in mice found that most of the silver remained on the catheter (approximately 16% of the silver was released after ten days). There was no accumulation of silver in the major organs. However, silver accumulated locally in the tissue surrounding the catheter, but its concentration remained far below the toxic level in humans [33]. Clinical results are also reassuring with no reported adverse effect linked to the accumulation of silver ions in organs and tissue [34].

5. Antibiotic Impregnated EVD

Two types of AI-EVD are used in clinical practice. Both are made of silicon elastomer. The first one is impregnated with a combination of rifampicin and minocycline (VentriClear™) and the second one is with a combination of rifampicin and clindamycin (Bactiseal™). The last one is the most studied device in the literature. Rifampicin has a broad-spectrum activity against GPC, some GNR, and intracellular bacteria. Its action is mediated through the inhibition of the bacterial DNA-dependent RNA polymerase [35]. It has also been widely used during biomaterial or implant infection because of its activity against Staphylococci-related biofilm. It is usually not advised to use it in monotherapy because of the associated risk with the emergence of resistance [36]. Clindamycin has activity against GPC (*Staphylococci, Streptococci*) and most anaerobes. It has no activity against aerobic GNR. Clindamycin inhibits bacterial protein synthesis by reversibly binding the 50S ribosomal subunits [37]. Its activity against biofilm is poor, experimentally [38]. Minocycline is a second-generation tetracycline antibiotic with activity against GPC, GNR (including *Acinetobacter baumanii*), and intracellular bacteria. Minocycline inhibits protein synthesis by binding the 30S ribosomal subunit. Its lipophilic properties allow it to easily cross cellular membranes and the blood-brain barrier [39].

6. In Vitro Assessment of Antibiotic Impregnated EVD Efficacy

In 1989, Bayston et al. described a process of impregnation of VS made in silicon with five different antibiotics previously selected because they adequately cross the silicone elastomer, resist the sterilization process, and exhibit optimal activity against GPC [40]. The impregnation process consisted of the immersion of the catheter in a solution of each antibiotic (single or in combination) and chloroform to give a concentration of 0.2% (w/v). Then, they were challenged three times with a 1 mL solution of *S. epidermidis*, at a concentration of 10^7 to 10^8 cfu/mL. The results showed that all of the catheters processed with a single antibiotic were ineffective to prevent bacterial colonization (with the exception of the rifampicin one). Catheters impregnated with rifampicin and clindamycin at a concentration of 0.2% (w/v) remained uncolonized after the three bacterial challenges [40]. Thus, they were selected for the development of the Bactiseal™ VS and EVD. The Bactiseal™ AI-EVD are impregnated with rifampicin and clindamycin at a concentration of 0.25 to 0.7 mg/g and 0.85 to 1.54 mg/g, respectively [41]. The total dose released by the catheter over time remained below the concentration measured after a single intravenous dose of each antibiotic limiting the risk of side effects [41]. In an in vitro model of antibiotics impregnated VS (AI-VS) was continuously perfused by a solution and challenged with a different strain of *Staphylococci* at a concentration of 10^8 cfu/mL (the exposition to bacteria range from 5 min to 1 h and was repeated every 14 days), Bayston et al. showed that AI-VS prevent bacterial colonization (assessed by standard culture) by up to 28 days and up to 42 days for *S. epidermidis* and *S. aureus* strains, respectively, while unprocessed catheters were colonized within 48 h. When a conditioning film (mimicking the in vivo protein deposit resulting from blood and CSF flow through the catheter) was applicated to the catheter layer, the effectiveness of AI-VS was similar [42]. They found that the continuous release of antibiotics does not prevent the adherence of bacteria but killed 100% of bacteria attached to the catheter in 50 h [41]. In a recent study, the in vitro activity of AI-EVD exposed to different mediums (air, saline, saline plus protein, and saline plus lipid medium) was tested. Antimicrobial activity drops quickly according to the medium dwelling notably in presence of a saline and lipid solution with no activity after day 28 for methicillin-resistant *Staphylococcus epidermidis* (MRSE), day 30 for MRSA and day 27 for methicillin-susceptible *Staphylococcus aureus* (MSSA) [43]. The decrease in antimicrobial activity could be related to the elution of the antibiotics according to the flow regiment crossing the catheter [41,44]. Indeed, when studying VentriClear™ EVD, Stevens et al. found that a significant amount of antibiotics were eluted in the fluid crossing the catheter [44].

7. In Vivo Assessment of Antimicrobial Activity of AI-EVD

7.1. Antimicrobial Activity

Mounier et al. studied in vivo durability of the antimicrobial activity of 65 Bactiseal™ AI-EVD [45]. They assessed the inhibition diameter of AI-EVD tips in contact with three different bacteria strains (MRSE, MRSA, and MSSA at a concentration of 10^4 and 10^5 cfu/mL) and the concentration of antibiotics remaining on the catheters after their removal. Inhibition diameters significantly decreased according to the duration of EVD placement and the volume of CSF drained (only for MRSE and MRSA). The activity dropped faster for the external side than the internal one, with no inhibition diameter in 23% (19/65) of all the AI-EVD for a median drainage time of 18 [15–25] days and a median volume of CSF drained of 2450 [1594–3246] mL. There was no correlation between clindamycin concentration and both duration of EVD catheterization and the volume of CSF drained, but the concentration decreased by about 80% only after five days of placement. Conversely, the rifampicin concentration dropped quickly according to duration and volume drained. For half of the AI-EVD analyzed for drug quantification, the concentration of rifampicin was less than one percent of the initial concentration [45]. In another in vivo study regarding AI-VS, authors found that the antimicrobial activity decreased by about 50% of the initial concentration within 10 days then it remained stable until approximately three months (97 days) [46].

7.2. Prevention of Bacterial Colonization with AI-EVD

When assessed through standard culture, most studies found that AI-EVD significantly decreases the rate of ventriculostomy-related colonization (VRC) [28,47]. However, standard culture has a poor sensibility to diagnosis catheter colonization when compared to other methods such as scanning electron microscopy (SEM) [48,49]. Ramirez et al. studied in vivo formation of biofilm on both unprocessed and Bactiseal™ AI-EVD with SEM. Biofilm was found on 86% of unprocessed EVD and 67% of AI-EVD ($p = 0.22$). After seven days the prevalence of biofilm was 89% and 88% for unprocessed and AI-EVD, respectively [50]. The culture of EVD tip was positive only in 37.5% (12/32). Similarly, observation of intraparenchymal intracranial pressure transducer (ICPT) on SEM showed a prevalence of mature biofilm of 73% while only five percent of the ICPT were positive in standard culture. The short median duration of the ICPT (4 [3–7] days) suggests that the formation of biofilm occurred quickly after insertion [48]. When studying CVC, similar results were found with a rate of biofilm formation of 100% after a median duration of ten days [51].

8. Efficacy of AI-EVD in Clinical Practice

The clinical literature shows conflicting results regarding the efficacy of AI-EVD to prevent VRI. Nonetheless, the use of AI-EVD is recommended in the IDSA guidelines [10]. Three RCTs tested the effectiveness of AI-EVD [47,52,53].

The first one compared the incidence of VRI between AI-EVD (impregnated with rifampicin and minocycline) and unprocessed EVD in 288 patients. VRI was defined as a positive CSF culture. All patients underwent prophylactic and then the maintenance of antibiotics during the time of EVD placement. The incidence of VRI was significantly lower in the AI-EVD than in the unprocessed EVD groups (1.3% and 9.4%; $p = 0.002$; respectively). Catheter colonization was significantly lower in the AI-EVD than in the standard EVD groups (17.9% and 36.9%; $p < 0.0012$, respectively). Interestingly all the microorganisms found in the AI-EVD culture were susceptible to minocycline and all the *staphylococcus* strains were susceptible to rifampicin [47].

The second one compared the incidence of VRI between Bactiseal™ AI-EVD and unprocessed EVD in 184 patients. VRI was defined as positive CSF culture associated with a CSF white cell count >10/mm^3 a CSF protein level > 0.8 g/L and a CSF/serum glucose ratio < 0.4. In the AI-EVD group, the patients received only prophylactic perioperative antibiotics (ampicillin-sulbactam and ceftriaxone) while in the standard group, the antibiotics were maintained during EVD placement. The incidence of nosocomial infection whatever the site concerned was similar between the two groups. The incidence of VRI was similar between AI-EVD and unprocessed EV (1% and 3%; $p = 0.282$, respectively) [52].

The last one compared the incidence of VRI between Bactiseal™ and unprocessed EVD in 357 patients. Proven VRI was defined as a CSF sample demonstrating a positive Gram stain and culture whereas suspected VRI was defined as a CSF sample demonstrating a negative Gram stain but a positive culture or CSF leukocytosis (White blood cell/Red blood cell CSF count > 0.02). The rate of proven VRI was similar in each group (2.3% and

2.8%; $p = 1.00$, respectively). The rate of suspected VRI was also similar in the two groups (17.5% and 20.4%; $p = 0.504$, respectively) [53].

In a large multicenter prospective cohort, analyzing 495 EVD (whose 188 were AI, 161 unprocessed, and 146 silver-impregnated EVD), there was no association between the catheter type and the incidence of VRI. Only the duration of EVD placement (>8 days; OR = 2.54; 95% CI [1.14–5.7]; $p = 0.02$) and regular sampling were significantly associated with the risk of VRI. The median time to infection was 9 [5–15] days [26]. However, a meta-analysis found a reduction of VRI with AI-EVD. In a pooled analysis of one RCT and three prospective observational studies (for a total of 2332 patients), Sonabend et al. found that AI-EVD significantly decreased the rate of VRI (RR = 0.19; 95% CI [0.07–0.52]; $p = 0.001$) [54]. Wang et al. also found that AI-EVD were associated with a significantly decreased risk of CSF infection in their meta-analysis of the three precited RCTs (for a total of 829 patients) (OR = 0.37; 95% CI [0.21–0.64]; $p = 0.0004$) [28]. Clinical results of AI compared to unprocessed EVD are resumed in Table 2.

Table 2. Effectiveness of antibiotic-impregnated compared to unprocessed EVD in clinical practice.

Comparison between Antibiotic Impregnated and Unprocessed EVD				
Years of Publication/Authors	**Type of Study**	**Catheter**	**Number of Patients**	**Incidence of VRI/Infection Rate According to the Type of EVD**
Zabramski et al., 2003 [47]	Randomized controlled trial	Minocycline/Rifampicin EVD vs. unprocessed EVD	288 patients (149 vs. 139)	1.3% vs. 9.4% $p = 0.0002$
Wong et al., 2010 [52]	Randomized controlled trial	Clindamycin/Rifampicin EVD vs. unprocessed EVD	184 patients (90 vs. 94)	1% vs. 3% $p = 0.282$
Pople et al., 2012 [53]	Randomized controlled trial	Clindamycin/Rifampicin EVD vs. unprocessed EVD	357 patients (176 vs. 181)	2.3% vs. 2.8% $p = 1.00$
Jamjoom et al., 2018 [26]	Prospective cohort	Clindamycin/Rifampicin EVD vs. unprocessed EVD	349 patients (188 vs. 161)	7.4% vs. 7.5% $p = 1.00$
Wang et al., 2013 [28]	Meta-analysis (3 RCT)	Clindamycin/Rifampicin Minocycline/Rifampicin EVD vs. unprocessed EVD	829 patients (415 vs. 414)	OR = 0.37; 95% CI [0.21–0.64]; $p = 0.0004$
Sonabend et al., 2015 [54]	Meta-analysis (1 RCT, 3 prospective studies)	Clindamycin/Rifampicin EVD vs. unprocessed EVD	2332 patients	RR = 0.19; 95% CI [0.07–0.52]; $p = 0.001$

EVD: External Ventricular Drain. OR: Odds Ratio. RCT: Randomized Controlled Trial. RR: Relative Risk.

Bacterial Selection and Adverse Effects

Currently there is no convincing evidence that AI-EVD or VS increase the risk of antibiotic resistance or bacterial selection [55]. However, in the first RCT, the organisms responsible for VRI isolated in the unprocessed EVD group were predominantly coagulase-negative *Staphylococci*, whereas those isolated in the AI-EVD group were MRSA, *Enterobacter aerogenes*, and *Enterococcus faecalis* [47]. Wong et al. found a similar rate of resistant opportunistic infection in each group [52]. In their meta-analysis, Konstantelias et al. found a higher risk of infection with MRSA and GNR in patients treated with AI-EVD compared to others [27]. However, another study found no difference in the incidence of MRSA with AI-VS compared to unprocessed shunts [56]. There was no reported local or systemic side effect with AI-EVD or VS currently in the literature [47,52,53]. However, no studies were designed to investigate the potential side effects of these antibiotic-impregnated biomaterials.

9. Discussion

EVD impregnated with antimicrobial agents are an attractive concept to prevent VRI. Until their development, prolonged systemic antibiotics were often used during the time of

EVD placement raising concerns about the risk of side effects and resistance selection. The IDSA guidelines recommend the use of AI-EVD although clinical studies showed contrasted results [10]. Interestingly, a large national cohort in the United Kingdom and Ireland (21 neurosurgical units for a total of 495 EVD) found that despite this recommendation only a third of the EVD implanted were AI-EVD [26]. Two of three RCTs found negative results regarding the efficacy of AI-EVD to prevent VRI [52,53]. In all RCT, the median time to infection was similar between each group and around day ten [47,52,53]. In the largest prospective study comparing AI-EVD with silver-impregnated and unprocessed EVD, there were no differences in the rate of VRI. The median time to infection was eleven, eight, and seven days for AI, unprocessed, and silver-impregnated EVD, respectively [26]. Results of the meta-analysis called for the use of AI-EVD despite an important heterogeneity in studies [28,54]. Their results should be interpreted with caution for the following reasons: First, the definition of VRI differs from one study to another and numerous cases of colonization could have been considered as infection. Second, the largest non-randomized study included in the meta-analysis which is in favor of AI-EVD is a before-after study comparing four different periods with the use of standardized protocol and different types of AI-EVD during each period [57]. Third, the largest prospective study showing negative results was not included (because its publication occurred later) [26].

In comparison, clinical studies on the effectiveness of antibiotics impregnated (AI)-CVC showed important discrepancies. A meta-analysis of 11 RCTs comparing AI versus unprocessed CVC showed no difference in the rate of CBSRI [58]. Conversely, a recent Cochrane Review including 60 studies showed a significantly lower rate of CBSRI when AI-CVC was used. However, there was no clinical benefice on mortality and clinical sepsis [59]. It is important to note that most of the studies comparing AI versus unprocessed devices used as primary endpoint a microbiological definition of infection without assessing the clinical status of the patients.

Although in vitro studies reported a prolonged activity of AI-EVD, in vivo results showed a quick drop in the antimicrobial activity related to the CSF flow and medium in which the catheter was inserted [45]. In some cases, there was no activity as soon as day ten, which corresponds to the pic of the incidence of VRI [11,12,45]. It could be linked to a continuous elution of the antibiotic through the flow crossing the catheter [44]. In addition, EVD are continuously challenged by a high density of bacteria composing the head skin microbiota despite optimal disinfection and dressing which only decrease bacterial density for a few hours [60]. This is a major difference with VS which are surgically internalized and thus, exposed to bacteria only during the time of surgery. The in vitro studies were primarily designed to assess the efficacy of AI-VS, but EVD are used in different clinical conditions (acute brain injury, acute local and systemic inflammation, other organs failure, prolonged stay in ICU, and use of systemic antibiotic among other) and the pathways of colonization and infection could be different. All these in vivo variables could explain the difference found in the durability of antimicrobial therapy and in the contrasted clinical results.

The primary goal of AI-EVD is to avoid bacterial adhesion and biofilm formation through a continuous release of antibiotics during the time of EVD placement. Colonization of the catheter is the preclude to infection. The standard diagnosis of VRC is based on the positive culture of EVD tip. Some clinical studies report a lower rate of VRC with AI compared to unprocessed EVD [28,47]. However, most of the bacteria involved in AI-EVD colonization were GPC suggesting a lack of efficacy of the impregnated antibiotics [12,27]. After the first phase of adhesion on biomaterial, bacteria rapidly organized themselves as a biofilm which are more difficult to detect by standard culture [61]. When observed through SEM, the rate of biofilm was almost constant and similar in both types of EVD only a few days after insertion [50]. Incidentally, Bayston et al. showed that AI-EVD cannot prevent bacteria adhesion on biomaterial and that the antimicrobial activity is exerted by killing the bacteria attached. In their protocol, the AI-EVD were challenged by bacterial strain intermittently (*Staphylococcus epidermidis* strain was added to the perfusate at day one, four, 10, and 21) [41]. Only a few minutes could be necessary to lead to an irreversible adhesion of the bacteria to the surface leading to the phenotypic and genotypic changes initiating biofilm formation, and as explicated above, EVD are constantly challenged by a high density of bacteria increasing the risk of adhesion and biofilm formation [62]. Observation of other transdermal devices (such as CVC impregnated or not and ICPT) found similar results [48,51]. It suggests that the natural history of such devices is to be colonized by the bacteria belonging to the skin microbiota and that other factors (not fully elucidated yet) triggered the transition to the infection. Regarding silver-impregnated EVD, the lack of effectiveness could be linked to the silver nanoparticles synthesis and impregnation process. As demonstrated in an experimental study the size of silver nanoparticles was too large (500 nm) and not uniformly distributed along the catheter limiting its biocidal activity [22]. Moreover, the inhibition of bacterial growth and biofilm formation is concentration dependent. When used in vivo numerous factors could influence the concentration of silver nanoparticles released over time, such as the deposit of proteins along or the flow regiment crossing the catheter.

The effect of impregnated EVD on bacterial selection and resistance remained unclear. Some studies suggested that silver and AI-EVD could be associated with a higher rate of VRI caused by GNR or MRSA [27,29,50]. However, a large prospective study found no difference in the rate of infection with GNR between the different types of EVD [26]. In the same way, a recent study found no evidence that impregnated EVD increases the risk of bacterial resistance [55].

10. Perspectives

The natural history of transdermal devices seems to be colonized by bacteria belonging to the skin microbiota. Factors triggering the transition from colonization to infection remained unclear. Some authors have raised an interesting hypothesis: Healthy biofilm as found on healthy skin, for example, could be protective whereas, an altered biofilm with a decreased bacterial diversity could lead to infection [63,64]. These new paradigms lead to the development of a new therapeutic approach consisting of replacing pathogenic biofilm with new healthy biofilm (i.e., fecal transplant) [63]. In the future, the identification of bacterial species composing biofilm implicated in VRI and its comparison to "healthy" biofilm will improve our understanding of infection pathways.

11. Conclusions

In vivo studies showed that colonization of EVD by bacterial biofilm is almost constant. Clinical studies failed to show a superiority of silver or AI-EVD versus unprocessed EVD to prevent VRI. A better understanding of biofilm dynamics on biomaterial will help us to improve our ability to develop new innovating ways to prevent VRI.

Funding: This research received no external funding.

Institutional Review Board Statement: Not applicable.

Informed Consent Statement: Not applicable.

Data Availability Statement: Not applicable.

Conflicts of Interest: The authors declare no conflict of interest.

References

1. Mojzsis, S.J.; Arrhenius, G.; McKeegan, K.D.; Harrison, T.M.; Nutman, A.P.; Friend, C.R. Evidence for life on Earth before 3800 million years ago. *Nature* **1996**, *384*, 55–59. [CrossRef]
2. Beraldi-Campesi, H. Early life on land and the first terrestrial ecosystems. *Ecol. Process.* **2013**, *2*, 1. [CrossRef]
3. Reysenbach, A.L.; Ehringer, M.; Hershberger, K. Microbial diversity at 83 degrees C in Calcite Springs, Yellowstone National Park: Another environment where the Aquificales and «Korarchaeota» coexist. *Extrem. Life Extrem. Cond.* **2000**, *4*, 61–67.
4. Brandwein, M.; Steinberg, D.; Meshner, S. Microbial biofilms and the human skin microbiome. *NPJ Biofilms Microbiomes* **2016**, *2*, 3. [CrossRef] [PubMed]

5. Cho, I.; Blaser, M.J. The human microbiome: At the interface of health and disease. *Nat. Rev. Genet.* **2012**, *13*, 260–270. [CrossRef]
6. Joung, Y.-H. Development of implantable medical devices: From an engineering perspective. *Int. Neurourol. J.* **2013**, *17*, 98. [CrossRef] [PubMed]
7. Potera, C. Forging a link between biofilms and disease. *Science* **1999**, *283*, 1837–1839. [CrossRef] [PubMed]
8. Bayston, R.; Lari, J. A study of the sources of infection in colonized shunts. *Dev. Med. Child Neurol.* **1974**, *16* (Suppl. S32), 16–22. [CrossRef]
9. Fried, H.I.; Nathan, B.R.; Rowe, A.S.; Zabramski, J.M.; Andaluz, N.; Bhimraj, A.; Guanci, M.M.; Seder, D.B.; Singh, J.M. The Insertion and Management of External Ventricular Drains: An Evidence-Based Consensus Statement: A Statement for Healthcare Professionals from the Neurocritical Care Society. *Neurocritical Care* **2016**, *24*, 61–81. [CrossRef]
10. Tunkel, A.R.; Hasbun, R.; Bhimraj, A.; Byers, K.; Kaplan, S.L.; Scheld, W.M.; van de Beek, D.; Bleck, T.P.; Garton, H.J.L.; Zunt, J.R. 2017 Infectious Diseases Society of America's Clinical practice guidelines for healthcare-associated ventriculitis and meningitis. *Clin. Infect. Dis.* **2017**, *64*, e34–e65. [CrossRef]
11. Lozier, A.P.; Sciacca, R.R.; Romagnoli, M.F.; Connolly, E.S., Jr. Ventriculostomy-related infections: A critical review of the literature. *Neurosurgery* **2002**, *51*, 170–181; discussion 81–82. [CrossRef] [PubMed]
12. Mounier, R.; Birnbaum, R.; Cook, F.; Jost, P.-H.; Martin, M.; Aït-Mamar, B.; Nebbad, B.; Couffin, S.; Tomberli, F.; Djedid, R.; et al. Natural history of ventriculostomy-related infection under appropriate treatment and risk factors of poor outcome: A retrospective study. *J. Neurosurg.* **2018**, *131*, 1052–1061. [CrossRef] [PubMed]
13. Mounier, R.; Lobo, D.; Cook, F.; Martin, M.; Attias, A.; Aït-Mamar, B.; Gabriel, I.; Bekaert, O.; Bardon, J.; Nebbad, B.; et al. From the skin to the brain: Pathophysiology of colonization and infection of external ventricular drain, a prospective observational study. *PLoS ONE* **2015**, *10*, e0142320. [CrossRef] [PubMed]
14. Wright, K.; Young, P.; Brickman, C.; Sam, T.; Badjatia, N.; Pereira, M.; Connolly, E.S.; Yin, M.T. Rates and determinants of ventriculostomy-related infections during a hospital transition to use of antibiotic coated external ventricular drains. *Neurosurg. Focus* **2013**, *34*, E12. [CrossRef]
15. Zschaler, R. Testing of the Antimicrobial Effect of Catheter Tubing with a Roll Culture Method. Available online: http://www.spiegelberg.de/home/documents/Zschaler.pdf (accessed on 18 January 2023).
16. Tapsoba, I.; Arbault, S.; Walter, P.; Amatore, C. Finding Out Egyptian Gods' Secret Using Analytical Chemistry: Biomedical Properties of Egyptian Black Makeup Revealed by Amperometry at Single Cells. *Anal. Chem.* **2010**, *82*, 457–460. [CrossRef]
17. Velusamy, P.; Su, C.-H.; Kumar, G.V.; Adhikary, S.; Pandian, K.; Gopinath, S.C.B.; Chen, Y.; Anbu, P. Biopolymers regulate silver nanoparticle under microwave irradiation for effective antibacterial and antibiofilm activities. *PLoS ONE* **2016**, *11*, e0157612. [CrossRef]
18. Nakamura, S.; Sato, M.; Sato, Y.; Ando, N.; Takayama, T.; Fujita, M.; Ishihara, M. Synthesis and application of silver nanoparticles (AG NPS) for the prevention of infection in healthcare workers. *Int. J. Mol. Sci.* **2019**, *20*, 3620. [CrossRef]
19. Burdușel, A.-C.; Gherasim, O.; Grumezescu, A.M.; Mogoantă, L.; Ficai, A.; Andronescu, E. Biomedical applications of silver nanoparticles: An up-to-date overview. *Nanomaterials* **2018**, *8*, 681. [CrossRef]
20. Savelyev, Y.; Gonchar, A.; Movchan, B.; Gornostay, A.; Vozianov, S.; Rudenko, A.; Rozhnova, R.; Travinskaya, T. Antibacterial polyurethane materials with silver and copper nanoparticles. *Mater. Today Proc.* **2017**, *4*, 87–94. [CrossRef]
21. Pino-Ramos, V.H.; Audifred-Aguilar, J.C.; Sánchez-Obregón, R.; Bucio, E. Antimicrobial polyurethane catheters synthesized by grafting-radiation method doped with silver nanoparticles. *React. Funct. Polym.* **2021**, *167*, 105006. [CrossRef]
22. Bayston, R.; Vera, L.; Mills, A.; Ashraf, W.; Stevenson, O.; Howdle, S.M. In vitro antimicrobial activity of silver-processed catheters for neurosurgery. *J. Antimicrob. Chemother.* **2009**, *65*, 258–265. [CrossRef] [PubMed]
23. Galiano, K.; Pleifer, C.; Engelhardt, K.; Brössner, G.; Lackner, P.; Huck, C.; Lass-Flörl, C.; Obwegeser, A. Silver segregation and bacterial growth of intraventricular catheters impregnated with silver nanoparticles in cerebrospinal fluid drainages. *Neurol. Res.* **2008**, *30*, 285–287. [CrossRef] [PubMed]
24. Keong, N.C.H.; Bulters, D.O.; Richards, H.K.; Farrington, M.; Sparrow, O.C.; Pickard, J.D.; Hutchinson, P.J.; Kirkpatrick, P.J. The silver (silver impregnated line versus EVD randomized trial). *Neurosurgery* **2012**, *71*, 394–404. [CrossRef] [PubMed]
25. Jakobs, M.; Klein, S.; Eigenbrod, T.; Unterberg, A.W.; Sakowitz, O.W. The siludrain trial: A prospective randomized controlled trial comparing standard versus silver-impregnated lumbar drains. *J. Neurosurg.* **2019**, *130*, 2040–2047. [CrossRef]
26. Jamjoom, A.A.; Joannides, A.J.; Poon, M.T.-C.; Chari, A.; Zaben, M.; Abdulla, M.A.; Roach, J.; Glancz, L.J.; Solth, A.; Duddy, J.; et al. Prospective, multicenter study of external ventricular drainage-related infections in the UK and Ireland. *J. Neurol. Neurosurg. Psychiatry* **2017**, *89*, 120–126. [CrossRef]
27. Konstantelias, A.A.; Vardakas, K.Z.; Polyzos, K.A.; Tansarli, G.S.; Falagas, M.E. Antimicrobial-impregnated and -coated shunt catheters for prevention of infections in patients with hydrocephalus: A systematic review and meta-analysis. *J. Neurosurg.* **2015**, *122*, 1096–1112. [CrossRef]
28. Wang, X.; Dong, Y.; Qi, X.-Q.; Li, Y.-M.; Huang, C.-G.; Hou, L.-J. Clinical review: Efficacy of antimicrobial-impregnated catheters in external ventricular drainage—A systematic review and meta-analysis. *Crit. Care* **2013**, *17*, 234. [CrossRef]
29. Atkinson, R.A.; Fikrey, L.; Vail, A.; Patel, H.C. Silver-impregnated external-ventricular-drain-related cerebrospinal fluid infections: A meta-analysis. *J. Hosp. Infect.* **2016**, *92*, 263–272. [CrossRef]
30. Chen, Y.-M.; Dai, A.-P.; Shi, Y.; Liu, Z.-J.; Gong, M.-F.; Yin, X.-B. Effectiveness of silver-impregnated central venous catheters for preventing catheter-related blood stream infections: A meta-analysis. *Int. J. Infect. Dis.* **2014**, *29*, 279–286. [CrossRef]

31. Lansdown, A.B.G. Silver in health care: Antimicrobial effects and safety in use. *Biofunctional Text. Skin* **2006**, *33*, 17–34.
32. McFadden, J.T. Tissue reactions to standard neurosurgical metallic implants. *J. Neurosurg.* **1972**, *36*, 598–603. [CrossRef] [PubMed]
33. Roe, D.; Karandikar, B.; Bonn-Savage, N.; Gibbins, B.; Roullet, J.-B. Antimicrobial surface functionalization of plastic catheters by silver nanoparticles. *J. Antimicrob. Chemother.* **2008**, *61*, 869–876. [CrossRef] [PubMed]
34. Winkler, K.M.; Woernle, C.M.; Seule, M.; Held, U.; Bernays, R.L.; Keller, E. Antibiotic-impregnated versus silver-bearing external ventricular drainage catheters: Preliminary results in a randomized controlled trial. *Neurocritical Care* **2013**, *18*, 161–165. [CrossRef]
35. Campbell, E.A.; Korzheva, N.; Mustaev, A.; Murakami, K.; Nair, S.; Goldfarb, A.; Darst, S.A. Structural mechanism for rifampicin inhibition of bacterial RNA polymerase. *Cell* **2001**, *104*, 901–912. [CrossRef] [PubMed]
36. Renz, N.; Trampuz, A.; Zimmerli, W. Controversy about the role of rifampin in biofilm infections: Is it justified? *Antibiotics* **2021**, *10*, 165. [CrossRef] [PubMed]
37. Guay, D. Update on clindamycin in the management of bacterial, fungal and protozoal infections. *Expert Opin. Pharmacother.* **2007**, *8*, 2401–2444. [CrossRef]
38. Smith, K.; Perez, A.; Ramage, G.; Gemmell, C.G.; Lang, S. Comparison of biofilm-associated cell survival following in vitro exposure of methicillin-resistant Staphylococcus aureus biofilms to the antibiotics: Clindamycin, daptomycin, linezolid, Tigecycline and Vancomycin. *Int. J. Antimicrob. Agents* **2009**, *33*, 374–378. [CrossRef]
39. Singh, S.; Khanna, D.; Kalra, S. Minocycline and doxycycline: More than antibiotics. *Curr. Mol. Pharmacol.* **2021**, *14*, 1046–1065. [CrossRef]
40. Bayston, R.; Grove, N.; Siegel, J.; Lawellin, D.; Barsham, S. Prevention of hydrocephalus shunt catheter colonisation in vitro by impregnation with antimicrobials. *J. Neurol. Neurosurg. Psychiatry* **1989**, *52*, 605–609. [CrossRef]
41. Bayston, R.; Ashraf, W.; Ortori, C. Does release of antimicrobial agents from impregnated external ventricular drainage catheters affect the diagnosis of ventriculitis? *J. Neurosurg.* **2016**, *124*, 375–381. [CrossRef]
42. Bayston, R.; Lambert, E. Duration of protective activity of cerebrospinal fluid shunt catheters impregnated with antimicrobial agents to prevent bacterial catheter-related infection. *J. Neurosurg.* **1997**, *87*, 247–251. [CrossRef]
43. Lang, E.; Hulin, A.; Egbeola-Martial, J.; Woerther, P.-L.; Drouard, L.; Roujansky, A.; Tomberli, F.; Bardon, J.; Schimpf, C.; Senova, S.; et al. In vitro study of factors influencing the duration of antimicrobial protection of antimicrobial-impregnated external ventricular drains. *J. Antimicrob. Chemother.* **2021**, *77*, 682–688. [CrossRef] [PubMed]
44. Stevens, E.A.; Palavecino, E.; Sherertz, R.J.; Shihabi, Z.; Couture, D.E. Effects of antibiotic-impregnated external ventricular drains on bacterial culture results: An in vitro analysis. *J. Neurosurg.* **2010**, *113*, 86–92. [CrossRef] [PubMed]
45. Mounier, R.; Lang, E.; Hulin, A.; Woerther, P.-L.; Lobo, D.; Martin, M.; Bitot, V.; Flores, L.; Cherruault, M.; Jost, P.-H.; et al. Durability of antimicrobial activity of antibiotic-impregnated external ventricular drains: A prospective study. *J. Antimicrob. Chemother.* **2019**, *74*, 3328–3336. [CrossRef] [PubMed]
46. Pattavilakom, A.; Kotasnas, D.; Korman, T.M.; Xenos, C.; Danks, A. Duration of in vivo antimicrobial activity of antibiotic-impregnated cerebrospinal fluid catheters. *Neurosurgery* **2006**, *58*, 930–935. [CrossRef]
47. Zabramski, J.M.; Whiting, D.; Darouiche, R.O.; Horner, T.G.; Olson, J.; Robertson, C.; Hamilton, A.J. Efficacy of antimicrobial-impregnated external ventricular drain catheters: A prospective, randomized, controlled trial. *J. Neurosurg.* **2003**, *98*, 725–730. [CrossRef]
48. Mounier, R.; Kapandji, N.; Gricourt, G.; Lobo, D.; Rodriguez, C.; Pons, S.; Djediat, C.; Woerther, P.-L.; Mellano, V.; Aït-Mamar, B.; et al. Assessment of bacterial colonization of intracranial pressure transducers: A prospective study. *Neurocritical Care* **2020**, *34*, 814–824. [CrossRef]
49. Hall-Stoodley, L.; Stoodley, P.; Kathju, S.; Høiby, N.; Moser, C.; Costerton, J.W.; Moter, A.; Bjarnsholt, T. Towards diagnostic guidelines for biofilm-associated infections. *FEMS Immunol. Med. Microbiol.* **2012**, *65*, 127–145. [CrossRef]
50. Ramírez, P.; Gordón, M.; Soriano, A.; Gil-Perotin, S.; Marti, V.; Gonzalez-Barbera, E.M.; Sanchez-Aguilar, M.T.; Simal, J.A.; Bonastre, J. Assessment of the in vivo formation of biofilm on external ventricular drainages. *Eur. J. Clin. Microbiol. Infect. Dis.* **2013**, *32*, 1437–1443. [CrossRef]
51. Raad, I.; Costerton, W.; Sabharwal, U.; Sadlowski, M.; Anaissie, E.; Bodey, G.P. Ultrastructural analysis of indwelling vascular catheters: A quantitative relationship between luminal colonization and duration of placement. *J. Infect. Dis.* **1993**, *168*, 400–407. [CrossRef]
52. Wong, G.K.; Ip, M.; Poon, W.S.; Mak, C.W.; Ng, R.Y. Antibiotics-impregnated ventricular catheter versus systemic antibiotics for prevention of nosocomial CSF and non-CSF infections: A prospective randomised clinical trial. *J. Neurol. Neurosurg. Psychiatry* **2010**, *81*, 1064–1067. [CrossRef] [PubMed]
53. Pople, I.; Poon, W.; Assaker, R.; Mathieu, D.; Iantosca, M.; Wang, E.; Zhang, L.W.; Leung, G.; Chumas, P.; Menei, P.; et al. Comparison of infection rate with the use of antibiotic-impregnated vs standard extraventricular drainage devices. *Neurosurgery* **2012**, *71*, 6–13. [CrossRef] [PubMed]
54. Sonabend, A.M.; Korenfeld, Y.; Crisman, C.; Badjatia, N.; Mayer, S.A.; Connolly, E.S. Prevention of ventriculostomy-related infections with prophylactic antibiotics and antibiotic-coated external ventricular drains: A systematic review. *Neurosurgery* **2011**, *68*, 996–1005. [CrossRef] [PubMed]
55. Talibi, S.; Tarnaris, A.; Shaw, S.A. Has the introduction of antibiotic-impregnated external ventricular drain catheters changed the nature of the microorganisms cultured in patients with drain-related infection? A single Neurosurgical Centre's experience. *Br. J. Neurosurg.* **2016**, *30*, 560–566. [CrossRef] [PubMed]

56. Parker, S.L.; Attenello, F.J.; Sciubba, D.M.; Garces-Ambrossi, G.L.; Ahn, E.; Weingart, J.; Carson, B.; Jallo, G.I. Comparison of shunt infection incidence in high-risk subgroups receiving antibiotic-impregnated versus standard shunts. *Childs Nerv. Syst.* **2009**, *25*, 77–83; discussion 85. [CrossRef]
57. Harrop, J.S.; Sharan, A.D.; Ratliff, J.; Prasad, S.; Jabbour, P.; Evans, J.J.; Veznedaroglu, E.; Andrews, D.W.; Maltenfort, M.; Liebman, K.; et al. Impact of a standardized protocol and antibiotic-impregnated catheters on ventriculostomy infection rates in cerebrovascular patients. *Neurosurgery* **2010**, *67*, 187–191. [CrossRef]
58. McConnell, S.A.; Gubbins, P.O.; Anaissie, E.J. Do antimicrobial-impregnated central venous catheters prevent catheter-related bloodstream infection? *Clin. Infect. Dis.* **2003**, *37*, 65–72. [CrossRef]
59. Chong, H.Y.; Lai, N.M.; Apisarnthanarak, A.; Chaiyakunapruk, N. Comparative efficacy of antimicrobial central venous catheters in reducing catheter-related bloodstream infections in adults: Abridged Cochrane Systematic Review and network meta-analysis. *Clin. Infect. Dis.* **2017**, *64* (Suppl. S2), S131–S140. [CrossRef]
60. Kramer, A.; Dissemond, J.; Kim, S.; Willy, C.; Mayer, D.; Papke, R.; Tuchmann, F.; Assadian, O. Consensus on wound antisepsis: Update 2018. *Skin Pharmacol. Physiol.* **2017**, *31*, 28–58. [CrossRef]
61. Donlan, R.M.; Costerton, J.W. Biofilms: Survival mechanisms of clinically relevant microorganisms. *Clin. Microbiol. Rev.* **2002**, *15*, 167–193. [CrossRef]
62. Carniello, V.; Peterson, B.W.; van der Mei, H.C.; Busscher, H.J. Physico-chemistry from initial bacterial adhesion to surface-programmed biofilm growth. *Adv. Colloid Interface Sci.* **2018**, *26*, 1–14. [CrossRef] [PubMed]
63. Darouiche, R.O. Device-Associated infections: A macroproblem that starts with microadherence. *Clin. Infect. Dis.* **2001**, *33*, 1567–1572. [CrossRef] [PubMed]
64. Relman, D.A. The human microbiome: Ecosystem resilience and health. *Nutr. Rev.* **2012**, *70*, S2–S9. [CrossRef] [PubMed]

International Journal of *Molecular Sciences*

MDPI

Article

Chitosan as a Promising Support of a CDH Activity Preservation System for Biomedical and Industrial Applications

Justyna Sulej [1,*], Monika Osińska-Jaroszuk [1], Magdalena Jaszek [1], Anna Olszewska [2], Anna Belcarz [3] and Wiktoria Piątek-Gołda [1]

1 Department of Biochemistry and Biotechnology, Institute of Biological Sciences, Maria Curie-Sklodowska University, Akademicka 19, 20-033 Lublin, Poland
2 Department of Human Physiology, Medical University of Lublin, 11 Radziwiłowska Street, 20-080 Lublin, Poland
3 Chair and Department of Biochemistry and Biotechnology, Medical University of Lublin, 1 Chodźki Street, 20-093 Lublin, Poland
* Correspondence: justyna.sulej@mail.umcs.pl; Tel.: +48-81-537-50-42

Abstract: Cellobiose dehydrogenase (CDH) is an extracellular hemoflavoprotein catalyzing the oxidation reaction of β-1,4-glycosidic-bonded sugars (lactose or cellobiose), which results in the formation of aldobionic acids and hydrogen peroxide as a byproduct. The biotechnological application of CDH requires the immobilization of the enzyme on a suitable support. As a carrier of natural origin used for CDH immobilization, chitosan seems to increase the catalytic potential of the enzyme, especially for applications as packaging in the food industry and as a dressing material in medical applications. The present study aimed to immobilize the enzyme on chitosan beads and determine the physicochemical and biological properties of immobilized CDHs obtained from different fungal sources. The chitosan beads with immobilized CDHs were characterized in terms of their FTIR spectra or SEM microstructure. The most effective method of immobilization in the proposed modification was the covalent bonding of enzyme molecules using glutaraldehyde, resulting in efficiencies ranging from 28 to 99%. Very promising results, compared to free CDH, were obtained in the case of antioxidant, antimicrobial, and cytotoxic properties. Summarizing the obtained data, chitosan seems to be a valuable material for the development of innovative and effective immobilization systems for biomedical applications or food packaging, preserving the unique properties of CDH.

Keywords: cellobiose dehydrogenase; immobilization; chitosan; antioxidant; antimicrobial; cytotoxic

Citation: Sulej, J.; Osińska-Jaroszuk, M.; Jaszek, M.; Olszewska, A.; Belcarz, A.; Piątek-Gołda, W. Chitosan as a Promising Support of a CDH Activity Preservation System for Biomedical and Industrial Applications. *Int. J. Mol. Sci.* **2023**, *24*, 4535. https://doi.org/10.3390/ijms24054535

Academic Editor: Helena Felgueiras

Received: 30 January 2023
Revised: 22 February 2023
Accepted: 23 February 2023
Published: 25 February 2023

1. Introduction

Fungi are a well-characterized group of producers of many biotechnologically important intracellular and extracellular enzymes. The best-known fungal enzymes include laccase, chitinase, esterase, manganese peroxidases, glucose oxidase, and cellobiose dehydrogenase [1–7]. These enzymes are applied in a wide range of industries, such as detergents, textiles, leather processing, forest products, baking, brewing, fermented products, pharmaceuticals, animal feed, fuel ethanol, and food [8,9]. In addition, fungal enzymes synthesize economically valuable compounds [10], the production of which with the use of metallic catalysts, although usually efficient, generates more byproducts and can be toxic to the environment. In contrast, the enzymes are non-toxic, biocompatible, and biodegradable, which eliminates pollution concerns [1]. In spite of many advantages, free fungal enzymes have some limitations, especially in difficult environmental conditions, such as strong acidity and/or high temperature, in which they can undergo denaturation decreasing their activity and affinity for substrates [11]. In order to improve the catalytic properties and stability of enzymes, the immobilization technique is used to optimize enzyme–substrate interactions and minimize non-specific interactions [12]. The use of immobilized enzymes in biotechnological processes allows multiple and repeated uses of a single batch of enzymes that are usually more stable than mobile enzymes. The catalyzed reaction can be controlled rapidly by removing the enzyme from the reaction solution, and contamination can be avoided by the easy separation of the enzyme from the product. The use of immobilized enzymes allows the development of multi-enzyme reaction systems, which are increasingly being used in industrial processes [13]. Various natural and synthetic matrices are used for immobilization adapted to the enzyme application. Organic and inorganic carriers such as alginate beads, glass and silica beads, activated carbon, and chitosan microsphere are most commonly used [14].

Chitosan, the product of chitin deacetylation, has gained a lot of interest as an attractive and cost-effective biomaterial for the immobilization of many enzymes used in a variety of fields, including medicine [15,16], wastewater treatment [17], agriculture [18], and biotechnology [19,20]. This polyaminosaccharide composed of β-(1–4)-linked *D*-glucosamine and *N*-acetyl-*D*-glucosamine units has several attractive properties, such as its natural origin, antibacterial activity, non-toxicity, easy modification, biodegradability, and low cost [21,22]. The presence of primary amino and hydroxyl groups in chitosan enhances the interaction between the biopolymer and enzymes, which allows the use of simple immobilization techniques, e.g., adsorption and covalent bonding [23]. The covalent bonding of the enzyme to the carrier is usually stronger and minimizes its leaching in the aqueous medium; it may also improve the biocatalyst stability and modify its catalytic properties [24].

Cellobiose dehydrogenase (CDH; EC1.1.99.18; CAZy: AA3.1) is an extracellular oxidoreductase (hemoflavoenzyme) secreted by fungi to assist lignocellulolytic enzymes in biomass degradation. CDH is a hemoflavoenzyme composed of a catalytic flavodehydrogenase (DH) domain linked to an *N*-terminal electron-transferring cytochrome (CYT) domain [25]. This enzyme is capable of catalyzing the oxidation of di- and oligosaccharides linked by β-1− 4-glycosidic bonds, including cellobiose and lactose, to the corresponding aldono-1,5-lactones, which spontaneously undergo hydrolysis to aldonic acids in an aqueous solution [26,27]. The cellobiose dehydrogenase (CDH) enzyme has been studied for many years for applications in biosensors and as an anode in biofuel cells due to its unique bioelectrochemical properties. Additionally, in the last decade, the antibiofilm, antimicrobial, and antioxidant properties of CDH have been associated with the presence of products of the CDH-catalyzed reaction consisting in the oxidation of lactose into lactobionic acid and such byproducts as hydrogen peroxide (H_2O_2), which are used in biomedicine as well as food and packaging industries [28]. The use of chitosan as a support for the immobilization of cellobiose dehydrogenases obtained by our research team is very reasonable given the application potential of these enzymes in both the medicine and food industries. Moreover, a very important issue concerning the application of immobilized enzymes, especially in these areas, is the potential cytotoxic effect. The use of a cross-linking agent, e.g., glutaraldehyde, can have a negative effect on healthy cells even if the matrix itself is non-toxic. This may affect the applicability of the immobilized enzyme system in a particular field.

Despite the huge potential of CDH, the research of the immobilized systems of this biocatalyst is a scientific novelty that is rare among research teams who work with this enzyme. Research on the use of cellobiose dehydrogenase immobilized on biocompatible matrices, such as chitosan, will allow expanding the application range of this enzyme. In the context of immobilization on the chitosan matrix, enzymes from *C. unicolor*, *P. sanguineus*, and *P. lindtnerii* have not been studied yet.

The aim of the present research was to apply chitosan beads as a support for the immobilization of four different fungal cellobiose dehydrogenases with varied application potential. By applying adsorption and the covalent immobilization of the enzymes and an enzyme/substrate system, the effectiveness of the immobilization process was determined. The chitosan beads with immobilized CDHs were characterized in terms of their FTIR spectra, SEM microstructure, and antioxidant, antimicrobial, and cytotoxic properties.

2. Results and Discussion

2.1. Immobilization of Cellobiose Dehydrogenase on Chitosan Beads

The use of immobilized enzymes instead of the free form of the biocatalyst has long been used for industrial purposes and offers additional advantages such as thermal stability and reusability. The immobilization of a protein with high application potential may increase the possibilities of its use in various industries, including those where it has not been used before.

There are two main methods of enzyme immobilization: physical and chemical. In the physical method, the enzyme interacts with the carrier (support) material without forming any formal chemical bonds. The protein can be adsorbed on the surface of the support material or entrapped within. The adsorption of the enzyme on the carrier is relatively simple, but the interaction between the carrier and the protein is minimal, and these materials are susceptible to leaching [29]. In the chemical method, the enzyme binds to the carrier through different covalent or non-covalent chemical interactions [30]. The most commonly used technique for enzyme immobilization is covalent immobilization using glutaraldehyde [31]. The reaction of glutaraldehyde with chitosan allows the activation of the support with aldehyde groups that may react with the enzyme through its amino groups (ε-amino group of lysine residues and terminal amino group) and sometimes with other surface functional groups (thiols, phenols, and imidazoles) [32]. Figure 1 shows the CDH immobilization scheme.

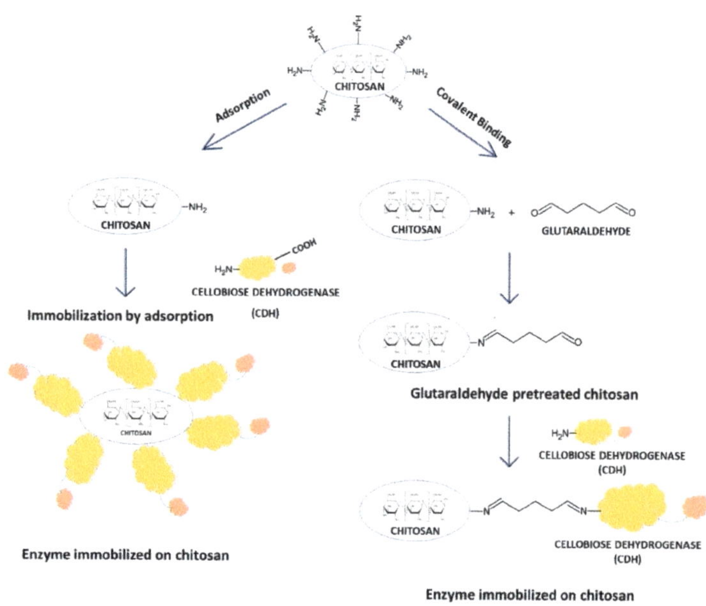

Figure 1. Immobilization of cellobiose dehydrogenase on chitosan through direct adsorption and covalent attachment using glutaraldehyde.

Natural biodegradable polymers such as alginate, carrageenan, or chitosan used as matrices for immobilization have found their special application in the engineering industry or medicine due to their ability to form hydrogels, fibers, films, or beads. In this study, we investigated the immobilization of cellobiose dehydrogenase on chitosan beads and compared the efficiency of the physical and chemical immobilization of CDH on chitosan. This polymer was used by Tegl et al. in their research on the biomedical application of CDH as an antibacterial agent [33].

2.1.1. Direct Adsorption

In the simplest immobilization technique, i.e., adsorption, the enzyme is bound to the carrier through hydrogen bonding, van der Waals forces, or hydrophobic interactions (physical adsorption) or through salt linkages (ionic binding) [34].

The chitosan carrier for CDH immobilization was prepared by precipitation in NaOH to form beads. The chitosan beads were immobilized by adsorption in the CDH-alone system and in the chitosan/CDH/substrate complex. Table 1 shows the differences in the protein binding efficiency and the CDH activity yield obtained from different fungi with or without the substrate. The highest activity yield of about 78% was obtained for CDH from the *C. unicolor*/lactose complex with protein binding of about 54%. Similarly, 60% efficiency of CDH immobilization on chitosan particles with the adsorption method was determined by Tegl et al. [33]. In our previous studies, we obtained higher CDH binding efficiency (from 60 to 100% depending on the enzyme tested) using the method of CDH entrapment in alginate beads [28].

Table 1. Efficiency of cellobiose dehydrogenase adsorption on the chitosan support (Ch) containing CDHs isolated from *P. lindtneri*: *Pl*CDH + Ch, with lactose *Pl*CDH + Ch + Lac, with cellobiose *Pl*CDH + Ch + Cel, from *C. unicolor*: *Cu*CDH + Ch, *Cu*CDH + Ch + Lac, *Cu*CDH + Ch + Cel, from *P. sanguineus*: *Ps*CDH + Ch, *Ps*CDH + Ch+ Lac, *Ps*CDH + Ch + Cel, and from *P. chrysosporium*: *Pch*CDH + Ch, *Pch*CDH + Ch + Lac, *Pch*CDH + Ch + Cel.

Sample Type	Applied Protein [mg/g Carrier]	Bound Protein [mg/g Carrier]	Protein Yield [%]	CDH Activity before Immobilization [U/g Carrier]	Activity of CDH Bound with Chitosan [U/g Carrier]	Activity Yield [%]
PlCDH + Ch	8.25 ± 0.02	1.76 ± 0.02	21.35	14.21 ± 0.03	0	0
PlCDH + Ch + Lac	8.23± 0.02	2.24 ± 0.02	27.21	15.27 ± 0.03	7.07 ± 0.04	46.29
PlCDH + Ch + Cel	9.61 ± 0.02	3.18 ± 0.01	33.14	15.28 ± 0.03	2.24 ± 0.02	14.64
CuCDH + Ch	6.49 ± 0.02	2.68 ± 0.02	41.27	37.53 ± 0.08	0.39 ± 0.001	1.04
CuCDH + Ch + Lac	5.76 ± 0.01	3.12 ± 0.03	54.12	40.17 ± 0.06	31.62 ± 0.08	78.74
CuCDH + Ch + Cel	6.18 ± 0.02	2.97 ± 0.03	48.04	36.93 ± 0.05	16.04 ± 0.05	43.44
PsCDH + Ch	8.19 ± 0.02	2.12 ± 0.02	25.91	17.10 ± 0.03	0	0
PsCDH + Ch + Lac	9.23 ± 0.03	4.77 ± 0.03	51.72	17.88 ± 0.03	6.60 ± 0.03	36.93
PsCDH + Ch + Cel	8.90 ± 0.02	6.12 ± 0.03	68.73	16.65 ± 0.03	2.54 ± 0.02	15.28
PchCDH + Ch	9.70 ± 0.03	5.89 ± 0.04	60.68	12.58 ± 0.02	1.56 ± 0.02	12.37
PchCDH + Ch + Lac	10.17 ± 0.04	2.59 ± 0.02	25.44	11.27 ± 0.03	2.11 ± 0.01	18.74
PchCDH + Ch + Cel	10.41 ± 0.04	3.27 ± 003	31.45	14.55 ± 0.03	2.66 ± 0.02	18.30

2.1.2. Covalent Immobilization Using Glutaraldehyde

The covalent immobilization of enzymes is one of the most widely used methods of attaching proteins to solid supports. This is associated with the stable nature of the bond preventing the enzyme from being released back into the solution [35]. However, such a strong attachment is a significant modification of the protein and can sometimes lead to changes in its hydrophilic/hydrophobic properties, which may significantly affect the efficiency of the immobilization process [29]. In our research, we used glutaraldehyde as a well-known cross-linking agent responsible for stiffening the enzyme structure and increasing its binding to the carrier [32]. A side effect of this procedure may be a remarkable decrease in the catalytic activity of the enzyme after immobilization. Nevertheless, from the practical point of view, it is much more important that biocatalyst preparations additionally crosslinked with glutaraldehyde exhibit significantly increased stability in processing conditions [36]. CDH was immobilized on chitosan beads activated with glutaraldehyde, which provided an excellent biocompatible surface since immobilization efficiency in the range of 28 to 99% for the different sources of the enzyme and enzyme/substrate complexes was achieved. Table 2 shows the immobilization yield of different variants of the immobilized enzyme. Compared to adsorption, the use of chitosan beads activated with the crosslinking agent significantly increased the efficiency of the immobilization process. Glutaraldehyde used as a crosslinking agent for the immobilization of enzymes on chitosan beads prevents the direct contact of the enzyme with the surrounding medium and enables the reagents to reach the catalytic site [37]. This technique has been used in many research studies to effectively immobilize such enzymes as lipase [38], β-galactosidase [37], or laccase [39] on chitosan beads The efficiency of the process of immobilization of various enzymes was high with values in the range of 84.7% [39] and 88.4 % [40] for laccase to 99.1% for lipase [41]. CDH immobilization has also been carried out on glutaraldehyde-modified magnetic chitosan spheres and chitosan particles used as carriers with activity yields of 61.54% [7] and 40% [42], respectively

Table 2. Efficiency of cellobiose dehydrogenase immobilization with glutaraldehyde on the chitosan support (ChGA) containing CDHs isolated from *P. lindtneri*: *Pl*CDH + ChGA, with lactose *Pl*CDH + ChGA + Lac, with cellobiose *Pl*CDH + ChGA + Cel, from *C. unicolor*: *Cu*CDH + ChGA, *Cu*CDH + ChGA + Lac, *Cu*CDH + ChGA + Cel, from *P. sanguineus*: *Ps*CDH + ChGA, *Ps*CDH + ChGA + Lac, *Ps*CDH + ChGA + Cel, and from *P. chrysosporium*: *Pch*CDH + ChGA, *Pch*CDH + ChGA + Lac, *Pch*CDH + ChGA + Cel).

Sample Type	Applied Protein [mg/g Carrier]	Bound Protein [mg/g Carrier]	Protein Yield [%]	CDH Activity before Immobilization [U/g Carrier]	Activity of CDH Bound with Chitosan [U/g Carrier]	Activity Yield [%]
PlCDH + ChGA	8.25 ± 0.02	3.58 ± 0.02	43.37	14.21 ± 0.03	4.41 ± 0.02	31.03
PlCDH + ChGA + Lac	8.23 ± 0.02	5.05 ± 0.03	61.36	15.27 ± 0.03	15.14 ± 0.06	99.18
PlCDH + ChGA + Cel	9.61 ± 0.02	6.85 ± 0.02	71.32	15.28 ± 0.03	13.64 ± 0.05	89.23
CuCDH + ChGA	6.49 ± 0.02	6.04 ± 0.04	93.06	37.53 ± 0.08	34.63 ± 0.08	92.28
CuCDH + ChGA + Lac	5.76 ± 0.01	3.57 ± 0.02	61.87	40.17 ± 0.06	29.60 ± 0.05	73.69
CuCDH + ChGA + Cel	6.18 ± 0.02	4.67 ± 0.01	75.49	36.93 ± 0.05	28.55 ± 0.04	77.31
PsCDH + ChGA	8.19 ± 0.02	3.95 ± 0.03	48.29	17.10 ± 0.03	4.89 ± 0.02	28.59
PsCDH + ChGA + Lac	9.23 ± 0.03	7.65 ± 0.05	82.87	17.88 ± 0.03	16.45 ± 0.03	92.02
PsCDH + ChGA + Cel	8.90 ± 0.02	7.07 ± 0.04	79.39	16.65 ± 0.03	12.91 ± 0.02	77.55
PchCDH + ChGA	9.70 ± 0.03	3.57 ± 0.02	36.76	12.58 ± 0.02	7.53 ± 0.03	59.84
PchCDH + ChGA + Lac	10.17 ± 0.04	7.76 ± 0.03	76.34	11.27 ± 0.03	10.13 ± 0.04	89.85
PchCDH + ChGA + Cel	10.41± 0.04	9.29 ± 0.03	89.28	14.55 ± 0.03	12.24 ± 0.04	84.16

2.2. Characterization of the Chitosan Beads

2.2.1. FTIR

The spectra of all the tested samples showed significant similarities (Figure 2). A broad band of overlapping peaks of O-H and N-H stretching was recorded at 3335–3353 cm^{-1} in all spectra, which is in agreement with observations of glutaraldehyde-crosslinked chitosan reported by another research team [43]. Two bands of asymmetric and symmetric -CH_3 and -CH_2 vibrations were found at 2936 and 2879 cm^{-1}, respectively. Li et al. (2013) found that the band of -CH_3 vibrations shifted from 2925 cm^{-1} in pure chitosan to 2937 cm^{-1} in glutaraldehyde-crosslinked chitosan, which may suggest successive crosslinking of the chitosan polymer in the case of this study. This may be supported by the observation of a small peak at 1404 cm^{-1} in all the spectra, indicating the presence of C-N stretching vibrations. Li et al. (2013) also observed the appearance of such a peak in glutaraldehyde-crosslinked chitosan in comparison with a peak at 1438 cm^{-1} for pure chitosan. These bands may indicate successful chitosan crosslinking by glutaraldehyde. On the other hand, a very small peak was observed at 1150 cm^{-1} in the spectra of all the samples [44]. As reported by Islam et al. (2019), this band indicated the presence of asymmetric stretching vibrations of the C-O-C bridge in the glycosidic bond in pure chitosan. Islam observed that this peak was shifted to 1110 cm^{-1} in the spectra of glutaraldehyde-crosslinked chitosan. Therefore, the presence of the minimal peak of 1150 cm^{-1} in these samples may suggest the presence of traces of uncrosslinked chitosan in all the tested samples [43].

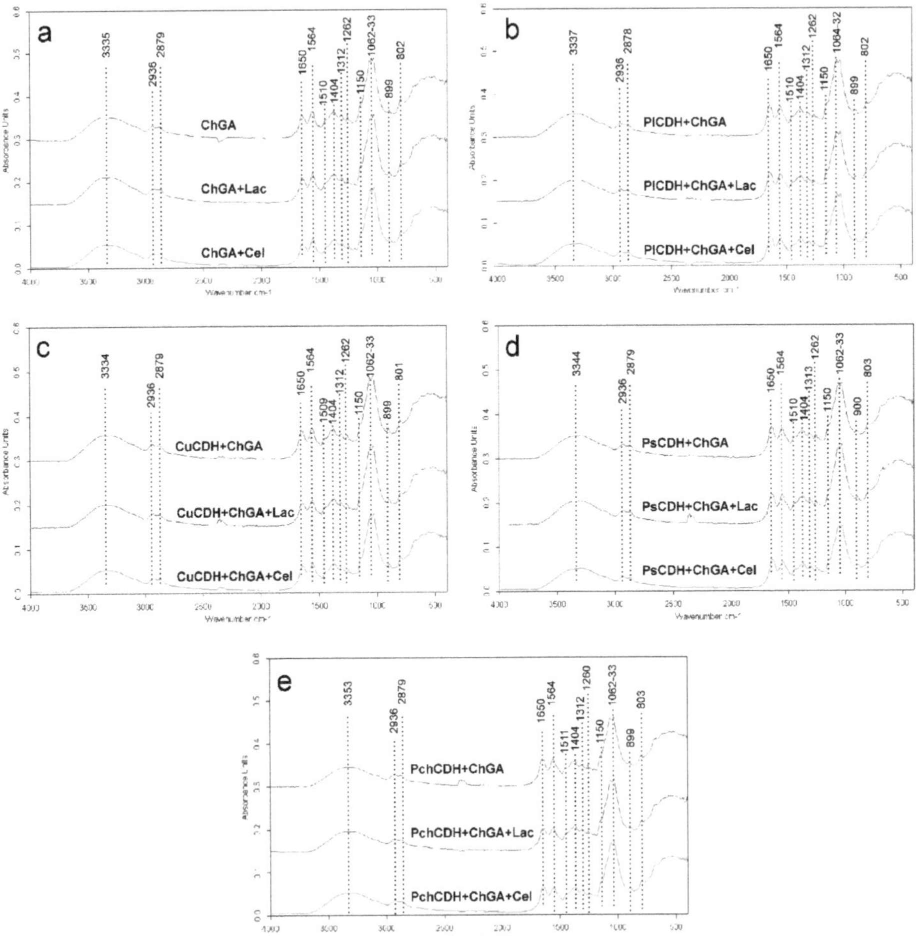

Figure 2. (**a**) FTIR spectra of chitosan with glutaraldehyde (ChGA) (C1-control), ChGA + Lac (C2-control), ChGA + Cel, (C3-control); (**b**) FTIR spectra chitosan with glutaraldehyde (ChGA) containing CDHs isolated from *P. lindtneri*: *Pl*CDH + ChGA, *Pl*CDH + ChGA + Lac, *Pl*CDH + ChGA + Cel; (**c**) FTIR spectra chitosan with glutaraldehyde (ChGA) containing CDHs isolated from *C. unicolor*: *Cu*CDH + ChGA, *Cu*CDH+ ChGA + Lac, *Cu*CDH + ChGA + Cel; (**d**) FTIR spectra chitosan with glutaraldehyde (ChGA) containing CDHs isolated from *P. sanguineus*: *Ps*CDH + ChGA, *Ps*CDH + ChGA + Lac, *Ps*CDH +ChGA + Cel; (**e**) FTIR spectra chitosan with glutaraldehyde (ChGA) containing CDHs isolated from *P. chrysosporium*: *Pch*CDH + ChGA, *Pch*CDH+ ChGA + Lac, *Pch*CDH + ChGA + Cel.

Two dominating sharp peaks at 1062 and 1033 cm^{-1} represent the O-C-O ring of chitosan [43,45]. The 1510 and 899 cm^{-1} bands may indicate the presence of N-H bending vibrations and C-C stretching vibrations (in the glucose ring), respectively [45]. The 1374 cm^{-1} band may indicate the presence of either CH_3 bending vibrations [44] or in-plane OH bending vibrations [38]. The 1262 cm^{-1} band can be assigned to the bending vibrations of hydroxyls present in chitosan [46] or to CH_2 wagging vibrations [47]. The 1312 cm^{-1} and 1201 cm^{-1} bands may indicate the presence of an amide III band (complex vibrations of NHCO group) [48] and OH plane deformation modes [49] observed in prawn-derived chitin.

The presence of 802 cm^{-1} bands in all the spectra is probably related to the presence of N-H wagging vibrations. Islam et al. (2019) and Vijayalakshmi et al. (2016) also observed this peak in glutaraldehyde-crosslinked chitosan but at a slightly different position (at 822 cm^{-1}). This peak shows slightly higher intensity in the spectrum of glutaraldehyde-crosslinked chitosan without any additives (sample ChGA) than in the spectra of the other samples. This may be caused by the increasing percentage of substances other than the crosslinked chitosan (namely sugars and proteins) in the spectra of these samples [43,45]. In all the tested samples, two significant peaks of 1650 and 1564 cm^{-1} were observed at positions recognized as amine I and amide II bands, respectively. The amide I band is assigned to C=O stretching vibrations of the peptide bond (about 80%) with a minor contribution from C-N stretching vibration (about 20%), whereas the amide II band derives mainly from in-plane NH bending and C-N stretching vibrations (40–60% and 18–40% of the potential energy, respectively) [50]. In samples ChGA (C1-control), ChGA + Lac (C2-control), and ChGA + Cel (C3-control) (without the enzyme addition), the amide II band dominates over the amide I band. However, when the spectrum of sample ChGA was compared with the spectra of samples *Pl*CDH + ChGA, *Cu*CDH + ChGA, *Ps*CDH + ChGA, or *Pch*CDH + ChGA (with the addition of protein), the height of the amide I band increased in the latter spectra. This suggested an increase in the C=O stretching vibration content in the overall signal intensity caused by the increase in the amount of peptide bonds in the tested samples. A similar relationship was observed for the spectrum of sample ChGA + Lac versus the spectra of samples *Pl*CDH + ChGA + Lac, *Cu*CDH + ChGA + Lac, *Ps*CDH + ChGA + Lac, and *Pch*CDH + ChGA + Lac (enriched with lactose) and for the spectrum of sample ChGA + Cel versus the spectra of samples *Pl*CDH + ChGA + Cel, *Cu*CDH + ChGA + Cel, *Ps*CDH + ChGA + Cel, and *Pch*CDH + ChGA + Cel (enriched with cellobiose). The ratio of the amide I to amide II band intensity ranged from 0.81 to 0.9 for chitosan without enzymes (ChGA, ChGA + Lac, and ChGA + Cel) but increased when the proteins were added (1.07–1.33 for chitosan containing *Cu*CDH and *Pch*CDH and 1.42–1.65 for chitosan containing *Pl*CDH and *Ps*CDH). This confirms the presence of proteins in these samples.

2.2.2. Examination of the Sample Surface Using the SEM Microscopy Technique

The study of the surface of many samples used in different biological systems (e.g., materials for the immobilization of cells or enzymes) requires the use of modern and precise tools, including microscopic techniques such as SEM [51]. The microscopic analysis of the surface of chitosan beads modified with glutaraldehyde, CDH (*Pl*CDH + ChGA, *Cu*CDH+ ChGA, *Ps*CDH + ChGA, *Pch*CDH + ChGA), lactose, and cellobiose evidenced morphological changes in all of the analyzed samples (Figure 3). Chitosan beads with glutaraldehyde (ChGA) and additionally bound with lactose (ChGA + Lac) and cellobiose (ChGA + Cel) were used as controls. During the experiments, it was observed that even in the case of the control without the CDH enzyme, the introduction of lactose or cellobiose significantly weakened the integrity of the surface of the tested biomaterials. Most likely as a result of the immobilization procedure, the beads exhibited clear cracks and a loss of shape (Figure 3a). Interestingly, the SEM detection showed that the introduction of *P. lindtneri* and *P. chrysosporium* CDH into the composition of the chitosan beads resulted in the smoothing of the biomaterial surface in comparison to the control, which suggests an improvement in the mechanical properties of the obtained samples (Figure 3b,c). It should be underlined that, in the case of *P. chrysosporium*, the presence of cellobiose additionally strengthened the surface of the chitosan beads, whereas no such effect was noted in the case of *P. lindtneri*. Contrarily, in the case of the *C. unicolor* and *P. sanguineus* enzymes, the modification of the composition of the chitosan beads by the immobilization of CDH and the addition of lactose and cellobiose caused negative changes in the mechanical properties of the tested biomaterials. The structure and shape of the surface of the modified beads were clearly disturbed, which may also suggest a reduction in their resistance to stress conditions present during their catalytic activity. In both cases, the presence of bound lactose further intensified this effect, and the bound cellobiose molecules seemed to weaken the effect by improving the integrity of the outer layer of the modified carrier (Figure 3b,c).

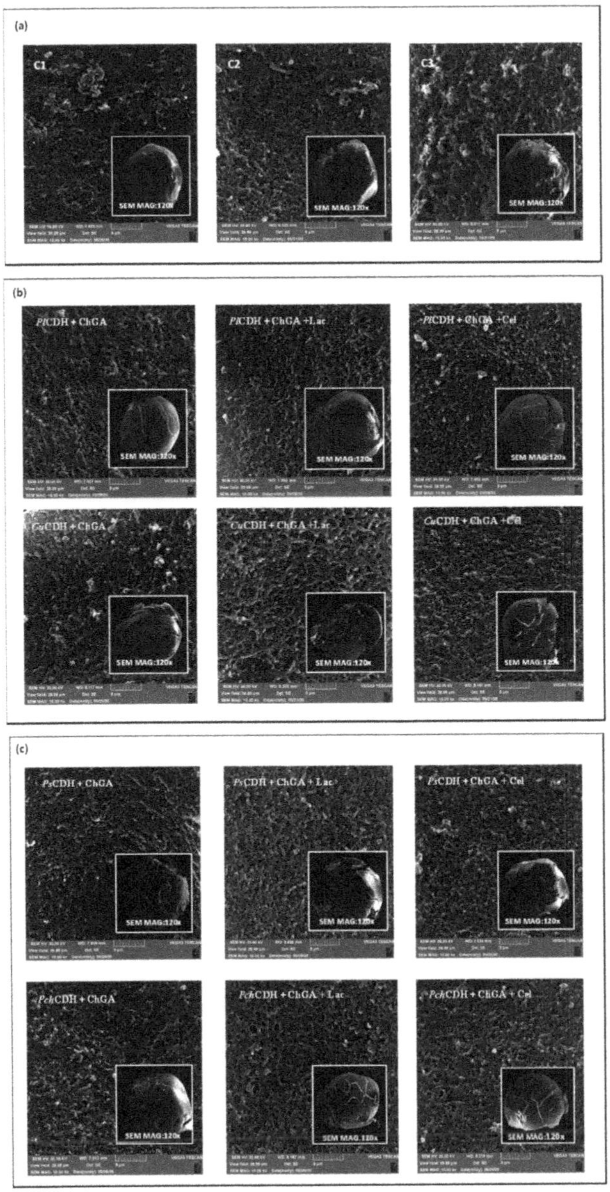

Figure 3. (**a**) SEM micrographs of the surface of chitosan with glutaraldehyde (ChGA) (C1-control), ChGA + Lac (C2-control), ChGA + Cel, (C3-control); (**b**) SEM micrographs of the surface of chitosan

with glutaraldehyde (ChGA) containing CDHs isolated from *P. lindtneri*: *Pl*CDH + ChGA (1a), *Pl*CDH + ChGA + Lac (1b), *Pl*CDH + ChGA + Cel (1c) and from *C. unicolor*: *Cu*CDH + ChGA (2a), *Cu*CDH+ ChGA + Lac (2b), *Cu*CDH + ChGA + Cel (2c); (**c**) SEM of the surface of chitosan with glutaraldehyde (ChGA) (containing CDHs isolated from *P. sanguineus*: *Ps*CDH + ChGA (3a), *Ps*CDH + ChGA + Lac (3b), *Ps*CDH + ChGA +Cel (3c) and from *P. chrysosporium*: *Pch*CDH + ChGA (4a), *Pch*CDH+ ChGA + Lac (4b), and *Pch*CDH + ChGA + Cel (4c).

Other papers examining the surface morphology of normal and activated chitosan beads as enzyme carriers found increased roughness after treatment with glutaraldehyde and enzyme immobilization [52,53]. Another study suggested that the porosity of the surface increased after the use of glutaraldehyde [54].

2.2.3. Antioxidant Properties

Fungi are known for the production of substances with potent antioxidant properties such as vitamins, carotenoids, polyphenols, polysaccharides, minerals, and enzymes. In our earlier studies, fungal cellobiose dehydrogenases have been reported to exhibit strong antioxidant activity in the presence of lactose and cellobiose substrates [55,56]. Although chitosan is known rather for its antibacterial properties, its antioxidant properties have been confirmed in some studies. Yen Ming-Tsung et al. observed antioxidant activities at a level of 58.3–70.2% for 1 mg/m chitosan from crab shells [57]. Additionally, the antioxidant activity of chitosan can be enhanced by synthesizing chitosan derivatives. For example, Chatterjee et al. showed that native chitosan with MW = 100 kDa exhibited no inhibition against DPPH in contrast to four new phenolic conjugates with significantly improved antioxidant activities (50 % inhibition of 2,2′-diphenyl-1-picrylhydrazyl radical and hydroxyl free radical) [58]. Simultaneously, the commercial chitosan carrier used in the immobilization process also showed antioxidant properties, which is confirmed by the results of our study presented in Figure 4.

In order to compare the antioxidant activities of CDHs immobilized on the chitosan carrier with the activities of native CDHs, the DPPH-scavenging method with Trolox and vitamin C as a control model system was used. Our result showed that, in the presence of lactose or cellobiose as substrate, all the native CDHs had very good antioxidant properties (about 80 to 90% of scavenging). As a result of the CDH immobilization on chitosan carriers, the appearance of antioxidant properties was also observed in the case of the enzyme without the presence of the substrate (about 30 to 78% of scavenging). This is probably related to the antioxidant properties of chitosan. In turn, a decrease in the antioxidant properties (against the control) was observed for the preparations with CDH immobilized on the chitosan carrier, especially in the presence of the cellobiose substrate (about 30 to 60 % of scavenging). The least significant decrease in the antioxidant activity after the immobilization process was observed for CDHs isolated from *Phlebia lindtneri* (*Pl*CDH) (Figure 4A). The decrease in the antioxidant activity of the studied enzymes against the control samples can be explained by the steric effect and structural changes in the protein conformation induced by the immobilization process [42]. These changes may have significant effects on the enzymatic activity of CDH. It should be emphasized that the structure of the enzyme molecule is the key of catalytic importance in this case for the production of lactobionic or cellobionic acid and hydrogen peroxide as a byproduct. In our previous studies, we demonstrated that CDHs without substrates (lactose and cellobiose) did not exhibit antioxidant properties. Reaction products were formed only in the presence of enzyme reaction substrates. These products are probably responsible for the antioxidant properties [28]. Thus, the antioxidant properties of CDH appear during the reaction with specific substrates. As a result of the immobilization process, the structure of the protein is probably changed or the active chemical groups responsible for the catalytic properties of the enzyme are partly blocked, which may explain the decrease in the antioxidant activity of the studied CDHs in comparison to the control samples [42].

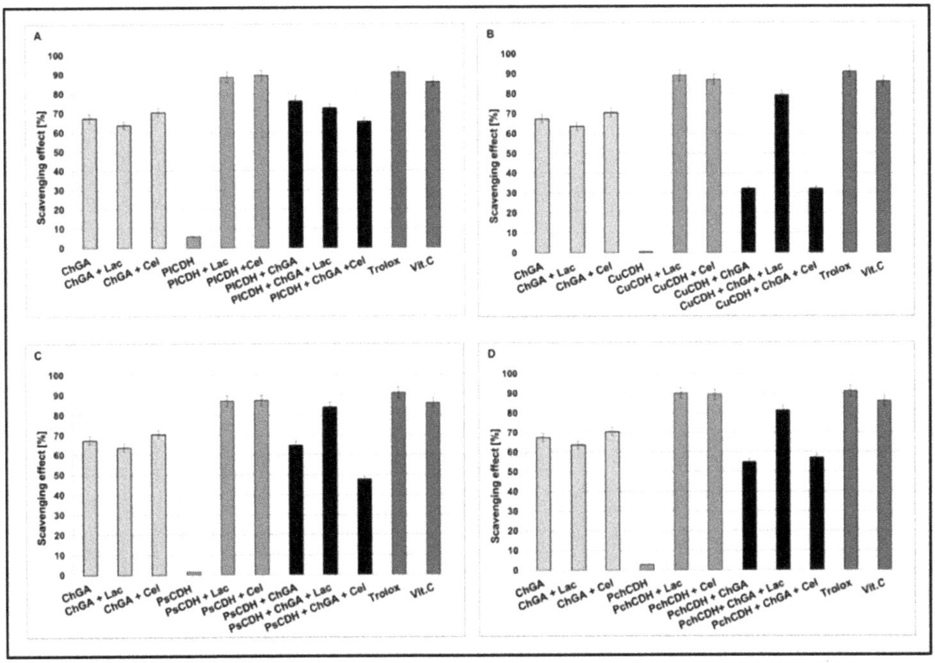

Figure 4. Scavenging effects of chitosan beads with glutaraldehyde (ChGA) (C1-control) with lactose ChGA + Lac (C2-control), or cellobiose ChGA + Cel (C3-control) containing CDHs isolated from *Phlebia lindtneri (Pl*CDH) (**A**), *Cerrena unicolor (Cu*CDH) (**B**), *Pycnoporus sanguineus* (*Ps*CDH) (**C**) and *Phanerochaete chrysosporium (Pch*CDH) (**D**). Data are mean ± SD for three measurements (n = 3).

2.2.4. Operational Stability of Chitosan Beads

Operational stability is an important parameter determining the usefulness of immobilization techniques in industry. The enzyme half-life ($t_{\frac{1}{2}}$) is a parameter characterizing the operational stability of preparations after immobilization. It is a quantity describing a period after which half of the initial catalytic activity of the enzyme is lost. In our study, we tested the antioxidant operational stability of CDH immobilized on chitosan beads. In all the tested samples, the best results were obtained for preparations containing lactose as a substrate, except chitosan beads containing CDH isolated from *Phlebia lindtneri (Pl*CDH). The most promising results in the three cycles of determination (about 60% and 58% of scavenging) were obtained for *Ps*CDH + ChGA + Lac and *Cu*CDH+ ChGA + Lac (Figure 5).

Our previous research on the immobilization of CDHs on alginate beads showed approx. only a 35% and 20% scavenging effect of CDH from *Phanerochaete chrysosporium* and *Phlebia lindtneri* [28]. In turn, Yang et al. immobilized cellobiose dehydrogenase from *Aspergillus fumigatus* (*Af*CDH) and laccase separately on magnetic chitosan spheres. In this multienzymatic system, they achieved 70 % relative catalytic efficiency after 10 cycles of reuse [7]. The loss of enzymatic activity after the immobilization process may depend on many factors, such as the elution of the biocatalyst from the support, blocking of the matrix pores, and destruction of the support material [59]. Good operational stability of enzymes is a very important parameter, as it facilitates multiple uses of the bound enzyme in the production process, which significantly reduces production costs. The ability of CDHs immobilized on chitosan beads to maintain enzymatic activity in several cycles of reuse is an advantage over the native form of this enzyme.

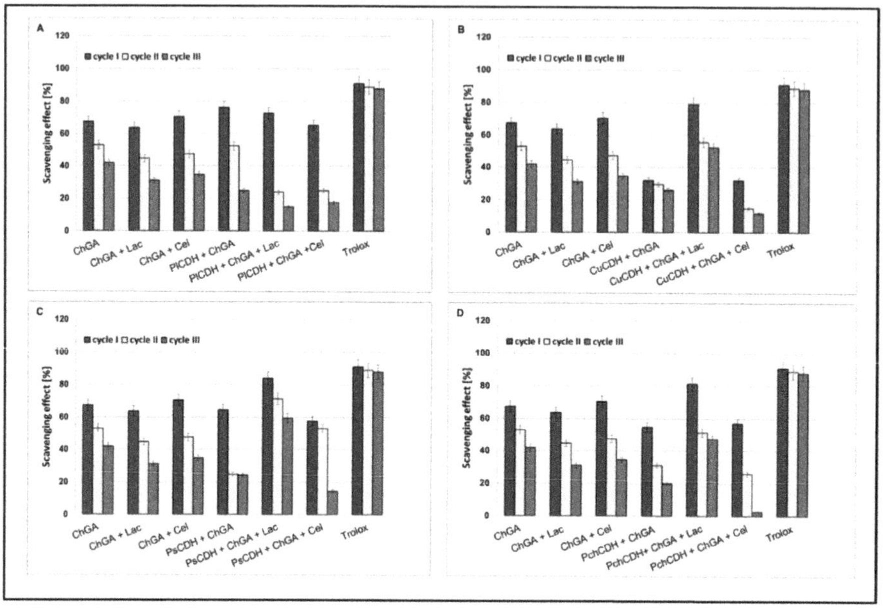

Figure 5. Operational stability of chitosan beads with glutaraldehyde (ChGA) (C1-control) with lactose, ChGA + Lac (C2-control), or cellobiose ChGA + Cel, (C3-control) containing CDHs isolated from *Phlebia lindtneri (Pl*CDH) (**A**), *Cerrena unicolor (Cu*CDH) (**B**), *Pycnoporus sanguineus* (*Ps*CDH) (**C**) and *Phanerochaete chrysosporium (Pch*CDH) (**D**). Data are mean ± SD for three measurements (n = 3).

2.2.5. In Vitro Cytotoxicity Assessment

The cytotoxic effect of chitosan beads with immobilized CDH and the CDH/substrate system was investigated using healthy mammalian Vero cells. To assess the toxicity of the beads with the different immobilized CDH systems (with or without the substrate) presented in this work, MTT viability cell assays were conducted in 24 h and 48 h incubation periods. The results obtained are shown in Figure 6. At first, the toxicity of chitosan beads without the addition of the enzyme activated and not activated with glutaraldehyde also in the presence of the substrates (lactose and cellobiose) was verified. Chitosan beads that were not activated with glutaraldehyde (Ch) exerted a strong proliferative effect on the cell line studied. On the other hand, the glutaraldehyde-activated beads showed low cytotoxicity after the 24 h incubation (less than 20% growth inhibition). After the 48 h incubation, cell viability decreased by another 20%. Similar levels of cell viability were obtained in the variant with the enzyme from *P. lindtnerii* in the presence of the substrates (PlCDH + ChGA + Lac and PlCDH + ChGA + Cel), while PlCDH without the substrate (PlCDH + ChGA) caused the proliferation of the cells. In the case of the other enzymes, there was a decrease (20–40%) in cell viability after 24 h of incubation. In addition, it was found that the enzyme from *C. unicolor* (CuCDH) had the most toxic effect on Vero cells after 48 h both in the presence and absence of the substrates. No research on the cytotoxicity of CDHs or products of enzymatic reactions (aldobionic acid and H_2O_2) has been performed so far. Our research team has done a series of such analyses, and they will be published shortly. However, research on the cytotoxicity of *N*, *N*, *N*-trimethyl chitosan/alginate beads containing gold nanoparticles was carried out by Martins et al. (2015). They showed that this material was slightly destructive for healthy VERO cells, with viability reduced to 75%. On the other hand, raw beads exhibited quite good biocompatibility with the healthy Vero line, as approximately 100% of the total cells remained viable after 24 h of incubation [60].

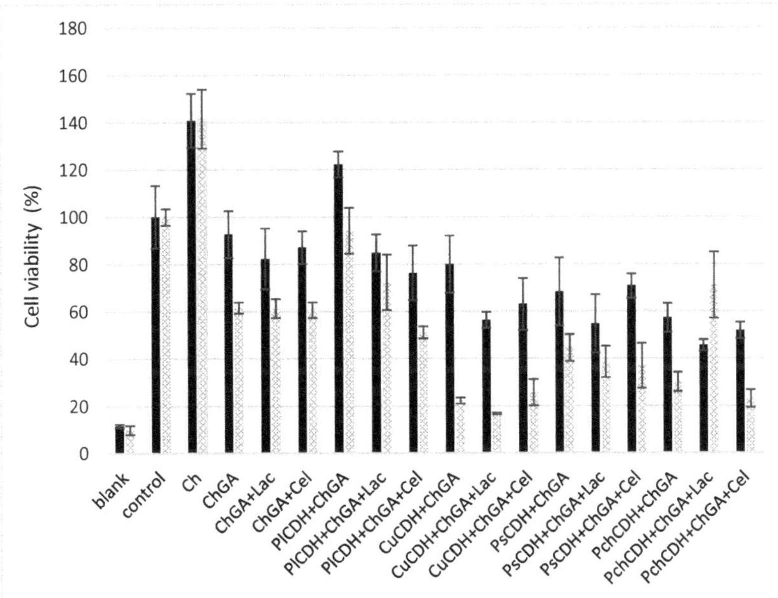

Figure 6. MTT cell viability assay of Vero cells incubated for 24 and 48 h with different chitosan beads: without glutaraldehyde (Ch), with glutaraldehyde (ChGA), with lactose (ChGA + Lac), or cellobiose (ChGA + Cel), containing CDHs isolated from *Phlebia lindtneri* (*Pl*CDH), *Cerrena unicolor* (*Cu*CDH), *Pycnoporus sanguineus* (*Ps*CDH), and *Phanerochaete chrysosporium* (*Pch*CDH). Data are mean ± SD for three measurements (n = 3).

The antimicrobial effects of CDHs immobilized on chitosan beads and the CDH/substrate systems from the same four fungi were tested against two strains of Gram-negative bacteria and two strains of Gram-positive bacteria. The assessment of the antibacterial activity of the tested CDHs and the CDH/substrate systems is presented in Table S1. The minimum inhibitory concentrations (MIC) of free CDHs (with and without the substrate) used for the immobilization process is presented in Table S2.

3. Materials and Methods

3.1. Microorganisms

Fungal strains producing cellobiose dehydrogenase (CDH) were taken from the collection at the Department of Biochemistry of Maria Curie, Sklodowska University (Poland) as described in previous publications [28].

3.2. Materials

The chitosan (medium molecular weight, degree of deacetylation 75–85%) and the glutaraldehyde (25% (*v*/*v*)) aqueous solution were obtained from Sigma Aldrich (Steinheim, Germany). Media components and other chemicals were purchased from Merck (Darmstadt, Germany), VWR (Vienna, Austria), BioRad (Warsaw, Poland), POCH (Gliwice, Polska), or BioMaxima (Lublin, Poland). All reagents and chemicals used in the work were of the highest available purity and analytical grade. Deionized water was used for the preparation of all aqueous solutions.

3.3. Enzyme Preparation

The preparation of semi-pure cellobiose dehydrogenase was carried out using a two-step purification strategy according to Sulej et al. (2019) [51]. Briefly, the crude enzyme was precipitated with ammonium sulfate fractionation in the saturation ranges of 40–90% (*P. sanguineus*), 30–50% (*P. lindtneri*), 15–85% (*C. unicolor*), and 20–80% (*P. chrysosporium*) at 0 °C. The precipitates were dissolved in deionized water and by diafiltration through centrifugal concentrators (Vivaspin Turbo 15) in polyethersulphone (PES) with a cut-off of 30 kDa (Sartorius, Göttingen, Germany). The diafiltrated sample was loaded onto a DEAE Sepharose (fast flow) column for anion exchange chromatography. Fractions with high CDH activity were pooled, desalted, and used as a partially purified enzyme for the immobilization experiments. The protein concentration during enzyme purification was monitored by ultraviolet (UV) absorbance at 280 nm.

3.4. Enzyme Activity Assay and Protein Determination

Cellobiose dehydrogenase activity was determined by the lactose-dependent reduction of 2,6-dichloroindophenol (DCIP) (Sigma Chemical Co., St. Louis, MO, USA) at 520 nm (ε_{520} = 6.8 $mM^{-1}\cdot cm^{-1}$), pH 4.5, and 30 °C. The method developed by Baminger et al. (1999) [61] with modifications described in our previous work was used [28]. The protein concentration was determined using the Bradford method [62] with crystalline bovine serum albumin (BSA) as a standard.

3.5. Immobilization of Cellobiose Dehydrogenase on Chitosan Beads

3.5.1. Preparation of Chitosan Beads

Chitosan powder (3 g) was dissolved in 100 mL of an acetic acid solution (2%, *v*/*v*) and mixed on a magnetic stirrer for 1 h at room temperature. Beads were formed by extruding the solution through a syringe needle into a coagulant bath of 1 M NaOH solution with constant stirring. The beads were kept in a 1 M NaOH solution for 24 h. After hardening, the pretreated chitosan beads were collected by filtration, washed with deionized water until neutrality, and stored at 4 °C.

3.5.2. Direct Adsorption

Chitosan beads (0.5 g) were added into 5 mL of a CDH solution (1 mg/mL). CDHs for immobilization were prepared by mixing with water and lactose or cellulose in a 1:1 ratio. The solution was stirred with chitosan beads at 25 °C, 100 rpm for 3.0 h and kept in the fridge overnight. In the next step, the chitosan beads with the adsorbed enzyme were separated and washed three times with 100 mM phosphate-citric acid buffer (pH 5.0) to remove unbound enzymes.

3.5.3. Covalent Immobilization Using Glutaraldehyde/Activation of Chitosan Beads

CDHs were immobilized on chitosan beads using glutaraldehyde as a linker. The chitosan beads (support) were activated by treatment thereof with 5% concentrations of glutaraldehyde and cross-linking (25 °C, 100 rpm, 1 h). After activation, the beads were washed thoroughly with distilled water to remove glutaraldehyde. The enzymes for immobilization were prepared in the same way as before. The immobilization process was carried out for 3 h at room temperature in a shaking incubator (100 rpm) and then the preparations were left overnight at 4 °C. The next day, the beads were washed three times with 100 mM phosphate-citric acid buffer (pH 5.0) to remove any unbound enzyme.

The percentage of immobilization was calculated as follows:

Total activity obtained in beads after immobilization/Total activity of the soluble enzyme loaded) × 100.

The scheme of preparation, activation, and immobilization of CDH on chitosan beads is shown in Figure 7.

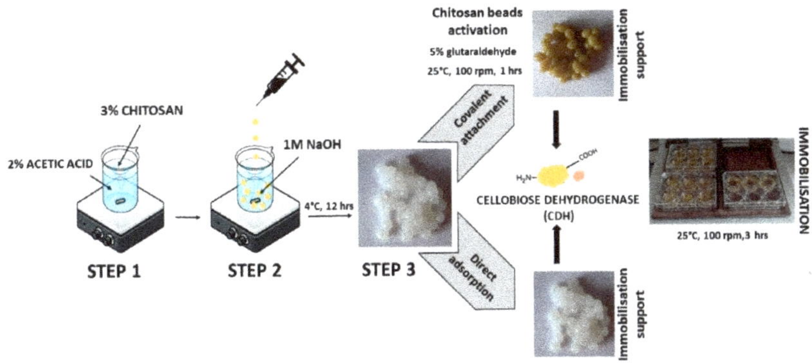

Figure 7. Steps in the immobilization of cellobiose dehydrogenase on chitosan beads.

3.6. Characterization of Chitosan Beads

3.6.1. FTIR

The FTIR-ATR analysis of the samples was performed using a Vertex 70 spectroscope (Bruker, Billerica, MA, USA) equipped with an ATR-diamond crystal within the range of 4000–370 cm with a 1 cm resolution. In total, 32 scans were collected for each sample. Prior to the measurement, all samples were dried in an exicator for 24 h to remove the excess unbound water. Spectra were analyzed using OPUS 7.0 program (Bruker, Billerica, MA, USA). The ratio of amide I to amide II bands was calculated in the same program using the values of 1650 cm^{-1} and 1564 cm^{-1} band intensity relative to the local baseline.

3.6.2. Scanning Microscopy (SEM) of Modified Chitosan Beads

The surface of the freeze-dried chitosan beads modified by immobilization of CDH isolated from the four fungal strains, lactose, and cellobiose was visualized using the scanning microscopy technique (SEM) at 30 kV accelerating voltage. Beads containing chitosan with glutaraldehyde (ChGA), ChGA with immobilized lactose, and ChGA with bounded cellobiose were the controls. All of the samples were coated with gold (Sigma-Aldrich, St. Louis, MO, USA) and their morphologies were observed using VEGA-3 LMU SEM (TESCAN, Brno, Czech Republic). The magnification ×10,000 was used to examine the surface of the control and the CDH-, laccase-, and cellobiose-modified samples. In order to check and confirm possible changes in the image of the samples, 10 separate areas with chitosan beads were selected.

3.6.3. Antioxidant Properties of CDH Immobilized on Chitosan Beads

The DPPH free radical scavenging activity of free and chitosan-bound cellobiose dehydrogenases (CDH) was determined with the method described by Paduch et al. (2008) [63] with some modification. The analyzed samples (0.1 mL or 0.1 mg chitosan beads, 6.25 to 800 μg/mL) were added to 0.1 mL of a DPPH· solution (0.2 mg/mL in ethanol). Trolox and vitamin C, the well-known standards with strong antioxidant activities, were used as positive controls. Absorbance at 515 nm was determined at room temperature after 15 min (optimal time of incubation). The percentage of inhibition of DPPH oxidation was calculated as follows:

$$DPPH\ scavenging\ effect\ (\%) = [(A_0 - A_1)/A_0] \times 100$$

where A_0 means the absorbance of the control sample and A_1 means the absorbance of the standard or the tested compound. The standard (Trolox and Vitamin C) calibration curves were prepared for a concentration range from 6.25 to 800 μg/mL.

3.6.4. Operational Stability

The antioxidant operational stability of the immobilized CDH was assessed by incubating chitosan beads (~0.1 mg) in 0.1 mL of a DPPH· solution (0.2 mg/mL in ethanol). After 15 min of incubation, the absorbance of the DPPH· solution was measured. The chitosan beads were washed five times with 0.1 M phosphate buffer pH 6.5 and suspended in a new solution of DPPH to begin the next cycle of activity measurement.

3.6.5. Cytotoxicity Evaluation

Cell viability was analyzed with the 3-(4,5-dimethylthiazol-2-yl)-2,5-diphenyltetrazolium bromide (MTT) method. The cytotoxicity of beads with immobilized CDHs and CDH/substrate systems was investigated as in Huber et al. (2017) [64] with several modifications. The beads were UV-sterilized for 1 h in sterile conditions. The sterile beads were then incubated for 24 h in DMEM medium at 37 °C and 5% CO_2 to allow leaching into the medium. Afterward, the medium was filtered through a 0.22-μm filter and the leach-out medium was used for cytotoxicity testing. The extracts were added into the wells of a 96-well plate containing 24 h Vero (African Green Monkey-kidney, ATCC, Kennett Square, PA, USA) cell line solutions (~1×10^4 cell/well) and incubated for 24 and 48 h. A control was performed by using untreated medium. After incubation, 10 μL of 5 mg/mL MTT (Sigma-Aldrich) was added to each well for 3 h incubation in darkness at room temperature, after which the medium was removed and formazan crystals within the cells were solubilized by 100 μL/well of DMSO (Chempur). The plates were shaken to fully dissolve the crystals. The optical density at 570 nm was detected with a Thermo Scientific Multiscan FC spectrophotometer. The results are presented as a percentage of the control values from three separate wells.

3.6.6. Statistical Analysis

The presented results are expressed as mean ± SD from three independent experiments (n = 3). The mean values and standard deviation were calculated using one-way ANOVA (Statgraphics Online); next, the means were compared using Tukey's multiple range test. The Excel program (Microsoft Office 365 package) was used for the calculation of the data. Values of $p \leq 0.05$ were considered statistically significant.

4. Conclusions

This paper presents a proposal for an innovative and efficient system for exploiting the catalytic potential of cellobiose dehydrogenase based on the immobilization of the enzyme on chitosan as an immobilization matrix. For the most complete demonstration of the possibilities of the proposed experimental setup, two different methods (adsorption and covalent binding) of immobilization of CDH on chitosan beads were tested. Although the adsorption method seems to be less invasive, the chemical bonding method using glutaraldehyde is definitely more efficient. As indicated by the obtained results, it is possible to obtain almost 90% binding efficiency while maintaining the most important catalytic properties. It should be emphasized that the physicochemical properties of the chitosan beads were confirmed by multi-plane analyses, including the FTIR and SEM techniques. It was shown that the immobilized enzyme in the proposed form of modification had good operational stability, which justifies its economical use in industry and biotechnology. It is particularly worth emphasizing that immobilized CDH maintains antioxidant, antibacterial, and cytotoxic properties at a level comparable to its free form. Based on the data obtained in this study, it can definitely be confirmed that chitosan is a promising and valuable material for the development of innovative immobilized CDH systems for biomedical applications or food packaging.

Supplementary Materials: The supporting information can be downloaded at: https://www.mdpi.com/article/10.3390/ijms24054535/s1. References [28,33,65–71] are cited in supplementary materials.

Author Contributions: Conceptualization, J.S. and M.O.-J.; methodology, J.S., M.O.-J., M.J., A.B. and A.O.; validation, M.O.-J., J.S. and W.P.-G.; investigation, J.S., M.O.-J., M.J., A.O., A.B. and W.P.-G.; resources, M.J.; writing—original draft preparation, J.S., M.O.-J. and M.J.; writing—review and editing, M.O.-J. and M.J.; visualization, W.P.-G. and A.O.; supervision, J.S.; funding acquisition, J.S. All authors have read and agreed to the published version of the manuscript.

Funding: This research was funded by the National Science Centre in Poland, grant number 2015/17/D/NZ9/02066.

Institutional Review Board Statement: Not applicable.

Informed Consent Statement: Not applicable.

Data Availability Statement: Not applicable.

Conflicts of Interest: The authors declare no conflict of interest.

References

1. Gao, Y.; Shah, K.; Kwok, I.; Wang, M.; Rome, L.H.; Mahendra, S. Immobilized fungal enzymes: Innovations and potential applications in biodegradation and biosynthesis. *Biotechnol. Adv.* **2022**, *57*, 107936. [CrossRef] [PubMed]
2. Janusz, G.; Pawlik, A.; Świderska-Burek, U.; Polak, J.; Sulej, J.; Jarosz-Wilkołazka, A.; Paszczyński, A. Laccase properties, physiological functions, and evolution. *Int. J. Mol. Sci.* **2020**, *21*, 966. [CrossRef] [PubMed]
3. Hartl, L.; Zach, S.; Seidl-Seiboth, V. Fungal chitinases: Diversity, mechanistic properties and biotechnological potential. *Appl. Microbiol. Biotechnol.* **2012**, *93*, 533–543. [CrossRef] [PubMed]
4. Tamayo-Cabezas, J.; Karboune, S. Optimizing immobilization and stabilization of feruloyl esterase from Humicola insolens and its application for the feruloylation of oligosaccharides. *Process Biochem.* **2020**, *98*, 11–20. [CrossRef]
5. Nousiainen, P.; Kontro, J.; Manner, H.; Hatakka, A.; Sipilä, J. Phenolic mediators enhance the manganese peroxidase catalyzed oxidation of recalcitrant lignin model compounds and synthetic lignin. *Fungal Genet. Biol.* **2014**, *72*, 137–149. [CrossRef]
6. Dubey, M.K.; Zehra, A.; Aamir, M.; Meena, M.; Ahirwal, L.; Singh, S.; Shukla, S.; Upadhyay, R.S.; Bueno-Mari, R.; Bajpai, V.K. Improvement strategies, cost effective production, and potential applications of fungal glucose oxidase (GOD): Current updates. *Front. Microbiol.* **2017**, *8*, 1032. [CrossRef]
7. Yang, J.; Xu, P.; Long, L.; Ding, S. Production of lactobionic acid using an immobilized cellobiose dehydrogenase/laccase system on magnetic chitosan spheres. *Process Biochem.* **2021**, *100*, 1–9. [CrossRef]
8. Østergaard, L.H.; Olsen, H.S. Industrial applications of fungal enzymes. In *Industrial Applications*; Springer: Berlin/Heidelberg, Germany, 2011; pp. 269–290.
9. Singh, R.; Kumar, M.; Mittal, A.; Mehta, P.K. Microbial enzymes: Industrial progress in 21st century. *3 Biotech.* **2016**, *6*, 174. [CrossRef]
10. McKelvey, S.M.; Murphy, R.A. Biotechnological use of fungal enzymes. *Fungi Biol. Appl.* **2017**, 201–225. [CrossRef]
11. Franssen, M.C.; Steunenberg, P.; Scott, E.L.; Zuilhof, H.; Sanders, J.P. Immobilised enzymes in biorenewables production. *Chem. Soc. Rev.* **2013**, *42*, 6491–6533. [CrossRef]
12. Homaei, A.A.; Sariri, R.; Vianello, F.; Stevanato, R. Enzyme immobilization: An update. *J. Chem. Biol.* **2013**, *6*, 185–205. [CrossRef] [PubMed]
13. Nguyen, H.H.; Kim, M. An overview of techniques in enzyme immobilization. *Appl. Sci. Converg. Technol.* **2017**, *26*, 157–163. [CrossRef]
14. Domínguez, A.; Gomez, J.; Lorenzo, M.; Sanromán, Á. Enhanced production of laccase activity by Trametes versicolor immobilized into alginate beads by the addition of different inducers. *World J. Microbiol. Biotechnol.* **2007**, *23*, 367–373. [CrossRef]
15. Wang, H.; Qian, J.; Ding, F. Recent advances in engineered chitosan-based nanogels for biomedical applications. *J. Mater. Chem. B* **2017**, *5*, 6986–7007. [CrossRef] [PubMed]
16. Pandey, A.R.; Singh, U.S.; Momin, M.; Bhavsar, C. Chitosan: Application in tissue engineering and skin grafting. *J. Polym. Res.* **2017**, *24*, 125. [CrossRef]
17. Desbrières, J.; Guibal, E. Chitosan for wastewater treatment. *Polym. Int.* **2018**, *67*, 7–14. [CrossRef]
18. Malerba, M.; Cerana, R. Recent advances of chitosan applications in plants. *Polymers* **2018**, *10*, 118. [CrossRef] [PubMed]
19. Krajewska, B. Application of chitin-and chitosan-based materials for enzyme immobilizations: A review. *Enzym. Microb. Technol.* **2004**, *35*, 126–139. [CrossRef]
20. Wang, K. Enzyme immobilization on chitosan-based supports. *Chitosan-Based Hydrogels* **2011**, *18*, 339–406.
21. Verma, M.L.; Kumar, S.; Das, A.; Randhawa, J.S.; Chamundeeswari, M. Enzyme immobilization on chitin and chitosan-based supports for biotechnological applications. *Sustain. Agric. Rev. 35* **2019**, 147–173. [CrossRef]
22. Zargar, V.; Asghari, M.; Dashti, A. A review on chitin and chitosan polymers: Structure, chemistry, solubility, derivatives, and applications. *ChemBioEng Rev.* **2015**, *2*, 204–226. [CrossRef]
23. Ribeiro, E.S.; de Farias, B.S.; Junior, T.R.S.A.C.; de Almeida Pinto, L.A.; Diaz, P.S. Chitosan–based nanofibers for enzyme immobilization. *Int. J. Biol. Macromol.* **2021**, *183*, 1959–1970. [CrossRef] [PubMed]

24. Hanefeld, U.; Gardossi, L.; Magner, E. Understanding enzyme immobilisation. *Chem. Soc. Rev.* **2009**, *38*, 453–468. [CrossRef] [PubMed]
25. Csarman, F.; Wohlschlager, L.; Ludwig, R. Cellobiose dehydrogenase. In *The Enzymes*; Elsevier: Amsterdam, The Netherlands, 2020; Volume 47, pp. 457–489.
26. Bao, W.; Renganathan, V. Cellobiose oxidase of Phanerochaete chrysosporium enhances crystalline cellulose degradation by cellulases. *FEBS Lett.* **1992**, *302*, 77–80. [CrossRef]
27. Henriksson, G.; Sild, V.; Szabó, I.J.; Pettersson, G.; Johansson, G. Substrate specificity of cellobiose dehydrogenase from Phanerochaete chrysosporium. *Biochim. Et Biophys. Acta (BBA)-Protein Struct. Mol. Enzymol.* **1998**, *1383*, 48–54. [CrossRef]
28. Sulej, J.; Osińska-Jaroszuk, M.; Jaszek, M.; Grąz, M.; Kutkowska, J.; Pawlik, A.; Chudzik, A.; Bancerz, R. Antimicrobial and antioxidative potential of free and immobilised cellobiose dehydrogenase isolated from wood degrading fungi. *Fungal Biol.* **2019**, *123*, 875–886. [CrossRef]
29. Imam, H.T.; Marr, P.C.; Marr, A.C. Enzyme entrapment, biocatalyst immobilization without covalent attachment. *Green Chem.* **2021**, *23*, 4980–5005. [CrossRef]
30. Romero-Fernández, M.; Paradisi, F. General overview on immobilization techniques of enzymes for biocatalysis. *Catal. Immobil. Methods Appl.* **2020**, 409–435.
31. Betancor, L.; López-Gallego, F.; Alonso-Morales, N.; Dellamora, G.; Mateo, C.; Fernandez-Lafuente, R.; Guisan, J.M. Glutaraldehyde in protein immobilization. In *Immobilization of Enzymes and Cells*; Springer: Berlin/Heidelberg, Germany, 2006; pp. 57–64.
32. Barbosa, O.; Ortiz, C.; Berenguer-Murcia, Á.; Torres, R.; Rodrigues, R.C.; Fernandez-Lafuente, R. Glutaraldehyde in bio-catalysts design: A useful crosslinker and a versatile tool in enzyme immobilization. *Rsc Adv.* **2014**, *4*, 1583–1600. [CrossRef]
33. Tegl, G.; Thallinger, B.; Beer, B.; Sygmund, C.; Ludwig, R.; Rollett, A.; Nyanhongo, G.S.; Guebitz, G.M. Antimicrobial cellobiose dehydrogenase-chitosan particles. *ACS Appl. Mater. Interfaces* **2016**, *8*, 967–973. [CrossRef]
34. Roig, M. *Immobilised Cells and Enzymes—A Practical Approach*; Woodward, J., Ed.; IRL Press: Oxford, UK, 1986; pp. 91, 177, 1985.£ 13; ISBN 947946-21-7.
35. Brena, B.M.; Batista-Viera, F. Immobilization of enzymes. In *Immobilization of Enzymes and Cells*, 2nd ed.; Guisan, J.M., Ed.; Humana Press Inc.: Totowa, NJ, USA, 2006; pp. 123–124.
36. Labus, K.; Wolanin, K.; Radosiński, Ł. Comparative study on enzyme immobilization using natural hydrogel matrices—Experimental studies supported by molecular models analysis. *Catalysts* **2020**, *10*, 489. [CrossRef]
37. Chen, H.; Zhang, Q.; Dang, Y.; Shu, G. The effect of glutaraldehyde cross-linking on the enzyme activity of immobilized β-galactosidase on chitosan bead. *Adv. J. Food Sci. Technol.* **2013**, *5*, 932–935. [CrossRef]
38. Gonawan, F.N.; Romli, M.; Zuhan, M.; Jaya, M. Immobilization of Candida Rugosa Lipase on the Glutaraldehyde-Activated Chitosan Beads. *J. Chem. Eng. Ind. Biotechnol.* **2022**, *8*, 33–41. [CrossRef]
39. Bilal, M.; Jing, Z.; Zhao, Y.; Iqbal, H.M. Immobilization of fungal laccase on glutaraldehyde cross-linked chitosan beads and its bio-catalytic potential to degrade bisphenol A. *Biocatal. Agric. Biotechnol.* **2019**, *19*, 101174. [CrossRef]
40. Mehandia, S.; Sharma, S.; Arya, S.K. Immobilization of laccase on chitosan-clay composite beads to improve its catalytic efficiency to degrade industrial dyes. *Mater. Today Commun.* **2020**, *25*, 101513. [CrossRef]
41. Helal, S.E.; Abdelhady, H.M.; Abou-Taleb, K.A.; Hassan, M.G.; Amer, M.M. Lipase from Rhizopus oryzae R1: In-depth characterization, immobilization, and evaluation in biodiesel production. *J. Genet. Eng. Biotechnol.* **2021**, *19*, 1–13. [CrossRef] [PubMed]
42. Zdarta, J.; Meyer, A.S.; Jesionowski, T.; Pinelo, M. A general overview of support materials for enzyme immobilization: Characteristics, properties, practical utility. *Catalysts* **2018**, *8*, 92. [CrossRef]
43. Islam, N.; Wang, H.; Maqbool, F.; Ferro, V. In vitro enzymatic digestibility of glutaraldehyde-crosslinked chitosan nanoparticles in lysozyme solution and their applicability in pulmonary drug delivery. *Molecules* **2019**, *24*, 1271. [CrossRef]
44. Li, B.; Shan, C.-L.; Zhou, Q.; Fang, Y.; Wang, Y.-L.; Xu, F.; Han, L.-R.; Ibrahim, M.; Guo, L.-B.; Xie, G.-L. Synthesis, characterization, and antibacterial activity of cross-linked chitosan-glutaraldehyde. *Mar. Drugs* **2013**, *11*, 1534–1552. [CrossRef]
45. Vijayalakshmi, K.; Devi, B.; Sudha, P.; Venkatesan, J.; Anil, S. Synthesis, characterization and applications of nanochitosan/sodium alginate/microcrystalline cellulose film. *J. Nanomed. Nanotechnol.* **2016**, *7*, 419. [CrossRef]
46. Song, C.; Yu, H.; Zhang, M.; Yang, Y.; Zhang, G. Physicochemical properties and antioxidant activity of chitosan from the blowfly Chrysomya megacephala larvae. *Int. J. Biol. Macromol.* **2013**, *60*, 347–354. [CrossRef] [PubMed]
47. Tsuboi, M. Infrared spectrum and crystal structure of cellulose. *J. Polym. Sci.* **1957**, *25*, 159–171. [CrossRef]
48. Pearson, F.; Marchessault, R.; Liang, C. Infrared spectra of crystalline polysaccharides. V. Chitin. *J. Polym. Sci.* **1960**, *43*, 101–116. [CrossRef]
49. Schwanninger, M.; Rodrigues, J.; Pereira, H.; Hinterstoisser, B. Effects of short-time vibratory ball milling on the shape of FT-IR spectra of wood and cellulose. *Vib. Spectrosc.* **2004**, *36*, 23–40. [CrossRef]
50. Krimm, S.; Bandekar, J. Vibrational spectroscopy and conformation of peptides, polypeptides, and proteins. *Adv. Protein Chem.* **1986**, *38*, 181–364. [PubMed]
51. Toldra, F.; Lequerica, J. Use of scanning electron microscopy techniques in the biocatalyst immobilization. *Food/Nahrung* **1991**, *35*, 739–744. [CrossRef]

52. Gür, S.D.; İdil, N.; Aksöz, N. Optimization of enzyme co-immobilization with sodium alginate and glutaraldehyde-activated chitosan beads. *Appl. Biochem. Biotechnol.* **2018**, *184*, 538–552. [CrossRef]
53. Singh, A.N.; Singh, S.; Suthar, N.; Dubey, V.K. Glutaraldehyde-activated chitosan matrix for immobilization of a novel cysteine protease, procerain B. *J. Agric. Food Chem.* **2011**, *59*, 6256–6262. [CrossRef]
54. Gilani, S.L.; Najafpour, G.D.; Moghadamnia, A.; Kamaruddin, A.H. Stability of immobilized porcine pancreas lipase on mesoporous chitosan beads: A comparative study. *J. Mol. Catal. B Enzym.* **2016**, *133*, 144–153. [CrossRef]
55. Sulej, J.; Janusz, G.; Osińska-Jaroszuk, M.; Małek, P.; Mazur, A.; Komaniecka, I.; Choma, A.; Rogalski, J. Characterization of cellobiose dehydrogenase and its FAD-domain from the ligninolytic basidiomycete Pycnoporus sanguineus. *Enzym. Microb. Technol.* **2013**, *53*, 427–437. [CrossRef]
56. Sulej, J.; Janusz, G.; Osińska-Jaroszuk, M.; Rachubik, P.; Mazur, A.; Komaniecka, I.; Choma, A.; Rogalski, J. Characterization of cellobiose dehydrogenase from a biotechnologically important Cerrena unicolor strain. *Appl. Biochem. Biotechnol.* **2015**, *176*, 1638–1658. [CrossRef] [PubMed]
57. Yen, M.-T.; Yang, J.-H.; Mau, J.-L. Physicochemical characterization of chitin and chitosan from crab shells. *Carbohydr. Polym.* **2009**, *75*, 15–21. [CrossRef]
58. Chatterjee, N.S.; Panda, S.K.; Navitha, M.; Asha, K.; Anandan, R.; Mathew, S. Vanillic acid and coumaric acid grafted chitosan derivatives: Improved grafting ratio and potential application in functional food. *J. Food Sci. Technol.* **2015**, *52*, 7153–7162. [CrossRef]
59. Podrepšek, G.H.; Primožič, M.; Knez, Ž.; Habulin, M. Immobilization of cellulase for industrial production. *Chem. Eng.* **2012**, *27*, 235–240.
60. Martins, A.F.; Facchi, S.P.; Monteiro, J.P.; Nocchi, S.R.; Silva, C.T.; Nakamura, C.V.; Girotto, E.M.; Rubira, A.F.; Muniz, E.C. Preparation and cytotoxicity of N, N, N-trimethyl chitosan/alginate beads containing gold nanoparticles. *Int. J. Biol. Macromol.* **2015**, *72*, 466–471. [CrossRef] [PubMed]
61. Baminger, U.; Nidetzky, B.; Kulbe, K.D.; Haltrich, D. A simple assay for measuring cellobiose dehydrogenase activity in the presence of laccase. *J. Microbiol. Methods* **1999**, *35*, 253–259. [CrossRef] [PubMed]
62. Bradford, M.M. A rapid and sensitive method for the quantitation of microgram quantities of protein utilizing the principle of protein-dye binding. *Anal. Biochem.* **1976**, *72*, 248–254. [CrossRef]
63. Paduch, R.; Matysik, G.; Wójciak-Kosior, M.; Kandefer-Szerszen, M.; Skalska-Kaminska, A.; Nowak-Kryska, M.; Niedziela, P. Lamium Album Extracts Express Free Radical Scavenging and Cytotoxic Activities. *Pol. J. Environ. Stud.* **2008**, *17*, 569–580.
64. Huber, D.; Tegl, G.; Mensah, A.; Beer, B.; Baumann, M.; Borth, N.; Sygmund, C.; Ludwig, R.; Guebitz, G.M. A dual-enzyme hydrogen peroxide generation machinery in hydrogels supports antimicrobial wound treatment. *ACS Appl. Mater. Interfaces* **2017**, *9*, 15307–15316. [CrossRef]
65. Augustin, M.; Ali-Vehmas, T.; Atroshi, F. Assessment of enzymatic cleaning agents and disinfectants against bacterial biofilms. *J. Pharm. Pharm. Sci.* **2004**, *7*, 55–64.
66. Ludwig, R.; Harreither, W.; Tasca, F.; Gorton, L. Cellobiose dehydrogenase: A versatile catalyst for electrochemical applications. *ChemPhysChem* **2010**, *11*, 2674–2697. [CrossRef] [PubMed]
67. Alt, E.; Leipold, F.; Milatovic, D.; Lehmann, G.; Heinz, S.; Schömig, A. Hydrogen peroxide for prevention of bacterial growth on polymer biomaterials. *Ann. Thorac. Surg.* **1999**, *68*, 2123–2128. [CrossRef] [PubMed]
68. Nyanhongo, G.S.; Thallinger, B.; Guebitz, G.M. Cellobiose dehydrogenase-based biomedical applications. *Process Biochem.* **2017**, *59*, 37–45. [CrossRef]
69. Thallinger, B.; Argirova, M.; Lesseva, M.; Ludwig, R.; Sygmund, C.; Schlick, A.; Nyanhongo, G.S.; Guebitz, G.M. Preventing microbial colonisation of catheters: Antimicrobial and antibiofilm activities of cellobiose dehydrogenase. *Int. J. Antimicrob. Agents* **2014**, *44*, 402–408. [CrossRef] [PubMed]
70. Öhlknecht, C.; Tegl, G.; Beer, B.; Sygmund, C.; Ludwig, R.; Guebitz, G.M. Cellobiose dehydrogenase and chitosan-based lysozyme responsive materials for antimicrobial wound treatment. *Biotechnol. Bioeng.* **2017**, *114*, 416–422. [CrossRef] [PubMed]
71. Wei, G.X.; Campagna, A.N.; Bobek, L.A. Effect of MUC7 peptides on the growth of bacteria and on *Streptococcus* mutans biofilm. *J. Antimicrob. Chemother.* **2006**, *57*, 1100–1109. [CrossRef] [PubMed]

International Journal of
Molecular Sciences

MDPI

Article

Structure and Microbiological Activity of 1*H*-benzo[*d*]imidazole Derivatives

Andrzej Olczak [1], Tomasz Pawlak [2,*], Sylwia Kałużyńska [1], Katarzyna Gobis [3], Izabela Korona-Głowniak [4], Katarzyna Suśniak [4], Marcin Zaborowski [5] and Małgorzata Szczesio [1]

1 Institute of General and Ecological Chemistry, Faculty of Chemistry, Lodz University of Technology, Zeromskiego 116, 90-924 Lodz, Poland
2 Centre of Molecular and Macromolecular Studies, Polish Academy of Science, Sienkiewicza 112, 90-363 Lodz, Poland
3 Department of Organic Chemistry, Faculty of Pharmacy, Medical University of Gdańsk, Gen. Hallera Ave. 107, 80-416 Gdańsk, Poland
4 Department of Pharmaceutical Microbiology, Faculty of Pharmacy, Medical University of Lublin, Chodźki Street 1, 20-093 Lublin, Poland
5 Dean Office's, Faculty of Chemistry, Lodz University of Technology, Zeromskiego 114, 90-543 Lodz, Poland
* Correspondence: tpawlak@cbmm.lodz.pl

Abstract: Three new crystal structures of 1*H*-benzo[*d*]imidazole derivatives were determined. In the structures of these compounds, an identical system of hydrogen bonds, C(4), was observed. Solid-state NMR was applied for testing the quality of the obtained samples. All of these compounds were tested for in vitro antibacterial activity against Gram-positive bacteria and Gram-negative bacteria, as well as antifungal activity, by checking their selectivity. ADME calculations indicate that the compounds can be tested as potential drugs.

Keywords: 1*H*-benzo[*d*]imidazole derivatives; antimicrobial activity; structure–activity relationship; X-ray; ADME; crossing the blood–brain barrier

Citation: Olczak, A.; Pawlak, T.; Kałużyńska, S.; Gobis, K.; Korona-Głowniak, I.; Suśniak, K.; Zaborowski, M.; Szczesio, M. Structure and Microbiological Activity of 1*H*-benzo[*d*]imidazole Derivatives. *Int. J. Mol. Sci.* **2023**, *24*, 3319. https://doi.org/10.3390/ijms24043319

Academic Editor: Dongho Kim

Received: 20 December 2022
Revised: 31 January 2023
Accepted: 5 February 2023
Published: 7 February 2023

1. Introduction

Many common infections today, such as pneumonia and urinary tract infection, do not respond to standard treatment. This extends treatment time and causes higher mortality. In the United States, drug-resistant bacteria cause over 2.8 million infections and approximately 35,000 deaths annually [1,2]. In Europe, AMR (antimicrobial resistance) causes approximately 33,000 deaths annually [3,4]. It is estimated that, in 2050, people will die more often from infections caused by antibiotic-resistant strains than from cancer [5]. The problem of drug resistance has been noticed by the CDC (Centers for Disease Control and Prevention) and the WHO (World Health Organization), as well as European and US governments. If coordinated global action is not taken immediately, we may face serious medical, social and economic complications in just a few years [1,2,4–6].

The extensive use, or even abuse, of antimicrobial drugs in medicine, animal husbandry and agriculture has resulted in strong selection pressure for the emergence and spread of various resistance mechanisms among bacteria [7]. According to a WHO report from 2022, the global problem of drug resistance is caused by bacteria that can cause both hospital and community-acquired infections. The most important ones among them are *Escherichia coli* (urinary tract infections, bacteremia/sepsis, nosocomial pneumonia and others)—resistant to third-generation cephalosporins and fluoroquinolones; *Klebsiella pneumoniae* (urinary tract infections, pneumonia, bacteraemia/sepsis and others)—resistant to third-generation cephalosporins and carbapenems; *Staphylococcus aureus* (bacteremia/sepsis, skin and soft tissue infections, osteoarthritis and others)—methicillin-resistant (MRSA) bacteria, resistant to all β-lactams except for ceftaroline and ceftobiprole);

Streptococcus pneumoniae (pneumonia, bacteremia/sepsis, meningitis and otitis media)—insensitive or resistant to penicillin; *Salmonella* (zoonotic salmonellosis)—resistant to fluoroquinolones; and *Shigella* (bacterial dysentery)—resistant to fluoroquinolones [6].

Drug-resistant strains of fungi, increasingly isolated in hospital environments, are also a problem. The fungus *Candida albicans* is most often responsible for infection, but more and more infections caused by other fungi are starting to appear [8]. It should be emphasized that, in recent years, the number of cases of fungemia caused by species other than *C. albicans* (*C. glabrata*, *C. krusei*, *C. parapsilosis*, *C. lusitaniae* and *C. tropicalis*) has increased by almost 2.5 times, which accounts for approx. 40% of fungi isolated from blood in ICUs (intensive care units) and up to 70% in hematology wards [9]. Unfortunately, the overuse of antifungal drugs, especially from the azole group, leads to the natural selection of resistant strains. Due to the fact that we do not have a very wide arsenal of agents to fight infections, their use should be thoughtful and prudent [10].

In the face of the spread of MDR (multidrug-resistant), XDR (extensively drug-resistant—sensitive to no more than two groups of antibiotics) and PDR (pandrug-resistant—resistant to all groups of antibiotics) microorganisms, it is necessary to implement a global strategy to prevent microbial drug resistance, mainly consisting in reducing the use of antimicrobial drugs; increasing the number of effective antimicrobial drugs active against resistant microbes, including the establishment of a global fund to support innovation in research on new drugs in early phases and non-commercial ones; promoting investment in new drugs and improving existing ones; and building a global coalition for action against drug resistance [4–6].

Therefore, we have undertaken the search for new compounds active against microorganisms resistant to antimicrobial drugs. Our interests include benzimidazoles, which exhibit multidirectional biological activity [11]. It was found, among others, that antibacterial and antifungal agents can be found in this group [12,13]. We focused in particular on benzimidazoles substituted at the C-2 position with cyclohexylethyl, cyclohexylpropyl and phenylpropyl moieties (Figure 1). These compounds were tested for tuberculostatic activity against *M. tuberculosis* [14]. These studies were conducted because drug-resistant and multidrug-resistant strains of *M. tuberculosis* have appeared [15,16]. The antitubercular activity of the **EJMCh-6** compound is very high, unlike that of others presented in this work. Additionally, this compound is not toxic to eukaryotic cells [14].

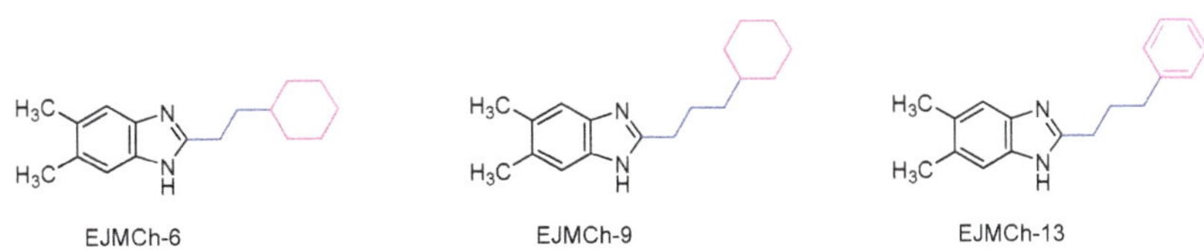

Figure 1. Structural formulas of the tested compounds.

The paper presents new crystal structures of three benzimidazoles substituted at the C-2 position with cyclohexylethyl, cyclohexylpropyl and phenylpropyl moieties (Figure 2). All of these compounds were tested for in vitro antibacterial activity against Gram-positive bacteria and Gram-negative bacteria, as well as antifungal activity, by checking their selectivity.

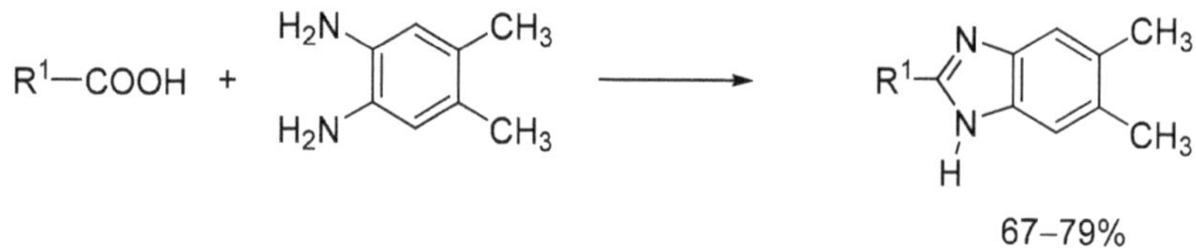

Figure 2. Synthesis of desired compounds.

2. Results and Discussion

2.1. Structure Analysis

Crystallographic data are included in Table 1, and the ORTEP representation of the structures is presented in Figure 3. Both **EJMCh-6** and **EJMCh-9** structures contain one molecule per asymmetric unit. Interestingly, they both show similar disorder in the cyclohexane ring (Figure 3). The ratios of the refined components are 0.79:0.21 and 0.84:0.16 for **EJMCh-6** and **EJMCh-9**, respectively. In contrast, the **EJMCh-13** compound contains three independent molecules in the asymmetric unit.

Table 1. Crystal data, data collection and refinement details of all compounds.

	EJMCh-6	EJMCh-9	EJMCh-13
Crystal data			
Chemical formula	$C_{17}H_{24}N_2$	$C_{18}H_{26}N_2$	$C_{18}H_{20}N_2$
M_r	256.38	270.41	264.36
Crystal system, space group	Monoclinic, *Cc*	Monoclinic, $P2_1/c$	Monoclinic, *Cc*
a, *b*, *c* (Å)	15.2276 (3), 12.7173 (3), 9.7501 (2)	14.2078 (4), 9.9210 (3), 11.5761 (4)	28.8210 (9), 15.4123 (5), 10.1758 (3)
β (°)	126.1959 (5)	104.7986 (9)	98.2484 (13)
V (Å^3)	1523.74 (6)	1577.59 (9)	4473.3 (2)
Z	4	4	12
Data collection			
No. of measured, independent and observed [$I > 2\sigma(I)$] reflections	15,809, 2939, 2936	3012, 3012, 2963	31367, 7814, 7731
R_{int}	0.020	0.024	0.022
$(\sin \theta/\lambda)_{max}$ (Å^{-1})	0.617	0.618	0.618
Refinement			
$R[F^2 > 2\sigma(F^2)]$, $wR(F^2)$, S	0.026, 0.072, 1.06	0.034, 0.092, 1.05	0.033, 0.090, 1.05
No. of reflections	2939	3012	7814
No. of parameters	234	240	547
No. of restraints	2	30	2
Δ_{max}, Δ_{min} (eÅ^{-3})	0.19, −0.12	0.22, −0.14	0.28, −0.20
Absolute structure	Flack x determined using 1439 quotients [(I+)-(I−)]/[(I+)+(I−)] [17].	-	Flack x determined using 3275 quotients [(I+)-(I−)]/[(I+)+(I−)] [17].
Absolute structure parameter	0.07 (5)	-	0.09 (8)

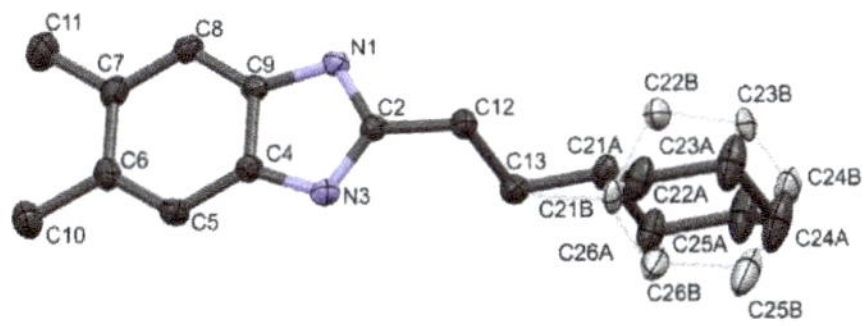

EJMCh-6

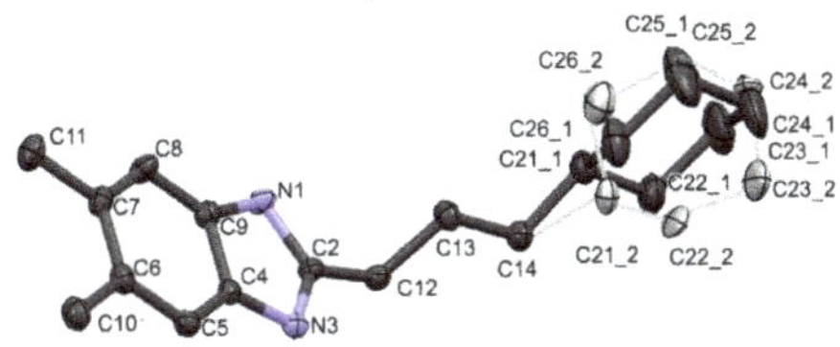

EJMCh-9

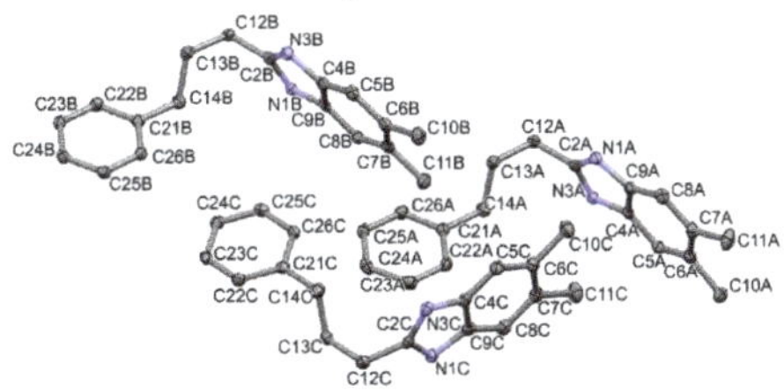

EJMCh-13

Figure 3. Molecular structures of all compounds showing atom-labeling schemes. Displacement ellipsoids are drawn at the 50% probability level. Drawings were prepared with Mercury software 2022.3.0 [18]. Light gray color (**EJMCh-6** and **EJMCh-9**) shows alternative conformations of disordered rings.

In all studied structures, the same pattern of hydrogen bonds of type N1-H1 … N3 is adopted (Figure 4). The system of hydrogen bonds observed in all structures is described as a chain, C(4), according to hydrogen-bond graph–set theory [19]. Interestingly, each of the independent molecules only forms a chain with its symmetrical counterparts, and all three chains align along the [001] crystallographic direction.

In the CSD (Cambridge Structural Database) [20], there is only one analogous structure, with code HIQXIR [21]. It has a phenyl end group and an ethyl linker (CH_2-CH_2). This structure, unlike the other three, is a hydrate in the crystal state. All the above-mentioned compounds with an ethyl linker have a stretched form (Figure 5), while the longer, three-carbon ones have a bent form. This is one of the factors that may affect the antibacterial activity of the tested compounds. A second factor affecting the activity may be the presence of an aromatic ring.

Sample Quality Verification via Solid-State NMR

The quality of the samples was tested by employing the solid-state NMR technique. Figure 6 shows the ^{13}C CP MAS NMR spectra of all studied samples. From the analysis of the experimental data, it is clear that all of them represent well-crystalline structures. Counting the number of isotropic resonance lines in the ^{13}C CP MAS NMR spectra of **EJMCh-6** and **EJMCh-9,** it is apparent that the number of molecules in the asymmetric unit is the same as that determined in the X-ray study. **EJMCh-13** should have a visible triplication of signals, at least for some positions. However, this is not the case, likely due to the fact that all three molecules in the asymmetric unit cell have very similar conformation and interatomic distances, resulting in no visible triplication of signals. Another reason could be the high broadening of the signals.

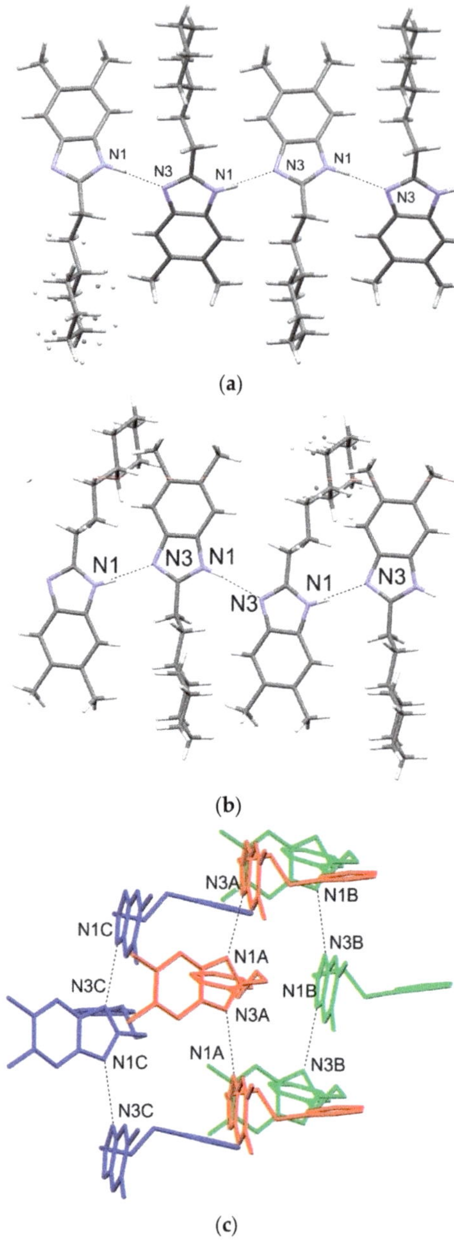

Figure 4. (**a**) Intermolecular hydrogen bonds in **EJMCh-6**. (**b**) Intermolecular hydrogen bonds in **EJMCh-9** (**c**) Intermolecular hydrogen bonds in **EJMCh-13**. The colors indicate three different molecules in the asymmetric unit and their crystallographic symmetry equivalents.

2.2. ADME Analysis

Bioavailability radars for all studied compounds were made (Figure 7). For drug-like properties, the compounds were found to have a good bioavailability score (0.55). Six physicochemical properties are considered on the bioavailability radar: lipophilicity, size, polarity, solubility, flexibility and saturation. Compound **EJMCh-13**, which contains the phenyl group instead of cyclohexane, has a better lipophilicity index, while the saturation index (fraction of carbons in the sp3 hybridization not less than 0.25) remains within a satisfactory range. All the compounds conform to the rules of Lipinski [22], Ghose [23], Egan [24], Veber [25] and Muegge [26] and thus are good drug candidates. The logKp values of the tested compounds range from −4.48 cm/s to −3.58 cm/s, and the more negative logKp is, the less the molecule penetrates the skin.

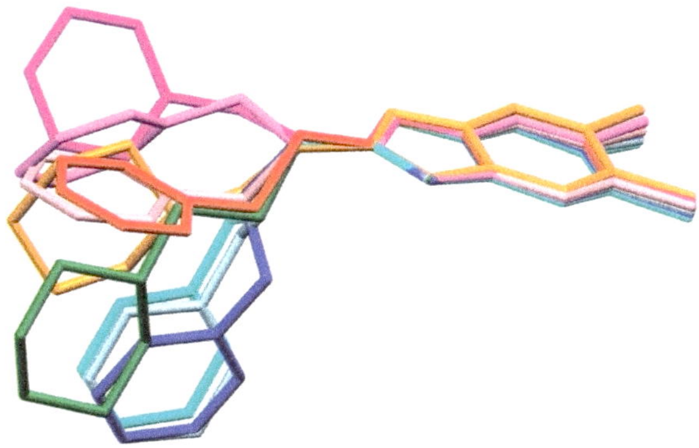

Figure 5. Superimposition of the studied molecules and HIQXIR (HIQXIR—red, pink, magenta and violet; **EJMCh-9**—green; **EJMCh-6**—orange; **EJMCh-13**—light blue, blue and cyan).

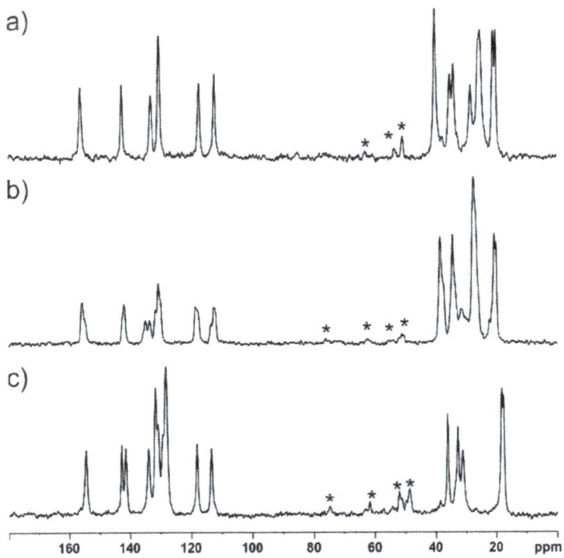

Figure 6. ^{13}C CP MAS spectra of **EJMCh-6** (**a**), **EJMCh-9** (**b**) and **EJMCh-13** (**c**) recorded at a spinning rate of 8 kHz. The recycle delay was 3 s. Asterisks indicate spinning sidebands.

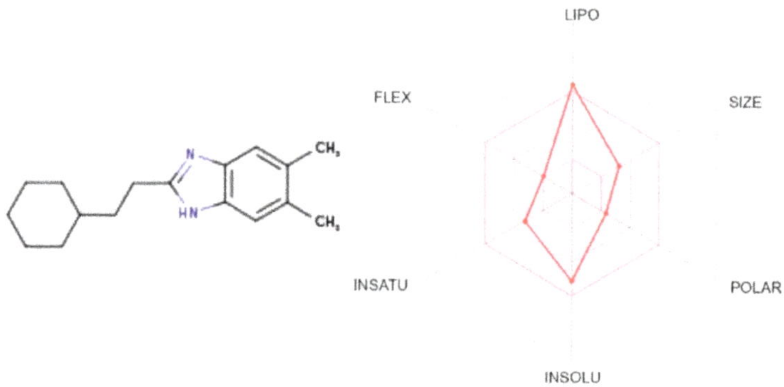

EJMCh-6

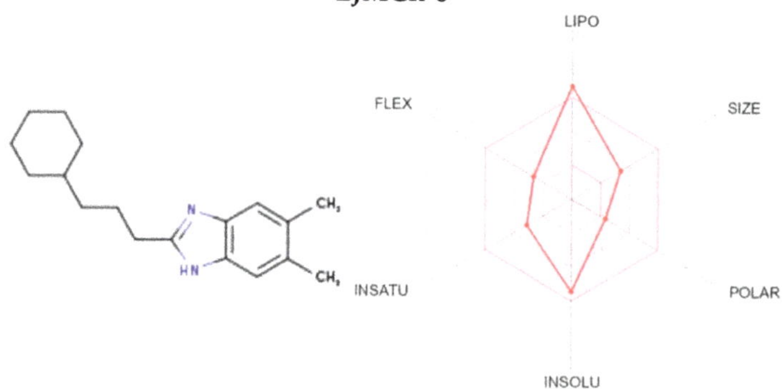

EJMCh-9

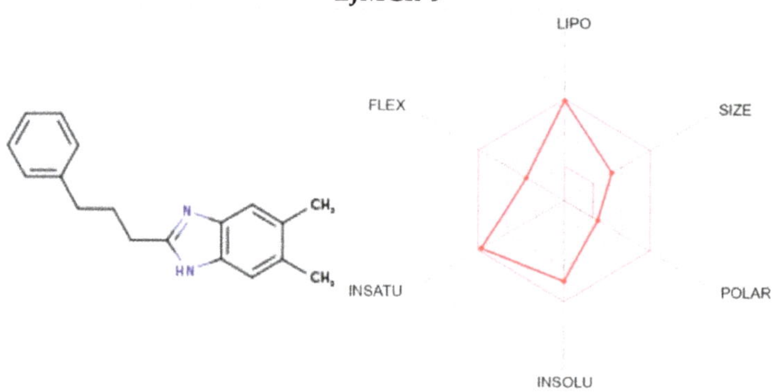

EJMCh-13

Figure 7. Bioavailability radars of all compounds.

The BOILED-Egg diagram (brain or intestinal estimated permeation predictive model) indicates that all compounds can be absorbed in the gastrointestinal tract, which may make them effective drugs (Figure 8). The analysis confirmed the effectiveness of the tested compounds in crossing the blood–brain barrier. This may be of particular importance in the treatment of tuberculous meningitis, which is the most severe and life-threatening form of tuberculosis. Despite the cure, more than half of patients suffer from permanent damage to the nervous system [27].

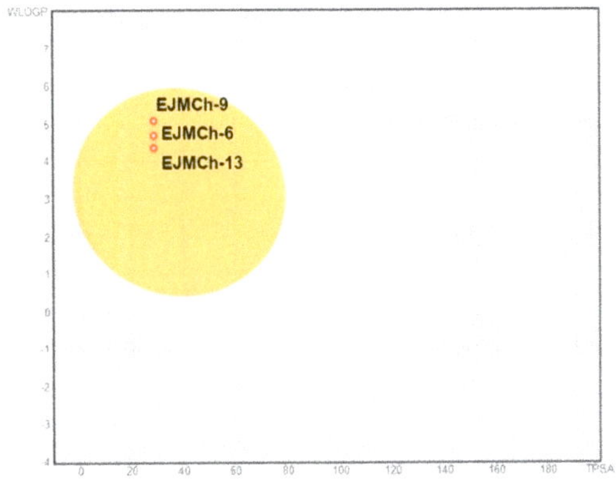

Figure 8. BOILED-Egg diagram for all compounds (lipophilicity (WLOGP) and polarity (tPSA)). Human intestinal absorption—white area; blood–brain barrier permeation—yellow area.

2.3. Antimicrobial Activity

The increase in and spread of antimicrobial resistance to available drugs necessitate a search for new molecules for the treatment of bacterial infections. Benzimidazole and its derivatives are the most potent classes of molecules against microorganisms. They were found as heterocyclic aromatic compounds with a variety of biological activities, such as anti-inflammatory, antiparasitic, antimalarial, antimycobacterial and antiviral activities, which was nicely reviewed by Tahlan et al. [28]. There are a number of derivatives evaluated for their antimicrobial activity against selected microbial species: 2-substituted-1H-benzimidazole derivatives, 2-mercaptobenzimidazole derivatives, 2-mercaptobenzimidazole and β-lactum segment derivatives containing –CONH–, new benzimidazoles bearing 2-pyridone, 3,4-dihydro triazino[1,2-a]benzimidazole compounds and novel substituted benzimidazole carboxamidine [28]. Most of these novel derivatives revealed good antimicrobial activity, ranging from MIC = 0.39–0.78 mg/L in the new class of benzimidazole and phenyl-substituted benzyl ethers or 0.12–0.5 mg/L in new bis-benzimidazole diamidine compounds to MIC = 100–250 mg/L in novel series of pyrido[1,2-a]benzimidazole derivatives [28].

The results of the antibacterial and antifungal activities of benzimidazoles substituted at the C-2 position with cyclohexylethyl, cyclohexylpropyl and phenylpropyl moieties tested in this study are presented in Table 2 as MICs, i.e., the lowest concentrations of compounds that prevent visible growth of the organism, and MBCs, i.e., the lowest concentrations that result in a ≥99.9% reduction in the microorganism inocula upon subculture with compound-free medium. Vancomycin, ciprofloxacin and nystatin were used as the standard drugs. **EJMCh-6** was found to be without bioactivity against the tested reference strains. Noticeably, the tested **EJMCh-13** and **EJMCh-9** showed no bioactivity against Gram-negative bacteria (*E. coli* ATCC 25922, S. Typhimurium ATCC 14028, *K. pneumoniae* ATCC 13883 and *P. aeruginosa* ATCC 9027; MIC >1000 mg/L) or mild bioactivity (*P. mirabilis* ATCC 12453) at minimal inhibition concentration (MIC = 250). The MIC values for Gram-positive reference bacteria indicate strong (MIC 15.6 mg/L) anti-staphylococcal (*S. aureus* ATCC 25923 and *S. epidermidis* ATCC 12228) and very strong anti–micrococcal (*M. luteus* ATCC 10240) activities of the **EJMCh-13** compound. The tested **EJMCh-9** and **-13** compounds also showed strong activity against enterococci and spore-forming pathogen *Bacillus cereus*. **EJMCh-9** presented good bioactivity against *S. aureus* strains (including MRSA) and very strong activity against reference *S. epidermidis* and *M. luteus* strains. The high values of the MBC/MIC ratio (8–16) for **EJMCh-9** and **EJMCh-13** indicate their bacteriostatic activities, except for the bactericidal activity of **EJMCh-13** against *S. epidermidis* and *B. cereus* reference strains (MBC/MIC 1). Surprisingly, the antifungal bioactivity (*C.*

albicans ATCC 2091, *C. glabrata* ATCC 90030 and *C. parapsilosis* ATCC 22019) of the tested compounds was not observed.

Table 2. Antimicrobial activity of tested chemicals presented as minimal inhibitory concentrations (MICs) and minimal bactericidal concentrations (MBCs) or minimal fungicidal concentrations (MFCs) against bacteria and yeasts, respectively.

Chemical / Microorganism	EJMCh-6		EJMCh-9		EJMCh-13		Standard Drug	
	mg/L						Vancomycin	
Gram-positive bacteria	MIC	MBC	MIC	MBC	MIC	MBC	MIC	MBC
S. aureus ATCC 25923	>1000	Nd	31.25	1000	15.6	125	0.98	0.98
S. aureus ATCC BAA-1707 *	>1000	Nd	31.25	1000	15.6	1000	0.98	0.98
S. epidermidis ATCC 12228	>1000	Nd	7.8	1000	15.6	15.6	0.98	0.98
M. luteus ATCC 10240	>1000	Nd	7.8	1000	7.8	62.5	0.12	0.12
E. faecalis ATCC 29212	>1000	Nd	15.6	1000	15.6	1000	1.95	3.9
B. cereus ATCC 10876	>1000	Nd	15.6	1000	15.6	15.6	0.98	15.6
Gram-negative bacteria							Ciprofloxacin	
S. typhimurium ATCC 14028	>1000	Nd	>1000	>1000	>1000	Nd	0.061	0.06
E. coli ATCC 25922	>1000	Nd	>1000	>1000	>1000	Nd	0.015	0.08
P. mirabilis ATCC 12453	>1000	Nd	250	1000	>1000	Nd	0.030	0.03
K. pneumoniae ATCC 13883	>1000	Nd	>1000	>1000	>1000	Nd	0.122	0.24
P. aeruginosa ATCC 9027	>1000	Nd	>1000	>1000	>1000	Nd	0.488	0.98
Yeasts	MIC	MFC	MIC	MFC	MIC	MFC	Nystatin	
C. albicans ATCC 102231	>1000	Nd	>1000	>1000	>1000	>1000	0.48	0.48
C. parapsilosis ATCC 22019	>1000	Nd	>1000	>1000	>1000	>1000	0.24	0.48
C. glabrata ATCC 90030	>1000	Nd	>1000	>1000	500	>1000	0.24	0.48

* Methicillin-resistant Staphylococcus aureus (MRSA); Nd—not detected.

The mechanism of action of the tested derivatives against Gram-positive and Gram-negative bacteria is not yet clearly understood. The biological activity against these two groups of bacteria could be different due to their cell wall structure differences and thus the difference in permeability. Peptidoglycan is a major component (90%) of the Gram-positive cell wall, whereas in Gram-negative bacteria, peptidoglycan, constituting 10% of the cell wall, lies between the cytoplasmic membrane and the outer lipid bilayer containing lipopolysaccharide, porins and adhesins, which create an additional barrier to overcome. The yeast cell wall is a characteristic structure of fungi and is mainly composed of glucans, chitin and glycoproteins, forming a different type of barrier. The dissimilar activity against Gram-positive bacteria of **EJMCh-9** and **EJMCh-13** may result from their different lipophilicity, stacking interactions and the possibility of hydrogen bonding.

3. Materials and Methods

3.1. Chemistry

The compounds were synthesized using simple synthesis from an appropriate diamine and carboxylic acid according to the methods previously described: Method A (**EJMCh-6**): carboxylic acid (1.5 equiv.), 5,6-dimethy-1,2-diaminobenzene (1 equiv.), and PPA (polyphosphoric acid) at 180–200 °C; $NaHCO_3/H_2O$. Method B (**EJMCh-9** and **EJMCh-13**): carboxylic acid (1.5 equiv.) and 5,6-dimethyl-1,2-dimanobenzene (1 equiv.) at 160–180 °C; $NaOH/H_2O$ (Figure 2) [14]. The IR spectra are shown in Supporting Information: Figure S1 (**EJMCh-6**), Figure S2 (**EJMCh-9**), Figure S3 (**EJMCh-13**).

3.2. X-ray Study

Single crystals of compounds suitable for X-ray diffraction were obtained with the slow evaporation of solvents at room temperature from methanol–DMF (1:1 v/v) solutions. The diffraction measurements were performed with Bruker SMART APEXII CCD Diffractometer (Bruker AXS Inc., Madison, WI, USA) at low temperature (100 K) and under CuK_α (1.54184 Å) radiation. The diffraction data were processed with SAINT ver. 8.34A, XPREP ver. 2014/2 and SADABS ver. 2014/4 for structures **EJMCh-6** and **EJMCh-13**, and TWINABS ver. 2008/4 for structure **EJMCh-9** (Bruker AXS Inc., Madison, WI, USA). Crystal structure solution and refinement were carried out with SHELX [29,30] and SHELXLE for visualization [31]. All H atoms (except for those engaged in hydrogen bonds) were geometrically optimized and considered riding atoms, with distances appropriate for 100 K temperatures and with $U_{iso}(H) = 1.2\ U_{eq}(C, N)$. The methyl H atoms were refined with $U_{iso}(H) = 1.5\ U_{eq}(C)$. CCDC 2231627, 2231628 and 2231629 contain the supplementary crystallographic data for this paper. The data are provided free of charge by The Cambridge Crystallographic Data Center at www.ccdc.cam.ac.uk/structures (accessed on 13 December 2022). The graphics of the crystal structures were obtained with Mercury [18]. PublCIF was used in data preparation [32]. In the aliphatic rings, structures **EJMCh-6** and **EJMCh-9** displayed some disorder, which was resolved with DSR [33].

3.3. NMR Study

^{13}C cross-polarization magic-angle spinning (CP MAS) NMR experiments were carried out with an Avance III spectrometer equipped with HX MAS probe heads of 4 mm, operating at 400.19 and 100.63 for ^1H and ^{13}C, respectively. The MAS frequency was 8 kHz. The CP MAS experiments were performed with the proton 90° pulse length of 3.4 μs and a contact time of 2 ms. For cross-polarization, the nutation frequency was 63 kHz with a ^1H ramp shape from 90% to 100%. A sample of U-^{13}C, ^{15}N-labeled histidine hydrochloride was used to set the Hartmann–Hahn condition. A total of 3.5k data points were acquired for the spectral width of 50 kHz. There were no special preparations of samples prior to the solid-state NMR study. The ^1H and ^{13}C NMR liquid-state spectra of all samples are shown in supplementary material information. All the spectra were recorded in DMSO-d_6 with a Neo Bruker 400 MHz spectrometer and are consistent with the results provided in [14], where we investigated the same set of samples. **EJMCh-6** ^1H NMR (Figure S4): δ 0.86–0.94 (m, 2H, CH_2), 1.10–1.23 (m, 4H, $2CH_2$), 1.60–1.75 (m, 6H $3CH_2$ and 1H CH), 2.27 (s, 6H, $2CH_3$), 2.75 (t, 2H, CH_2), 7.14 (s, 1H, ArH), 2.25 (s, 1H, ArH), 11.8 (s, 1H, NH); ^{13}C NMR (Figure S5): δ 20.4, 26.2, 26.5, 26.6, 33.0. 35.7, 37.1, 111.3, 118.7, 129.1, 129.9, 133.2, 142.4, 154.8. **EJMCh-9** ^1H NMR (Figure S6): δ 0.81 (br m, 2H, CH_2), 1.17 (br m, 6H, $3CH_2$), 1.63–1.70 (br m, 7H, $3CH_2$ and 1H CH), 2.24 (s, 6H, $2CH_3$), 2.68 (br m, 2H, CH_2), 3.33 (br m, 2H, CH_2), 7.12 (s, 1H, ArH), 7.23 (s, 1H, ArH), 11.84 (s, 1H, NH); ^{13}C NMR (Figure S7): δ 21.9, 27.1, 27.9, 28.2, 30.9, 34.8, 38.6, 38.8, 112.8, 120.3, 130.6, 131.5, 134.8, 144.0, 156.2. **EJMCh-13** ^1H NMR (Figure S8): δ 2.04 (m, 2H, CH_2), 2.27 (s, 6H, $2CH_3$), 2.64 (t, 2H, CH_2), 2.76 (t, 2H, CH_2), 7.15–7.17 (m, 2H, ArH), 7.19–7.23 (m, 3H, ArH), 7.30 (s, 2H, ArH), 11.90 (s, 1H, NH); ^{13}C NMR (Figure S9): δ 20.4, 28.5, 29.8, 35.1, 111.3, 118.8, 126.3, 128.8, 128.9, 129.2, 130.1, 133.2, 142.1, 142,4, 154.2.

3.4. ADME

The compounds were analyzed for their pharmacokinetic properties, drug likeness and absorption. ADME (absorption, distribution, metabolism and excretion) analysis was performed using the SwissADME service 2022 [34,35] and BOILED-Egg—to predict the gastrointestinal absorption and brain penetration of molecules [36].

3.5. In Vitro Antibacterial Activity Assay

All target compounds were screened for antibacterial and antifungal activities using the 2-fold micro-dilution broth method. The minimal inhibitory concentration (MIC) of the tested compounds was evaluated for a panel of reference Gram-positive bacteria (*Staphy-*

lococcus aureus ATCC 25923, *S. aureus* ATCC BAA-1707, *Staphylococcus epidermidis* ATCC 12228, *Micrococcus luteus* ATCC 10240 and *Bacillus cereus* ATCC 10876) and Gram-negative bacteria (*Salmonella* Typhimurium ATCC 14028, *Escherichia coli* ATCC 25922, *Proteus mirabilis* ATCC 12453, *Klebsiella pneumoniae* ATCC 13883, *Pseudomonas aeruginosa* ATCC 9027 and yeasts: *Candida albicans* ATCC 102231, *Candida parapsilosis* ATCC 22019 and *Candida glabrata* ATCC 90030). The procedure for conducting antimicrobial activity testing was previously described in detail [37]. The solutions of the tested compounds in dimethylosulfoxide (DMSO) were suspended in Mueller–Hinton broth for bacteria or Mueller–Hinton broth with 2% glucose for fungi. Then, series of two-fold dilutions were carried out in sterile 96-well polystyrene microtitrate plates (Nunc, Roskilde, Denmark), obtaining concentrations from 1000 to 7.8 μg/mL in the appropriate medium. Simultaneously, the inocula of 24 h cultures of microorganisms in sterile physiological saline (0.5 McFarland standard density) were prepared and added to each well, obtaining the final density of 1.5 × 105 CFU/mL for bacteria and 5 × 10^4 CFU/mL for yeasts; CFU—colony forming units. Proper positive (inoculum without tested compound) and negative (compound without inoculum) controls were added to each microplate. Vancomycin, ciprofloxacin and nystatin were used as the standard reference reagents.

After incubation (35 °C, 24 h) the growth of microorganisms was measured spectrophotometrically at 600 nm. MICs were marked at the lowest dilution of 1H-benzo[d]imidazole derivatives without the growth of bacteria or yeasts.

Then, 5 μL of the suspension from each well, including controls, was subcultured on the agar plates in order to determine the minimal bactericidal concentration (MBC) or minimal fungicidal concentration (MFC). The plates were incubated at 35 °C for 24 h. The MBC/MFC was determined at the lowest concentration of derivatives inhibiting the growth of microbes. The MBC/MIC index was also calculated to show the bacteriostatic or bactericidal effect of the tested compounds.

4. Conclusions

Studies on the selectivity of the biological activity of compounds with known antitubercular activity were carried out. The tuberculostatic activity of **EJMCh-6** is very high, unlike that of other compounds presented in this work. On the other hand, in **EJMCh-9** and **EJMCh-13**, significant activity against other Gram-positive bacteria can be observed. The presence of an odd propyl linker may cause an increase in activity against other bacteria. The molecules of this compound are bent, unlike the molecules with an ethyl linker, which are more stretched. It seems that the decisive factor regarding the biological activity of the tested compounds is the length of the linker. Compounds with a longer linker have more conformational freedom, which can make them more flexible to fit the binding site. Despite the conformational differences, compounds with propyl and ethyl linkers form the same arrangement of hydrogen-bond chains in the crystal. The in silico ADME analysis shows that all three tested 1H-benzo[*d*]imidazole derivatives are good candidates as potential drugs.

Supplementary Materials: The following supporting information can be downloaded at: https://www.mdpi.com/article/10.3390/ijms24043319/s1.

Author Contributions: Conceptualization, A.O., T.P., K.G., I.K.-G. and M.S.; synthesis, K.G.; antimicrobial activity, I.K.-G., S.K. and K.S.; NMR methodology, T.P.; X-ray and ADME methodology, A.O., S.K., M.Z. and M.S.; writing—original draft preparation, A.O., T.P., K.G., I.K.-G. and M.S.; visualization, T.P., K.G., M.Z. and M.S. All authors have read and agreed to the published version of the manuscript.

Funding: This research received no external funding.

Institutional Review Board Statement: Not applicable.

Informed Consent Statement: Not applicable.

Data Availability Statement: Data are contained within the article.

Conflicts of Interest: The authors declare no conflict of interest.

References

1. Centers for Disease Control and Prevention. Antibiotic Resistance Threats in the United States. 2019. Available online: https://www.cdc.gov/drugresistance/pdf/threats-report/2019-ar-threats-report-508.pdf (accessed on 20 December 2022).
2. President's Council of Advisors on Science and Technology. Report to the President on Combating Antibiotic Resistance. Available online: https://obamawhitehouse.archives.gov/sites/default/files/microsites/ostp/PCAST/pcast_carb_report_sept2014.pdf (accessed on 20 December 2022).
3. Sriram, A.; Kalanxhi, E.; Kapoor, G.; Craig, J.; Balasubramanian, R.; Brar, S.; Criscuolo, N.; Hamilton, A.; Klein, E.; Tseng, K.; et al. State of the World's Antibiotics, 2021. Washington, DC: Center for Disease Dynamics, Economics, and Policy. 2021. Available online: https://onehealthtrust.org/wp-content/uploads/2021/02/The-State-of-the-Worlds-Antibiotics-in-2021.pdf (accessed on 20 December 2022).
4. European Centre for Disease Prevention and Control. Antimicrobial Resistance in the EU/EEA (EARS-Net). Annual Epidemiological Report for 2019. Available online: https://www.ecdc.europa.eu/sites/default/files/documents/AER-EARS-Net%202019-final.pdf (accessed on 20 December 2022).
5. O'Neill, J. Tackling Drug-Resistant Infections Globally: Final Report and Recommendations. Review on Antimicrobial Resistance, May 2016. Available online: https://amr-review.org/sites/default/files/160518_Final%20paper_with%20cover.pdf (accessed on 20 December 2022).
6. World Health Organization: Antimicrobial Resistance: Global Report on Surveillance. 2022. Available online: https://www.who.int/publications/i/item/9789240062702 (accessed on 20 December 2022).
7. Andersson, D.I.; Hughes, D. Evolution of antibiotic resistance at non-lethal drug concentrations. *Drug Resist. Update.* **2012**, *15*, 162–172. [CrossRef] [PubMed]
8. Hammarskjöld, F.; Mernelius, S.; Andersson, R.E.; Berg, S.; Hanberger, H.; Löfgren, S.; Malmvall, B.-E.; Petzold, M.; Matussek, A. Possible transmission of Candida albicans on an intensive care unit: Genotype and temporal cluster analyses. *J. Hosp. Infect.* **2013**, *85*, 60–65. [CrossRef] [PubMed]
9. Pfaller, M.A.; Diekema, D.J. Epidemiology of invasive candidiasis: A persistent public health problem. *Clin. Microbiol. Rev.* **2007**, *20*, 133–163. [CrossRef] [PubMed]
10. Siscar-Lewin, S.; Hube, B.; Brunke, S. Emergence and evolution of virulence in human pathogenic fungi. *Trends Microbiol.* **2022**, *30*, 693–704. [CrossRef]
11. Pathare, B.; Bansode, T. Review- biological active benzimidazole derivatives. *Results Chem.* **2021**, *3*, 100200. [CrossRef]
12. Dokla, E.M.E.; Abutaleb, N.S.; Milik, S.N.; Li, D.; El-Baz, K.; Shalaby, M.-A.W.; Al-Karaki, R.; Nasr, M.; Klein, C.D.; Abouzid, K.A.M.; et al. Development of benzimidazole-based derivatives as antimicrobial agents and their syn-ergistic effect with colistin against gram-negative bacteria. *Eur. J. Med. Chem.* **2020**, *186*, 111850. [CrossRef]
13. Huang, D.; Qiu, F.; Zhang, Z.; Shi, L.; Cao, C.; Ke, S. Synthesis and antifungal activity of substituted 2-Aryl benzimidaz-oles derivatives. *J. Het. Chem.* **2019**, *56*, 2494–2498. [CrossRef]
14. Gobis, K.; Foks, H.; Serocki, M.; Augustynowicz-Kopeć, E.; Napiórkowska. Synthesis and evaluation of in vitro antimycobacterial activity of novel 1H-benzo[d]imidazole derivatives and analogues. *Eur. J. Med. Chem.* **2015**, *89*, 13–20. [CrossRef]
15. *Global Tuberculosis Report 2022*; World Health Organization: Geneva, Switzerland, 2022; Available online: https://www.who.int/publications/i/item/9789240061729 (accessed on 20 December 2022).
16. Yew, W.-W. Management of multidrug-resistant tuberculosis and extensively drug-resistant tuberculosis: Current status and future prospects. *Kekkaku* **2011**, *86*, 9–16.
17. Parsons, S.; Flack, H.D.; Wagner, T. Use of intensity quotients and differences in absolute structure refinement. *Acta Cryst.* **2013**, *B69*, 249–259. [CrossRef]
18. Macrae, C.F.; Sovago, I.; Cottrell, S.J.; Galek, P.T.A.; McCabe, P.; Pidcock, E.; Platings, M.; Shields, G.P.; Stevens, J.S.; Towler, M.; et al. Mercury 4.0: From visualization to analysis, design and prediction. *J. Appl. Cryst.* **2020**, *53*, 226–235. [CrossRef] [PubMed]
19. Bernstein, J.; Davis, R.E.; Shimoni, L.; Chang, N.-L. Patterns in Hydrogen Bonding Functionality and Graph Set Analysis in Crystals. *Angew. Chem. Int. Ed. Engl.* **1995**, *34*, 1555–1573. [CrossRef]
20. Groom, C.R.; Bruno, I.J.; Lightfoot, M.P.; Ward, S.C. The Cambridge Structural Database. *Acta Cryst.* **2016**, *B72*, 171–179. [CrossRef] [PubMed]
21. Główka, M.L.; Kałużyńska, S.; Krause, M.; Gobis, K.; Foks, H.; Szczesio, M.; Olczak, A. The structures of benzimidazole de-rivatives and their potential as tuberculostatics. *Acta Cryst.* **2018**, *C74*, 1684–1691.
22. Lipinski, C.A.; Lombardo, F.; Dominy, B.W.; Feeney, P.J. Experimental and computational approaches to estimate solubility and permeability in drug discovery and development settings. *Adv. Drug Deliv. Rev.* **2001**, *46*, 3–26. [CrossRef]
23. Ghose, A.K.; Viswanadhan, V.N.; Wendolski, J.J. A knowledge-based approach in designing combinatorial or medicinal chemistry libraries for drug discovery. 1. A qualitative and quantitative characterization of known drug databases. *J. Comb. Chem.* **1999**, *1*, 55–68. [CrossRef]
24. Egan, W.J.; Merz, K.M.; Baldwin, J.J. Prediction of drug absorption using multivariate statistics. *J. Med. Chem.* **2000**, *3*, 3867–3877. [CrossRef]

25. Veber, D.F.; Johnson, S.R.; Cheng, H.Y.; Smith, B.R.; Ward, K.W.; Kopple, K.D. Molecular properties that influence the oral bioavailability of drug candidates. *J. Med. Chem.* **2002**, *45*, 2615–2623. [CrossRef]
26. Muegge, I.; Heald, S.L.; Brittelli, D. Simple selection criteria f or drug-like chemical matter. *J. Med. Chem.* **2001**, *44*, 1841–1946. [CrossRef]
27. Huynh, J.; Donovan, J.; Hoan Phu, N.; Dang Trung Nghia, H.; Thuy Thuong Thuong, N.; Thwaites, G.E. Tuberculous men-ingitis: Progress and remaining questions. *Lancet Neurol.* **2022**, *21*, 450–464. [CrossRef]
28. Tahlan, S.; Kumar, S.; Narasimhan, B. Antimicrobial potential of 1H-benzo[d]imidazole scaffold: A review. *BMC Chem.* **2019**, *13*, 18–40. [CrossRef] [PubMed]
29. Sheldrick, G.M. SHELXT—Integrated space-group and crystal-structure determination. *Acta Cryst.* **2015**, *C71*, 3–8. [CrossRef] [PubMed]
30. Sheldrick, G.M. A short history of SHELX. *Acta Cryst.* **2008**, *A64*, 112. [CrossRef] [PubMed]
31. Hübschle, C.B.; Sheldrick, G.M.; Dittrich, B. ShelXle: A Qt graphical user interface for SHELXL. *J. Appl. Cryst.* **2011**, *44*, 1281–1284. [CrossRef] [PubMed]
32. Westrip, S.P. publCIF: Software for editing, validating and formatting crystallographic information files. *J. Apply. Cryst.* **2010**, *43*, 920–925. [CrossRef]
33. Kratzert, D.; Krossing, I. Recent improvements in DSR. *J. Appl. Cryst.* **2018**, *51*, 928–934. [CrossRef]
34. Daina, A.; Michielin, O.; Zoete, V. SwissADME: A free web tool to evaluate pharmacokinetics, drug-likeness and medicinal chemistry friendliness of small molecules. *Sci. Rep.* **2017**, *7*, 42717. [CrossRef]
35. Daina, A.; Michielin, O.; Zoete, V. iLOGP: A simple, robust, and efficient description of n-octanol/water partition coefficient for drug design using the GB/SA approach. *J. Chem. Inf. Model.* **2014**, *54*, 3284–3301. [CrossRef]
36. Daina, A.; Zoete, V. A BOILED-Egg To Predict Gastrointestinal Absorption and Brain Penetration of Small Molecules. *ChemMedChem* **2016**, *11*, 1117–1121. [CrossRef]
37. Zesławska, E.; Korona-Głowniak, I.; Szczesio, M.; Olczak, A.; Zylewska, A.; Tejchman, W.; Malm, A. Structural analysis and antimicrobial activity of 2[1H]-pyrimidinethione/selenone derivatives. *J. Mol. Struct.* **2017**, *1142*, 261–266. [CrossRef]

International Journal of *Molecular Sciences*

MDPI

Article

Design, Synthesis and Antimicrobial Evaluation of New *N*-(1-Hydroxy-1,3-dihydrobenzo[*c*][1,2]oxaborol-6-yl)(hetero)aryl-2-carboxamides as Potential Inhibitors of Mycobacterial Leucyl-tRNA Synthetase

Petr Šlechta [1], Adam Anthony Needle [1], Ondřej Jand'ourek [2], Pavla Paterová [3], Klára Konečná [2], Pavel Bárta [4], Jiří Kuneš [5], Vladimír Kubíček [4], Martin Doležal [1] and Marta Kučerová-Chlupáčová [1,*]

1 Department of Pharmaceutical Chemistry and Pharmaceutical Analysis, Faculty of Pharmacy in Hradec Králové, Charles University, 500 05 Hradec Králové, Czech Republic
2 Department of Biological and Medical Sciences, Faculty of Pharmacy in Hradec Králové, Charles University, 500 05 Hradec Králové, Czech Republic
3 Department of Clinical Microbiology, University Hospital, 500 05 Hradec Králové, Czech Republic
4 Department of Biophysics and Physical Chemistry, Faculty of Pharmacy in Hradec Králové, Charles University, 500 05 Hradec Králové, Czech Republic
5 Department of Organic and Bioorganic Chemistry, Faculty of Pharmacy in Hradec Králové, Charles University, 500 05 Hradec Králové, Czech Republic
* Correspondence: kucerom@faf.cuni.cz

Abstract: Tuberculosis remains a serious killer among infectious diseases due to its incidence, mortality, and occurrence of resistant mycobacterial strains. The challenge to discover new antimycobacterial agents forced us to prepare a series of *N*-(1-hydroxy-1,3-dihydrobenzo[*c*][1,2]oxaborol-6-yl)(hetero)aryl-2-carboxamides 1–19 via the acylation of 6-aminobenzo[*c*][1,2]oxaborol-1(3*H*)-ol with various activated (hetero)arylcarboxylic acids. These novel compounds have been tested in vitro against a panel of clinically important fungi and bacteria, including mycobacteria. Some of the compounds inhibited the growth of mycobacteria in the range of micromolar concentrations and retained this activity also against multidrug-resistant clinical isolates. Half the maximal inhibitory concentrations against the HepG2 cell line indicated an acceptable toxicological profile. No growth inhibition of other bacteria and fungi demonstrated selectivity of the compounds against mycobacteria. The structure–activity relationships have been derived and supported with a molecular docking study, which confirmed a selectivity toward the potential target leucyl-tRNA synthetase without an impact on the human enzyme. The presented compounds can become important materials in antimycobacterial research.

Keywords: antimicrobial; antimycobacterial; benzoxaborole; cytotoxicity; molecular docking; multidrug-resistant tuberculosis; tuberculosis

Citation: Šlechta, P.; Needle, A.A.; Jand'ourek, O.; Paterová, P.; Konečná, K.; Bárta, P.; Kuneš, J.; Kubíček, V.; Doležal, M.; Kučerová-Chlupáčová, M. Design, Synthesis and Antimicrobial Evaluation of New N-(1-Hydroxy-1,3-dihydrobenzo [c][1,2]oxaborol-6-yl)(hetero)aryl-2-carboxamides as Potential Inhibitors of Mycobacterial Leucyl-tRNA Synthetase. *Int. J. Mol. Sci.* **2023**, *24*, 2951. https://doi.org/10.3390/ijms24032951

Academic Editor: Helena Felgueiras

Received: 31 December 2022
Revised: 31 January 2023
Accepted: 1 February 2023
Published: 2 February 2023

1. Introduction

Tuberculosis (TB), a life-threatening infectious disease, is one of the leading causes of death around the world. TB was ranked as the second cause of death from a single infectious agent in 2021 and even surpassed HIV. TB is caused by the bacillus *Mycobacterium tuberculosis* (*Mtb*), which is spread via droplets from infected people. An estimated 10.6 million people fell ill in 2021 with TB, which is a slight increase from 10.1 million in 2020 [1]. The current treatment for TB requires 6–9 months of long combinational therapy consisting of four first-line drugs such as rifampicin, isoniazid, pyrazinamide, and ethambutol. Unfortunately, emerging drug resistance reveals drug-resistant TB strains that continue to be a health threat; multidrug-resistant TB (MDR-TB) [2] and extensively drug-resistant TB (XDR-TB) [3] are especially causing most of the first-line and second-line anti-TB drugs to be ineffective. This creates an urgent need for the development of new antimycobacterial agents with a unique mechanism of action.

(3*H*)-Benzo[*c*][1,2]oxaborol-1-ol (Figure 1, often abbreviated in publications as benzoxaborole) is a derivative of boronic acid combining a cyclic hemiester moiety with a free hydroxyl group. Boron in the molecule represents a strong Lewis acidic center and the benzoxaboroles are considered more acidic than arylboronic acids [4]. Benzoxaboroles show chemical stability and high hydrolytic resistance of the boron–carbon bond. Boron-containing compounds are able either to create an ester bond or a dative bonding to the active site of enzymes [5]. Synthesis of the unsubstituted (3*H*)-benzo[*c*][1,2]oxaborol-1-ol was published in 1957 [6] and it was originally called boronophthalide [7]. Although it was synthesized 65 years ago, it had not been employed in drug design and discovery until the last two decades. The biological activities of benzoxaboroles have been reviewed in several publications [4,8–10]. Tavaborole (Figure 1) is the first clinically used drug in the field of antimicrobial therapy approved for the treatment of onychomycoses by the US Food and Drug Administration in 2014 [11]. The mechanism of action is the inhibition of leucyl-tRNA synthetase (LeuRS) via the oxaborole tRNA-trapping (OBORT) mechanism. The OBORT mechanism utilizes the ability of the boron atom to bond to the *cis*-diols of the 3′-terminal adenosine nucleotide Ade76 of $tRNA^{Leu}$. The resulting covalent adduct traps the 3′ end of $tRNA^{Leu}$ in the editing site of LeuRS and creates a nonproductive complex. This causes inhibition of leucylation and, thus, the inhibition of protein synthesis [12]. In recent years, a series of 3-aminomethylbenzoxaboroles targeting *Mtb* LeuRS by the OBORT mechanism has been identified. Thorough research focused on these derivatives led to detailed structure–activity relationship (SAR) studies and the identification of a potent *Mtb* LeuRS inhibitor with oral bioavailability and in vivo efficacy, with the 3-aminomethylbenzoxaborole derivative GSK656 (Figure 1, IC_{50} against LeuRS = 0.20 μM) submitted into clinical trials [13,14], currently in Phase 2 [15]. However, the majority of the candidates struggled with potential toxicity issues caused by the inhibition of mammalian cytoplasmic LeuRS [13]. In the work of Hu and colleagues [16], a distinctive structural difference between the editing domain of bacterial (*Streptococcus pneumoniae*) and human cytosolic LeuRS was mentioned as well as the fact that the human cytosolic LeuRS editing domain is more compact due to the presence of four additional eukaryote-specific insertion [17]. One of these additional insertions covers the opening of the editing pocket and causes the binding of more bulky compounds to be difficult [13]. We expected that a similar difference could be found between the editing domain of mycobacterial LeuRS and human cytosolic LeuRS and should serve as valuable information for the design of new selective derivatives. Different biological targets have been also described for benzoxaboroles in bacteria and fungi, e.g., β-lactamases, D,D-carboxypeptidase [4], or NADH dehydrogenase [18]; however, LeuRS is the most often mentioned and is probably the only well-described target in mycobacteria [8]. Benzoxaborole-derived compounds with excellent antimycobacterial activity [19–22] and concurrently LeuRS inhibitory activity [23–26] have been published. Many compounds have been patented and some of them reached clinical trials. Benzoxaboroles might be also suitable for surface functionalization modification or employment into co-polymers with PEG or polyacrylamide [4]. General inhibitors of microbial LeuRS have been summarized in a review article by Zhang and Ma in 2019 [27].

The design of our compounds started with the experience of the above-mentioned research group [13]. The bidentate covalent adduct of benzoxaborole with Ade76 of tRNA may be mimicked by an adduct with adenosine monophosphate (AMP) [28,29]. Under the proper inspection of the crystallographic structure of *Mtb* H37Rv LeuRS co-crystallized with a confirmed inhibitor in the form of the spiro adduct with AMP (PDB ID: 5AGR, see Figure 2), it is visible that most of the H-bond interactions occur between the merged ribose part of the adduct and amino acid residues T336 and T337 of the threonine-rich region. There are several other H-bond interactions and hydrophobic contacts between L432 and Y435 of the AMP binding loop and AMP. Finally, the research group described three additional key interactions of the amino group of (*S*)-3-(aminomethyl)-

7-ethoxybenzo[c][1,2]oxaborol-1(3*H*)-ol (Figure 2) to carboxylic acid side chains of D447, D450, and the carbonyl group of M441. The 7-ethoxy substitution was important for packing with the Ade76 ribose, thus further stabilizing the boron-tRNA adduct. The research group concluded an extensive SAR study on the interaction with *Mtb* LeuRS using isothermal titration calorimetry. Halogen substitution (Cl/Br) in position 4 of benzoxaborole can enhance its affinity to the *Mtb* LeuRS editing domain [13]. In the following study, the authors concluded that the positions C-6 and C-7 are solvent-exposed and may be used for further structural modification to improve the selectivity. Large aromatic substitution in position 7 led to a decrease in antimycobacterial activity. For the modification of C-6, only aliphatic substitution was applied [14].

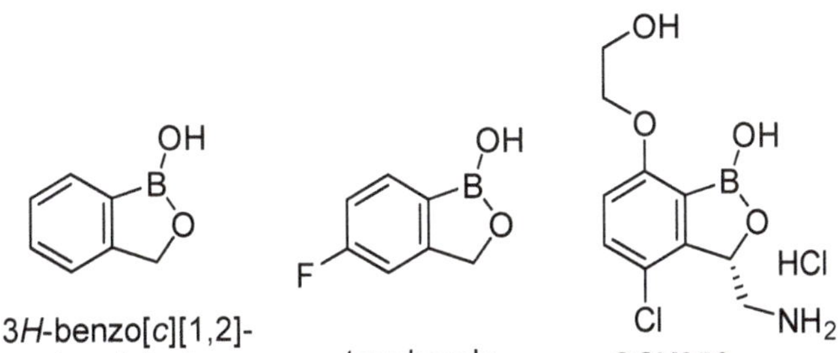

Figure 1. 3*H*-benzo[*c*][1,2]oxaborol-1-ol and its important derivatives with antimicrobial activity.

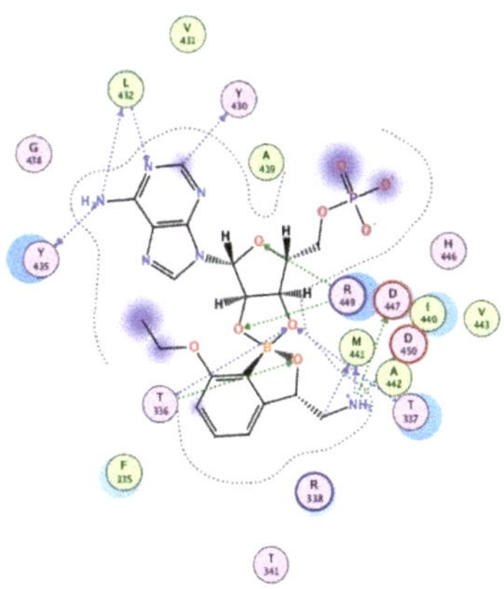

Figure 2. Ligand interaction diagram of *Mtb* H37Rv LeuRS and (*S*)-3-(aminomethyl)-7-ethoxybenzo [c][1,2]-oxaborol-1(3*H*)-ol in the form of the bivalent adduct with AMP (PDB ID: 5AGR).

2. Results and Discussion

2.1. In Silico Studies

To explore the hypothesis on the difference between human cytosolic LeuRS and *Mtb* H37Rv LeuRS, we first superimposed crystallographic structures of the complex of *Mtb* H37Rv LeuRS (mtbLeuRS) editing domain (PDB ID: 5AGR deposited by Palencia et al. [13]) and human cytosolic LeuRS (hLeuRS) editing domain (PDB ID: 2WFD deposited by Seiradake et al. [17]). For the reason of the significant difference in the sequences of human and mycobacterial editing domains, the superimposition of the enzymes was performed according to the structural motifs and then the alignment of amino acid sequences was performed to match the superimposition, otherwise the superimposition was not successful. From the superimposition (Figure 3a), a significant difference in the binding sites of these two enzymes was observed. The binding pocket of the human cytosolic enzyme was closed over by an additional alfa-helix (R457EKLAEAKEKIYLKGFYE474), which was missing in the mycobacterial enzyme.

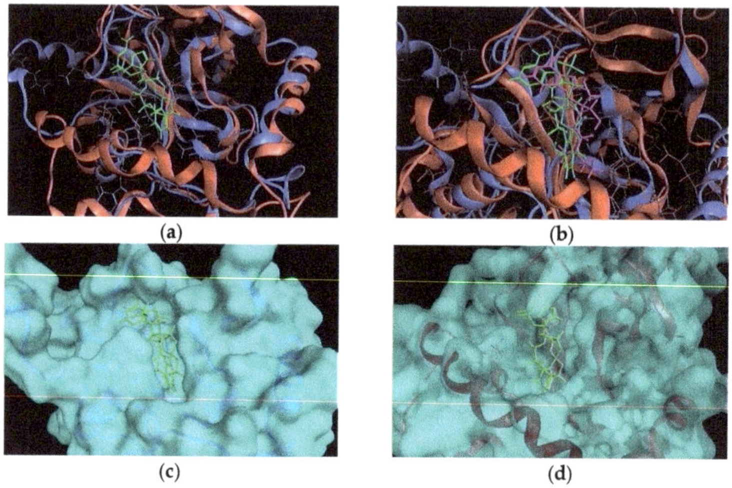

Figure 3. (**a**) Superimposition of 5AGR mtbLeuRS (blue color) with co-crystallized adduct (*S*)-3-(aminomethyl)-7-ethoxybenzo[*c*][1,2]oxaborol-1(3*H*)-ol-AMP as a ligand (light green color) and 2WFD hLeuRS (pink color) with highlighted additional alfa-helix (red color). (**b**) Superimposition of 5AGR mtbLeuRS (blue color) with ligand (green color) and 6LPF hLeuRS (pink color) with co-crystallized 2′(L-norvalyl)amino-2′-deoxyadenosine and 5′-*O*-(L-leucylsulfamoyl)adenosine as a ligand (purple color) and highlighted additional alfa-helix (red color). (**c**) 5AGR mtbLeuRS with a calculated surface area of the enzyme (light blue color) and co-crystallized adduct (light green color). (**d**) 2WFD hLeuRS with a calculated surface area of the enzyme (light blue color) and co-crystallized adduct from 5AGR (light green color).

Due to the poor resolution of the crystallographic structure of 2WFD and poor sequence similarity of mtbLeuRS and hLeuRS, we decided to superimpose the 5AGR with another crystallographic structure of human cytosolic LeuRS co-crystallized with 2′-(L-norvalyl)amino-2′-deoxyadenosine and 5′-*O*-(L-leucylsulfamoyl)adenosine (PDB ID: 6LPF deposited by Liu et al. [30]) to check whether the position of the adenine core remains in the same position within the binding pocket of hLeuRS. The superimposition of 5AGR with 6LPF had to be performed according to the structural motifs as well. The superimposition of mtbLeuRS and the other human enzyme 6LPF (Figure 3b) revealed an analogic position of ligands in both enzymes and, again, an additional alfa-helix (R457EKLAEAKEKIYLKGFYE474) that closes over the pocket of human LeuRS was detected.

To better rendering the binding pocket, the surface area of the enzymes was calculated, which is depicted in Figure 3c for the mtbLeuRS 5AGR and Figure 3d for the hLeuRS 2WFD.

With the aim to investigate if the planned compounds (**1**–**19**) (Table 1) would fit in the mtbLeuRS and whether the large substitution would provide important additional interactions with mycobacterial enzyme, we ran docking studies. As our compounds were expected to act via the OBORT mechanism [12], we assumed that they should form a bivalent adduct with Ade76 and bound to the enzyme as the co-crystalized ligand does to 5AGR. Therefore, a library of bivalent adducts of the compounds with AMP was created with the stereochemistry corresponding to the co-crystalized ligand to mimic Ade76. To keep the interaction between the AMP moiety of the adduct and the enzyme unchanged, template docking was set with the AMP core and induced fit receptor. The score values for every unique molecule were in the range from −10.318074 to −8.267931 (Table S1).

Table 1. Calculated lipophilicity of the prepared compounds (log*P*), comparison of their antimycobacterial activity (MIC) with standards, and their cytotoxicity against HepG2 (IC_{50}).

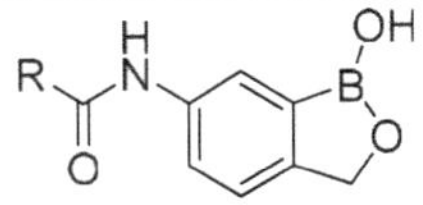

		Log*P*	MIC						IC_{50}
Comp.	**R**		***Mtb* H37Ra (μM)**	***Mtb* H37Rv (μM)**	***M. kansasii* (μM)**	***M. avium* (μM)**	***M. aurum* (μM)**	***M. smeg.* (μM)**	**HepG2 (μM)**
1	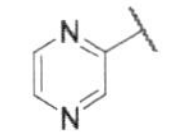	0.42	122.53	49.01	30.62	122.53	15.33	30.62	>1000
2	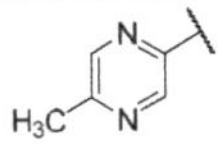	0.92	58.07	46.46	58.07	58.07	14.53	29.03	>500
3	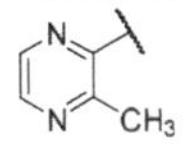	0.58	>1851.4	92.91	58.07	>1851.4	14.53	14.53	>500
4	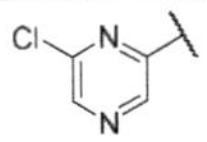	1.16	107.95	86.36	**26.98**	**26.98**	13.51	26.98	>1000
5	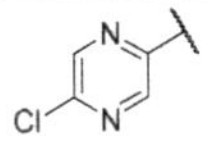	1.16	**26.98**	**21.59**	107.95	53.98	13.51	53.98	82.44
6	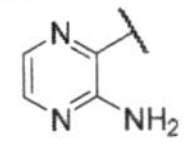	0.67	>1851.47	>370.29	>1851.47	>1851.47	14.48	462.87	>100
7	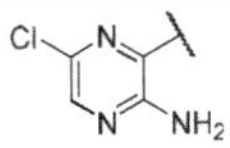	1.40	>1851.47	>328.41	>1851.47	410.51	>1851.47	410.51	>100
8	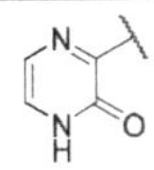	−0.61	230.59	92.24	230.59	57.65	28.82	28.82	>1000
9	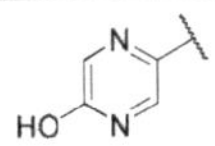	1.05	115.30	184.48	115.30	230.59	230.59	57.65	>1000
10	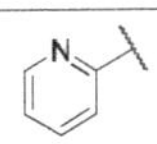	1.20	61.50	49.20	30.74	61.50	15.39	61.50	>1000
11	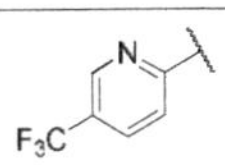	2.19	97.03	**9.72**	**12.14**	97.03	24.25	24.25	>100
12	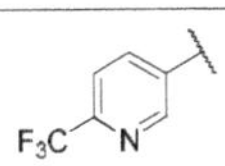	1.84	194.07	155.26	194.07	388.14	**12.14**	24.25	>250
13	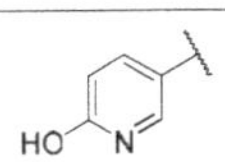	1.19	462.88	>370.30	115.72	462.88	115.72	115.72	>1000
14	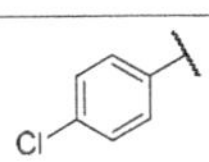	2.52	434.77	43.48	217.39	434.77	13.60	217.39	>1000

Table 1. *Cont.*

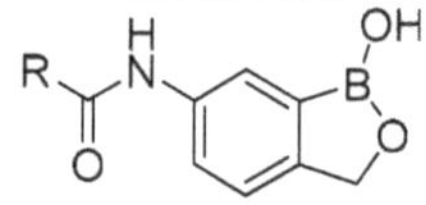

Comp.	R	Log*P*	MIC						IC$_{50}$
			Mtb H37Ra (μM)	*Mtb* H37Rv (μM)	*M. kansasii* (μM)	*M. avium* (μM)	*M. aurum* (μM)	*M. smeg.* (μM)	HepG2 (μM)
15	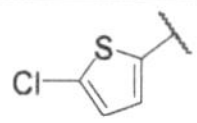	2.27	53.23	42.59	106.46	425.85	53.23	851.71	210.5
16	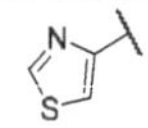	1.06	120.16	96.13	30.03	240.32	15.03	30.03	>1000
17	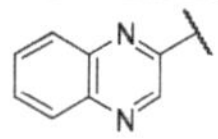	2.01	>819.40	**20.49**	**25.60**	102.43	**12.82**	**12.82**	>250
18	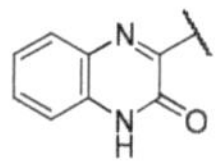	0.98	>1851.47	>311.43	>1851.47	48.66	24.32	24.32	243.7
19	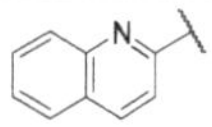	2.59	205.52	82.21	205.52	**25.68**	25.68	25.68	466.7
	INH	-	3.65	2.84	45.57	7291.87	28.51	227.87	na
	RIF	-	0.00759	na	0.06	0.08	0.47	15.19	na
	CIP	-	1.51	na	0.75	1.51	0.05	0.19	na
	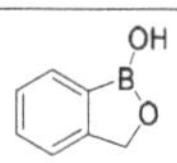 *		na	55.99 *	na	na	na	na	na
	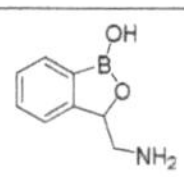 *		na	11.04 *	na	na	na	na	na

Used standards: INH = isoniazid, RIF = rifampicin, CIP = ciprofloxacin; * compounds synthesized and evaluated by Palencia et al. [13]; na = not available.

The retrieved poses matched the thesis, no steric clashes were observed that would indicate any problems with the binding mode of the mtbLeuRS (5AGR) with the compounds. Some pyrazine- and pyridine-substituted derivatives, that were bearing the nitrogen atom in the neighboring position in the (hetero)aryl moiety bound to the carboxamide group, exhibited specific H-bond interaction between the nitrogen atom of the heterocycle and D447 through a water molecule. The same amino acid residue formed one of the crucial interactions with the aminomethyl group in derivatives described by Palencia et al. [13] (Figure 4). In some cases, the aromatic substitution also exhibited an additional arene–H interaction with the hydrogen atom from T476.

To complete our thesis, the results from the template-docking into mtbLeuRS were compared with the hLeuRS (PDB ID: 2WFD) in a superimposition. It seems that the additional alfa-helix (R^{457}EKLAEAKEKIY^{468}LKGFYE^{474}) in the human enzyme would cause serious steric clashes with the largely substituted compounds (Figure 5a), especially due to the presence of Y468, which is more obvious in the rendering with the calculated surface of 2WFD (Figure 5b).

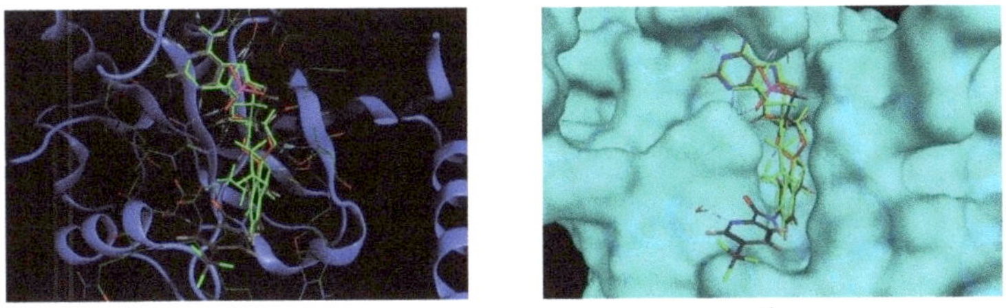

(**a**) (**b**)

Figure 4. (**a**) The pose of one of the template-docked compounds, compound **11** (grey) into mtbLeuRS (PDB ID: 5AGR, the original co-crystallized ligand in green) interacting with D447. (**b**) The same pose with the calculated surface.

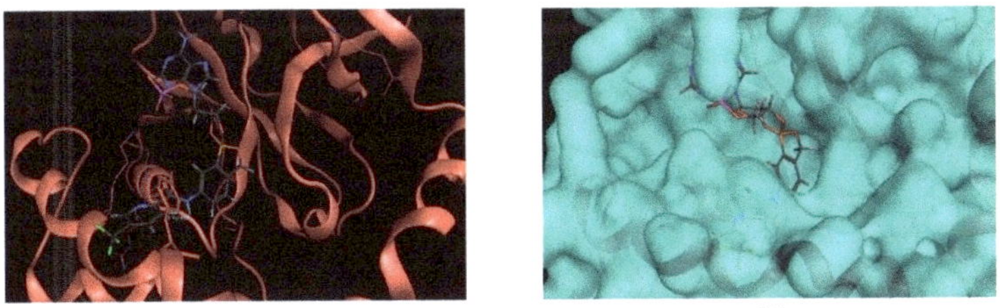

(**a**) (**b**)

Figure 5. Superimposition of the hLeuRS (PDB ID: 2WFD) visualized with docking results of compound **11** (grey color). (**a**) 2WFD hLeuRS (pink color) visualized without calculated surface with highlighted Y468 (red color). (**b**) 2WFD hLeuRS visualized with the calculated surface.

Based on the computational studies, it can be concluded that the large aromatic substitution in position 6 of the benzoxaborole moiety should be beneficial for the selectivity of the compounds toward LeuRS of *Mtb* H37Rv and may provide some additional interactions, whereas the planned compounds should not be able to bind to the human LeuRS, as they should not be able to fit in the relatively small binding pocket.

2.2. Synthesis of the Compounds

The final compounds **1–19** (Table 1) were prepared by the acylation of 6-aminobenzo[*c*][1,2]oxaborol-1(3*H*)-ol hydrochloride with the corresponding activated (hetero)arylcarboxylic acid in an overall yield of 19–98%. The compounds have been characterized using melting points, ^{1}H-NMR, ^{13}C-NMR, ^{11}B-NMR, IR, and MS spectra. Their purity was checked using elemental analysis or HPLC. In the ^{13}C NMR spectra, the carbon attached to boron was usually not observed due to the quadrupolar relaxation of ^{10}B and ^{11}B nuclei. The ^{13}C NMR spectrum of compound **8** was not obtained in sufficient quality due to the low solubility of the compound. The shift of the boron signal in the ^{11}B-NMR spectra was found within the range of 30.36–35.07 ppm, which is in accordance with published data (32 ppm [4]).

In the IR spectra, a sharp absorption band was observed for valence vibration of the O-H bond at the frequencies 3460–3298 cm^{-1}, which is comparable to the literature data 3400–3200 cm^{-1} [7] and 3450–3350 cm^{-1} [31] for benzoxaborole. For some compounds, a broad absorption band for valence vibration N-H bond in amide appeared in the range of 3376–3298 cm^{-1}. Typical absorption bands were detected for all the compounds for the deformation vibration of carbonyl at the frequencies 1706–1631 cm^{-1} and for the deformation vibration of the N-H bond in amide at the frequencies 1541–1523 cm^{-1}. Apart from the typical absorption band of the hydroxy group attached to benzoxaborole, there are two other typical bands, which are the absorbance of the C-O bond producing an intensive band of valency vibration at the frequencies 1001–956 cm^{-1} comparable to the literature value of 985–970 cm^{-1} [7] and 1005–970 cm^{-1} [31] and the absorbance of the

B-O bond presenting at frequencies 1394–1361 cm^{-1} comparable to literature values of 1475–1375 cm^{-1} [7] and 1380–1340 cm^{-1} [31].

2.3. *In Vitro Antimycobacterial Activity*

2.3.1. Antimycobacterial Activity against Collection Strains of Mycobacteria

All the synthesized compounds **1–19** were evaluated on their antimycobacterial activity using microplate Alamar Blue assay (MABA, [32]) against pathogenic *Mycobacterium tuberculosis* H37Rv CNCTC My 331/88 (ATCC 27294), avirulent strain *Mycobacterium tuberculosis* H37Ra ITM-M006710 (ATCC 9431), fast-growing *M. smegmatis* DSM 43,465 (ATCC 607), *M. aurum* DSM 43,999 (ATCC 23366), and non-tuberculous mycobacteria *M. avium* DSM 44,156 (ATCC 25291), *M. kansasii* DSM 44,162 (ATCC 12478). The most attractive compounds were also tested against multi-drug-resistant (MDR) clinical isolates of *Mtb*. The minimum inhibitory concentration (MIC) values were measured in μg/mL and then were recalculated into μM. The fast-growing *M. smegmatis* and *M. aurum* [33] (recently reclassified into genus *Mycolicibacterium* [34]) are commonly used as valid and safe non-pathogenic models for antimycobacterial research as an alternative to *Mtb* H37Rv [35–37].

Aryl or heteroaryl substitution bound in position 6 of benzoxaborole via the amidic linker-produced compounds active against the virulent strain *Mtb* H37Rv (Table 1) with few exceptions (compounds **6**, **7**, **13**, and **18**). Most of the compounds (**1–9**) were substituted with a pyrazine ring mimicking the clinically used anti-TB drug pyrazinamide and supported by our previous long-term experience with the synthesis of pyrazine-based derivatives [38–40]. However, other ring analogs with a pyridine ring (compounds **10–13**), a benzene ring (compound **14**), or even with a five-member aromatic ring (compounds **15** and **16**), and bicyclic aromatic systems (compounds **17–19**) have been prepared.

The most active compound among the pyrazine derivatives was the 5-chloro substituted compound **5** ($MIC_{\mathit{Mtb}H37Rv}$ 21.59 μM). The high antimycobacterial activity of compound **5** can be explained by the higher lipophilicity due to the chlorine substitution and, thus, the better penetration through the mycolic cell wall of mycobacteria. The pyrazine ring should also have the ability to form H-bond interaction with D447 through a water molecule. In comparison to the 5-chloro benzene analog, the inhibitory activity was retained only against *M. aurum* (MIC around 15 μM). The change of the six-member aromatic ring for a five-member aromatic ring did not bring an increase in activity against *Mtb* H37Rv. Among the derivatives bearing aromatic nitrogen bicycles, the quinoxaline substitution produced compound **17** with MIC = 20.49 μM against *Mtb* H37Rv. The most active derivative against *Mtb* H37Rv in the whole series was 5-trifluoro-2-pyridyl derivative (**11**) with MIC = 9.72 μM, ruling the positional 2-trifluoro-5-pyridyl isomer (**12**, MIC = 155.26 μM) and the non-substituted 2-pyridyl derivative (**10**, MIC = 49.20 μM). Notably, compounds **1**, **2**, **5**, **10**, **11**, **14**, **15**, and **17** exhibited comparable or better activity against *Mtb* H37Rv than unsubstituted benzoxaborole moiety and compound **11** exhibited comparable inhibition to the activity of 3-aminomethylbenzoxaborole (both included in Table 1 and extracted from the publication by Palencia et al. [13]). Nevertheless, the activity of compound **11** did not overtake the MIC value of isoniazid (2.84 μM).

Regarding the comparison of inhibitory activity against *Mtb* H37Rv with the one against *Mtb* H37Ra, there was a satisfactory correlation with the exception of a few compounds (**3**, **8**, **11**, **14**, **17**, and **19**).

As for the SAR of the nitrogen-containing six-member ring, it can be concluded that a halogenated substitution in position 5 is favorable. 3-Amino substitution of the pyrazine ring led to a complete loss of inhibitory activity against *Mtb*. It can be generally stated that a lipophilic substitution is more convenient than any hydrophilic substitution (amino, hydroxy). However, no direct correlation has been found between the inhibitory activity against *Mtb* and the calculated lipophilicities (Table 1).

As for other tested mycobacterial strains, there is a good correlation between the activity of the compounds against *Mtb* H37Rv and *M. kansasii*. The most active derivatives against *M. kansasii* were the 5-trifluoro-2-pyridyl derivative (compound **11**, MIC 12.14 μM) and the quinoxalin-2-yl derivative (compound **17**, MIC 25.60 μM). An inverse effect was observed for the chlorinated pyrazine derivative. The more convenient location for chloro substitution was position 5 in the case of *Mtb* H37Rv inhibition, whereas position 6 (compound **4**) is more convenient for the inhibition of *M. kansasii* (MIC = 26.98 μM) and *M. avium* (MIC = 26.98 μM).

M. avium turned out to be the least susceptible strain to the tested compounds. The most active compound against *M. avium* proved to be the derivative with quinolin-2-yl substitution (compound **19**, MIC = 25.58 μM).

The rapidly growing mycobacteria *M. aurum* and *M. smegmatis* appeared to be the most susceptible strains. Compound **17** with the quinolin-2-yl substitution was the most active compound inhibiting *M. aurum* and *M. smegmatis* with identical MIC =12.82 μM. Generally, it can be stated that most of the compounds exhibited a broad spectrum of antimycobacterial activity also against non-tuberculous strains of mycobacteria in several cases even better than standard isoniazid. Compounds **3** and **17** proved comparable inhibition as rifampicin against *M. smegmatis*.

Compound **10** does not seem to show the best growth inhibition, but it should be pointed out from the series, as it inhibited the growth of all mycobacterial strains (MIC below 62 μM).

Some SAR phenomena can be explained by docking studies. Pyrazine and pyridine derivatives were able to bind to a molecule of water and the D447 residue to form an H-bond interaction between the heterocyclic nitrogen and an editing domain of mycobacterial LeuRS. This interaction could not be observed for the benzene derivatives, which is in accordance with the lower activity of these derivatives. Focusing on the two position isomers **11** and **12**, the higher activity of 5-trifluoro-2-pyridyl (**11**) can be explained by the sufficient proximity of nitrogen for hydrogen bond formation, whereas the nitrogen in 2-trifluoro-5-pyridyl derivative (**12**) is too far for the formation of the additional hydrogen bond. The same correlation has been found for the pair of compounds **9** (4-hydroxypyrazin-2-yl derivative) and **13** (2-hydroxy-5-pyridyl derivative). Only compound **9** was able to form the hydrogen bond, which was confirmed in the docking studies, and it also proved higher inhibitory activity against *Mtb* H37Rv (MIC = 184.48 μM) compared to compound **13** (MIC > 370.30 μM).

2.3.2. Antimycobacterial Activity against Multi-Drug-Resistant Clinical Isolates of *M. tuberculosis*

The promising candidates, namely compounds **4**, **5**, **10**, and **17**, were also evaluated against multidrug-resistant (MDR) clinical isolates of *Mtb*. The used MDR isolates were resistant to streptomycin and the majority of first-line anti-TB drugs such as isoniazid, rifampicin, and pyrazinamide. In Table 2, the consistent results of the prepared compounds against *Mtb* H37Rv and MDR strains are demonstrated. There is just a slight decrease in the activity against MDR strains (one dilution difference), which indicates a unique mechanism of action that is not related to any of the first-line anti-TB drugs. The complete resistance profile of the MDR strains can also be found in Section 3.3.1 (antimycobacterial screening against *Mtb* H37Rv and MDR strains of *Mtb*).

Table 2. Comparison of the antimycobacterial activity of selected compounds against collection strain of *Mtb* H37Rv and the clinical isolates of MDR strains of *Mtb* (labeled IZAK and MATI) including standards.

Comp.	MIC *Mtb* H37Rv (μM)	MIC *Mtb* IZAK (μM)	MIC *Mtb* MATI (μM)
4	86.36	86.36	86.36
5	21.59	43.18	43.18
10	49.20	98.41	98.41
17	20.49	40.97	40.97
INH	2.84	91.15	>91.15
CPX	0.6	0.6	0.6
EMB	1.91	7.64	7.64

2.4. In Vitro Antibacterial and Antifungal Evaluation

Benzoxaborole moiety itself exerts a broad spectrum of inhibitory activity against several strains of fungi [41]. Its simple derivatives may affect the viability of several strains of bacteria. So, the problem in designing new benzoxaborole derivatives is not just in obtaining compounds with high activity but also obtaining compounds with enough selectivity. Therefore, the series of title compounds were evaluated on activities against a set of sixteen microorganisms, eight bacterial strains, and eight fungal strains. A microdilution method according to EUCAST [42–44] was used. The MIC values were expressed in μM. The derivatives **7**, **11**, **14**, **18**, and **19** were tested up to the concentration of 125 μM, while the rest were tested up to the concentration of 500 μM. For the methodology and the complete list of tested strains, please see the Supplementary Materials.

None of the tested compounds proved any inhibitory activity at the tested concentrations against the included strains of fungi and yeasts (Table S4). This may be explained by the difference in the structure between prokaryotic and eukaryotic LeuRS and by the steric clash with the extra alfa helix discussed herein in the computational studies. Most of the compounds did not exert any inhibitory activity against Gram-negative strains of bacteria (Table S2) besides compound **16**, which proved very weak inhibition (MIC = 500 μM) after 24 h incubation and no inhibition after 48 h incubation against *Escherichia coli* and *Klebsiella pneumoniae*. In the case of a few compounds, very weak inhibitory activity was determined against Gram-positive bacteria with MIC values in the concentration range of 125–250 μM. This led us to the conclusion that the title compounds selectively affect only strains of mycobacteria and mycolicibacteria.

2.5. Cytotoxicity Screening

The cytotoxicity of the most promising compounds was evaluated using standard human liver cancer cell line HepG2 in a commercial CellTiter 96 assay. The determined IC_{50} values allowing a quantitative comparison of the toxicity among the tested compounds are presented in Table 1. Several established anti-tuberculosis drugs are known to manifest a risk of hepatotoxic behavior [45] and the cytotoxic effect on the hepatic cell line is thus a relevant surrogate.

The majority of the tested compounds were not toxic and no significant decrease in cell viability was observed in the highest tested concentrations (IC_{50} > 1000 μM). For seven compounds with solubility issues in the incubation medium, the IC_{50} values were determined in the highest concentrations possible. The derivatives **2** and **3** were determined up to the concentration of 500 μM, compounds **12** and **17** up to the concentration of 250 μM, and compounds **6**, **7**, and **11** up to the concentration of 100 μM. All appeared to be non-toxic up to their highest tested concentrations.

The selectivity of the most active compounds Is expressed as a selectivity index (SI = IC_{50} (μM)/$MIC_{Mtb H37Rv}$ (μM)) (Table 3). For the majority of compounds, the SI value is above 10, which can be considered a reasonable starting point for further optimization. Only compounds **5** and **15** proved to have SI values below 5.

Table 3. Cytotoxicity of the selected compounds against HepG2 cell line (IC_{50}) and selectivity indices (SI) in relation to their antimycobacterial activity against *Mtb* H37Rv (MIC).

Compound	HepG2 IC_{50} (μM)	*Mtb* H37Rv MIC (μM)	SI
1	>1000	49.01	>20.40
2	>500	46.46	>10.76
5	82.44	21.59	3.82
10	>1000	49.20	>20.33
11	>100	9.72	>10.29
14	>1000	43.48	>23.00
15	210.5	42.59	4.94
17	>250	20.49	>12.20

3. Materials and Methods

3.1. In Silico Study

Docking studies were performed using Dock utility in MOE 2022.02 (Chemical Computing Group, Montreal, QC, Canada), using the AMBER10:EHT force field. The editing domain of mtbLeuRS (PDB ID: 5AGR), the editing domain of hLeuRS (PDB ID: 2WFD), and human cytosolic LeuRS (hLeuRS2, PDB ID: 6LPF) were downloaded from the RCSB PDB database. 5AGR was firstly adjusted by removing artifacts from crystallography, namely Met-Leu and EDO (1,2-ethanediol) in the sequence editor. Then, solvent molecules were erased, besides the ones that formed H-bond interaction with amino acid residues. 2WFD was adjusted by the deletion of the additional B chain in the sequence editor. 6LPF was adjusted in the sequence editor by removing the A chain of the dimer, several ligands (LSS, GOL), and solvent molecules. The following enzymes were superposed: 5AGR with 2WFD and 5AGR with 6LPF. The position of 5AGR was firstly fixed and then the superposition of enzymes according to their structural motifs and subsequently, the alignment of the sequences was performed to match the superposition, otherwise, the superposition was not successful. All the proteins or complexes were prepared for subsequent docking using the function QuickPrep (with default settings). The bivalent adducts of synthesized compounds and AMP were drawn in ChemDraw 20.0 (PerkinElmer Informatics, Waltham, MA, USA). After import into MOE, the structures were prepared (choice of protonation state at pH 7, calculation of partial charges, and energy minimization until RMS = 0.00001 kcal.mol^{-1}Å^{-1}). The adducts of synthesized compounds were docked in the 5AGR using the template docking utility with the AMP moiety defined as the template. The superposed hLeuRS1 and hLeuRS2 were inactivated during docking. The thirty best poses from placement were refined to the final two poses for every entry (minimized inside the induced fit receptor) and scored using the GBVI/WSA dG scoring function. The resulting poses were sorted according to the ascending score value and the poses were analyzed visually. The score values for every unique molecule are involved in Table S1.

3.2. Synthesis of the Compounds

3.2.1. General Information

All the reagents and solvents (unless stated otherwise) were purchased from Sigma-Aldrich (St. Louis, MO, USA) or Fluorochem (Hadfield, Derbyshire, UK) and used without further purification. The reaction progress and purity of the products were monitored using thin-layer chromatography (TLC) using Silica gel 60 F_{254} sheets (Merck, Darmstadt, Germany). The flash chromatography of some final compounds was performed using PuriFlash XS 420+ (Interchim, Montluçon, France) with original columns (spherical silica, particle size 30 µm) with detection using UV-VIS detection at 254 nm and 280 nm. The elution ran in isocratic mode using a mobile phase consisting of hexane (Hex), ethyl acetate (EtOAc), and methanol (MeOH) at a volume ratio of 40:55:5. The NMR spectra were recorded using a Varian VNMR S500 (Varian, Palo Alto, CA, USA) at 500 MHz for ^{1}H, 126 MHz for ^{13}C, and 160 MHz for ^{11}B, and the ^{1}H and ^{13}C spectra of compounds **3, 9–14, 16, 18,** and **19** using Jeol JNM-ECZ600R (Jeol Ltd., Tokyo, Japan) at 600 MHz for ^{1}H and 151 MHz for ^{13}C. The spectra were measured in DMSO-*d6* at ambient temperature. The chemical shifts reported as δ values in ppm indirectly referred to tetramethylsilane (TMS) via the solvent signal (2.49 for ^{1}H and 39.7 for ^{13}C in DMSO-*d6*). The IR spectra were recorded using an NICOLET 6700 FT-IR spectrophotometer (Thermo Scientific, Waltham, MA, USA) using the ATR-Ge method. The elemental analyses were measured using a Vario Micro Cube Elemental Analyzer (Elementar Analysensysteme, Hanau, Germany) and the values are provided in percentage. The yields are provided in percentage and refer to the amount of pure product after all the purification steps. The melting points were determined in open capillary on a Stuart SMP30 melting point apparatus (Bibby Scientific Limited, Staffordshire, UK) and are uncorrected. The purities of the studied compounds were determined using Agilent Technologies 1200 SL liquid chromatograph (Agilent Technologies, Santa Clara, CA, USA) consisting of a vacuum microdegasser,

1200 SL binary pump, 1200 SL plus autosampler, TCC Infinity 1290 column thermostat, and 1200 SL diode-array detector. The chromatographic system was controlled using an Agilent ChemStation (Agilent Technologies, Santa Clara, CA, USA), version B.04.02 extended by a spectral module. The area percentage method at the wavelength 250 nm was applied where it was proven to have >95% purity. The Log*P* values were calculated using ChemDraw 20.0 part of ChemOffice 2020 package (PerkinElmer Informatics, Waltham, MA, USA). The mass spectra in both positive and negative mode (APCI-MS) were measured using the Expression® Compact Mass Spectrometer (Advion, Ithaca, NY, USA) with a single-quad detector. The samples were applied as solids by the ASAP probe method (Advion, Ithaca, NY, USA).

3.2.2. General Synthetic Procedure

In the first step, the corresponding (hetero)arylcarboxylic acid (2 mmol) was activated with 1,1′-carbonyldiimidazole (CDI; 2.2 mmol) in DMSO (5 mL) for 1 h. After initial heating, it proceeded at room temperature. Then, 6-aminobenzo[*c*][1,2]oxaborol-1(3*H*)-ol hydrochloride (2 mmol) dissolved in DMSO (10 mL) was added and the reaction mixture was stirred at room temperature overnight. The working-up was accomplished by the addition of 1M HCl (10 mL). Within one hour of stirring, precipitation appeared. The precipitate was filtrated off and washed with water. All the final compounds were dried over anhydrous phosphorus pentoxide under reduced pressure at room temperature before characterization and purity check. Compounds **10** and **15** were not completely clean so we further purified them using Flash chromatography using isocratic elution with the mobile phase consisting of Hex:EtOAc:MeOH = 4:5.5:0.5. The compounds demonstrated R_f values of 0.85 and 0.9, respectively, in this mobile phase.

3.2.3. Characterization and Purity of the Compounds **1–19**

In the ^{1}H-NMR spectra, the following abbreviations have been used: PzH—pyrazine hydrogen, PyH—pyridine hydrogen, QxH—quinoxaline hydrogen, TzH—thiazole hydrogen, and LmH—hydrogen involved in lactam-lactim tautomerism. In the ^{13}C-NMR spectra, the following abbreviations have been used: PyC—pyridine carbon. The ^{1}H-NMR, ^{13}C-NMR, and ^{11}B-NMR spectra of the title compounds with solvent signals omitted are involved.

***N*-(1-hydroxy-1,3-dihydrobenzo[*c*][1,2]oxaborol-6-yl)pyrazine-2-carboxamide (1).** Light yellow solid. Yield: 79%. mp 261.3–262.3 °C. ^{1}H NMR (500 MHz, dmso) δ 10.75 (s, 1H, CONH), 9.30 (d, *J* = 1.5 Hz, 1H, PzH), 9.26 (s, 1H, OH), 8.92 (d, *J* = 2.5 Hz, 1H, PzH), 8.80 (dd, *J* = 2.5, 1.5 Hz, 1H, PzH), 8.28 (d, *J* = 2.0 Hz, 1H, ArH), 7.85 (dd, *J* = 8.3, 2.0 Hz, 1H, ArH), 7.42–7.37 (m, 1H, ArH), 4.97 (s, 2H, CH_2). ^{13}C NMR (126 MHz, dmso) δ 161.89, 150.01, 147.86, 145.38, 144.25, 143.45, 137.14, 131.15, 124.08, 122.72, 121.77, 69.96. ^{11}B NMR (160 MHz, dmso) δ 31.62. IR (ATR-Ge, cm^{-1}): 3317 (ν O-H), 1672 (δ, C=O), 1535 (δ CONH), 1369 (ν B-O), 990 (ν C-O ether). Analysis calculated for $C_{12}H_{10}BN_3O_3$ (M_r 255.04): C, 56.51; H, 3.95; N, 16.48. Found: C, 56.30; H, 3.65; N, 16.48. MS: $[M + H]^+$ = 255.9 (exact mass 255.08).

***N*-(1-hydroxy-1,3-dihydrobenzo[*c*][1,2]oxaborol-6-yl)-5-methylpyrazine-2-carboxamide (2).** Light yellow solid. Yield: 72%. mp 215.0–216.1 °C. ^{1}H NMR (500 MHz, dmso) δ 10.67 (s, 1H, CONH), 9.24 (s, 1H, OH), 9.15 (d, *J* = 1.5 Hz, 1H, PzH), 8.68 (d, *J* = 1.5 Hz, 1H, PzH), 8.28 (d, *J* = 2.0 Hz, 1H, ArH), 7.84 (dd, *J* = 8.2, 2.0 Hz, 1H, ArH), 7.41–7.35 (m, 1H, ArH), 4.96 (s, 2H, CH_2), 2.62 (s, 3H, CH_3). ^{13}C NMR (126 MHz, dmso) δ 162.26, 157.58, 150.12, 143.46, 143.18, 142.86, 137.47, 131.38, 124.29, 122.91, 121.97, 70.20, 21.88. ^{11}B NMR (160 MHz, dmso) δ 33.75. IR (ATR-Ge, cm^{-1}): 3319 (ν O-H), 1673 (δ, C=O), 1537 (δ CONH), 1367 (ν B-O), 993 (ν C-O ether). Analysis calculated for $C_{13}H_{12}BN_3O_3$ (M_r 269.07): C, 58.03; H, 4.50; N, 15.62. Found: C, 57.80; H, 4.23; N, 15.60. MS: $[M + H]^+$ = 269.9 (exact mass 269.10).

***N*-(1-hydroxy-1,3-dihydrobenzo[*c*][1,2]oxaborol-6-yl)-3-methylpyrazine-2-carboxamide (3).** Light white solid. Yield: 72%. mp 262.9–264.0 °C. ^{1}H NMR (600 MHz, dmso) δ 10.66 (s, 1H, CONH), 9.24 (brs, 1H, OH), 8.72 (d, *J* = 2.5 Hz, 1H, PzH), 8.60 (dd, *J* = 2.5, 0.8 Hz, 1H, PzH), 8.26 (d, *J* = 2.0 Hz, 1H, ArH), 7.76 (dd, *J* = 8.2, 2.0 Hz, 1H, ArH), 7.39 (dd, *J* = 8.2, 0.9 Hz, 1H, ArH), 4.96 (s, 2H, CH_2), 2.76 (d, *J* = 0.6 Hz, 3H, CH_3). ^{13}C NMR (151 MHz, dmso) δ 164.39, 153.49, 150.10, 146.18, 146.12, 141.33, 137.87, 131.54, 123.83, 122.47, 122.15, 70.27, 22.93. ^{11}B NMR (160 MHz, dmso) δ 32.51. IR (ATR-Ge, cm^{-1}): 3344 (ν O-H), 3068 (ν C-H arom.), 1683 (δ, C=O), 1525 (δ CONH), 1373 (ν B-O), 978 (ν C-O ether). Analysis calculated for $C_{13}H_{12}BN_3O_3$ (M_r 269.07): C, 58.03; H, 4.5; N, 15.62. Found: C, 57.58; H, 4.22; N, 15.18. MS: $[M + H]^+$ = 270.0 (exact mass 269.10).

6-chloro-*N*-(1-hydroxy-1,3-dihydrobenzo[*c*][1,2]oxaborol-6-yl)pyrazine-2-carboxamide (4). White solid. Yield: 66%. mp 250.9–252.4 °C. ^{1}H NMR (500 MHz, dmso) δ 10.68 (s, 1H, CONH), 9.26 (s, 1H, OH), 9.23 (s, 1H, PzH), 9.05 (s, 1H, PzH), 8.24 (d, *J* = 2.0 Hz, 1H, ArH), 7.82 (dd, *J* = 8.2, 2.0 Hz, 1H, ArH), 7.40 (d, *J* = 8.2 Hz, 1H, ArH), 4.97 (s, 2H, CH_2). ^{13}C NMR (126 MHz, dmso) δ 160.85, 150.23, 147.56, 147.10, 145.46, 142.52, 136.91, 131.14, 124.29, 123.04, 121.77, 69.96.^{11}B NMR (160 MHz, dmso) δ 32.52. IR (ATR-Ge, cm^{-1}): 3368 (ν O-H), 3057 (ν C-H arom.), 1681 (δ, C=O), 1523 (δ CONH), 1367 (ν B-O), 992 (ν C-O ether). Analysis calculated for $C_{12}H_9BClN_3O_3$ (M_r 289.48): C, 49.79; H, 3.13; N, 14.52. Found: C, 49.32; H, 2.87; N, 14.38. MS: $[M + H]^+$ = 289.9 (exact mass 289.04).

5-chloro-*N*-(1-hydroxy-1,3-dihydrobenzo[*c*][1,2]oxaborol-6-yl)pyrazine-2-carboxamide (5). White solid. Yield: 41%. mp 233.8–235.3 °C. ^{1}H NMR (500 MHz, dmso) δ 10.77 (s, 1H, CONH), 9.25 (brs, 1H, OH), 9.11 (d, *J* = 1.4 Hz, 1H, PzH), 8.92 (d, *J* = 1.4 Hz, 1H, PzH), 8.26 (d, *J* = 2.1 Hz, 1H, ArH), 7.83 (dd, *J* = 8.2, 2.1 Hz, 1H, ArH), 7.39 (d, *J* = 8.2 Hz, 1H, ArH), 4.96 (s, 2H, CH_2). ^{13}C NMR (126 MHz, dmso) δ 161.34, 151.32, 150.37, 144.47, 144.41, 143.39, 137.29, 124.41, 123.08, 122.02, 70.21. ^{11}B NMR (160 MHz, dmso) δ 31.95. IR (ATR-Ge, cm^{-1}): 3355 (ν O-H), 1693 (δ, C=O), 1542 (δ CONH), 1369 (ν B-O), 978 (ν C-O ether). Analysis calculated for $C_{12}H_9BClN_3O_3$ (M_r 289.48): C, 49.79; H, 3.13; N, 14.52. Found: C, 50.16; H, 2.96; N, 14.41. MS: $[M + H]^+$ = 289.9 (exact mass 289.04).

3-amino-*N*-(1-hydroxy-1,3-dihydrobenzo[*c*][1,2]oxaborol-6-yl)pyrazine-2-carboxamide (6). Yellow solid. Yield: 98%. mp 246.6–249.4 °C. ^{1}H NMR (500 MHz, dmso) δ 10.52 (s, 1H, CONH), 9.23 (s, 1H, OH), 8.27 (d, *J* = 2.3 Hz, 1H, PzH), 8.20 (d, *J* = 2.1 Hz, 1H, ArH), 7.92 (d, *J* = 2.3 Hz, 1H, PzH), 7.77 (dd, *J* = 8.2, 2.1 Hz, 1H, ArH), 7.59 (brs, 2H, NH_2), 7.37 (d, *J* = 8.2 Hz, 1H, ArH), 4.96 (s, 2H, CH_2). ^{13}C NMR (126 MHz, dmso) δ 164.77, 155.67, 149.71, 147.51, 137.13, 131.22, 125.67, 124.09, 122.58, 121.73, 69.97. ^{11}B NMR (160 MHz, dmso) δ 32.21. IR (ATR-Ge, cm^{-1}): 3409 (ν O-H), 3337 and 3299 (NH_2), 1672 (δ, C=O), 1542 (δ CONH), 1367 (ν B-O), 986 (ν C-O ether). Analysis calculated for $C_{12}H_{11}BN_4O_3$ (M_r 270.06): C, 53.37; H, 4.11; N, 20.75. Found: C, 52.93; H, 3.73; N, 20.39. MS: $[M + H]^+$ = 270.9 (exact mass 270.09).

3-amino-6-chloro-*N*-(1-hydroxy-1,3-dihydrobenzo[*c*][1,2]oxaborol-6-yl)pyrazine-2-carboxamide (7). Yellow solid. Yield: 53%. mp 242.9–244.4 °C. ^{1}H NMR (500 MHz, dmso) δ 10.33 (s, 1H, CONH), 9.21 (s, 1H, OH), 8.37 (s, 1H, PzH), 8.15 (d, *J* = 2.1 Hz, 1H, ArH), 7.76 (dd, *J* = 8.2, 2.1 Hz, 1H, ArH), 7.72 (brs, 2H, NH_2), 7.38 (d, *J* = 8.2 Hz, 1H, ArH), 4.96 (s, 2H, CH_2). ^{13}C NMR (126 MHz, dmso) δ 164.02, 154.72, 150.19, 147.07, 137.12, 132.19, 124.64, 124.50, 123.31, 121.94, 70.22. ^{11}B NMR (160 MHz, dmso) δ 32.03. IR (ATR-Ge, cm^{-1}): 3440 (ν O-H), 3348 and 3325 (NH_2), 1670 (δ, C=O), 1535 (δ CONH), 1362 (ν B-O), 977 (ν C-O ether). Analysis calculated for $C_{12}H_{10}BClN_4O_3$ (M_r 304.5): C, 47.33; H, 3.31; N, 18.4. Found: C, 47.42; H, 3.05; N, 18.39. MS: $[M + H]^+$ = 305.0 (exact mass 304.05).

***N*-(1-hydroxy-1,3-dihydrobenzo[*c*][1,2]oxaborol-6-yl)-3-oxo-3,4-dihydropyrazine-2-carboxamide (8).** Light green solid. Yield: 53%. mp 326.4–328.3 °C (decomp.). ^{1}H NMR (500 MHz, dmso) δ 13.23 (brs, 1H, LmH), 11.46 (s, 1H, CONH), 9.25 (brs, 1H, OH), 8.12 (d, *J* = 2.0 Hz, 1H, PzH), 7.82 (s, 1H, ArH), 7.75 (dd, *J* = 8.2, 2.1 Hz, 1H, ArH), 7.73–7.69 (m, 1H, PzH), 7.39 (d, *J* = 8.2 Hz, 1H, ArH), 4.96 (s, 2H, CH_2). ^{11}B NMR (160 MHz, dmso) δ 33.88. IR (ATR-Ge, cm^{-1}): 3460 (ν O-H), 3040 (ν C-H arom.), 1694 (δ, C=O), 1652 (δ CONH), 1368 (ν B-O), 956 (ν C-O ether). Analysis calculated for $C_{12}H_{10}BN_3O_4$ (M_r 271.04): C, 53.18; H, 3.72; N, 15.50. Found: C, 52.79; H, 3.39; N, 15.36. MS: $[M + H]^+$ = 271.8 (exact mass 271.08).

5-hydroxy-*N*-(1-hydroxy-1,3-dihydrobenzo[*c*][1,2]oxaborol-6-yl)pyrazine-2-carboxamide (9). Light brown solid. Yield: 93%. mp 338.3–340.1 °C. ^{1}H NMR (600 MHz, dmso) δ 12.84 (brs, 1H, OH), 10.15 (s, 1H, CONH), 9.18 (brs, 1H, OH), 8.20 (d, *J* = 3.7 Hz, 1H, ArH), 8.09 (m, 1H, PzH), 8.05–8.01 (m, 1H, PzH), 7.80–7.74 (m, 1H, ArH), 7.37–7.32 (m, 1H, ArH), 4.96–4.93 (m, 2H, CH_2). ^{13}C NMR (151 MHz, dmso) δ 161.68, 156.80, 149.82, 146.83, 137.77, 131.26, 127.04, 124.21, 122.76, 121.96, 70.26. ^{11}B NMR (160 MHz, dmso) δ 32.73. IR (ATR-Ge, cm^{-1}): 3357 (ν O-H), 2796 (NH, Lactam), 1677 (δ, C=O), 1659 (CO, CONH), 1529 (δ CONH), 1367 (ν B-O), 968 (ν C-O ether). Calculated for $C_{12}H_{10}BN_3O_4$ (M_r 271.04): C, 53.18; H, 3.72; N, 15.5. Found: C, 52.74; H, 3.45; N, 15.10. MS: $[M + H]^+$ = 271.8 (exact mass 271.08).

***N*-(1-hydroxy-1,3-dihydrobenzo[*c*][1,2]oxaborol-6-yl)picolinamide (10).** Yellow solid. Yield: 80%. mp 183.7–186.1 °C. ^{1}H NMR (600 MHz, dmso) δ 10.64 (s, 1H, CONH), 9.23 (s, 1H, OH), 8.73 (m, 1H, PyH), 8.30 (d, *J* = 2.0 Hz, 1H, ArH), 8.16 (m, 1H, PyH), 8.06 (m, 1H, PyH), 7.85 (dd, *J* = 8.2, 2.0 Hz, 1H, ArH), 7.67 (m, 1H, PyH), 7.42–7.35 (m, 1H, ArH), 4.97 (s, 2H, CH_2). ^{13}C NMR (151 MHz, dmso) δ 163.00, 150.54, 150.03, 148.98, 138.66, 137.68, 131.50, 127.41, 124.16, 122.90, 122.68, 122.08, 70.29. ^{11}B NMR (160 MHz, dmso) δ 31.61. IR (ATR-Ge, cm^{-1}): 3311 (ν O-H), 1677 (δ, C=O), 1540 (δ CONH), 1366 (ν B-O), 980 (ν C-O ether). Calculated for $C_{13}H_{11}BN_2O_3$ (M_r 254.05): C, 61.46; H, 4.36; N, 11.03. Found: C, 61.19; H, 3.99; N, 10.99. MS: $[M + H]^+$ = 255.0 (exact mass 254.09); R_f (Hex:EtOAc:MeOH = 4:5.5:0.5) = 0.85.

***N*-(1-hydroxy-1,3-dihydrobenzo[*c*][1,2]oxaborol-6-yl)-5-(trifluoromethyl)picolinamide (11).** Light yellow solid. Yield: 82%. mp 188.1–188.9 °C. ^{1}H NMR (600 MHz, dmso) δ 10.79 (s, 1H, CONH), 9.24 (s, 1H, OH), 9.10 (m, 1H, PyH), 8.50–8.45 (m, 1H, PyH), 8.34 (m, 1H, PyH), 8.30 (d, *J* = 2.0 Hz, 1H, ArH), 7.86 (dd, *J* = 8.2, 2.0 Hz, 1H, ArH), 7.40 (dd, *J* = 8.2, 0.8 Hz, 1H, ArH), 4.97 (s, 2H, CH_2). ^{13}C NMR (151 MHz, dmso) δ 161.95, 154.13, 150.38, 145.83, 137.42, 136.31, 131.52, 127.92 (q, *J* = 32.5 Hz, 1C, PyC), 124.37, 123.92 (q, *J* = 273.8 Hz, 1C, CF_3), 123.30, 122.98, 122.11, 70.29. ^{11}B NMR (160 MHz, dmso) δ 30.36. IR (ATR-Ge, cm^{-1}): 3343 (ν O-H), 1682 (δ, C=O), 1531 (δ CONH), 1366 (ν B-O), 978 (ν C-O ether). HPLC purity 95.1 %. MS: $[M + H]^+$ = 323.1 (exact mass 322.07).

***N*-(1-hydroxy-1,3-dihydrobenzo[*c*][1,2]oxaborol-6-yl)-6-(trifluoromethyl)nicotinamide (12).** Light yellow solid. Yield: 87%. mp 227.3–229.4 °C. ^{1}H NMR (600 MHz, dmso) δ 10.67 (s, 1H, CONH), 9.26 (s, 1H, OH), 9.25 (d, *J* = 2.1 Hz, 1H, PyH), 8.57 (dd, *J* = 8.1, 2.1 Hz, 1H, PyH), 8.17 (d, *J* = 2.0 Hz, 1H, ArH), 8.09 (dd, *J* = 8.1, 0.9 Hz, 1H, PyH), 7.78 (dd, *J* = 8.2, 2.0 Hz, 1H, ArH), 7.43–7.39 (m, 1H, ArH), 4.98 (s, 2H, CH_2). ^{13}C NMR (151 MHz, dmso) δ 163.38, 150.35, 149.85, 148.67 (q, *J* = 34.1 Hz, 1C, PyC), 138.40, 137.88, 134.38, 131.57, 124.13, 122.98, 122.16, 121.95 (q, *J* = 273.9 Hz, 1C, CF_3), 121.14, 70.29. ^{11}B NMR (160 MHz, dmso) δ 32.24. IR (ATR-Ge, cm^{-1}): 3298 (ν CONH), 1649 (δ, C=O), 1524 (δ CONH), 1372 (ν B-O), 985 (ν C-O ether). HPLC purity 95.0 %. MS: $[M + H]^+$ = 323.1 (exact mass 322.07).

6-hydroxy-*N*-(1-hydroxy-1,3-dihydrobenzo[*c*][1,2]oxaborol-6-yl)nicotinamide (13). Beige solid. Yield: 28%. mp 304.8–306.9 °C. ^{1}H NMR (600 MHz, dmso) δ 12.07 (brs, 1H, OH), 10.00 (s, 1H, CONH), 9.20 (s, 1H, OH), 8.19 (d, *J* = 2.7 Hz, 1H, PyH), 8.07 (d, *J* = 2.0 Hz, 1H, ArH), 7.97 (dd, *J* = 9.6, 2.7 Hz, 1H, PyH), 7.69 (dd, *J* = 8.2, 2.0 Hz, 1H, ArH), 7.35 (d, *J* = 8.2 Hz, 1H, ArH), 6.40 (d, *J* = 9.6 Hz, 1H, PyH), 4.95 (s, 2H, CH_2). ^{13}C NMR (151 MHz, dmso) δ 163.10, 162.90, 149.60, 139.90, 138.52, 138.34, 131.37, 124.09, 122.85, 121.95, 119.69, 113.17, 70.25. ^{11}B NMR (160 MHz, dmso) δ 33.55. IR (ATR-Ge, cm^{-1}): 3457 (ν O-H), 3131 (ν C-H arom.), 2918 (ν CH_2), 2868 (NH, Lactam), 1662 (δ, C=O), 1556 (δ CONH), 1394 (ν B-O), 1001 (ν C-O ether). Calculated for $C_{13}H_{11}BN_2O_4$ (M_r 270.05): C, 57.82; H, 4.11; N, 10.37. Found: C, 57.39; H, 4.35; N, 9.95. MS: $[M + H]^+$ = 270.9 (exact mass 270.08).

4-chloro-*N*-(1-hydroxy-1,3-dihydrobenzo[*c*][1,2]oxaborol-6-yl)benzamide (14). White solid. Yield: 92%. mp 225.5–228.1 °C. ^{1}H NMR (600 MHz, dmso) δ 10.34 (s, 1H, CONH), 9.22 (s, 1H, OH), 8.16 (d, *J* = 2.0 Hz, 1H, ArH), 8.02–7.97 (m, 2H, ArH), 7.75 (dd, *J* = 8.1, 2.0 Hz, 1H, ArH), 7.62–7.57 (m, 2H, ArH), 7.38 (d, *J* = 8.1 Hz, 1H, ArH), 4.96 (s, 2H, CH_2). ^{13}C NMR (151 MHz, dmso) δ 164.95, 149.93, 138.27, 136.88, 134.23, 131.43, 130.17, 128.99,

124.26, 123.04, 122.00, 70.28. ^{11}B NMR (160 MHz, dmso) δ 35.07. IR (ATR-Ge, cm^{-1}): 3298 (ν O-H), 1638 (δ, C=O), 1536 (δ CONH), 1366 (ν B-O), 982 (ν C-O ether). Calculated for $C_{14}H_{11}BClNO_3$ (M_r 287.51): C, 58.49; H, 3.86; N, 4.87. Found: C, 58.43; H, 3.64; N, 4.85. MS: $[M + H]^+$ = 287.9 (exact mass 287.05).

5-chloro-*N*-(1-hydroxy-1,3-dihydrobenzo[*c*][1,2]oxaborol-6-yl)thiophene-2-carboxamide (15). Light yellow solid. Yield: 21%. mp 234.9–237.1 °C. ^{1}H NMR (500 MHz, dmso) δ 10.35 (s, 1H, CONH), 9.24 (s, 1H, OH), 8.07 (d, *J* = 2.0 Hz, 1H, ArH), 7.92 (d, *J* = 4.1 Hz, 1H, TfH), 7.71 (dd, *J* = 8.2, 2.0 Hz, 1H, ArH), 7.39 (dd, *J* = 8.2, 0.8 Hz, 1H, ArH), 7.26 (d, *J* = 4.1 Hz, 1H, TfH), 4.96 (s, 2H, CH_2). ^{13}C NMR (126 MHz, dmso) δ 159.27, 150.03, 139.70, 137.54, 134.27, 129.50, 128.74, 124.15, 123.00, 122.06, 70.20. ^{11}B NMR (160 MHz, dmso) δ 30.49. IR (ATR-Ge, cm^{-1}): 3309 (ν O-H), 1631 (δ, C=O), 1541 (δ CONH), 1365 (ν B-O), 982 (ν C-O ether). Analysis calculated for $C_{12}H_9BClNO_3S$ (M_r 293.53): C, 49.10; H, 3.09; N, 4.77; S, 10.92. Found: C, 49.02; H, 2.95; N, 4.77; S, 11.26. MS: $[M + H]^+$ = 293.9 (exact mass 293.01); R_f (Hex:EtOAc:MeOH = 4:5.5:0.5) = 0.9.

***N*-(1-hydroxy-1,3-dihydrobenzo[*c*][1,2]oxaborol-6-yl)thiazole-4-carboxamide (16).** Light yellow solid. Yield: 86%. mp 225.7–227.9 °C. ^{1}H NMR (600 MHz, dmso) δ 10.33 (s, 1H, CONH), 9.26 (d, *J* = 2.1 Hz, 1H, ArH), 9.22 (brs, 1H, OH), 8.49 (d, *J* = 1.8 Hz, 1H, TzH), 8.22 (d, *J* = 1.8 Hz, 1H, TzH), 7.80 (dd, *J* = 8.2, 2.1 Hz, 1H, ArH), 7.37 (d, *J* = 8.2 Hz, 1H, ArH), 4.96 (s, 2H, CH_2). ^{13}C NMR (151 MHz, dmso) δ 159.67, 155.58, 151.33, 149.99, 137.78, 131.42, 125.93, 124.35, 122.96, 122.00, 70.28. ^{11}B NMR (160 MHz, dmso) δ 33.03. IR (ATR-Ge, cm^{-1}): IR 3373 (ν O-H), 3019 (ν C-H arom.), 2918 and 2868 (ν CH_2), 1664 (δ, C=O), 1538 (δ CONH), 1361 (ν B-O), 977 (ν C-O ether). Analysis calculated for $C_{11}H_9BN_2O_3S$ (M_r 260.07): C, 50.8; H, 3.49; N, 10.77; S, 12.33. Found: C, 50.45; H, 3.47; N, 10.62; S, 12.48. MS: $[M + H]^+$ = 260.9 (exact mass 260.04).

***N*-(1-hydroxy-1,3-dihydrobenzo[*c*][1,2]oxaborol-6-yl)quinoxaline-2-carboxamide (17).** Yellow solid. Yield: 19%. mp 231.4–233.0 °C. ^{1}H NMR (500 MHz, dmso) δ 10.86 (s, 1H, CONH), 9.55 (s, 1H, QxH), 9.28 (s, 1H, OH), 8.33 (d, *J* = 2.1 Hz, 1H, ArH), 8.29 (m, 1H, QxH), 8.25–8.18 (m, 1H, QxH), 8.01 (m, 2H, QxH), 7.91 (dd, *J* = 8.2, 2.1 Hz, 1H, ArH), 7.43 (d, *J* = 8.2 Hz, 1H, ArH), 4.99 (s, 2H, CH_2). ^{13}C NMR (126 MHz, dmso) δ 162.19, 150.03, 145.01, 144.17, 143.12, 139.87, 137.15, 132.27, 131.57, 129.76, 129.32, 124.01, 122.70, 121.83, 69.98. ^{11}B NMR (160 MHz, dmso) δ 31.35. IR (ATR-Ge, cm^{-1}): 3376 (ν CONH), 3320 (ν O-H), 2965 (ν CH_2), 1680 (δ, C=O), 1539 (δ CONH), 1370 (ν B-O), 983 (ν C-O ether). Analysis calculated for $C_{16}H_{12}BN_3O_3$ (M_r 305.1): C, 62.99; H, 3.96; N, 13.77. Found: C, 63.46; H, 3.79; N, 13.51. MS: $[M + H]^+$ = 306.0 (exact mass 305.10).

***N*-(1-hydroxy-1,3-dihydrobenzo[*c*][1,2]oxaborol-6-yl)-3-oxo-3,4-dihydroquinoxaline-2-carboxamide (18).** Yellow solid. Yield: 71%. mp 348.9–350.2 °C (decomp.). ^{1}H NMR (600 MHz, dmso) δ 12.92 (s, 1H, LmH), 11.10 (s, 1H, CONH), 9.28 (s, 1H, OH), 8.12 (d, *J* = 2.1 Hz, 1H, ArH), 7.92–7.87 (m, 1H, ArH), 7.75 (dd, *J* = 8.2, 2.1 Hz, 1H, ArH), 7.68–7.62 (m, 1H, ArH), 7.44–7.37 (m, 3H, ArH), 4.97 (s, 2H, CH_2). ^{13}C NMR (151 MHz, dmso) δ 162.06, 154.43, 152.31, 150.11, 137.75, 133.05, 132.52, 131.77, 129.86, 124.58, 123.08, 122.44, 121.90, 116.24, 70.29. ^{11}B NMR (160 MHz, dmso) δ 31.28. IR (ATR-Ge, cm^{-1}): 3328 (ν O-H), 1706 (δ, C=O), 1550 (δ CONH), 1369 (ν B-O), 973 (ν C-O ether). Analysis calculated for $C_{16}H_{12}BN_3O_4$ (M_r 321.1): C, 59.85; H, 3.77; N, 13.09. Found: C, 59.38; H, 3.68; N, 12.73. MS: $[M + H]^+$ = 322.0 (exact mass 321.09).

***N*-(1-hydroxy-1,3-dihydrobenzo[*c*][1,2]oxaborol-6-yl)quinoline-2-carboxamide (19).** Light yellow solid. Yield: 87%. mp 201.0–203.1 °C. ^{1}H NMR (600 MHz, dmso) δ 10.76 (s, 1H, CONH), 9.26 (brs, 1H, OH), 8.62 (dd, *J* = 8.6, 0.8 Hz, 1H, ArH), 8.34 (d, *J* = 2.0 Hz, 1H, ArH), 8.25 (d, *J* = 8.5 Hz, 2H, ArH), 8.11 (dd, *J* = 8.2, 1.4 Hz, 1H, ArH), 7.95–7.88 (m, 2H, ArH), 7.75 (ddd, *J* = 8.1, 6.8, 1.2 Hz, 1H, ArH), 7.43 (dd, *J* = 8.1, 0.8 Hz, 1H, ArH), 4.99 (s, 2H, CH_2). ^{13}C NMR (151 MHz, dmso) δ 163.21, 150.69, 150.10, 146.45, 138.74, 137.65, 131.60, 131.21, 129.89, 129.47, 128.88, 128.68, 124.08, 122.69, 122.17, 119.30, 70.33. ^{11}B NMR (160 MHz, dmso) δ 32.53. IR (ATR-Ge, cm^{-1}): 3355 (ν O-H), 3319 (ν CONH), 1675 (δ, C=O), 1528 (δ CONH), 1361 (ν B-O), 976 (ν C-O ether). Analysis calculated for $C_{17}H_{13}BN_2O_3$ (M_r

304.11): C, 67.14; H, 4.31; N, 9.21. Found: C, 66.88; H, 4.04; N, 9.18. MS: $[M + H]^+ = 305.0$ (exact mass 304.10).

3.3. Biological Studies

3.3.1. Antimicrobial Screening

Antimycobacterial Screening against Mtb H37Ra, *M. smegmatis*, *M. aurum*, *M. avium*, and *M. kansasii*

The antimycobacterial assay was performed using fast-growing *Mycolicibacterium smegmatis* DSM 43,465 (ATCC 607), *Mycolicibacterium aurum* DSM 43,999 (ATCC 23366); non-tuberculous (atypical) mycobacteria, namely *Mycobacterium avium* DSM 44,156 (ATCC 25291) and *Mycobacterium kansasii* DSM 44,162 (ATCC 12478), purchased from German Collection of Microorganisms and Cell Cultures (Braunschweig, Germany); and an avirulent strain of *Mycobacterium tuberculosis* H37Ra ITM-M006710 (ATCC 9431) purchased from Belgian Co-ordinated Collections of Microorganisms (Antwerp, Belgium). The technique used for activity determination was the microdilution broth panel method using 96-well microtitration plates [32]. The culture medium was Middlebrook 7H9 broth (Merck, Darmstadt, Germany) enriched with glycerol (Merck, Darmstadt, Germany) and Middlebrook OADC growth supplement (Himedia, Mumbai, India) according to manufacturer instructions.

The mycobacterial strains were cultured on supplemented Middlebrook 7H9 agar and suspensions were prepared in supplemented Middlebrook 7H9 broth. The final density was adjusted to value 1.0 according to the McFarland scale and diluted in the ratio 1:20 (for fast-growing mycobacteria) or 1:10 (for the rest of mycobacteria) with broth.

The tested compounds were dissolved in DMSO (Sigma-Aldrich, St. Louis, MO, USA), then Middlebrook 7H9 broth was added to obtain a concentration of 2000 μg/mL. The standards used for activity determination were isoniazid (INH), rifampicin (RIF), and ciprofloxacin (CIP) (Merck). The final concentrations were reached using binary dilution and the addition of mycobacterial suspension and were set as 500, 250, 125, 62.5, 31.25, 15.625, 7.81, and 3.91 μg/mL. The final concentration of DMSO in any well did not exceed 2.5% (*v*/*v*) and did not affect the growth of mycobacteria. Positive (broth, DMSO, bacteria) and negative (broth, DMSO) growth controls were included.

The plates with slow-growing mycobacteria were sealed with polyester adhesive film and all the plates were incubated in the dark at 37 °C without agitation. The addition of 0.01% solution of resazurin sodium salt followed after 48 h of incubation for *M. smegmatis*, after 72 h of incubation for *M. aurum*, after 96 h of incubation for *M. avium* and *M. kansasii*, and after 120 h of incubation for *Mtb* H37Ra, respectively. The microtitration panels were then incubated for a further 2.5 h for the determination of activity against *M. smegmatis*, 4 h for *M. aurum*, 5–6 h for *M. avium* and *M. kansasii*, and 18 h for *Mtb* H37Ra, respectively. The antimycobacterial activity was expressed as the minimum inhibitory concentration (MIC). The MIC (in μg/mL) was determined on the basis of stain color change (blue color—active; pink color—not active). All the experiments were conducted in duplicates.

Antimycobacterial Screening against Mtb H37Rv and MDR Strains of Mtb

The microdilution method based on Microplate Alamar Blue Assay (MABA) was applied [32]. The tested strain *Mtb* H37Rv CNCTC My 331/88 (ATCC 27294) was obtained from the Czech National Collection of Type Cultures (CNCTC), National Institute of Public Health (Prague, Czech Republic). Multi-drug resistant strains of *Mtb* laboratory ID designation IZAK and MATI were obtained from the Department of Clinical Microbiology, University Hospital Hradec Králové from Dr. Pavla Paterová. The Middlebrook 7H9 broth of declared pH 6.6 (Sigma-Aldrich, St. Louis, MO, USA) enriched with 0.4% of glycerol (Sigma-Aldrich, St. Louis, MO, USA) and 10% of OADC growth supplement (Himedia, Mumbai, India) was used for cultivation.

Tested compounds were dissolved and diluted in DMSO and mixed with broth (25 μL of DMSO solution in 2.475 mL of broth) and placed (100 μL) into microplate wells. The mycobacterial inocula were suspended in isotonic saline solution and the density was adjusted to 0.5–1.0 according to McFarland scale. These suspensions were diluted by 10^{-1} and used to inoculate the testing wells, adding 100 µL of mycobacterial suspension per well. The final concentrations of tested compounds in wells were 100, 50, 25, 12.5, 6.25, 3.13, and 1.56 µg/mL. INH was used as the standard (inhibition of growth). The positive control (visible growth) consisted of broth plus mycobacterial suspension plus DMSO. A total of 30 µL of Alamar Blue working solution (1:1 mixture of 0.02% resazurin sodium salt (aq. sol.) and 10% Tween 80) was added after five days of incubation. The results were then determined after 24 h of incubation. The MIC (in µg/mL) was determined as the lowest concentration that prevented the from blue to pink color change. All the experiments were conducted in duplicates.

Susceptibility Profiles for the Used MDR Mtb Strains

Mtb laboratory ID IZAK, isolated from a 63-year-old man from bronchial aspirate in 2020, was tested and interpreted according to CLSI (Clinical and Laboratory Standards Institute) breakpoints in 2020.

Mtb laboratory ID MATI, isolated from a 23-year-old man from sputum in 2021, was tested and interpreted according to CLSI breakpoints in 2021. The drug susceptibility is summarized in Table 4.

Table 4. Susceptibility profiles of tested MDR *Mtb* strains.

Laboratory ID	Drug	Concentration (µM)	Susceptibility
IZAK	STM	6.88	resistant
	INH	29.17	resistant
	RIF	>9.72	resistant
	EMB	2.45	sensitive
	PZA	>129.96	resistant
MATI	STM	>27.51	resistant
	INH	>58.33	resistant
	RIF	>9.72	resistant
	EMB	2.45	sensitive
	PZA	>1039.70	resistant

STM = streptomycin; INH = isoniazid; RIF = rifampicin; EMB = ethambutol; PZA = pyrazinamide.

3.3.2. Cytotoxicity Screening

The human hepatocellular liver carcinoma cell line HepG2 purchased from Health Protection Agency Culture Collections (ECACC, Salisbury, UK) was cultured in DMEM (Dulbecco's Modified Eagle's Medium–high glucose) (Sigma-Aldrich, St. Louis, MO, USA) supplemented with 10% fetal bovine serum (PAA Laboratories GmbH, Pasching, Austria), 1% L-glutamine solution (Sigma-Aldrich, St. Louis, MO, USA), and non-essential amino acid solution (Sigma-Aldrich, St. Louis, MO, USA) in a humidified atmosphere containing 5% CO_2 at 37 °C. For subculturing, the cells were harvested after trypsin/EDTA (Sigma-Aldrich, St. Louis, MO, USA) treatment at 37 °C. For cytotoxicity evaluation, the cells treated with the tested substances were used as experimental groups. Untreated HepG2 cells served as controls.

The cells were seeded in a density of 10,000 cells per well in a 96-well plate 24 h prior to the experiment. The next day, the cells were treated with each of the tested substances dissolved in DMSO. The tested substances were prepared at different incubation concentrations (1–1000 µM) in triplicates according to their solubility. Concurrently, the controls representing 100% cell viability, 0% cell viability (the cells treated with 10% DMSO), no cell control, and vehiculum controls were prepared in triplicates. After 24 h of incubation in a humidified atmosphere containing 5% CO_2 at 37 °C, the reagent from the kit CellTiter 96 AQueous One Solution Cell Proliferation Assay (CellTiter 96, PROMEGA, Fitchburg, WI, USA) was added. After 2 h of incubation at 37 °C, the absorbance of the samples was recorded at 490 nm (TECAN, Infinita M200, Grödig, Austria). A standard toxicological parameter IC_{50} was calculated using nonlinear regression from a semilogarithmic plot of incubation concentration versus the percentage of absorbance (log(inhibitor) vs. normalized response model, least squares fit) relative to untreated controls using GraphPad Prism 9 software, (GraphPad Software, San Diego, CA, USA).

4. Conclusions

Within this work, nineteen derivatives of *N*-(1-hydroxy-1,3-dihydrobenzo[*c*][1,2] oxaborol-6-yl)(hetero)aryl-2-carboxamides have been prepared and fully characterized. In the in vitro biological antimicrobial screening, the compounds showed selective growth inhibition of the tested mycobacteria in comparison to no growth inhibition of the representatives of Gram-positive and Gram-negative bacteria, yeasts, and fungi. The selected compounds retained the biological activity also against resistant strains of clinical isolates of *Mtb* H37Rv. The majority of the most active compounds proved to possess a beneficial selectivity index concerning the negligible inhibition of the HepG2 cell line. This series of compounds provides a point for further research of new antimycobacterial compounds.

Supplementary Materials: The following supporting information can be downloaded at: https://www.mdpi.com/article/10.3390/ijms24032951/s1.

Author Contributions: Conceptualization, M.K.-C. and M.D.; Methodology, P.Š., M.K.-C., O.J., P.P., K.K., P.B., J.K. and V.K., Validation, O.J., P.P., K.K., P.B., J.K. and V.K.; Formal Analysis, P.Š.; Investigation, P.Š., A.A.N., O.J., P.P., K.K., P.B., J.K. and V.K.; Resources, P.Š., O.J., P.P., K.K., P.B., J.K. and V.K.; Data Curation, P.Š.; Writing—Original Draft Preparation, P.Š. and M.K.-C.; Writing—Review and Editing, M.K.-C., O.J., P.B., K.K. and M.D.; Visualization, P.Š. and M.K.-C.; Supervision, M.K.-C.; Project Administration, P.Š.; Funding Acquisition, M.D. All authors have read and agreed to the published version of the manuscript.

Funding: This research was funded by the Ministry of Health of the Czech Republic, grant no. NU21-05-00482 and by Charles University project SVV 260 547 (Czech Republic).

Institutional Review Board Statement: Not applicable.

Informed Consent Statement: Not applicable.

Data Availability Statement: Data are contained within the article or Supplementary Materials. All original data presented in this study are available from the authors upon request.

Acknowledgments: The authors acknowledge, Ing. Jana Vacková, Ida Dufková, and Iva Vencovská for their technical support.

Conflicts of Interest: The authors declare no conflict of interest.

References

1. WHO. *Global Tuberculosis Report 2022*; WHO: Geneve, Switzerland, 2022; p. 68. ISBN 978-92-4-006172-9.
2. Coker, R.J. Review: Multidrug-resistant tuberculosis: Public health challenges. *Trop. Med. Int. Health* **2004**, *9*, 25–40. [CrossRef] [PubMed]
3. Esposito, S.; D'Ambrosio, L.; Tadolini, M.; Schaaf, H.S.; Luna, J.C.; Centis, B.R.; Dara, M.; Matteelli, A.; Blasi, F.; Migliori, G.B. ERS/WHO Tuberculosis Consilium assistance with extensively drug-resistant tuberculosis management in a child: Case study of compassionate delamanid use. *Eur. Respir. J.* **2014**, *44*, 811–815. [CrossRef] [PubMed]
4. Adamczyk-Wozniak, A.; Borys, K.M.; Sporzynski, A. Recent Developments in the Chemistry and Biological Applications of Benzoxaboroles. *Chem. Rev.* **2015**, *115*, 5224–5247. [CrossRef]
5. Adamczyk-Wozniak, A.; Cyranski, M.K.; Zubrowska, A.; Sporzynski, A. Benzoxaboroles-Old compounds with new applications. *J. Organomet. Chem.* **2009**, *694*, 3533–3541. [CrossRef]
6. Snyder, H.R.; Reedy, A.J.; Lennarz, W.J. Synthesis of aromatic boronic acids. aldehydo boronic acids and a boronic acid analog of tyrosine1. *J. Am. Chem. Soc.* **1958**, *80*, 835–838. [CrossRef]
7. Lennarz, W.J.; Snyder, H.R. Arylboronic Acids.4. Reactions of Boronophthalide. *J. Am. Chem. Soc.* **1960**, *82*, 2172–2175. [CrossRef]
8. Baker, S.J.; Ding, C.Z.; Akama, T.; Zhang, Y.K.; Hernandez, V.; Xia, Y. Therapeutic potential of boron-containing compounds. *Future Med. Chem.* **2009**, *1*, 1275–1288. [CrossRef]
9. Fernandes, G.F.S.; Denny, W.A.; Dos Santos, J.L. Boron in drug design: Recent advances in the development of new therapeutic agents. *Eur. J. Med. Chem.* **2019**, *179*, 791–804. [CrossRef]
10. Dhawan, B.; Akhter, G.; Hamid, H.; Kesharwani, P.; Alam, M.S. Benzoxaboroles: New emerging and versatile scaffold with a plethora of pharmacological activities. *J. Mol. Struct.* **2022**, *1252*, 21. [CrossRef]

11. Coghi, P.S.; Zhu, Y.H.; Xie, H.M.; Hosmane, N.S.; Zhang, Y.J. Organoboron Compounds: Effective Antibacterial and Antiparasitic Agents. *Molecules* **2021**, *26*, 26. [CrossRef]
12. Rock, F.L.; Mao, W.M.; Yaremchuk, A.; Tukalo, M.; Crepin, T.; Zhou, H.C.; Zhang, Y.K.; Hernandez, V.; Akama, T.; Baker, S.J.; et al. An antifungal agent inhibits an aminoacyl-tRNA synthetase by trapping tRNA in the editing site. *Science* **2007**, *316*, 1759–1761. [CrossRef] [PubMed]
13. Palencia, A.; Li, X.F.; Bu, W.; Choi, W.; Ding, C.Z.; Easom, E.E.; Feng, L.; Hernandez, V.; Houston, P.; Liu, L.; et al. Discovery of Novel Oral Protein Synthesis Inhibitors of Mycobacterium tuberculosis That Target Leucyl-tRNA Synthetase. *Antimicrob. Agents Chemother.* **2016**, *60*, 6271–6280. [CrossRef] [PubMed]
14. Li, X.F.; Hernandez, V.; Rock, F.L.; Choi, W.; Mak, Y.S.L.; Mohan, M.; Mao, W.M.; Zhou, Y.; Easom, E.E.; Plattner, J.J.; et al. Discovery of a Potent and Specific M-tuberculosis Leucyl-tRNA Synthetase Inhibitor: (S)-3-(Aminomethyl)-4-chloro-7-(2hydroxyethoxy)benzo c 1,2 oxaborol-1(3 H)-ol (GSK656). *J. Med. Chem.* **2017**, *60*, 8011–8026. [CrossRef]
15. Edwards, B.D.; Field, S.K. The Struggle to End a Millennia-Long Pandemic: Novel Candidate and Repurposed Drugs for the Treatment of Tuberculosis. *Drugs* **2022**, *82*, 1695–1715. [CrossRef] [PubMed]
16. Hu, Q.H.; Liu, R.J.; Fang, Z.P.; Zhang, J.; Ding, Y.Y.; Tan, M.; Wang, M.; Pan, W.; Zhou, H.C.; Wang, E.D. Discovery of a potent benzoxaborole-based anti-pneumococcal agent targeting leucyl-tRNA synthetase. *Sci. Rep.* **2013**, *3*, 2475. [CrossRef]
17. Seiradake, E.; Mao, W.; Hernandez, V.; Baker, S.J.; Plattner, J.J.; Alley, M.R.K.; Cusack, S. Crystal Structures of the Human and Fungal Cytosolic Leucyl-tRNA Synthetase Editing Domains: A Structural Basis for the Rational Design of Antifungal Benzoxaboroles. *J. Mol. Biol.* **2009**, *390*, 196–207. [CrossRef]
18. Korkegian, A.; O'Malley, T.; Xia, Y.; Zhou, Y.S.; Carter, D.S.; Sunde, B.; Flint, L.; Thompson, D.; Ioerger, T.R.; Sacchettini, J.; et al. The 7-phenyl benzoxaborole series is active against Mycobacterium tuberculosis. *Tuberculosis* **2018**, *108*, 96–98. [CrossRef]
19. Alam, M.A.; Arora, K.; Gurrapu, S.; Jonnalagadda, S.K.; Nelson, G.L.; Kiprof, P.; Jonnalagadda, S.C.; Mereddy, V.R. Synthesis and evaluation of functionalized benzoboroxoles as potential anti-tuberculosis agents. *Tetrahedron* **2016**, *72*, 3795–3801. [CrossRef]
20. Patel, N.; O'Malley, T.; Zhang, Y.K.; Xia, Y.; Sunde, B.; Flint, L.; Korkegian, A.; Ioerger, T.R.; Sacchettini, J.; Alley, M.R.K.; et al. A Novel 6-Benzyl Ether Benzoxaborole Is Active against Mycobacterium tuberculosis In Vitro. *Antimicrob. Agents Chemother.* **2017**, *61*, 3. [CrossRef]
21. Gumbo, M.; Beteck, R.M.; Mandizvo, T.; Seldon, R.; Warner, D.F.; Hoppe, H.C.; Isaacs, M.; Laming, D.; Tam, C.C.; Cheng, L.W.; et al. Cinnamoyl-Oxaborole Amides: Synthesis and Their in Vitro Biological Activity. *Molecules* **2018**, *23*, 13. [CrossRef]
22. Adamczyk-Wozniak, A.; Komarovska-Porokhnyavets, O.; Misterkiewicz, B.; Novikov, V.P.; Sporzynski, A. Biological activity of selected boronic acids and their derivatives. *Appl. Organomet. Chem.* **2012**, *26*, 390–393. [CrossRef]
23. Si, Y.Y.; Basak, S.; Li, Y.; Merino, J.; Iuliano, J.N.; Walker, S.G.; Tonge, P.J. Antibacterial Activity and Mode of Action of a Sulfonamide-Based Class of Oxaborole Leucyl-tRNA-Synthetase Inhibitors. *ACS Infect. Dis.* **2019**, *5*, 1231–1238. [CrossRef] [PubMed]
24. Ganapathy, U.S.; del Rio, R.G.; Cacho-Izquierdo, M.; Ortega, F.; Lelievre, J.; Barros-Aguirre, D.; Lindman, M.; Dartois, V.; Gengenbacher, M.; Dick, T. A Leucyl-tRNA Synthetase Inhibitor with Broad-Spectrum Antimycobacterial Activity. *Antimicrob. Agents Chemother.* **2021**, *65*, 13. [CrossRef] [PubMed]
25. Ganapathy, U.S.; Gengenbacher, M.; Dick, T. Epetraborole Is Active against Mycobacterium abscessus. *Antimicrob. Agents Chemother.* **2021**, *65*, 5. [CrossRef]
26. Kim, T.; Hanh, B.T.B.; Heo, B.; Quang, N.; Park, Y.; Shin, J.; Jeon, S.; Park, J.W.; Samby, K.; Jang, J. A Screening of the MMV Pandemic Response Box Reveals Epetraborole as A New Potent Inhibitor against Mycobacterium abscessus. *Int. J. Mol. Sci.* **2021**, *22*, 5936. [CrossRef]
27. Zhang, P.P.; Ma, S.T. Recent development of leucyl-tRNA synthetase inhibitors as antimicrobial agents. *MedChemComm* **2019**, *10*, 1329–1341. [CrossRef]
28. Lincecum, T.L.; Tukalo, M.; Yaremchuk, A.; Mursinna, R.S.; Williams, A.M.; Sproat, B.S.; Van Den Eynde, W.; Link, A.; Van Calenbergh, S.; Grotli, M.; et al. Structural and mechanistic basis of pre- and posttransfer editing by leucyl-tRNA synthetase. *Mol. Cell* **2003**, *11*, 951–963. [CrossRef]
29. Palencia, A.; Crepin, T.; Vu, M.T.; Lincecum, T.L.; Martinis, S.A.; Cusack, S. Structural dynamics of the aminoacylation and proofreading functional cycle of bacterial leucyl-tRNA synthetase. *Nat. Struct. Mol. Biol.* **2012**, *19*, 677–684. [CrossRef]
30. Liu, R.J.; Long, T.; Li, H.; Zhao, J.H.; Li, J.; Wang, M.Z.; Palencia, A.; Lin, J.Z.; Cusack, S.; Wang, E.D. Molecular basis of the multifaceted functions of human leucyl-tRNA synthetase in protein synthesis and beyond. *Nucleic Acids Res.* **2020**, *48*, 4946–4959. [CrossRef]
31. Cummings, W.M.; Cox, C.H.; Snyder, H.R. Arylboronic acids. Medium-size ring-containing boronic ester groups. *J. Org. Chem.* **1969**, *34*, 1669–1674. [CrossRef]
32. Franzblau, S.G.; Witzig, R.S.; McLaughlin, J.C.; Torres, P.; Madico, G.; Hernandez, A.; Degnan, M.T.; Cook, M.B.; Quenzer, V.K.; Ferguson, R.M.; et al. Rapid, low-technology MIC determination with clinical Mycobacterium tuberculosis isolates by using the microplate Alamar Blue assay. *J. Clin. Microbiol.* **1998**, *36*, 362–366. [CrossRef] [PubMed]
33. Sundarsingh, T.J.A.; Ranjitha, J.; Rajan, A.; Shankar, V. Features of the biochemistry of Mycobacterium smegmatis, as a possible model for Mycobacterium tuberculosis. *J. Infect. Public Health* **2020**, *13*, 1255–1264. [CrossRef]
34. Gupta, R.S.; Lo, B.; Son, J. Phylogenomics and Comparative Genomic Studies Robustly Support Division of the Genus Mycobacterium into an Emended Genus Mycobacterium and Four Novel Genera. *Front. Microbiol.* **2018**, *9*, 67. [CrossRef] [PubMed]

35. Namouchi, A.; Cimino, M.; Favre-Rochex, S.; Charles, P.; Gicquel, B. Phenotypic and genomic comparison of Mycobacterium aurum and surrogate model species to Mycobacterium tuberculosis: Implications for drug discovery. *BMC Genom.* **2017**, *18*, 530. [CrossRef]
36. Chaturvedi, V.; Dwivedi, N.; Tripathi, R.P.; Sinha, S. Evaluation of Mycobacterium smegmatis as a possible surrogate screen for selecting molecules active against multi-drug resistant Mycobacterium tuberculosis. *J. Gen. Appl. Microbiol.* **2007**, *53*, 333–337. [CrossRef]
37. Heinrichs, M.T.; May, R.J.; Heider, F.; Reimers, T.; Sy, S.K.B.; Peloquin, C.A.; Derendorf, H. Mycobacterium tuberculosis Strains H37ra and H37rv have Equivalent Minimum Inhibitory Concentrations to Most Antituberculosis Drugs. *Int. J. Mycobacteriol.* **2018**, *7*, 156–161. [CrossRef]
38. Ambrozkiewicz, W.; Kucerova-Chlupacova, M.; Jand'ourek, O.; Konecna, K.; Paterova, P.; Barta, P.; Vinsova, J.; Dolezal, M.; Zitko, J. 5-Alkylamino-N-phenylpyrazine-2-carboxamides: Design, Preparation, and Antimycobacterial Evaluation. *Molecules* **2020**, *25*, 21. [CrossRef]
39. Nawrot, D.; Suchankova, E.; Jandourek, O.; Konecna, K.; Barta, P.; Dolezal, M.; Zitko, J. N-pyridinylbenzamides: An isosteric approach towards new antimycobacterial compounds. *Chem. Biol. Drug Des.* **2021**, *97*, 686–700. [CrossRef]
40. Konecna, K.; Diepoltova, A.; Holmanova, P.; Jand'ourek, O.; Vejsova, M.; Voxova, B.; Barta, P.; Maixnerova, J.; Trejtnar, F.; Kucerova-Chlupacova, M. Comprehensive insight into anti-staphylococcal and anti-enterococcal action of brominated and chlorinated pyrazine-based chalcones. *Front. Microbiol.* **2022**, *13*, 14. [CrossRef]
41. Baker, S.J.; Zhang, Y.K.; Akama, T.; Lau, A.; Zhou, H.; Hernandez, V.; Mao, W.M.; Alley, M.R.K.; Sanders, V.; Plattner, J.J. Discovery of a new boron-containing antifungal agent, 5-fluoro-1,3-dihydro-1-hydroxy-2,1-benzoxaborole (AN2690), for the potential treatment of onychomycosis. *J. Med. Chem.* **2006**, *49*, 4447–4450. [CrossRef]
42. European Committee for Antimicrobial Susceptibility Testing (EUCAST) of the European Society for Clinical Microbiology and Infectious Diseases (ESCMID). Eucast Discussion Document E. Dis 5.1: Determination of minimum inhibitory concentrations (MICs) of antibacterial agents by broth dilution. *Clin. Microbiol. Infect.* **2003**, *9*, 9–15. [CrossRef]
43. Arendrup, M.C.; Meletiadis, J.; Mouton, J.W.; Lagrou, K.; Hama, P.; Guinea, J.; Afst-Eucast. Eucast Definitive Document E. Def 7.3.1. Method for the Determination of Broth Dilution Minimum Inhibitory Concentrations of Antifungal Agents for Yeasts. 2017. Available online: https://www.eucast.org/fileadmin/src/media/PDFs/EUCAST_files/AFST/Files/EUCAST_E_Def_7_3_1_Yeast_testing__definitive.pdf (accessed on 20 December 2022).
44. Arendrup, M.C.; Meletiadis, J.; Mouton, J.W.; Lagrou, K.; Hamal, P.; Guinea, J.; Afst-Eucast. Eucast Definitive Document E.Def 9.3.1. Method for the Determination of Broth Dilution Minimum Inhibitory Concentrations of Antifungal Agents for Conidia Forming Moulds. 2017. Available online: https://www.eucast.org/fileadmin/src/media/PDFs/EUCAST_files/AFST/Files/EUCAST_E_Def_9_3_1_Mould_testing__definitive.pdf (accessed on 20 December 2022).
45. Ramappa, V.; Aithal, G.P. Hepatotoxicity Related to Anti-tuberculosis Drugs: Mechanisms and Management. *J. Clin. Exp. Hepatol.* **2013**, *3*, 37–49. [CrossRef] [PubMed]

International Journal of
Molecular Sciences

MDPI

Article

Antimicrobial Performance of Innovative Functionalized Surfaces Based on Enamel Coatings: The Effect of Silver-Based Additives on the Antibacterial and Antifungal Activity

Hamza Ur Rehman [1], Francesca Russo [2], Massimo Calovi [2], Orietta Massidda [1,*] and Stefano Rossi [2,*]

1 Department of Cellular, Computational and Integrative Biology, University of Trento, 38123 Trento, Italy
2 Department of Industrial Engineering, University of Trento, 38123 Trento, Italy
* Correspondence: orietta.massidda@unitn.it (O.M.); stefano.rossi@unitn.it (S.R.); Tel.: +39-0461-282442 (S.R.)

Abstract: Frequently touched surfaces (FTS) that are contaminated with pathogens are one of the main sources of nosocomial infections, which commonly include hospital-acquired and healthcare-associated infections (HAIs). HAIs are considered the most common adverse event that has a significant burden on the public's health worldwide currently. The persistence of pathogens on contaminated surfaces and the transmission of multi-drug resistant (MDR) pathogens by way of healthcare surfaces, which are frequently touched by healthcare workers, visitors, and patients increase the risk of acquiring infectious agents in hospital environments. Moreover, not only in hospitals but also in high-traffic public places, FTS play a major role in the spreading of pathogens. Consequently, attention has been devoted to developing novel and alternative methods to tackle this problem. This study planned to produce and characterize innovative functionalized enameled coated surfaces supplemented with 1% $AgNO_3$ and 2% $AgNO_3$. Thus, the antimicrobial properties of the enamels against relevant nosocomial pathogens including the Gram-positive *Staphylococcus aureus* and the Gram-negative *Escherichia coli* and the yeast *Candida albicans* were assessed using the ISO:22196:2011 norm.

Keywords: vitreous enamel; antimicrobial coatings; silver; fungicidal activity

Citation: Rehman, H.U.; Russo, F.; Calovi, M.; Massidda, O.; Rossi, S. Antimicrobial Performance of Innovative Functionalized Surfaces Based on Enamel Coatings: The Effect of Silver-Based Additives on the Antibacterial and Antifungal Activity. *Int. J. Mol. Sci.* **2023**, *24*, 2364. https://doi.org/10.3390/ijms24032364

Academic Editor: Helena Felgueiras

Received: 30 December 2022
Revised: 20 January 2023
Accepted: 21 January 2023
Published: 25 January 2023

1. Introduction

With the increasing number of infectious diseases, their management has become a growing challenge. According to the Centers for Diseases Control and Prevention (CDC) and the World Health Organization (WHO), infectious diseases are a primary global concern [1].

Contaminated surfaces within healthcare and public environments play a significant role as a potential driver in the transmission of infectious diseases [2–4]. Since January 2020, with the emergence and spread of the SARS-CoV-2 pandemic, hygiene and precautionary measures such as sanitizing hands and disinfection of surfaces have been implemented worldwide at all levels [5,6].

Pathogenic microorganisms can survive for a prolonged time on frequently touched surfaces (FTS) and are the leading cause of outbreaks of hospital-acquired infections and healthcare-associated infections (commonly named HAIs) [7]. HAIs are the most frequent affliction [8] that affect between 5% and 10% of patients admitted to hospitals in developed countries, of which 15–40% end up in intensive care units (ICU). The risk of contracting HAIs is much higher in developing countries [9]. According to the European Center for Disease Prevention and Control (ECDC), in Europe HAIs are acquired by more than 4 million people annually compared to 1.7 million in the US [10]. This is because pathogens typically shed by healthcare workers and patients' hands contaminate the surrounding surfaces for days, increasing the risk of transmission and acquisition by healthy individuals and other patients [11]. Therefore, along with the routine practice of hand hygiene with sanitizers and alcohol-based hand rub, environmental hygiene and disinfection is another crucial aspect in controlling these infections [12,13].

Ethanol is the most common alcohol ingredient used routinely as a disinfectant and appears to have both effective antibacterial and antiviral potential. However, ethanol does not seem very effective against acquired antibiotic-resistant bacteria and, most importantly, must be used repeatedly in medical settings [14]. In addition, or as an alternative to the use of disinfectants, novel strategies are required [11,15], as the microorganisms causing HAIs also show resistance due to biofilm formation, which renders them phenotypically less susceptible to disinfectants [1] and to a large number of antimicrobials used in therapy [16]. Furthermore, unhygienic environmental surfaces elevate the resistance of pathogens to disinfectants. In parallel to FTS at a nosocomial level that is responsible for HAIs, the role of surfaces in the transmission of pathogenic microorganisms in public settings, including public transport [17], public toilets [18], and even kitchens and washrooms in a home [19] should not be neglected where doorknobs, lift buttons and light switches serve as a reservoir of a large number of pathogens [20]. The persistency of pathogens on FTS for months [21] makes it essential to adopt and carry out routine cleaning of these surfaces through non-alcohol-based disinfectants [22], that persist for more extended periods, or through the development of functionalized surfaces/materials with antimicrobial activity to counteract the spread of pathogens.

Antimicrobial coatings represent an effective and advancing strategy against microbial growth and probably the development of multidrug-resistant bacteria to reduce HAIs [20,23,24]. Antimicrobial materials create self-disinfecting surfaces by applying coatings with antimicrobial activity. Antimicrobial materials are divided into anti-biofouling surfaces (also known as non-adhesive surfaces) that prevent microbial adhesion and actively antimicrobial surfaces, which are further grouped into biocide-releasing surfaces that contain active eluting agents (such as ions silver, copper, zinc, or antibiotics, chloride, iodine), contact killing surfaces that become active upon contact of microbes (such as quaternary ammonium polymers or peptides), and light-activated antimicrobial surfaces that require photosensitizers such as TiO_2 photocatalyst thin films [23,25]. In addition to chemical additions, the surface topographical properties (e.g., the durability of material and prevention of microbial retention over time) should always be considered, which can considerably affect hygienic properties [26].

The property of high durability is particularly appreciated in vitreous enamel [27]. Vitreous enamel is an inorganic coating that possesses excellent physical and mechanical properties, such as resistance to light, scratches, chemicals, intense temperature, abrasion, corrosion, optimal ease of cleaning, and hygiene. These properties make it a suitable candidate for use in public healthcare and high-traffic environments where frequent cleaning and disinfection are compulsory [28]. Moreover, despite being based on a relatively ancient production technique, vitreous enamel coatings meet very modern criteria of environmental friendliness and recyclability. At the end of life, enameled metallic components can be treated as metal scraps, without the need to separate the coating from the metal substrate, as enamel compounds become smelt waste that separates from the metal and does not alter the properties of the smelt itself [29]. In addition to that, many recent technological advances have allowed the improvement of closed-loop waste management in the enameling process. In this context, reducing harmful chemicals and introducing new deposition methods have allowed the full recycling of effluents and the recycling of wet and dry formulations with a consequent reduction in over spraying [30,31]. Thus, the production of enamel coatings is perfectly compatible with the efforts to promote the circular economy and the closed-loop management plan.

Silver-based antimicrobial enamel coatings have been recently exploited [20]: these materials could represent a solution to prevent the retention, spread, and transmission of pathogenic microorganisms at the nosocomial level and in public high-traffic environments.

Ag^+ ions show the highest antimicrobial potential that has been demonstrated in a wide range of applications against bacteria, fungi, protozoa, and certain viruses [32]. The mechanism of action of Ag^+ ions against bacteria still needs to be understood entirely. However, several mechanisms have been proposed and accepted (Figure 1).

1. As bacterial cell walls are made up of negatively charged peptidoglycan (PG), the positively charged Ag^+ ions will bind to the cell walls, the cell membrane, and the membranal transport proteins, thus blocking the transport of essential substances in and out of the bacterial cells [33–35].
2. The Ag^+ ions may destroy the energy production of the bacterial cell by binding to the cellular proteins and enzymes through sulfhydryl groups (-SH), which leads to the inactivation of proteins and enzymes [33,35–37]. Subsequently, the bacterial cell will burst, leading to bacteria destruction [37].
3. Once inside the cell, Ag^+ ions will bind to bacterial DNA [33] and subsequently block replication and cell division [33,38,39].

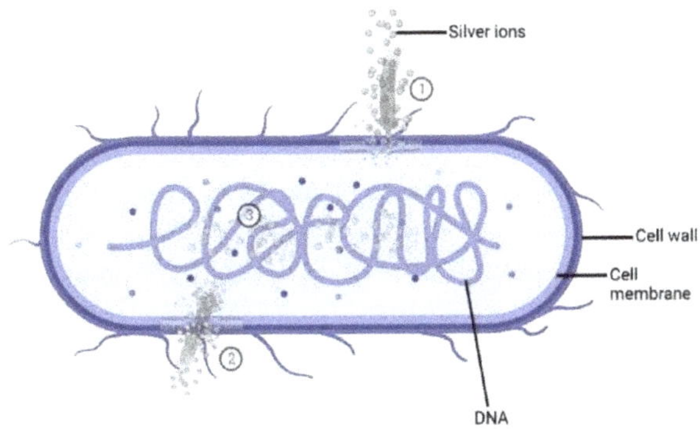

Figure 1. Schematic representation of the mode of action of silver ions (Ag^+). ①. The binding of Ag^+ to bacterial cell walls and membranes blocks the transport of essential substances. ②. Cellular uptake of Ag^+ leads to the binding and inactivation of enzymes. ③. The interaction of Ag^+ with bacterial DNA and blocking its cell division. Created with BioRender.com.

Silver has been widely used in the fabrication of antimicrobial surfaces/materials. It is considered one of the best alternatives in antimicrobial coatings due to its substantial antimicrobial effects even at low concentrations [40]. Its addition to polymers allows a persistent control of the environmental pathogenic bacteria [41]. It has been added to plastics [42] and synthetic fabrics [43], with a strong reduction in bacterial contamination. Moreover, some works have reported also the antifungal activity of silver, and silver nanoparticles have been shown to exert activity against dermatophytes [44]. It has been also shown that the conjugation of silver nanoparticles with multipurpose contact lenses and amphotericin B, an antifungal drug, significantly reduces corneal infections caused by *Acanthamoeba castellanii* [45,46]. Furthermore, it has been suggested that Ag^+ ions may act against fungi either by destroying the integrity of the cell membrane and halting the budding process [44] or through increased production of reactive oxygen species (ROS) after exposure in fungi [47], supporting the notion that silver nanoparticles with different sizes and shapes possess different levels of effectiveness, which in turn vary depending on the fungus and incubation time [48].

This work aimed to test the antimicrobial activities of newly developed innovative antimicrobial functionalized surfaces based on enamel coatings with the addition of silver-based additives ($AgNO_3$) that have previously shown efficient antibacterial activity [20]. The antibacterial activity of the composite coatings was tested in a time-course experiment against *Escherichia coli* and *Staphylococcus aureus*, as two main representative causative agents of HAIs within Gram-negative and Gram-positive bacteria, respectively [49]. In addition, the enamel coatings with silver-based additives were also tested against *Candida albicans*, as a model organism for fungal pathogens in hospital settings [50].

2. Results

2.1. Coatings Characterization

Figure 2 depicts the appearance of the three types of samples. The image illustrates a distinct variation in the surface hue of the different enamels, owing to the different concentrations of silver in the top layers of sample E1 and sample E2, containing 1 wt.% and 2 wt.% of $AgNO_3$, respectively. As a result, adding silver alters the visual features of the coatings, changing the upper shade of the materials.

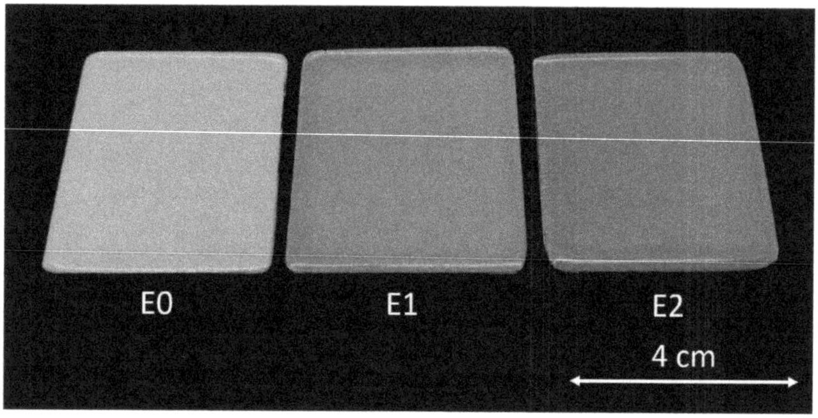

Figure 2. Top view of the enameled samples (40 mm × 40 mm). E0 (reference samples), E1 (admixed with 1% wt. $AgNO_3$), and E2 (admixed with 2% wt. $AgNO_3$).

Table 1 shows the roughness evaluation of the samples acquired with the PS1 roughness gauge (5 measurements on five samples—a total of 25 measurements). As demonstrated in Table 1, the surface roughness of each sample has shallow values (comparable among all the samples), and the addition of silver does not show a discernable impact on the topography of the samples that are typical of suitable manufactured enamel coatings.

Table 1. Surface roughness measurements (in-plane measurements).

Sample Type	Ra (μm)	Rz (μm)
E0	0.15 ± 0.01	0.74 ± 0.05
E1	0.15 ± 0.02	0.77 ± 0.16
E2	0.11 ± 0.02	0.63 ± 0.09

Cross-sectional analyses were carried out via optical and SEM microscopes to measure the thickness of three different layers (ground coat, white coat, and silver top-coat) and the internal microstructure of the test samples. The measurements in Table 2 relative to the thickness of the different layers are the average value of 5 measurements carried out on five samples (25 measurements in total) per series. The samples were cut and cold-mounted with an epoxy bi-component resin. Then, the samples were ground and mechanically polished from a #180 SiC grinding paper until a 1 μm polishing diamond paste was ready to be observed under optical and SEM microscopes.

Figure 3 represents the optical cross-sectional micrographs of the samples. The discrimination between the three different layers of the test samples is shown in the images. The dark bottom layer is the ground-coat layer and is often placed to ensure proper adhesion between the metal substrate and the enamel coating [51]. The ground-coat layer thickness is approximately 39–51 μm for all samples, as shown in Table 2. The thickness of the white-coat layer displays some inhomogeneity, ranging from 120–151 μm for each sample. This thickness variance is mostly attributed to the inherent unpredictability of the selected deposition technique, which does not guarantee complete reproducibility as it is heavily impacted by operator expertise. Finally, the pink silver-based top-coat represents the outer layer, with an estimated thickness of 30 μm and 24 μm for samples E1 and E2, respectively.

Table 2. Sample thickness measurements.

Sample Type	Total Thickness (μm)	Ground Coat (μm)	White Coat (μm)	Silver Coat (μm)
E0	171 ± 1	51 ± 7	120 ± 6	N/A
E1	217 ± 3	50 ± 5	137 ± 1	30 ± 2
E2	214 ± 1	39 ± 7	151 ± 3	24 ± 3

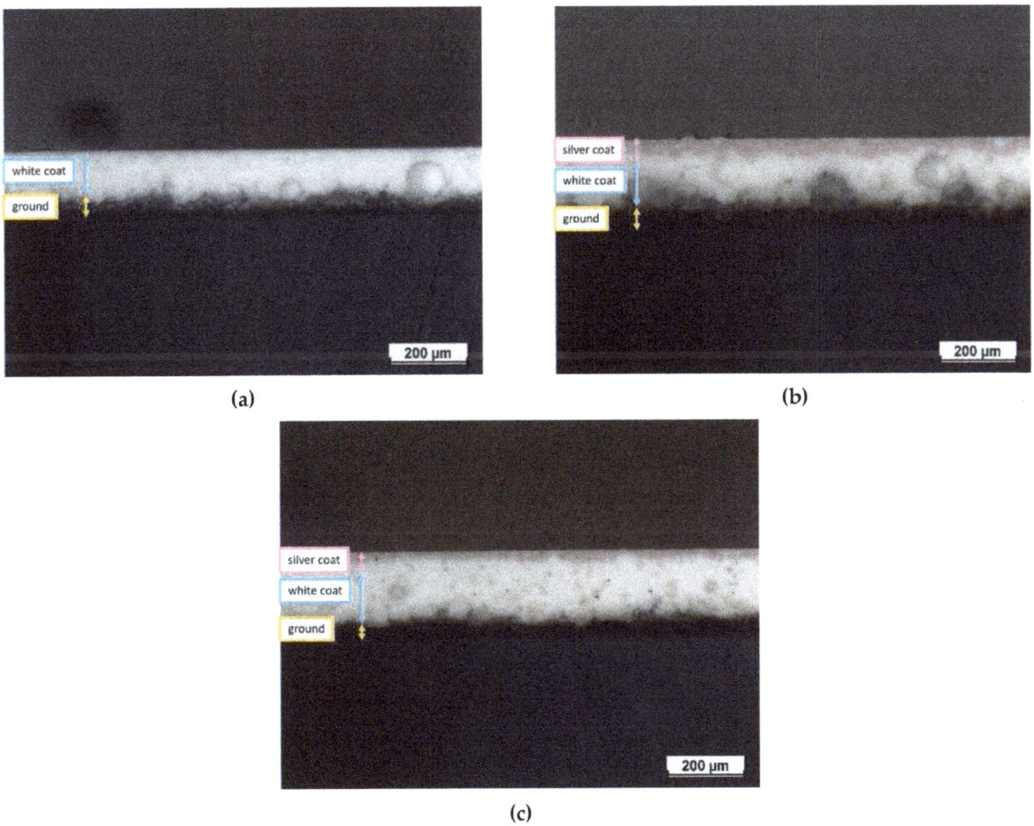

Figure 3. Optical cross-sectional micrographs of (**a**) reference sample E0, (**b**) sample E1, and (**c**) sample E2.

Similarly, Figure 4 represents the SEM micrographs of the three series of samples. The compact layers of the enamels exhibit the typically closed porosity structure which characterizes this type of protective coatings. The concentration of porosity decreases going up from the ground layer to the top-coat. The same bubbles appear larger (about 50 μm) in the ground and then decrease in size in the overlaying two layers.

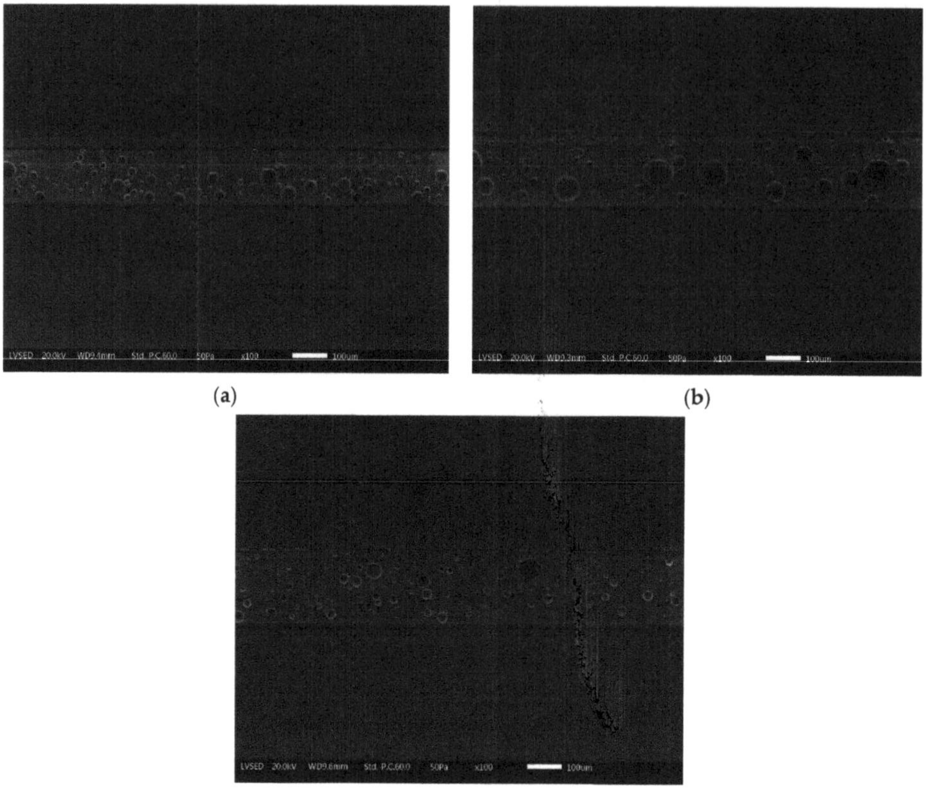

Figure 4. SEM cross-sectional micrographs of (**a**) reference sample E0, (**b**) sample E1, and (**c**) sample E2.

2.2. In Vitro Analysis of Coatings Antimicrobial Effectiveness

2.2.1. Antibacterial Efficacy

The treated enameled surfaces reduced the number of both *E. coli* and *S. aureus* strains compared to their respective untreated controls after 2, 8, and 24 h of exposure.

Figure 5 shows the reduction of the CFU/mL after the exposure of the silver-treated contaminated surfaces with *E. coli*. While the reference sample EE0 showed a relatively low decrease in CFU/mL over time, the presence of silver in the two samples, EE1, and EE2, caused a clear drop in the CFU/mL during the time course of the experiment. This result was more evident at higher silver concentrations, with no *E. coli* cells recovered in the EE2 sample already after 8 h of exposure.

To better highlight the inhibitory activity of silver, Figure 6 shows the percentage of recovered bacteria over time, normalized with respect to each untreated reference sample taken at the same time. The treated samples EE1 reduced the growth of *E. coli* to 30% and 98%, after an exposure time of 2 and 8 h, respectively, while the treated samples EE2 completely reduced the number of *E. coli* cells after 8 h. Both silver concentrations, however, resulted in complete inhibition of *E. coli* after 24 h, in agreement with our previous report [20]. The antibacterial activity (R) of the Ag-coated enameled samples against *E. coli* at 24 h, calculated based on the ISO:22196 formula, was of 3.55 for both EE1 and EE2.

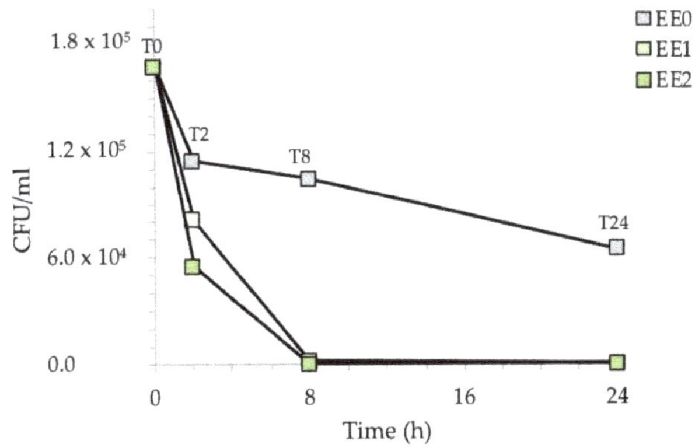

Figure 5. CFU/mL recovered after the exposure of the silver-treated contaminated surfaces to *E. coli* with respect to the untreated control. The results are representative of three independent determinations.

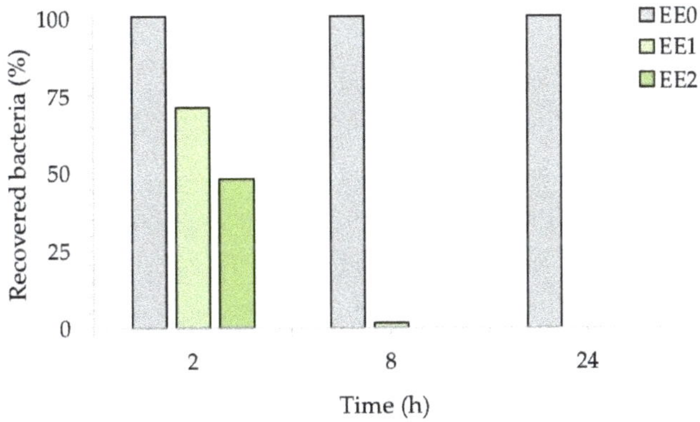

Figure 6. Antibacterial activity of the test surfaces against *E. coli* after, 2, 8 and 24 h of exposure.

Similarly, Figure 7 shows the CFU/mL recovered after the exposure of the contaminated surfaces with *S. aureus*. In this case, the antibacterial activity of silver was not particularly evident before 8 h of exposure for any of the silver-treated samples. However, sample SE2, containing a higher amount of Ag, showed a more rapid inhibitory behavior than sample SE1.

Figure 8, which illustrates the percentage of recovered bacteria over time, normalized with respect to each untreated reference sample taken at the same time, clearly points out this. Accordingly, the number of *S. aureus* cells was reduced in sample SE1 by 6% and 20% after 2 and 8 h of exposure, respectively. On the other hand, sample SE2 showed a 68% reduction compared to the untreated samples SE0 already after 8 h of exposure. Both silver concentrations, however, resulted in complete inhibition of *S. aureus* after 24 h, in agreement with our previous report [20]. The antibacterial activity (R) of the Ag-coated enameled samples against *S. aureus* at 24 h, calculated based on the ISO:22196 formula, was of 3.14 for both SE1 and SE2.

2.2.2. Antifungal Activity

A protocol similar to that used to evaluate the antibacterial activity of silver enamel against *E. coli* and *S. aureus* was used to analyze the antifungal potential of the treated enameled surfaces against *C. albicans*. *C. albicans* is an opportunistic dimorphic fungus that can grow as both yeast and filamentous cells forms; it has a commensal relationship with humans, but it is also responsible for about 70% of oral infections in humans and about 75% of genital infections in women [49,52]. Figure 9 shows the typical morphology of *C. albicans* cells, observed with the optical microscope after 24 h of contact with the surface of the reference sample E0.

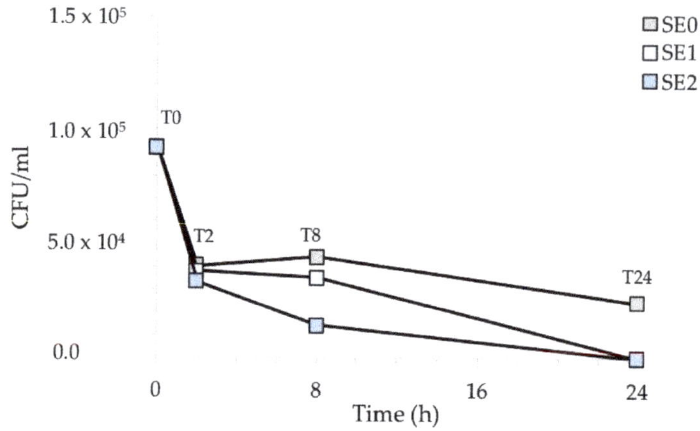

Figure 7. CFU/mL recovered after the exposure of the silver-treated contaminated surfaces to *S. aureus* with respect to the untreated control. The results are representative of three independent determinations.

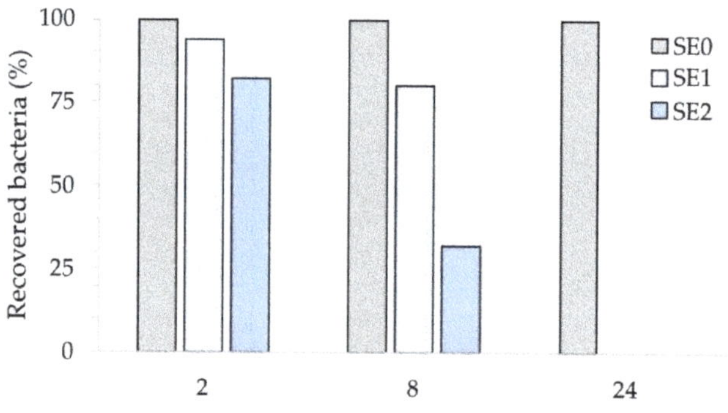

Figure 8. Antibacterial activity of the test surfaces against *S. aureus* after 2, 8 and 24 h of exposure.

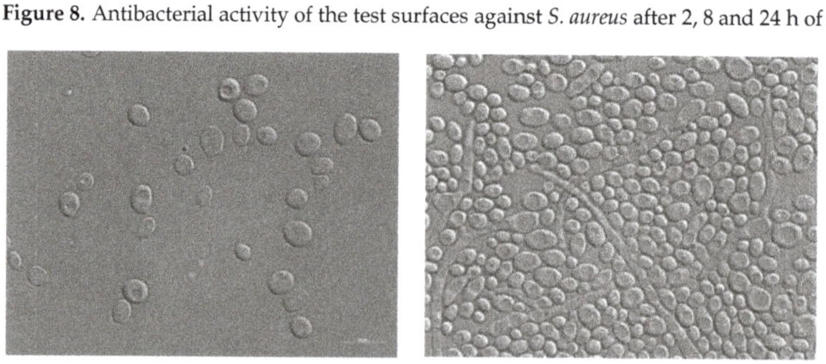

(a) **(b)**

Figure 9. Differential interference contrast (DIC) micrographs showing *C. albicans* at T0 (**a**) and T24 (**b**) of contact with the surface of the untreated enameled coating E0. Size bar, 10 μm.

Figure 10 reveals the reduction of the number of CFU/mL recovered during the exposure of the samples to *C. albicans*. In this case, due to the slower growth of *C. albicans* with respect to bacteria, the exposure was monitored every 24 h up to 72 h, to better observe the antifungal effect of silver, which became more evident over time.

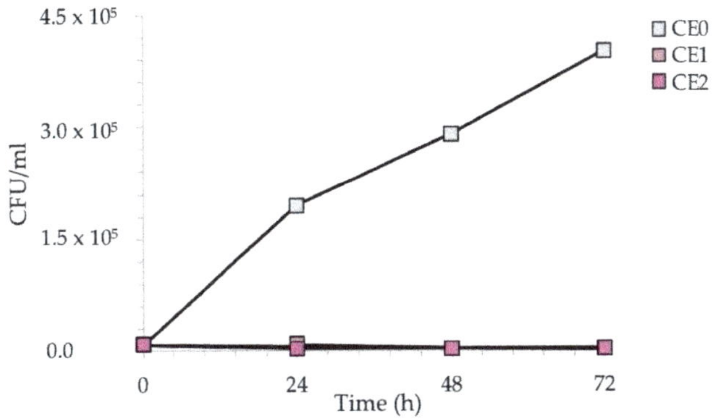

Figure 10. CFU/mL recovered after the exposure of the contaminated silver-treated enameled surfaces to *C. albicans*. The results are representative of three independent determinations.

As expected, the CFU/mL of the reference sample CE0 increased over time, as *C. albicans* requires longer incubation times for growth with respect to bacteria. Conversely, silver causes a sharp decrease in CFU/mL after 48 h in the case of sample CE1, and already after 24 h in the case of sample CE2. Again, the amount of silver seemed to play a role in the antimicrobial kinetics, anticipating a more efficient antifungal activity at higher Ag concentrations.

Figure 11 shows the percentage (%) of recovered fungal cells from three surfaces of each sample at each time point, normalized with respect to each untreated reference sample taken at the same time. Both CE1 and CE2 showed a reduction in the recovery of *C. albicans*. After 24 h of contact with CE1 and CE2, *C. albicans* showed a reduction of 80% and 98%, respectively. Longer exposure (48 h and 72 h) with CE1 and CE2 led to a larger inhibitory effect on *C. albicans* with a 99.9% reduction compared to the untreated surface CE0. Consistent with these results, the R activity for CE1 and CE2, based on the ISO:22196 formula, was of 1.53 and 1.88, respectively, at 24 h, of 2.19 and 3.03, respectively, at 48 h to reach 3.68 and 3.71, respectively, at 72 h.

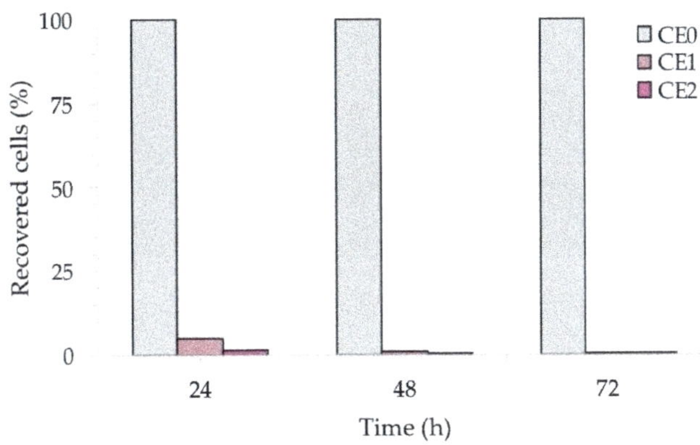

Figure 11. Antifungal activity of the test surfaces against *C. albicans* after 24, 48 and 72 h of exposure.

3. Discussion

The deposition of the silver layer in the most superficial layers of the coating proved to be able to guarantee an effective antimicrobial activity for longer exposure times [20]. The EDXS analyses carried out on these outer layers in previous work [20] demonstrated that sample E1 and sample E2 contain 0.5 wt.% and 1.5 wt.% of silver, respectively. Furthermore, from the SEM images, which highlight the internal morphology of the coatings, the presence of silver in the top-coat does not alter the structure of the enameled layer, nor does it preclude its correct deposition.

The results of the current work confirm that the functionalized surfaces, such as enameled coated surfaces supplemented with 1% wt. $AgNO_3$ (E1) and 2% wt. $AgNO_3$ (E2), possesses a remarkable antibacterial activity against the Gram-negative *E. coli* and the Gram-positive *S. aureus* and demonstrated that this activity occurs rapidly over time, dependent on $AgNO_3$ concentrations. Indeed, both E1 and E2 showed a consistent reduction in the number of *E. coli* after 2 h of exposure to the surfaces. Duration of 8 h and 24 h of contact with the treated surfaces led to complete inhibition of *E. coli*. On the other hand, *S. aureus* showed somewhat more resistance to both E1 and E2 after 2 and 8 h of contact. However, the complete absence of recovery of *S. aureus*, observed after 24 h of contact with the treated surfaces, is consistent with the antibacterial activity observed in a recent work published by Russo et al. [20]. As sample E2 possesses a higher concentration of Ag, it exhibits a more rapid antibacterial effect against both the Gram-negative bacteria and Gram-positive tested in this work and is consistent with the norm for these studies [53–55].

The antibacterial activity of both $AgNO_3$ concentrations was very consistent in all the experiments performed with *E. coli*, whereas the experiments performed with *S. aureus* showed more fluctuation. It has been shown that Gram-negative bacteria are more vulnerable to silver ions than Gram-positive bacteria [55], which is in agreement with the results in the current work. This may be mainly because the cells wall of Gram-positives and Gram-negatives differ significantly. The layers of PG in the cell wall of Gram-positives are thicker than those surrounding the Gram-negatives [56], likely enforcing a mechanism of protection against silver ions' entry into the cytoplasm [36]. However, no bacterial growth or multiplication was observed during continuous cultivation by inoculating the silver-treated cells into a fresh liquid LB media according to a work published by Feng et al. [36].

According to the concept of fungistatic and fungicidal action [57], both silver-based enameled coatings were shown to have a fungistatic effect after 24 h and a likely fungicidal effect after 48 and 72 h against *C. albicans*.

A few studies have demonstrated the antifungal properties of AgNPs against *C. albicans* [47,58–61], but no data on the antifungal properties of silver-based enameled coatings are available. The finding of the present work is consistent with the finding of previous studies that the effect of silver against fungi is dependent on its concentration [62] and incubation time [63], which explains why E1 is less active after 24 h and 48 h than E2. The current understanding of the mode of action of silver against fungi is poor but ultrastructure analysis of *C. albicans* while exposed to AgNPs revealed particles cluster outside the fungal cells, slowly releasing the silver ions and driving cell death via the reduction process by the contact of cell components with ionic silver [58]. As previously mentioned, a 99.9% reduction in the viability of *C. albicans* was observed after 48 h of contact time by both E1 and E2. The results were similar after 72 h, indicating an antifungal effect that fails to reach 100% efficacy in such a short time.

4. Material and Methods

4.1. Samples Deposition and Morphological Characterization

The samples were produced according to the previous work [20]. DC04EK cold-rolled steel panels (40 mm × 40 mm × 1.5 mm) were used as substrates for the coatings deposition. The steel panels were degreased in a hot detergent solution at 40 °C for 10 min, to remove possible contaminants and activate the surface for the deposition step. The frit and the samples were developed and deposited at Emaylum Italia s.r.l (Chignolo d'Isola, BG, Italy).

Silver nitrate ($AgNO_3$ > 99.0%, Sigma-Aldrich, St. Louis, MO, USA) was employed as the source of silver. Three series of samples were produced by the wet spraying method: the reference sample, free of silver, and the sample with the addition of 1 wt.% $AgNO_3$, and the sample with the addition of 2 wt.% $AgNO_3$. The reference sample was deposited by a 2A/2F (2 applications—2 firings) cycle: a "ground coat", was fired at 850 °C for 3–4 min, and subsequently, a "white top-coat", fired at 800 °C for 2–3 min, was deposited on it. The firing steps were carried out in a furnace in an air atmosphere. In order to limit the interpenetration of the two layers during the second firing step, the top-coat was fired at a lower temperature. The two samples containing silver were deposited by a 3A/3F application cycle: the three layers—namely, ground coat, white top-coat, and silver-based coat—were deposited and fired in separate steps. The silver-based layer was fired at 800 °C for 2–3 min. The formulation of the single layers can be found in the previous work [20]. Table 3 shows the nomenclature of the three series of samples.

Table 3. Samples nomenclature.

	Reference Sample	1 wt.% $AgNO_3$	2 wt.% $AgNO_3$
Sample nomenclature	E0	E1	E2
Number of layers	2	3	3

As a first step, the topological parameters of the enamel coatings, such as surface roughness and thickness, were characterized. The roughness of the surfaces was characterized using the MAHR Marsurf PS1 (MAHR GmbH, Göttingen, Germany) roughness gauge as per ISO 4287:1997 standard [64]. The Ra and Rz values exhibited in the results represent the averages of 25 measurements performed on five different surfaces from each sample. The thickness of the samples, including the thickness of different layers (ground-coat, white-coat, and silver-coat layers), was analyzed with the stereotypical optical microscope (NIKON SMZ25, Nikon Instruments, Amstelveen, the Netherlands). A total of 50 measurements were taken on five surfaces from each sample. The internal microstructure of the antimicrobial coatings was exploited using the stereotypical optical microscope (NIKON SMZ25, Nikon Instruments, Amstelveen, the Netherlands) and scanning electron microscope (SEM JEOL IT300, JEOL, Akishima, Tokyo, Japan) in low-vacuum mode at 20 kV.

4.2. Microbial Strains and Culture Media

The microorganisms tested in this work were the Gram-positive *S. aureus* ATCC 6538 strain, the Gram-negative *E. coli* ATCC 8739 strain, and the dimorphic fungus *C. albicans*. Nutrient agar (NA), Nutrient broth (NB), Sabouraud dextrose agar (SDA), and Plate count agar (PCA) were purchased from Microbial S.n.c (Uta, Cagliari, Italy). Tryptone soya agar (TSA) and Tryptone soya broth (TSB) were purchased from Oxoid (Oxoid Ltd. Basingstoke, Hants, UK). These media were used for bacterial and fungal stock culture preparation, pre-culture, inoculum preparation and to calculate the viable counts in all antimicrobial assays. Phosphate-buffered saline (PBS 1×), obtained from Gibco (Life Sciences, Waltham, MA, USA), was used for bacterial and fungal cell recovery from the test specimens and for 10-fold serial dilutions.

Bacterial stock cultures were kept in TSB at −80 °C with 15% glycerol for Gram-positives and with 30% glycerol for Gram-negatives. Fungal stock cultures were stored in sterile distilled water with 20% glycerol. All media were prepared as per the manufacturer's instructions.

4.3. Evaluation of Antimicrobial Activity

The antimicrobial efficacy of silver-based enameled coated surfaces at different interval points was evaluated using the ISO 22196:2011 standard [65], with some modifications when necessary. The steps carried out during this study are summarized in Figure 12.

1. Sterilization of test specimens and cover films: The test samples (reference, 1% $AgNO_3$ coated, and 2% $AgNO_3$ coated surface) and cover films were sterilized following the protocol described by Calovi et al. [66]. The polypropylene (PP) films (3.6 × 3.6 cm) were used to cover the suspension of test microbes dispersed on coated surfaces (4.0 × 4.0 cm) to prevent evaporation of microbial suspension. PP films and test specimens were kept in 70% ethanol for 10 min and then exposed to ultraviolet (UV) radiation for 1 h per sample side inside the laminar flow hood for sterilization.
2. Preparation of the test *inoculum*: The microbial strains were sub-cultured on appropriate agar plates (NA for *E. coli*, TSA for *S. aureus*, and SDA for *C. albicans*) at 37 °C for 24–48 h, according to the microbial strain, to obtain fresh cultures. Isolated colonies of the test microorganisms from the fresh cultures were resuspended in 10 mL of 1:500 NB (*E. coli* and *S. aureus*) or TSB (*C. albicans*) using a sterile loop. Cell number was adjusted to ~1.5×10^8 cells/mL, corresponding to 0.5 McFarland standard, and the optical density (OD) at 600 nm (OD600), ranging between 0.175 and 0.180 for *E. coli*, 0.210 for *S. aureus* and 0.250 for *C. albicans*, was determined with a UV-Vis spectrophotometer (Biochrom, Cambridge, USA). The suspensions were then diluted in PBS 1× to obtain the respective inoculum with a final concentration ranging from 3.9×10^5 cells/mL to 1.6×10^6 cells/mL (target concentration of 9.4×10^5 cells/mL). Microbial counts were determined by measuring the colony-forming units per mL (CFU/mL), following 10-fold dilution in PBS 1× and inoculated onto agar plates using an L-shaped loop and incubated for 24–48 h at 37 °C.
3. Inoculation of the test specimens: The sterilized enameled surfaces were placed in the middle of the petri dish, with the coated surface facing up. The test inoculum was pipetted onto each surface of the test samples. The contaminated surfaces were covered with a PP film, gently pressing down to spread the test inoculum on the coated surfaces but not to leak beyond the edges of the PP films. Then, the contaminated surfaces were incubated for different time intervals at 35 ± 1 °C, keeping the relative humidity above 90%. Reference samples used as controls to represent the test inoculum were processed instantly after microbial suspension at the T0.
4. Recovery of microbial cells from test specimens: Microorganisms were recovered from each sample after different inoculation times by adding 10 mL of PBS 1× to the Petri dishes containing the contaminated samples. Upon adding 10 mL of PBS 1×, the Petri dishes containing the contaminated samples were placed on Major Science Orbital Shaker (Saratoga, CA, USA) for 5 min at 125 revolutions per minute (rpm), to ensure the dissociation of the microorganisms from the surfaces. Then, the contaminated surfaces were washed by collecting and releasing the PBS 1× using a pipette at least four times. After this step, 10-fold serial dilutions in PBS 1× were performed and, finally, 1 mL of the undiluted sample and of each dilution was spread over the Petri dishes, and 15 mL of PCA was included. The colonies grown on PCA plates were counted according to the microbial strain incubation time at 35 ± 1 °C.
5. Determination of antimicrobial activity (R) of enamel surfaces contaminated for 24 h: The antimicrobial efficacy of enameled surfaces with 1% wt. $AgNO_3$ and 2% wt. $AgNO_3$ was evaluated by assessing the recovered microbial cells in contact with the enameled surfaces included for 24 h. The general formula for calculating the recovered cells/cm^2 from each specimen as per ISO 22,196 standard is as follows:

$$N\ (\text{cells/cm}^2) = (100 * C * D * V)/A \quad (1)$$

where N is the number of viable counts recovered per cm^2, C is the colony count, D is the dilution factor of the counted plate, V is the volume of the PBS added to the test samples and A is the surface area (mm^2) of the PP cover films. As per ISO standard accordance, three conditions should be satisfied, respectively, to have a valid test.

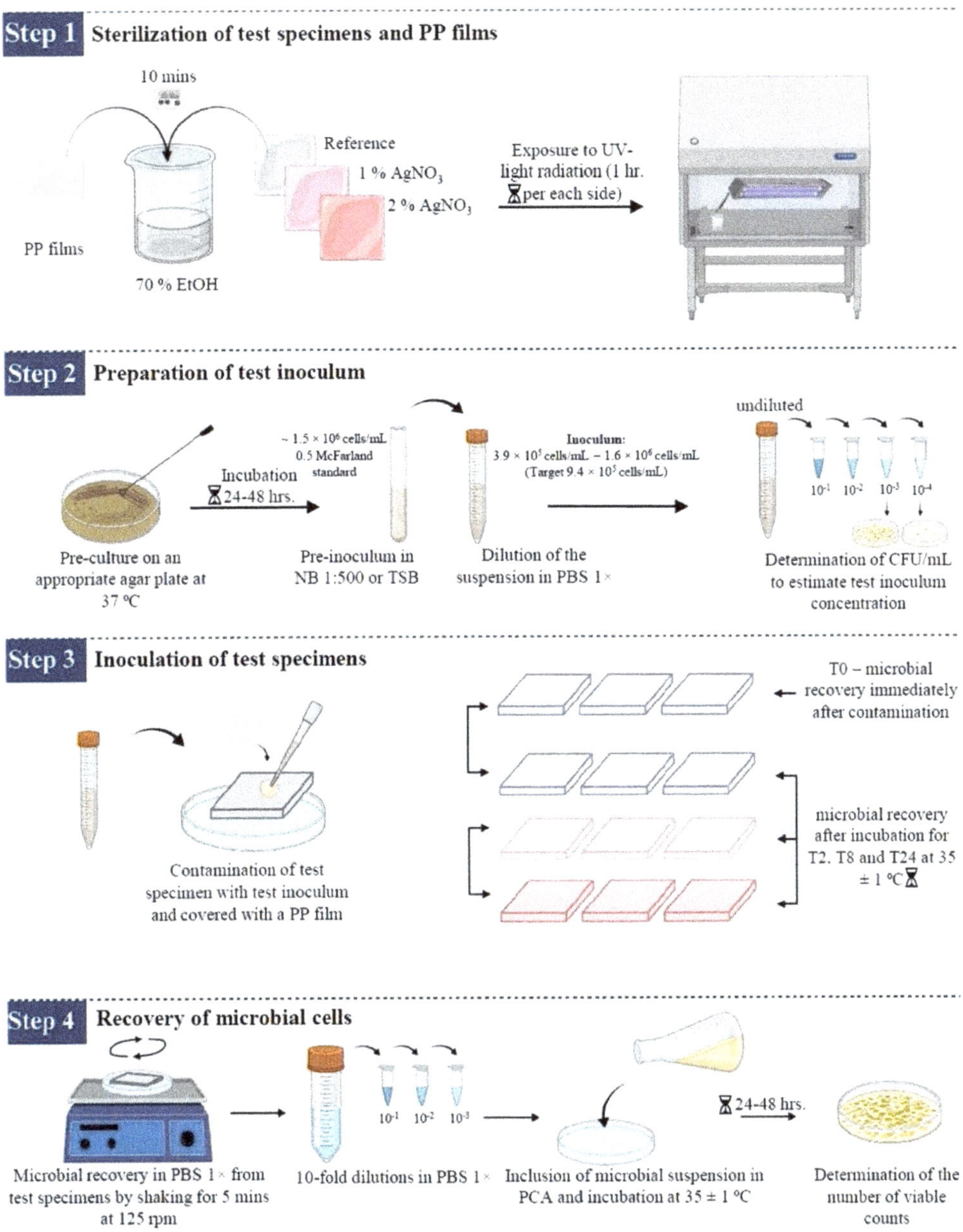

Figure 12. Schematic representation of the procedure used to determine the antimicrobial activity of the enameled surfaces. The image was taken and modified from [20].

I. The average number of viable counts recovered at T0 from the control samples should range between 6.2×10^3 cells/cm^2 and 2.5×10^4 cells/cm^2.
II. The logarithmic value of the number of viable counts recovered from control samples immediately after inoculation must meet the following criteria:

$$(L_{max} - L_{min})/L_{mean} \leq 0.2 \quad (2)$$

where L_{max} is log10 of the maximum number of viable microbial counts, L_{min} is log10 of the minimum number of viable counts, and L_{mean} is the mean number of viable counts recovered from control samples.

III. The number of viable microbial counts recovered from each reference sample at T24 should be more than 6.2×10 cells/cm^2.

When the conditions mentioned above are satisfied, the antimicrobial activity of test specimens is calculated by the following formula:

$$R = (U_t - U_0) - (A_t - U_0) = U_t - A_t \quad (3)$$

where R represents antimicrobial activity, U_0, and at log10 of the number of viable microbial counts recovered from control samples at T0, Ut represents an average of the log10 of the number of viable counts recovered at T24 from control samples, and A_t is the average of the log10 of the number of viable counts recovered from treated samples at T24.

The residual microbial counts at the different time points (T0, T2, T8, and T24 for *E. coli* and *S. aureus* and T24, T48, and T72 for *C. albicans*) were determined by calculating the CFU/mL.

5. Conclusions

The persistence of nosocomial pathogens on inanimate surfaces in hospital and health-care settings is a major source of transmission of HAIs. The highest likelihood of lengthy persistence is provided by a large inoculum, a high relative humidity (such as more than 70%), and low temperatures [67], which are addressed through this study with the ISO:22196 standard.

The functionalized enameled surfaces were supplemented with 1% wt. $AgNO_3$ and 2% $AgNO_3$ have significant antimicrobial activity. The experiments carried out in the present study confirmed that the enameled coated surfaces containing Ag have a complete antibacterial effect after 24 h of exposure against the Gram-positive *S. aureus* and the Gram-negative *E. coli*. Moreover, the treated enameled surfaces E2, containing more Ag, were more effective in reducing the viability of both bacteria than the treated enameled surfaces E1 at earlier time intervals. With respect to *E. coli*, *S. aureus* showed less reduction in the number of bacteria during the time course, likely due to its thicker cell wall. Nevertheless, longer exposure to silver-based coated surfaces led to complete inhibition of *S. aureus*. Furthermore, this work proves an antifungal activity (fungistatic or fungicidal) of the silver-based enameled coatings against the opportunistic pathogenic fungus *C. albicans*.

In the future, these silver-based enameled coatings could be tested also against other nosocomial clinical isolates that are multi-drug resistant (MDR). Due to their unique properties, such as non-toxicity, hygiene, ease of cleaning, resistance to corrosion and abrasion, surface smoothness, hardness, and durability, silver-based enameled coatings can be applied in the field of architecture and design, and in the medical, chemical and pharmaceutical areas that can benefit from the above-mentioned properties.

Author Contributions: Conceptualization, H.U.R., F.R., M.C., O.M. and S.R.; methodology, H.U.R., F.R., M.C., O.M. and S.R.; investigation, H.U.R., F.R., M.C. and O.M.; data curation, H.U.R., F.R, M.C. and O.M.; writing—original draft preparation, H.U.R., F.R. and M.C.; writing—review and editing, S.R. and O.M.; supervision, S.R. and O.M. All authors have read and agreed to the published version of the manuscript.

Funding: This research was funded by the RIVID project granted by the University of Trento in 2020 in the context of "COVID19" call awarded to S.R. (coordinator) and O.M. (collaborator).

Institutional Review Board Statement: Not applicable.

Informed Consent Statement: Not applicable.

Data Availability Statement: The data presented in this study are available on request from the corresponding author. The data are not publicly available due to the absence of an institutional repository.

Acknowledgments: The authors thank Attilio Compagnoni (Emaylum Italia, Chignolo d'Isola, BG, Italy) for the supply of samples and useful suggestions. The authors would like to thank Berenice Furlan (Univerisity of Trento) for her support in the antibacterial assays.

Conflicts of Interest: The authors declare no conflict of interest.

References

1. Mahira, S.; Jain, A.; Khan, W.; Domb, A.J. Antimicrobial materials—An overview. In *Antimicrobial Materials for Biomedical Applications*, 1st ed.; Abraham, J., Domb, K.R.K., Farah, S., Eds.; Royal Society of Chemistry: London, UK, 2019; pp. 1–37.
2. Bures, S.; Fishbain, J.T.; Uyehara, C.F.; Parker, J.M.; Berg, B.W. Computer keyboards and faucet handles as reservoirs of nosocomial pathogens in the intensive care unit. *Am. J. Infect. Control* **2000**, *28*, 465–471. [CrossRef] [PubMed]
3. Weber, D.J.; Rutala, W.A. Role of environmental contamination in the transmission of vancomycin-resistant enterococci. *Infect. Control Hosp. Epidemiol.* **1997**, *18*, 306–309. [CrossRef] [PubMed]
4. Otter, J.A.; Yezli, S.; Salkeld, J.A.; French, G.L. Evidence that contaminated surfaces contribute to the transmission of hospital pathogens and an overview of strategies to address contaminated surfaces in hospital settings. *Am. J. Infect. Control* **2013**, *41*, S6–S11. [CrossRef]
5. Moore, L.D.; Robbins, G.; Quinn, J.; Arbogast, J.W. The impact of COVID-19 pandemic on hand hygiene performance in hospitals. *Am. J. Infect. Control* **2021**, *49*, 30–33. [CrossRef] [PubMed]
6. Bhat, S.A.; Sher, F.; Kumar, R.; Karahmet, E.; Haq, S.A.U.; Zafar, A.; Lima, E.C. Environmental and health impacts of spraying COVID-19 disinfectants with associated challenges. *Environ. Sci. Pollut. Res.* **2021**, *29*, 85648–85657. [CrossRef]
7. Cheng, V.; Chau, P.; Lee, W.; Ho, S.; Lee, D.; So, S.; Wong, S.; Tai, J.; Yuen, K. Hand-touch contact assessment of high-touch and mutual-touch surfaces among healthcare workers, patients, and visitors. *J. Hosp. Infect.* **2015**, *90*, 220–225. [CrossRef]
8. Burke, J.P. Infection control-a problem for patient safety. *N. Engl. J. Med.* **2003**, *348*, 651–656. [CrossRef]
9. Pittet, D.; Donaldson, L. Clean care is safer care: A worldwide priority. *Lancet* **2005**, *366*, 1246–1247. [CrossRef]
10. Adlhart, C.; Verran, J.; Azevedo, N.F.; Olmez, H.; Keinänen-Toivola, M.M.; Gouveia, I.; Melo, L.F.; Crijns, F. Surface modifications for antimicrobial effects in the healthcare setting: A critical overview. *J. Hosp. Infect.* **2018**, *99*, 239–249. [CrossRef] [PubMed]
11. Dancer, S.J. Controlling hospital-acquired infection: Focus on the role of the environment and new technologies for decontamination. *Clin. Microbiol. Rev.* **2014**, *27*, 665–690. [CrossRef] [PubMed]
12. Khan, A.; Dancer, S.; Humphreys, H. Priorities in the prevention and control of multidrug-resistant Enterobacteriaceae in hospitals. *J. Hosp. Infect.* **2012**, *82*, 85–93. [CrossRef]
13. Datta, R.; Platt, R.; Yokoe, D.S.; Huang, S.S. Environmental cleaning intervention and risk of acquiring multidrug-resistant organisms from prior room occupants. *Arch. Intern. Med.* **2011**, *171*, 491–494. [CrossRef]
14. Gold, N.A.; Mirza, T.M.; Avva, U. Alcohol sanitizer. 2018. Available online: https://europepmc.org/article/nbk/nbk513254 (accessed on 20 September 2022).
15. Choi, H.; Chatterjee, P.; Lichtfouse, E.; Martel, J.A.; Hwang, M.; Jinadatha, C.; Sharma, V.K. Classical and alternative disinfection strategies to control the COVID-19 virus in healthcare facilities: A review. *Environ. Chem. Lett.* **2021**, *19*, 1945–1951. [CrossRef]
16. Avershina, E.; Shapovalova, V.; Shipulin, G. Fighting antibiotic resistance in hospital-acquired infections: Current state and emerging technologies in disease prevention, diagnostics and therapy. *Front. Microbiol.* **2021**, *12*, 2044. [CrossRef] [PubMed]
17. Goscé, L.; Johansson, A. Analysing the link between public transport use and airborne transmission: Mobility and contagion in the London underground. *Environ. Health* **2018**, *17*, 1–11. [CrossRef]
18. Vardoulakis, S.; Oyarce, D.A.E.; Donner, E. Transmission of COVID-19 and other infectious diseases in public washrooms: A systematic review. *Sci. Total Environ.* **2022**, *803*, 149932. [CrossRef]
19. Kagan, L.J.; Aiello, A.E.; Larson, E. The role of the home environment in the transmission of infectious diseases. *J. Community Health* **2002**, *27*, 247–267. [CrossRef] [PubMed]
20. Russo, F.; Furlan, B.; Calovi, M.; Massidda, O.; Rossi, S. Silver-based vitreous enamel coatings: Assessment of their antimicrobial activity towards *Escherichia coli* and *Staphylococcus aureus* before and after surface degradation. *Surf. Coat. Technol.* **2022**, *445*, 128702. [CrossRef]
21. Kramer, A.; Schwebke, I.; Kampf, G. How long do nosocomial pathogens persist on inanimate surfaces? A systematic review. *BMC Infect. Dis.* **2006**, *6*, 1–8. [CrossRef]
22. Disinfection & Sterilization Guidelines | Guidelines Library | Infection Control | CDC. 2019. Available online: https://www.cdc.gov/infectioncontrol/guidelines/disinfection/index.html (accessed on 30 September 2022).
23. Page, K.; Wilson, M.; Parkin, I.P. Antimicrobial surfaces and their potential in reducing the role of the inanimate environment in the incidence of hospital-acquired infections. *J. Mater. Chem.* **2009**, *19*, 3819–3831. [CrossRef]
24. Crijns, F.R.; Keinänen-Toivola, M.M.; Dunne, C.P. Antimicrobial coating innovations to prevent healthcare-associated infection. *J. Hosp. Infect.* **2017**, *95*, 243–244. [CrossRef]

25. Chen, X.; Hirt, H.; Li, Y.; Gorr, S.-U.; Aparicio, C. Antimicrobial GL13K peptide coatings killed and ruptured the wall of Streptococcus gordonii and prevented formation and growth of biofilms. *PLoS ONE* **2014**, *9*, e111579. [CrossRef]
26. Whitehead, K.A.; Verran, J. The effect of surface topography on the retention of microorganisms. *Food Bioprod. Process.* **2006**, *84*, 253–259. [CrossRef]
27. Rossi, S.; Russo, F.; Calovi, M. Durability of vitreous enamel coatings and their resistance to abrasion, chemicals, and corrosion: A review. *J. Coat. Technol. Res.* **2021**, *18*, 39–52. [CrossRef]
28. Savvova, O. Effect of zinc and tin oxides on the bactericidal properties of glass enamel coatings. *Glass Ceram.* **2014**, *71*, 254–257. [CrossRef]
29. Shippy, G. Reclamation of Scrap Frit. In *Proceedings of the 41st Porcelain Enamel Technical Forum: Ceramic Engineering and Science Proceedings (1979)*; Smothers, W.J., Ed.; John Wiley & Sons, Inc.: Hoboken, NJ, USA, 1980; pp. 208–210.
30. Elliott, R.H. Zero Discharge, Zero Pollution, and Source Reduction. In *Proceedings of the 50th Porcelain Enamel Institute Technical Forum: Ceramic Engineering and Science Proceedings*; Wachtman, J.B., Ed.; John Wiley & Sons, Inc.: Hoboken, NJ, USA, 1989; p. 513.
31. Cameron, D.S. Waste Minimization. In *Proceedings of the 50th Porcelain Enamel Institute Technical Forum: Ceramic Engineering and Science Proceedings (1989)*; Wachtman, J.B., Ed.; John Wiley & Sons, Inc.: Hoboken, NJ, USA, 1989.
32. Silvestry-Rodriguez, N.; Sicairos-Ruelas, E.E.; Gerba, C.P.; Bright, K.R. Silver as a disinfectant. *Rev. Environ. Contam. Toxicol.* **2007**, *191*, 23–45. [PubMed]
33. Thurman, R.B.; Gerba, C.P.; Bitton, G. The molecular mechanisms of copper and silver ion disinfection of bacteria and viruses. *Crit. Rev. Environ. Sci. Technol.* **1989**, *18*, 295–315. [CrossRef]
34. Slawson, R.; Lee, H.; Trevors, J. Bacterial interactions with silver. *Biol. Met.* **1990**, *3*, 151–154. [CrossRef] [PubMed]
35. Slawson, R.M.; Van Dyke, M.I.; Lee, H.; Trevors, J.T. Germanium and silver resistance, accumulation, and toxicity in microorganisms. *Plasmid* **1992**, *27*, 72–79. [CrossRef]
36. Feng, Q.L.; Wu, J.; Chen, G.Q.; Cui, F.; Kim, T.; Kim, J. A mechanistic study of the antibacterial effect of silver ions on *Escherichia coli* and *Staphylococcus aureus*. *J. Biomed. Mater. Res.* **2000**, *52*, 662–668. [CrossRef] [PubMed]
37. Liau, S.; Read, D.; Pugh, W.; Furr, J.; Russell, A. Interaction of silver nitrate with readily identifiable groups: Relationship to the antibacterialaction of silver ions. *Lett. Appl. Microbiol.* **1997**, *25*, 279–283. [CrossRef]
38. Klueh, U.; Wagner, V.; Kelly, S.; Johnson, A.; Bryers, J. Efficacy of silver-coated fabric to prevent bacterial colonization and subsequent device-based biofilm formation. *J. Biomed. Mater. Res.* **2000**, *53*, 621–631. [CrossRef] [PubMed]
39. Richards, R. Antimicrobial action of silver nitrate. *Microbios* **1981**, *31*, 83–91.
40. Marambio-Jones, C.; Hoek, E. A review of the antibacterial effects of silver nanomaterials and potential implications for human health and the environment. *J. Nanopar. Res.* **2010**, *12*, 1531–1551. [CrossRef]
41. Brady, M.J.; Lisay, C.M.; Yurkovetskiy, A.V.; Sawan, S.P. Persistent silver disinfectant for the environmental control of pathogenic bacteria. *Am. J. Infect. Control* **2003**, *31*, 208–214. [CrossRef] [PubMed]
42. Silver, S. Bacterial silver resistance: Molecular biology and uses and misuses of silver compounds. *FEMS Microbiol. Rev.* **2003**, *27*, 341–353. [CrossRef] [PubMed]
43. Takai, K.; Ohtsuka, T.; Senda, Y.; Nakao, M.; Yamamoto, K.; Matsuoka, J.; Hirai, Y. Antibacterial properties of antimicrobial-finished textile products. *Microbiol. Immunol.* **2002**, *46*, 75–81. [CrossRef]
44. Kim, K.-J.; Sung, W.-S.; Moon, S.-K.; Choi, J.-S.; Kim, J.-G.; Lee, D.-G. Antifungal effect of silver nanoparticles on dermatophytes. *J. Microbiol. Biotechnol.* **2008**, *18*, 1482–1484.
45. Hendiger, E.B.; Padzik, M.; Sifaoui, I.; Reyes-Batlle, M.; López-Arencibia, A.; Zyskowska, D.; Grodzik, M.; Pietruczuk-Padzik, A.; Hendiger, J.; Olędzka, G. Silver Nanoparticles Conjugated with Contact Lens Solutions May Reduce the Risk of Acanthamoeba Keratitis. *Pathogens* **2021**, *10*, 583. [CrossRef]
46. Anwar, A.; Siddiqui, R.; Hussain, M.A.; Ahmed, D.; Shah, M.R.; Khan, N.A. Silver nanoparticle conjugation affects antiacanthamoebic activities of amphotericin B, nystatin, and fluconazole. *Parasitol. Res.* **2018**, *117*, 265–271. [CrossRef]
47. Lee, B.; Lee, M.J.; Yun, S.J.; Kim, K.; Choi, I.-H.; Park, S. Silver nanoparticles induce reactive oxygen species-mediated cell cycle delay and synergistic cytotoxicity with 3-bromopyruvate in *Candida albicans*, but not in *Saccharomyces cerevisiae*. *Int. J. Nanomed.* **2019**, *14*, 4801. [CrossRef] [PubMed]
48. González-Fernández, S.; Lozano-Iturbe, V.; Menéndez, M.F.; Ordiales, H.; Fernández-Vega, I.; Merayo, J.; Vazquez, F.; Quirós, L.M.; Martín, C. A Promising Antifungal and Antiamoebic Effect of Silver Nanorings, a Novel Type of AgNP. *Antibiotics* **2022**, *11*, 1054. [CrossRef] [PubMed]
49. Poolman, J.T.; Anderson, A.S. *Escherichia coli* and *Staphylococcus aureus*: Leading bacterial pathogens of healthcare associated infections and bacteremia in older-age populations. *Expert Rev. Vaccines* **2018**, *17*, 607–618. [CrossRef] [PubMed]
50. Kabir, M.A.; Hussain, M.A.; Ahmad, Z. *Candida albicans*: A model organism for studying fungal pathogens. *ISRN Microbiol.* **2012**, *2012*, 1–15. [CrossRef] [PubMed]
51. Voges, D.L. Porcelain Enamel: Properties and Applications. In Proceedings of the 58th Porcelain Enamel Institute Technical Forum: Ceramic Engineering and Science Proceedings, Nashville, TN, USA, 12–16 May 1996; pp. 22–26.
52. Aboualigalehdari, E.; Ghafourian, S.; Sadeghifard, N.; Sekawi, Z. Is *Candida albicans* a cause of nosocomial infection in Iran? *Rev. Med. Microbiol.* **2013**, *24*, 85–88. [CrossRef]

53. Nabipour, Y.S.; Rostamzad, A. Comparing the antimicrobial effects of silver and copper nanoparticles against pathogenic and resistant bacteria of *Klebsiella pneumonia, Pseudomonas aeruginosa* and *Staphylococcus aureus*. *Cumhur. Üniversitesi Fen Edeb. Fakültesi Fen Bilim. Derg.* **2015**, *36*, 2541–2546.
54. Schrand, A.M.; Rahman, M.F.; Hussain, S.M.; Schlager, J.J.; Smith, D.A.; Syed, A.F. Metal-based nanoparticles and their toxicity assessment. *Wiley Interdiscip. Rev. Nanomed.* **2010**, *2*, 544–568. [CrossRef]
55. Lara, H.H.; Ayala-Núnez, N.V.; Ixtepan Turrent, L.d.C.; Rodríguez Padilla, C. Bactericidal effect of silver nanoparticles against multidrug-resistant bacteria. *World J. Microbiol. Biotechnol.* **2010**, *26*, 615–621. [CrossRef]
56. Silhavy, T.J.; Kahne, D.; Walker, S. The bacterial cell envelope. *Cold Spring Harb. Perspect. Biol.* **2010**, *2*, a000414. [CrossRef]
57. Graybill, J.; Burgess, D.; Hardin, T. Key issues concerning fungistatic versus fungicidal drugs. *Eur. J. Clin. Microbiol.* **1997**, *16*, 42–50. [CrossRef]
58. Vazquez-Muñoz, R.; Avalos-Borja, M.; Castro-Longoria, E. Ultrastructural analysis of *Candida albicans* when exposed to silver nanoparticles. *PLoS ONE* **2014**, *9*, e108876. [CrossRef] [PubMed]
59. Ahamad, I.; Bano, F.; Anwer, R.; Srivastava, P.; Kumar, R.; Fatma, T. Antibiofilm Activities of Biogenic Silver Nanoparticles Against *Candida albicans*. *Front. Microbiol.* **2021**, *12*, 4093. [CrossRef] [PubMed]
60. Kim, K.-J.; Sung, W.S.; Suh, B.K.; Moon, S.-K.; Choi, J.-S.; Kim, J.G.; Lee, D.G. Antifungal activity and mode of action of silver nano-particles on *Candida albicans*. *Biometals* **2009**, *22*, 235–242. [CrossRef]
61. Lara, H.H.; Romero-Urbina, D.G.; Pierce, C.; Lopez-Ribot, J.L.; Arellano-Jiménez, M.J.; Jose-Yacaman, M. Effect of silver nanoparticles on *Candida albicans* biofilms: An ultrastructural study. *J. Nanobiotechnol.* **2015**, *13*, 1–12. [CrossRef] [PubMed]
62. Deyá, C.; Bellotti, N. Biosynthesized silver nanoparticles to control fungal infections in indoor environments. *Adv. Nat. Sci. Nanosci. Nanotechnol.* **2017**, *8*, 025005. [CrossRef]
63. Ishida, K.; Cipriano, T.F.; Rocha, G.M.; Weissmüller, G.; Gomes, F.; Miranda, K.; Rozental, S. Silver nanoparticle production by the fungus *Fusarium oxysporum*: Nanoparticle characterisation and analysis of antifungal activity against pathogenic yeasts. *Mem. Inst. Oswaldo Cruz* **2013**, *109*, 220–228. [CrossRef]
64. *ISO 4287:1997*; Geometrical Product Specifications (GPS)–Surface Texture: Profile Method–Terms, Definitions and Surface Texture Parameters. International Organization for Standardization: Geneva, Switzerland, 1997.
65. Cashera, A.; Porosa, L. ISO 22196: 2011 Measurement of Antibacterial Activity on Plastics and Other Non-Porous Surfaces "Modified Large Droplet Inoculation Method". Available online: http://www.skinguard24.com/data/SG24_Ryerson.pdf (accessed on 30 September 2022).
66. Calovi, M.; Furlan, B.; Coroneo, V.; Massidda, O.; Rossi, S. Facile Route to Effective Antimicrobial Aluminum Oxide Layer Realized by Co-Deposition with Silver Nitrate. *Coatings* **2022**, *12*, 28. [CrossRef]
67. Donkor, E.S. Nosocomial pathogens: An in-depth analysis of the vectorial potential of cockroaches. *Trop. Med. Infect. Dis.* **2019**, *4*, 14. [CrossRef] [PubMed]

International Journal of
Molecular Sciences

MDPI

Article

The In Vitro Effect of Laser Irradiation (Er:YAG and CO_2) and Chemical Reagents (Hydrogen Peroxide, Sodium Hypochlorite, Chlorhexidine, or Sodium Fluoride) Alone or in Combination on Reducing Root Caries Bacteria

Nitya Reddy [1], Janina Golob Deeb [1], Todd Kitten [2], Caroline K. Carrico [3] and Kinga Grzech-Leśniak [1,4,*]

1 Department of Periodontics, School of Dentistry, Virginia Commonwealth University, Richmond, VA 23298-0566, USA
2 Department of Oral and Craniofacial Molecular Biology, Philips Institute for Oral Health Research, School of Dentistry, Virginia Commonwealth University, Richmond, VA 23298-0566, USA
3 Department of Dental Public Health and Policy, School of Dentistry, Virginia Commonwealth University, Richmond, VA 23298-0566, USA
4 Laser Laboratory, Department of Oral Surgery, Wroclaw Medical University, 50-425 Wroclaw, Poland
* Correspondence: kgl@periocare.pl

Citation: Reddy, N.; Golob Deeb, J.; Kitten, T.; Carrico, C.K.; Grzech-Leśniak, K. The In Vitro Effect of Laser Irradiation (Er:YAG and CO_2) and Chemical Reagents (Hydrogen Peroxide, Sodium Hypochlorite, Chlorhexidine, or Sodium Fluoride) Alone or in Combination on Reducing Root Caries Bacteria. *Int. J. Mol. Sci.* **2022**, *23*, 15732. https://doi.org/10.3390/ijms232415732

Academic Editor: Helena Felgueiras

Received: 9 October 2022
Accepted: 8 December 2022
Published: 12 December 2022

Abstract: (1) Lasers have been used for the treatment of dentinal hypersensitivity and bacterial reductions in periodontology. The purpose of this in vitro study was to evaluate the effect of Carbon Dioxide (CO_2) and Erbium-doped Yttrium Aluminum Garnet (Er:YAG) lasers with chlorhexidine (CHX), hydrogen peroxide (H_2O_2), sodium hypochlorite (NaOCl), or sodium fluoride (NaF) on the viability of oral bacteria associated with root caries. (2) *Streptococcus mutans*, *Streptococcus sanguinis*, and *Enterococcus faecalis* were grown in Brain Heart Infusion (BHI) broth, diluted to an OD660 of 0.5, and treated with antiseptics with or without simultaneous irradiation with the Er:YAG and CO_2 lasers for 30 s repeated three times. The treatment groups consisted of 1: no treatment, 2: 0.5% H_2O_2 alone, 3: 0.5% NaOCl alone, 4: 0.12% CHX alone, 5: 2% NaF alone, 6: laser alone, 7: laser with 0.5% H_2O_2, 8: laser with 0.5% NaOCl, 9: laser with 0.12% CHX, and 10: laser with 2% NaF for both lasers. The microbial viability was determined through plating and viable colonies were counted, converted into CFU/mL, and transformed into log form. The statistical analysis was performed using a two-tailed paired t-test. (3) The use of CO_2 and Er:YAG lasers alone failed to show statistically significant antibacterial activity against any of the bacteria. The only effective monotreatment was CHX for *S. mutans*. The combined treatment of 0.5% NaOCl with Er:YAG produced the greatest reduction in overall viability. (4) The combination of the Er:YAG laser with 0.5% NaOCl resulted in the largest reduction in bacterial survival when compared to monotherapies with antimicrobial solutions or lasers.

Keywords: CO_2 laser; *Enterococcus faecalis*; erbium laser; oral pathogens; *Streptococcus mutans*; *Streptococcus sanguinis*

1. Introduction

Infection of the oral cavity is an important global and public health problem. In particular, root caries has become a significant concern with the increasing lifespan of humans and the longevity of dentition with a reported pooled prevalence of 41.5% [1]. Gingival recession can occur in patients with periodontal disease, resulting in local attachment loss and hypersensitivity of dentin [2]. It can also occur in patients with aggressive tooth brushing habits, thin gingival phenotype, and thin bone of alveolar housing and has been associated with aberrant frenal attachment, mucogingival deficiencies, orthodontic therapy, and positional characteristics of teeth, as well as aging [3,4]. These exposed root surfaces create new environments for microbial colonization and are more susceptible to caries due to better bacterial adhesion to the dentin surface and hypersensitivity of the exposed root surface leading to reduced toothbrushing. Demineralization is twice as rapid on the root surface as it is on the enamel because of its lower mineral and higher organic content compared to enamel [5]. Restorative root caries treatment with adhesive materials can be very challenging due to compromised isolation and access, adhesive properties of the root surface, and the lack of retention in preparations due to root form and anatomy [4].

The main etiology for the onset and progression of root caries is the presence of bacteria and fermentable carbohydrates on the root surface [6]. While the demineralization process for enamel starts at the pH of 5.5, root caries initiation and progression start considerably faster at a pH of 6.4 due to the lower degree of mineralization of cementum and dentin covering root surfaces [7]. Untreated caries causes increasing pain complicated by inflammations of gingiva and pulp and can lead to tooth extraction affecting ability to chew and have a direct impact on oral health, which is related to decreased quality of life [8]. Early bacterial colonizers of the root surface implicated in the etiology of root caries include Gram-positive non-mutans streptococci species such as *Streptococcus sanguinis*, *Actinomyces* spp., *Streptococcus mutans*, *Enterococcus faecalis*, *Rothia*, *Veillonella*, anaerobic Gram-negative bacteria, and the yeast, *Candida albicans* [9]. The microflora associated with root surfaces and carious lesions have considerable bacterial diversity and complexity. The presence of *Enterococcus faecalis*, *Streptococcus mutans*, and *Pseudoramibacter alactolyticus* were found in levels higher than 40% in elderly subjects with root caries. As these bacteria were commonly present in dentinal samples from root caries while rarely present or absent in other categories, *Enterococcus faecalis* was implicated as a bacterium associated with root caries [9].

Laser light affects microbial cell integrity including Gram-positive and Gram-negative bacteria, *C. albicans*, and other periopathogens [10]. The erbium lasers, including the Erbium-doped Yttrium Aluminum Garnet (Er:YAG) laser, coincide with the most optimal absorption peak of water and has reportedly had good bactericidal effects at low energy outputs and have thus been used as an effective alternative method to traditional periodontal scaling and root planning (SRP) [11]. It has been shown to remove the smear layer on the root surfaces without any apparent heat damage [10]. The use of erbium lasers as an adjunct to SRP is favorable [12,13]. When compared to mechanical scaling alone, laser treatment decreases endotoxins and lipopolysaccharides on the root surfaces and increases the growth and adherence of gingival fibroblasts, which may aid in the reattachment of gingival tissues to the root surface [11]. Erbium lasers can also be used for other applications such as deepithelialization, debonding ceramic restorations, and soft and hard tissue surgery [14,15].

The Carbon Dioxide (CO_2) laser with 10,600 nm is commonly used in dentistry and widely considered the best surgical laser for coagulation during and after surgery [16]. The chromophore for the CO_2 laser is water, which is similar to the Er:YAG, but differs in that the CO_2 laser targets water inside soft tissues and does not rely on an external water source as does the Er:YAG. The absorbed CO_2 laser energy causes the water within the tissue to vaporize and removes tissue through ablation [17].

Oral rinses are used in dentistry for their antimicrobial properties. Chlorhexidine (CHX) has bacteriostatic effects and is used pre-operatively and post-operatively due to its effective reduction in bacterial load [18,19]. Hydrogen peroxide (H_2O_2) rinses have been used for many years to help control plaque and oral infections and provide high antiplaque efficacy, decreased gingival inflammation, and oral bacteria counts with no side effects at 1.5% concentration [20]. At 3% concentration, available over the counter, with a maximum contact time of seven minutes, no mucosal irritation was noted in an animal model. The hydroxyl radicals generated by H_2O_2 photolysis provide a powerful oxidizing agent capable of inducing oxidative damage to oral bacteria [21].

There has been limited research examining the effect of chemical irrigants in conjunction with lasers on the bacterial counts of species implicated in root caries. The purpose of this in vitro study is to evaluate the effectiveness of using chemical irrigants, H_2O_2, NaOCl,

CHX, and NaF, in conjunction with the Er:YAG and CO_2 laser on the bacterial viability of *Streptococcus mutans*, *Streptococcus sanguinis*, and *Enterococcus faecalis* in an effort to identify if they could be used as adjunctive therapies in the treatment of root caries. We hypothesize that the combination of an irrigant with a laser will yield a greater bactericidal effect than either the irrigant or laser alone.

2. Results

Under the herein applied parameters, the use of CO_2 and Er:YAG lasers alone had limited antibacterial effects, but there were significant differences in bacterial survival based on the use of irrigant, laser, and bacteria for all combinations. A summary of the models is presented in Table 1. Table 2 presents the pairwise comparisons for the use of irrigants with and without the laser (laser + irrigant vs. irrigant alone). Table 3 presents the pairwise comparisons for the use of each laser with and without the irrigant (i.e., laser + irrigant vs. laser alone).

Table 1. Model results.

	CO_2		Er:YAG	
	F	*p*-Value	F	*p*-Value
Bacteria	8.09	0.0005	22.64	<0.0001
Irrigant	96.47	<0.0001	132.87	<0.0001
Laser (Y/N)	41.55	<0.0001	42.1	<0.0001
Bacteria*Irrigant	23.74	<0.0001	22.87	<0.0001
Bacteria*Irrigant*Laser (Y/N)	7.62	<0.0001	11.25	<0.0001

Note: Bacteria*Irrigant and Bacteria*Irrigant*Laser (Y/N) indicate the two and three-way interaction terms in the model.

Table 2. Pairwise comparisons for the effect of irrigants with and without a laser.

Laser Comparison	Bacterium	Irrigant	Δ	SE	*p*-Value	Adj *p*
CO_2 vs. No Laser	*E. faecalis*	CHX	−0.06	0.78	0.9341	1
		H_2O_2	−0.81	0.78	0.2983	1
		NaF	−1.22	0.78	0.1175	0.9993
		NaOCl	−5.74	0.78	<0.0001	<0.0001
		None	−1.59	0.78	0.0419	0.9655
	S. mutans	CHX	0.57	0.78	0.4637	1
		H_2O_2	−0.64	0.78	0.4103	1
		NaF	−0.85	0.78	0.276	1
		NaOCl	−2.20	0.78	0.0052	0.5227
		None	−1.09	0.78	0.1616	0.9999
	S. sanguinis	CHX	0.21	0.78	0.7832	1
		H_2O_2	1.57	0.78	0.0451	0.9713
		NaF	−0.56	0.78	0.4721	1
		NaOCl	−6.41	0.78	<0.0001	<0.0001
		None	−0.55	0.78	0.4783	1
Er:YAG vs. No Laser	*E. faecalis*	CHX	0.63	0.76	0.407	1
		H_2O_2	0.09	0.76	0.9033	1
		NaF	−0.93	0.76	0.2249	1
		NaOCl	−5.74	0.76	<0.0001	<0.0001
		None	−0.97	0.76	0.2074	1
	S. mutans	CHX	−0.26	0.76	0.7378	1
		H_2O_2	−0.58	0.76	0.4449	1
		NaF	−0.36	0.76	0.639	1
		NaOCl	−6.63	0.76	<0.0001	<0.0001
		None	−0.28	0.76	0.7128	1
	S. sanguinis	CHX	0.17	0.76	0.8237	1
		H_2O_2	1.68	0.76	0.0288	0.9233
		NaF	−0.19	0.76	0.8048	1
		NaOCl	−5.77	0.76	<0.0001	<0.0001
		None	−0.03	0.76	0.9688	1

SE = Standard Error; Adj *p* is Tukey's adjusted *p*-value.

Table 3. Pairwise comparisons for effect of laser with and without an irrigant.

Laser	Bacterium	Irrigant		Δ	SE	p-Value	Adj p
CO_2	*E. faecalis*	CHX	vs. No Irrigant	0.03	0.95	0.9753	1
		H_2O_2		−0.20	0.95	0.8376	1
		NaF		0.42	0.95	0.6595	1
		NaOCl		−7.61	0.95	<0.0001	<0.0001
	S. mutans	CHX		−7.00	0.95	<0.0001	<0.0001
		H_2O_2		−0.55	0.95	0.5619	1
		NaF		0.24	0.95	0.8045	1
		NaOCl		−3.39	0.95	0.0005	0.1051
	S. sanguinis	CHX		−2.65	0.95	0.0059	0.5554
		H_2O_2		−0.30	0.95	0.75	1
		NaF		0.00	0.95	0.9972	1
		NaOCl		−8.10	0.95	<0.0001	<0.0001
Er:YAG	*E. faecalis*	CHX	vs. No Irrigant	0.10	0.93	0.9145	1
		H_2O_2		0.08	0.93	0.9315	1
		NaF		0.09	0.93	0.9275	1
		NaOCl		−8.24	0.93	<0.0001	<0.0001
	S. mutans	CHX		−8.64	0.93	<0.0001	<0.0001
		H_2O_2		−1.31	0.93	0.1639	0.9999
		NaF		−0.08	0.93	0.9278	1
		NaOCl		−8.64	0.93	<0.0001	<0.0001
	S. sanguinis	CHX		−3.22	0.93	0.0007	0.1465
		H_2O_2		−0.71	0.93	0.4481	1
		NaF		−0.15	0.93	0.8689	1
		NaOCl		−7.99	0.93	<0.0001	<0.0001

2.1. CO_2 Laser

The CO_2 laser demonstrated a synergistic effect with NaOCl by killing significantly more *E. faecalis* than NaOCl alone (−5.7 logs, adj p-value < 0.0001) and significantly more than the CO_2 laser alone (−7.6, adj p-value = <0.0001). None of the other laser and irrigant combinations were effective at reducing *E. faecalis*.

The CO_2 laser also demonstrated a synergistic effect with NaOCl by killing significantly more *S. sanguinis* than with NaOCl alone (−6.4, p-value < 0.0001) or the CO_2 laser alone (−8.1, p-value < 0.0001). None of the other laser and irrigant combinations were effective at reducing *S. sanguinis*.

For *S. mutans*, the only effective treatment was chlorhexidine and there was no additional benefit with the CO_2 laser (adjusted p-value = 1.00) (Figure 1).

2.2. Er:YAG Laser

The Er:YAG laser demonstrated a synergistic effect with NaOCl by killing significantly more *E. faecalis* than with NaOCl alone (−6.6 logCFU, adj p-value < 0.0001) and significantly more than with the Er:YAG laser alone (−8.2 logCFU, adj. p-value < 0.0001).

There was also a synergistic effect between the Er:YAG laser and NaOCl for *S. mutans*, with the combination killing significantly more *S. mutans* than with NaOCl alone (−6.6 logCFU, adj p-value < 0.0001) or the laser alone (−8.6 logCFU, adj p-value < 0.0001). The chlorhexidine reduced *S. mutans* to undetectable levels independent of the Er:YAG laser (−0.3 logCFU, adj p-value = 1).

For *S. sanguinis*, there was a synergistic effect between the Er:YAG laser and NaOCl with significantly less *S. sanguinis* growth combined than with the laser alone (−8.0 logCFU, adj p-value < 0.0001) or with NaOCl alone (−5.8 logCFU; adj p-value < 0.0001) (Figure 2).

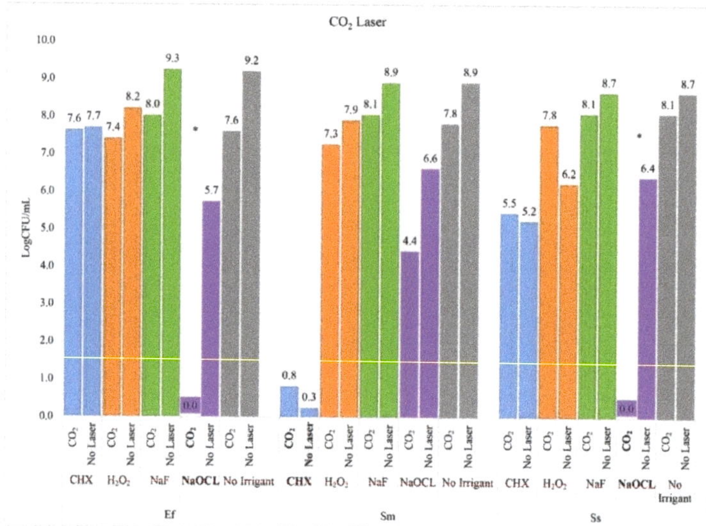

Figure 1. CO_2 Laser. Asterisk indicates a statistically significant difference between the two groups (i.e., Laser + Irrigant vs. Irrigant Alone) based on Tukey's adjusted pairwise comparisons.

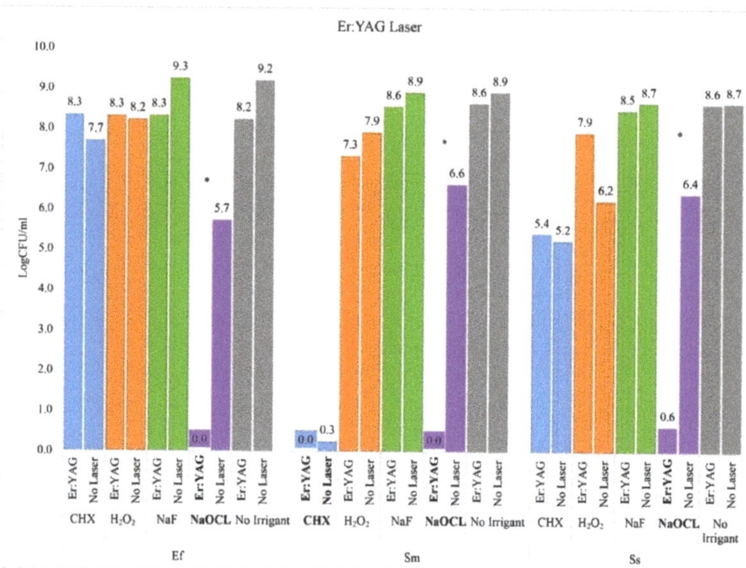

Figure 2. Er:YAG Laser. Asterisk indicates a statistically significant difference between the two groups (i.e., Laser + Irrigant vs. Irrigant Alone) based on Tukey's adjusted pairwise comparisons.

2.3. Chemical Reagents

Fluoride did not have any effect alone or in combination with any laser on bacterial viability for *S. sanguinis*, *S. mutans*, or *E. faecalis*. Hydrogen peroxide was only statistically significant in reducing bacterial growth for *S. mutans*, but still did not eliminate all bacteria (−3.2 logCFU, adj p-value = 0.0240). CHX was effective as a monotherapy for *S. mutans* reducing the bacterial count to undetectable levels with and without lasers (adj p-values = 1) and was statistically significantly more effective on Ss (−5.2 logCFU, adj *p*-value < 0.0001) and *E. faecalis* (−5.2 logCFU, adj *p*-value < 0.0001) when used in combination with the Er:YAG or CO_2 laser. Sodium hypochlorite 0.5%, proved to have the most extensive effect. NaOCl in combination with Er:YAG resulted in a synergistic effect for all three bacterial species. NaOCl with the CO_2 laser was effective for *E. faecalis* (−5.7 logCFU, adj *p*-value < 0.0001), and *Ss* (−6.4 logCFU, adj *p*-value < 0.0001) but not *S. mutans* (−2.2 logCFU, adj *p*-value = 0.5227).

2.4. Bacterial Strains

The most effective treatment for *E. faecalis* was NaOCl in combination with CO_2, or the Er:YAG laser, which resulted in undetectable amounts of *E. faecalis* after treatment. Either treatment, NaOCl or the laser alone, was unable to achieve the same level of bacterial reduction as the combination showing the synergistic effects of an antimicrobial with laser treatment. For the CO_2 laser, NaOCl (5.7, adj *p*-value < 0.0001) or the laser alone (7.6 logCFU, adj *p*-value < 0.0001) resulted in significantly more growth than the combination. The results were similar for the Er:YAG laser and NaOCl combination as compared to NaOCl (5.7 logCFU, adj *p*-value < 0.0001) or the laser alone (8.2 logCFU, adj *p* value < 0.0001).

The most effective treatment for *S. mutans* was CHX alone or combination Er:YAG with NaOCl. The Er:YAG with NaOCl resulted in significantly less bacterial growth than NaOCl alone (−6.6 logCFU, adj *p*-value < 0.0001) and Er:YAG laser alone (−8.6 logCFU, adj *p*-value < 0.0001). While the Er:YAG and CO_2 lasers provided a synergistic effect in combination with NaOCl, CHX by itself also produced the same result as compared to CHX with Er:YAG with CHX (0.26 logCFU, adj *p*-value = 1) or CO_2 with CHX (0.57 logCFU, adj *p*-value = 1).

The most effective treatment for *S. sanguis* was CHX or combination NaOCl with the Er:YAG and combination NaOCl with the CO_2 laser. All these combinations resulted in undetectable levels of *Ss* after treatment. While the Er:YAG and CO_2 lasers provided a synergistic effect in combination with NaOCl, CHX by itself also produced the same result as compared to the Er:YAG with CHX (0.17 logCFU, adj *p*-value = 1) or CO_2 with CHX (0.21 logCFU, adj *p*-value = 1).

3. Discussion

This study was able to show the effectiveness of the CO_2 and Er:YAG lasers in conjunction with antimicrobials in reducing the bacterial viability of *E. faecalis*, *S. mutans*, and *S. sanguis*. A combination of NaOCl with Er:YAG laser has shown in vitro synergistic bactericidal effects in reducing the viability and growth of these three bacterial species to undetectable levels. Both lasers as singular treatments did not reduce the bacterial load for any of the three tested species.

Previous studies have confirmed the bactericidal and synergistic effect of the Er:YAG laser when combined with a low concentration of antimicrobials, such as 0.5% NaOCl, 0.5% H_2O_2, or 0.03% CHX, on reducing periodontal pathogens, specifically *P. gingivalis*, *F. nucleatum*, and *S. gordonii* [22,23]. Similar laser settings were used in this investigation that could be employed as a supportive periodontal treatment protocol and root caries prevention strategy. The use of 3% hydrogen peroxide with 980 nm diode laser irradiation has been shown clinically effective in reducing the total bacteria count including periopathogens immediately and sustained for 6 weeks, 12 weeks, and 6 months following treatment in >5 mm and >6 mm periodontal pockets with a reduction in red and orange bacterial complexes [24,25]. The results from this investigation support the concept of a non-invasive supportive treatment protocol that can be implemented during periodontal maintenance in high caries risk patients with attachment loss following periodontal disease or therapy.

The lasers used in this study have different clinical indications. When mechanical periodontal therapy using SRP is compared to a combination of both Er:YAG and Nd:YAG lasers, a significant reduction in red and orange complexes was observed for the laser groups [26]. Using the Er:YAG laser resulted in an 85% reduction in periopathogens, whereas combinations of Er:YAG and Nd:YAG observed a 93% reduction and best improvements in clinical attachment level gain, compared to only 46% reduction in the SRP alone group [26]. The current study builds upon the already established evidence of the benefits of lasers in improving microbiological and clinical outcomes in non-surgical periodontal therapy.

The present study did not find any effect of NaF when used as a monotherapy or in conjunction with lasers on bacterial reduction. Previous studies examining the effect of various concentrations of fluorides on *P. gingivalis* and *S. mutans* cultured on titanium disks also did not find a significant decrease in bacterial growth [27] and even reported a slight increase in bacteria growth when a concentration of 1% gel was used [28]. The application of NaF to the surface of the tooth reduces demineralization of the enamel and promotes remineralization. Fluoride-treated teeth exhibit higher pH values by inhibiting bacterial acid production, which contributes to its antimicrobial effect, rather than having a direct bactericidal effect [29].

This study evaluated three bacteria associated with root caries. *S. mutans* and *S. sanguinis* are frequently found in root caries, whereas *E. faecalis* is more commonly associated with endodontic infections [30]. However, a growing number of studies have observed it in carious lesions, periodontitis, and peri-implantitis [31]. Studies have found *Enterococci* species at much higher rates in carious lesions (55.8%) versus caries-free lesions (7%) and have specifically found *E. faecalis* at a much higher frequency than any other *Enterococci* species [32]. *E. faecalis* appears to be associated with root caries disease processes as it was frequently observed in samples from carious roots, but was rarely present or absent in non-carious dentinal root samples [9].

Periodontally treated patients are more susceptible to caries formation due to root exposure and a high incidence of root caries has been observed at 4, 8, 12, and 14 years of maintenance periodontal therapy with approximately two-thirds of patients developing root caries during the first four years [33–36]. Besides oral hygiene instructions, nutrition counseling, and educating patients, fluoride can be used as an adjunctive treatment to aid in the remineralization of tooth surfaces despite lacking bactericidal effects.

Root caries prevention is of significant importance to periodontists due to its high prevalence and risk rates in 20% of patients referred for periodontal treatment [37]. The findings of the present study suggest that the use of a Er:YAG laser with low concentration of NaOCl can be an effective treatment protocol during maintenance appointments in high caries risk periodontal patients, particularly since there is established use of laser treatments with the present settings and use of chemical irrigants in dentistry [38].

There are several limitations to the present study. The determination of the bactericidal effects of the laser with chemical irrigants was performed in vitro, therefore, its clinical significance in periodontal therapy remains unclear. Additionally, only three bacterial species were evaluated with a small sample size.

4. Materials and Methods

4.1. Culture Conditions

Three oral bacterial species implicated in the etiology of root caries were used in this study: *S. mutans*, *S. sanguinis*, and *E. faecalis*. They were individually grown but treated in parallel. Freezer stocks (−80 °C) of each bacterial species were obtained and five microliters of a single-use aliquot were used to inoculate 5 mls of Brain Heart Infusion (BHI) broth (Becton–Dickinson; Franklin Lakes, NJ, USA). The inoculum was incubated overnight in an aerobic environment at 37 °C. The optical densities (OD) of the cultures were measured using a spectrophotometer (Genesys 150, Thermo Fisher Scientific, Waltham, MA, USA) at 660 nm (OD660) and normalized to an OD of 0.5. Ten aliquots of 150 μL of each bacterial strain were then transferred to non-adjacent wells of a 96-well plate to facilitate ten treatment groups per experiment. All the treatments were performed in a sterile biological safety cabinet. The chemicals used were added in the form of concentrated stock solutions (3%H_2O_2; 5.25% NaOCl; 2%CHX; 75% NaF) directly to the bacterial cultures in the plate. The stock solutions were diluted with sterile water to the concentrations specified per study group. The irradiations with each laser were performed on the appropriate samples. The chemical treatments and laser irradiation were performed one well at a time so that each bacterial culture was in contact with the chemical agent and/or laser for 60 s. Following laser irradiation, each treated sample was diluted 50-fold into fresh BHI broth. Additional dilutions were performed and spread onto BHI plates using an Eddy Jet 2 spiral plater (Neutec Group Inc.; Farmingdale, NY, USA). The plates were incubated for 24–48 h anaerobically at 37 °C. The viable colonies were counted for each plate, converted into CFU/mL, and transformed into log form for statistical analysis.

4.2. Laser Irradiation

The Er:YAG and CO_2 laser was set to normal periodontal clinical settings. The samples were irradiated by a Er:YAG laser at 2940 nm (LightWalker, Fotona, Slovenia), using a 400 µm Varian fiber tip of cylindrical quartz at parameters: 40 mJ; 40 Hz; 1.6 W for 30 s with the 300 µs short pulse duration in contact mode (SP).

An UltraSpeed Smart CO_2 laser at 10,600 nm (DEKA, Implant Direct, Thousand Oaks, CA, USA) irradiated using a contra-angle attachment and tapered tip at parameters: 50 Hz for 30 s with 0.3 s pulses in direct contact mode. The irradiation was performed with a disinfected aluminum foil barrier to isolate the treated wells from contamination.

4.3. Study Group

Group 1a–c: bacteria alone (*S. mutans*, *S. sanguinis*, or *E. faecalis* alone)
Group 2a–c: bacteria (*S. mutans*, *S. sanguinis*, or *E. faecalis*) + H_2O_2 (0.5%)
Group 3a–c: bacteria (*S. mutans*, *S. sanguinis*, or *E. faecalis*) + NaOCl (0.5%)
Group 4a–c: bacteria (*S. mutans*, *S. sanguinis*, or *E. faecalis*) + CHX (0.12%)
Group 5a–c: bacteria (*S. mutans*, *S. sanguinis*, or *E. faecalis*) + NaF (2%)
Group 6a–i: bacteria (*S. mutans*, *S. sanguinis*, or *E. faecalis*) + laser (Er:YAG or CO_2)
Group 7a–i: bacteria (*S. mutans*, *S. sanguinis*, or *E. faecalis*) + laser (Er:YAG or CO_2) + H_2O_2 (0.5%)
Group 8a–i: bacteria (*S. mutans*, *S. sanguinis*, or *E. faecalis*) + laser (Er:YAG or CO_2) + NaOCl (0.5%)
Group 9a–i: bacteria (*S. mutans*, *S. sanguinis*, or *E. faecalis*) + laser (Er:YAG or CO_2) + CHX (0.12%)
Group 10a–i: bacteria (*S. mutans*, *S. sanguinis*, or *E. faecalis*) + laser (Er:YAG or CO_2) + NaF (2%)

Each experiment was performed three times.

4.4. Statistical Analysis

The number of surviving colonies of bacterial species were analyzed using ANOVA models. The models evaluated the effects of the combinations of lasers and irrigants on various bacteria species using interaction terms. The post hoc pairwise comparisons were adjusted using Tukey's adjustment. The significance level was set at 0.05.

5. Conclusions

Under the conditions applied herein, lasers combined with low concentrations of commonly used dental chemical agents can provide additional benefits by reducing the bacteria associated with root caries. The use of 0.5% NaOCl in combination with an Er:YAG laser irradiation resulted in the greatest reduction in bacterial viability when compared to monotherapies with antimicrobial solutions or lasers alone. The laser alone was the most effective means for reducing *S. mutans* colonies. NaF had no effect alone or when combined with any of the lasers on bacterial reduction.

Author Contributions: Conceptualization, J.G.D., T.K. and K.G.-L.; methodology, J.G.D., T.K. and K.G.-L.; software, C.K.C.; validation, C.K.C. and J.G.D.; formal analysis, C.K.C.; investigation, N.R., resources, N.R.; data curation, N.R., J.G.D. and T.K. writing—original draft preparation, N.R.; J.G.D.,

T.K. and K.G.-L.; writing—review and editing, N.R., J.G.D., T.K. and K.G.-L.; supervision, J.G.D., T.K., K.G.-L. and C.K.C. project administration, J.G.D., T.K. and K.G.-L. All authors have read and agreed to the published version of the manuscript.

Funding: Nitya Reddy received funding from the VCU School of Dentistry Alexander Fellowship for this study. No financial relationships exist between any of the authors and any commercial firms that may pose a conflict of interest.

Institutional Review Board Statement: Not applicable.

Informed Consent Statement: Not applicable.

Conflicts of Interest: The authors declare no conflict of interest.

References

1. Pentapati, K.C.; Siddiq, H.; Yeturu, S.K. Global and regional estimates of the prevalence of root caries—Systematic review and meta-analysis. *Saudi Dent. J.* **2019**, *31*, 3–15. [CrossRef] [PubMed]
2. Kinane, D.F. Causation and pathogenesis of periodontal disease: Causation and pathogenesis of periodontal disease. *Periodontol. 2000* **2001**, *25*, 8–20. [CrossRef]
3. Chapple, I.L.C.; Mealey, B.L.; Van Dyke, T.E.; Bartold, P.M.; Dommisch, H.; Eickholz, P.; Geisinger, M.L.; Genco, R.J.; Glogauer, M.; Goldstein, M.; et al. Periodontal health and gingival diseases and conditions on an intact and a reduced periodontium: Consensus report of workgroup 1 of the 2017 World Workshop on the Classification of Periodontal and Peri-Implant Diseases and Conditions. *J. Periodontol.* **2018**, *89*, S74–S84. [CrossRef] [PubMed]
4. Burgess, J.O.; Gallo, J.R. Treating root-surface caries. *Dent. Clin. N. Am.* **2002**, *46*, 385–404. [CrossRef] [PubMed]
5. Banting, D.W.; Ellen, R.P.; Fillery, E.D. Prevalence of root surface caries among institutionalized older persons. *Community Dent. Oral Epidemiol.* **1980**, *8*, 84–88. [CrossRef]
6. Clarkson, B.H.; Krell, D.; Wefel, J.S.; Crall, J.; Feagin, F.F. In Vitro caries-like lesion production by Streptococcus mutans and Actinomyces viscosus using sucrose and starch. *J. Dent. Res.* **1987**, *66*, 795–798. [CrossRef]
7. Melberg, J.R. Demineralization and remineralization of root surface caries. *Gerodontology* **1986**, *5*, 25–31. [CrossRef]
8. Petersen, P.E. Sociobehavioural risk factors in dental caries—International perspectives. *Community Dent. Oral Epidemiol.* **2005**, *33*, 274–279. [CrossRef]
9. Preza, D.; Olsen, I.; Aas, J.A.; Willumsen, T.; Grinde, B.; Paster, B.J. Bacterial Profiles of Root Caries in Elderly Patients. *J. Clin. Microbiol.* **2008**, *46*, 2015–2021. [CrossRef]
10. Cobb, C.M. Lasers and the treatment of periodontitis: The essence and the noise. *Periodontol. 2000* **2017**, *75*, 205–295. [CrossRef]
11. Bader, C.; Krejci, I. Indications and limitations of Er:YAG laser applications in dentistry. *Am. J. Dent.* **2006**, *19*, 178–186.
12. Crespi, R.; Romanos, G.E.; Barone, A.; Sculean, A.; Covani, U. Er:YAG Laser in Defocused Mode for Scaling of Periodontally Involved Root Surfaces: An In Vitro Pilot Study. *J. Periodontol.* **2005**, *76*, 686–690. [CrossRef]
13. Feist, I.S.; Micheli, G.D.; Carneiro, S.R.S.; Eduardo, C.P.; Miyagi, S.P.H.; Marques, M.M. Adhesion and Growth of Cultured Human Gingival Fibroblasts on Periodontally Involved Root Surfaces Treated by Er:YAG Laser. *J. Periodontol.* **2003**, *74*, 1368–1375. [CrossRef] [PubMed]
14. Ishikawa, I.; Aoki, A.; Takasaki, A.A. Clinical application of erbium:YAG laser in periodontology. *J. Int. Acad. Periodontol.* **2008**, *10*, 22–30. [PubMed]
15. ALBalkhi, M.; Swed, E.; Hamadah, O. Efficiency of Er:YAG laser in debonding of porcelain laminate veneers by contact and non-contact laser application modes (in vitro study). *J. Esthet. Restor. Dent.* **2018**, *30*, 223–228. [CrossRef] [PubMed]
16. Garg, N.; Verma, S.; Chadha, M.; Rastogi, P. Use of carbon dioxide laser in oral soft tissue procedures. *Natl. J. Maxillofac. Surg.* **2015**, *6*, 84. [CrossRef]
17. Monteiro, L.; Delgado, M.L.; Garcês, F.; Machado, M.; Ferreira, F.; Martins, M.; Salazar, F.; Pacheco, J.J. A histological evaluation of the surgical margins from human oral fibrous-epithelial lesions excised with CO_2 laser, Diode laser, Er:YAG laser, Nd:YAG laser, electrosurgical scalpel and cold scalpel. *Med. Oral Patol. Oral Y Cir. Bucal* **2019**, *24*, e271. [CrossRef] [PubMed]
18. Veksler, A.E.; Kayrouz, G.A.; Newman, M.G. Reduction of Salivary Bacteria by Pre-Procedural Rinses With Chlorhexidine 0.12%. *J. Periodontol.* **1991**, *62*, 649–651. [CrossRef]
19. Sanz, M.; Newman, M.G.; Anderson, L.; Matoska, W.; Otomo-Corgel, J.; Saltini, C. Clinical Enhancement of Post-Periodontal Surgical Therapy by a 0.12% Chlorhexidine Gluconate Mouthrinse. *J. Periodontol.* **1989**, *60*, 570–576. [CrossRef]
20. Muniz, F.W.M.G.; Cavagni, J.; Langa, G.P.J.; Stewart, B.; Malheiros, Z.; Rösing, C.K. A Systematic Review of the Effect of Oral Rinsing with H_2O_2 on Clinical and Microbiological Parameters Related to Plaque, Gingivitis, and Microbes. *Int. J. Dent.* **2020**, *2020*, 8841722. [CrossRef]
21. Sato, H.; Niwano, Y.; Nakamura, K.; Mokudai, T.; Ikai, H.; Kanno, T.; Egusa, H. Efficacy and safety of a therapeutic apparatus using hydrogen peroxide photolysis to treat dental and periodontal infectious diseases. *J. Toxicol. Sci.* **2016**, *41*, 793–799. [CrossRef] [PubMed]

22. Golob Deeb, J.; Smith, J.; Belvin, B.R.; Lewis, J.; Grzech-Leśniak, K. Er:YAG Laser Irradiation Reduces Microbial Viability When Used in Combination with Irrigation with Sodium Hypochlorite, Chlorhexidine, and Hydrogen Peroxide. *Microorganisms* **2019**, *7*, 612. [CrossRef] [PubMed]
23. Grzech-Leśniak, K.; Belvin, B.R.; Lewis, J.P.; Golob Deeb, J. Treatment with Nd:YAG Laser Irradiation Combined with Sodium Hypochlorite or Hydrogen Peroxide Irrigation on Periodontal Pathogens: An In Vitro Study. *Photobiomodul. Photomed. Laser Surg.* **2021**, *39*, 46–52. [CrossRef] [PubMed]
24. El Mobadder, M.; Nammour, S.; Namour, M.; Namour, A.; Grzech-Leśniak, K. Disinfection Potential of 980 nm Diode Laser and Hydrogen Peroxide (3%) in Critical Probing Depths Periodontal Pockets: Retrospective Study. *Life* **2022**, *12*, 370. [CrossRef]
25. Nammour, S.; El Mobadder, M.; Maalouf, E.; Namour, M.; Namour, A.; Rey, G.; Matamba, P.; Matys, J.; Zeinoun, T.; Grzech-Leśniak, K. Clinical Evaluation of Diode (980 nm) Laser-Assisted Nonsurgical Periodontal Pocket Therapy: A Randomized Comparative Clinical Trial and Bacteriological Study. *Photobiomodul. Photomed. Laser Surg.* **2021**, *39*, 10–22. [CrossRef]
26. Grzech-Leśniak, K.; Sculean, A.; Gašpirc, B. Laser reduction of specific microorganisms in the periodontal pocket using Er:YAG and Nd:YAG lasers: A randomized controlled clinical study. *Lasers Med. Sci.* **2018**, *33*, 7–1461. [CrossRef]
27. Barrak, I.; Urbán, E.; Turzó, K.; Nagy, K.; Braunitzer, G.; Stájer, A. Short- and Long-Term Influence of Fluoride-Containing Prophylactics on the Growth of Streptococcus mutans on Titanium Surface. *Implant Dent.* **2015**, *24*, 6–675. [CrossRef]
28. Stájer, A.; Urbán, E.; Pelsőczi, I.; Mihalik, E.; Rakonczay, Z.; Nagy, K.; Turzó, K.; Radnai, M. Effect of caries preventive products on the growth of bacterial biofilm on titanium surface. *Acta Microbiol. Immunol. Hung.* **2012**, *59*, 51–61. [CrossRef]
29. Ishiguro, T.; Mayanagi, G.; Azumi, M.; Otani, H.; Fukushima, A.; Sasaki, K.; Takahashi, N. Sodium fluoride and silver diamine fluoride-coated tooth surfaces inhibit bacterial acid production at the bacteria/tooth interface. *J. Dent.* **2019**, *84*, 30–35. [CrossRef]
30. Blancas, B.; Lanzagorta, M.L.; Jiménez-Garcia, L.F.; Lara, R.; Molinari, J.L.; Fernández, A.M. Study of the ultrastructure of Enterococcus faecalis and Streptococcus mutans incubated with salivary antimicrobial peptides. *Clin. Exp. Dent. Res.* **2021**, *7*, 365–375. [CrossRef]
31. Komiyama, E.Y.; Lepesqueur, L.S.; Yassuda, C.G.; Samaranayake, L.P.; Parahitiyawa, N.B.; Balducci, I.; Koga-Ito, C.Y. Enterococcus Species in the Oral Cavity: Prevalence, Virulence Factors and Antimicrobial Susceptibility. *PLoS ONE* **2016**, *11*, e0163001. [CrossRef] [PubMed]
32. Kouidhi, B.; Zmantar, T.; Mahdouani, K.; Hentati, H.; Bakhrouf, A. Antibiotic resistance and adhesion properties of oral Enterococci associated to dental caries. *BMC Microbiol.* **2011**, *11*, 155. [CrossRef] [PubMed]
33. Ravald, N.; Hamp, S.E. Prediction of root surface caries in patients treated for advanced periodontal disease. *J. Clin. Periodontol.* **1981**, *8*, 400–414. [CrossRef] [PubMed]
34. Ravald, N.; Johansson, C.S. Tooth loss in periodontally treated patients: A long-term study of periodontal disease and root caries. *J. Clin. Periodontol.* **2012**, *39*, 73–79. [CrossRef]
35. Ravald, N.; Hamp, S.E.; Birkhed, D. Long-term evaluation of root surface caries in periodontally treated patients. *J. Clin. Periodontol.* **1986**, *13*, 758–767. [CrossRef]
36. Ravald, N.; Birkhed, D.; Hamp, S.E. Root caries susceptibility in periodontally treated patients. Results after 12 years. *J. Clin. Periodontol.* **1993**, *20*, 124–129. [CrossRef]
37. Fadel, H.; Al Hamdan, K.; Rhbeini, Y.; Heijl, L.; Birkhed, D. Root caries and risk profiles using the Cariogram in different periodontal disease severity groups. *Acta Odontol. Scand.* **2011**, *69*, 118–124. [CrossRef]
38. Aoki, A.; Mizutani, K.; Schwarz, F.; Sculean, A.; Yukna, R.A.; Takasaki, A.A.; Romanos, G.E.; Taniguchi, Y.; Sasaki, K.M.; Zeredo, J.L.; et al. Periodontal and peri-implant wound healing following laser therapy. *Periodontol. 2000* **2015**, *68*, 217–269. [CrossRef]

International Journal of
Molecular Sciences

MDPI

Article

Ultrasensitive Detection of Multidrug-Resistant *Mycobacterium tuberculosis* Using SuperSelective Primer-Based Real-Time PCR Assays

Anshika Narang [1], Salvatore A. E. Marras [1,*], Natalia Kurepina [2], Varsha Chauhan [3], Elena Shashkina [2], Barry Kreiswirth [2], Mandira Varma-Basil [3], Christopher Vinnard [4] and Selvakumar Subbian [1,*]

1 Public Health Research Institute, New Jersey Medical School, Rutgers University, Newark, NJ 07103, USA
2 Center for Discovery and Innovation, Nutley, NJ 07110, USA
3 Department of Microbiology, Vallabhbhai Patel Chest Institute, University of Delhi, Delhi 110021, India
4 New Jersey Medical School, Rutgers University, Newark, NJ 07103, USA
* Correspondence: marrassa@njms.rutgers.edu (S.A.E.M.); subbiase@njms.rutgers.edu (S.S.)

Abstract: The emergence of drug-resistant tuberculosis is a significant global health issue. The presence of heteroresistant *Mycobacterium tuberculosis* is critical to developing fully drug-resistant tuberculosis cases. The currently available molecular techniques may detect one copy of mutant bacterial genomic DNA in the presence of about 1–1000 copies of wild-type *M. tuberculosis* DNA. To improve the limit of heteroresistance detection, we developed SuperSelective primer-based real-time PCR assays, which, by their unique assay design, enable selective and exponential amplification of selected point mutations in the presence of abundant wild-type DNA. We designed SuperSelective primers to detect genetic mutations associated with *M. tuberculosis* resistance to the anti-tuberculosis drugs isoniazid and rifampin. We evaluated the efficiency of our assay in detecting heteroresistant *M. tuberculosis* strains using genomic DNA isolated from laboratory strains and clinical isolates from the sputum of tuberculosis patients. Results show that our assays detected heteroresistant mutations with a specificity of 100% in a background of up to 10^4 copies of wild-type *M. tuberculosis* genomic DNA, corresponding to a detection limit of 0.01%. Therefore, the SuperSelective primer-based RT-PCR assay is an ultrasensitive tool that can efficiently diagnose heteroresistant tuberculosis in clinical specimens and contributes to understanding the drug resistance mechanisms. This approach can improve the management of antimicrobial resistance in tuberculosis and other infectious diseases.

Keywords: drug resistance; molecular diagnosis; anti-tuberculosis therapy; real-time PCR; isoniazid; rifampicin; genomic DNA; mutation

Citation: Narang, A.; Marras, S.A.E.; Kurepina, N.; Chauhan, V.; Shashkina, E.; Kreiswirth, B.; Varma-Basil, M.; Vinnard, C.; Subbian, S. Ultrasensitive Detection of Multidrug-Resistant *Mycobacterium tuberculosis* Using SuperSelective Primer-Based Real-Time PCR Assays. *Int. J. Mol. Sci.* **2022**, *23*, 15752. https://doi.org/10.3390/ijms232415752

Academic Editor: Helena Felgueiras

Received: 19 October 2022
Accepted: 9 December 2022
Published: 12 December 2022

1. Introduction

Tuberculosis (TB) is currently the second leading infectious killer disease worldwide after COVID-19, with a disease burden shared across all countries, genders, and age groups. In 2020, an estimated 10 million people fell ill with TB, and 1.5 million people died [1]. Multidrug-resistant TB (MDR-TB), defined as resistance to isoniazid (INH) and rifampin (RIF), remains a serious public health threat worldwide. Preventing the emergence of bacterial drug resistance is a critical component of TB elimination efforts, with the World Health Organization (WHO) aiming to reduce 95% of TB-attributable deaths by 2035 [1].

The standard regimen of directly observed therapy short course (DOTS) strategy for global TB control includes 4 months of treatment with INH, RIF, pyrazinamide (PZA), and ethambutol (EMB), followed by 2 months of INH and RIF [1]. Single-nucleotide polymorphisms (SNPs) at a handful of genetic loci (known as hot spots) are responsible for the development of the majority of drug resistance in *M. tuberculosis* [2]. MDR-TB is thought to develop first with the selection of INH-resistant mutant *M. tuberculosis* strains, followed by the selection of RIF resistance among these mutants [3]. Treatment of MDR-TB requires the use of more toxic, more expensive, and less effective second-line drugs administered for a prolonged time (more than a year) [4], although shorter MDR treatment regimens have been endorsed by WHO [5].

Acquired drug resistance in TB occurs via the selection of spontaneous mutants in a wild-type (WT) background with a spontaneous mutation rate of 10^{-8} to 10^{-9} per round of *M. tuberculosis* replication [6]. With a sufficiently large *M. tuberculosis* population in infected tissue (greater than 10^9 organisms) [7], spontaneous mutants with resistance to a single drug are likely to be present, which could then be selected by inadequate drug pressure [3]. Heteroresistant TB is defined as a disease in which drug-resistant and drug-susceptible *M. tuberculosis* strains co-exist in the same patient [8–10]. In the conventional proportional drug sensitivity tests (pDST), heteroresistance is defined as 1–99% bacterial colony growth on drug-containing media [11].

Current TB diagnostic methods are insensitive in detecting low levels of mutant *M. tuberculosis* in a WT background, with lower limits of quantification ranging between 1 and 10% for standard methods, such as line probe assays [12,13] and mycobacteria growth indicator tubes (MGIT) [14]. Critically, the major limitation of traditional polymerase chain reaction (PCR) methodology is the inability to detect rare mutant sequences in a background of abundant WT DNA. With recent advances, such as digital PCR [10] and next-generation sequencing [15], the limit of detection of rare mutants may approach the 0.1% threshold (corresponding to the ratio of mutant DNA/WT DNA), which remains far above the level required to detect the baseline presence of heteroresistant strains in clinical samples or dynamic changes in the emergence of drug-resistant *M. tuberculosis* [7]. Even with the limitations of current tools, the clinical impact of heteroresistant TB is increasingly recognized [8,11,14,16]. A recent study reported the weighted pooled prevalence of INH heteroresistance (from 19 studies) to be 5% and RIF heteroresistance (from 17 studies) to be 7% from 2001 to 2020 [17]. Among South African TB patients, heteroresistant infections identified by the mycobacterial interspersed repetitive units variable number of tandem repeat (MIRU-VNTR) genotyping were associated with a 90% increase in the odds of delayed sputum conversion after 2 months of treatment [18], a critical treatment endpoint for TB control programs (and related to the risk of ongoing transmission in the community). Heteroresistant *M. tuberculosis* infection can lead to a false negative rapid nucleic acid test for RIF resistance [19,20] and contributes to variable treatment responses observed in different anatomic regions of infected lungs [21]. Most worrisome, patients with clinically undetected heteroresistance can progress to MDR-TB in the face of first-line treatment pressure [22], either due to formulation differences [23] or intra-individual pharmacokinetic variability [24]. All of these mechanisms contribute to the failure of TB control efforts through inadequate treatment and the further spread of drug-resistant *M. tuberculosis* strains [3].

In this report, we propose a revolutionary new approach in PCR primer design, called "SuperSelective PCR primers" [25], for ultrasensitive detection of drug-resistant *M. tuberculosis* bacilli in an abundant background of wild-type *M. tuberculosis*. Our innovation in PCR primer design and demonstration of these primers in RT-PCR assays provide a significant technological advancement with clinical application toward detecting rare drug-resistant mutants in TB and other infectious diseases. SuperSelective primers innovate by separating a relatively long "anchor" sequence targeting the gene of interest- and a short "foot" sequence selective for the point mutation (i.e., the interrogating nucleotide). In between, there is a "bridge" sequence, which is non-binding to the target DNA sequence. When the SuperSelective primer is hybridized to the template molecule, the bridge sequence in the primer and the corresponding template sequence make a "bubble" region that functionally separates the efficient formation of the anchor hybrid with the formation of the foot hybrid [25]. Both the anchor (corresponding to the target gene) and foot (with the interrogating nucleotide) sequences must bind to the DNA fragment for amplification to proceed. Use of SuperSelective PCR primers in RT-PCR overcomes the limitations of other primer-based approaches, like amplification refractory mutation system (ARMS) primers [26], dual-priming oligonucleotide (DPO) primers [27], hairpin primers [28], and PlexPrimers [29], which are either not sufficiently sensitive to detect extremely rare mutants, not compatible with real-time PCR due to the presence of unnatural nucleotides in their sequence, or have not been shown to enable quantitative determinations in multiplex real-time PCR assays when different target mutations occur in the same codon. By relying on standard PCR equipment and reagents, SuperSelective PCR methods can be readily adopted by any laboratory equipped to perform conventional PCR, supporting implementation in settings with medium and high burdens of TB disease.

2. Results

2.1. SuperSelective Primer Design

We first designed a prototype SuperSelective primer to detect *inhA* promoter mutation -15C→T and optimized the primer configuration and PCR conditions to achieve maximum sensitivity for mutant detection. A SuperSelective primer 18-14/14-6:1:0 (inh15_SSP3) was paired with a conventional reverse primer and tested in a RT-PCR assay to detect one copy of -15C→T mutant in a background of 10^5 copies of WT DNA. Linearized pUC plasmid containing a cloned sequence carrying *inhA* promoter mutation -15C→T was used as a template for these experiments, and its WT counterpart was used as background. Results show that our primer design efficiently amplifies as low as one copy of mutant DNA (in three out of six replicates) in a background of 10^5 copies of WT DNA. Amplification of mutant in the control set of the experiment (10^5 copies of mutant DNA without the WT background) down to one copy confirmed that the amplification signal for mutant from low copy numbers was specific and not a product of the non-specific binding of primers to the background WT pDNA. Although this SuperSelective primer design also resulted in the amplification of WT DNA (one out of six replicates), the WT amplification peaks were seen occasionally at PCR cycle (Ct) > 40, which overlapped with the amplification peaks with low copy numbers (ten and one) of mutant. The non-suppression of the unintended WT target amplification using this SuperSelective primer could be attributed to the high GC content in *M. tuberculosis* gDNA [30], leading to a firm binding of anchor and/or foot regions to the WT template.

The main objective of our investigation was to design primers adequate for early amplification of mutant (ideally Ct 19–25 for 10^5 copies) to attain amplification down to 1 copy within a short RT-PCR cycle (ideally 50 cycles), with complete suppression of WT background. Therefore, to determine optimal parameters for this assay, we tweaked the SuperSelective primer design in different parts of the PCR (owing to different functions), making one to two variations at a time, and tested these modified designs under different annealing temperatures (Tm) and assay conditions (Table S1).

2.2. Standardization of Primer Foot Length

To strengthen the binding of SuperSelective primers with the mutant DNA template, we tested SuperSelective primer 18-14/13-7:1:0 (inh15_SSP4) with a longer foot length (8 nt instead of 7 nt) in the RT-PCR assay described as above. The resulting Ct values of 10^5 copies of mutant and 10^5 copies of WT DNA with this longer foot primer were plotted against the Ct values obtained with primer 18-14/14-6:1:0 (inh15_SSP3). Results show that the extended foot length (8 bp) in primer inh15_SSP4 helped in the early amplification of mutant DNA template (3–4 cycles early) but also enhanced the binding of SuperSelective primers with WT DNA (Figure S1a). These results demonstrate that a shorter foot length may delay achieving the threshold cycle, leading to enhanced selectivity, whereas a longer foot length may reduce the selectivity of SuperSelective primers. From a thermodynamic standpoint, the improved selectivity with shorter foot lengths can be due to the higher ratio of the equilibrium abundance of perfectly complementary mutant foot hybrids compared to the equilibrium abundance of mismatched WT foot hybrids [25].

2.3. Standardization of the Size and Symmetry of Intervening Loop (Bubble)

In order to inhibit the amplification of WT gene copy or delay the amplification beyond 50 cycles, which would avoid overlap with amplification of low copies of mutant, we tested SuperSelective primers 18-14/21-6:1:0 (inh15_SSP2) with increased lengths of the intervening sequence (from 14 to 21 bp) (forming an asymmetric bubble) in RT-PCR assays. The resulting Ct values of 10^5 copies of mutant and 10^5 copies of WT with this primer were plotted against the Ct values obtained with 18-14/14-6:1:0 (inh15_SSP3), which has a shorter length of the intervening sequence (equal to the length of the bridge, forming a symmetric bubble) (Figure S1b). Two reactions were carried out with each of the primers, of which 18-14/21-6:1:0 initiated the amplification of WT in one reaction and completely suppressed the amplification in the other sample. In contrast, the 18-14/14-6:1:0 primer amplified the targets of both WT replicate reactions. The asymmetric bubble with 18-14/21-6:1:0 also resulted in a Ct delay (~5 cycles) in the amplification of the mutant target and lower fluorescence signal. These results signify that the asymmetricity of the bubble, created by the intervening sequence, may significantly affect the binding efficiency of a SuperSelective primer with its target and, thus, the amplification threshold, resulting in the formation of fewer amplicons.

We also tested SuperSelective primers with a variation (15/17 nt) in bridge length of the primer design 18-14/13-7:1:0 (inh15_SSP4). In all other aspects, the composition of the primers was the same, i.e., 18-15/13-7:1:0 (inh15_SSP5) and 18-17/13-7:1:0 (inh15_SSP6). We did not find a significant difference in Ct values for 10^5 copies of mutant, or 10^5 copies of WT obtained with primers inh15_SSP4 and inh15_SSP5, which have bridge lengths of 14 and 15 bp, respectively. Primer inh15_SSP6 with a bridge length of 17 bp formed a bigger and more asymmetric bubble, delaying the amplification of 10^5 copies of mutant or 10^5 copies of WT by 1 cycle. The window of discrimination (ΔCt) between the Ct for the mutant and the WT was similar for these two primers.

Furthermore, we tested different bridge-intervening sequence length combinations—20/8, 8/8, and 8/20 (inh15_SSP7-9)—in the SuperSelective primer designs. SuperSelective primers with a long bridge but short intervening sequence 18-20/8-7:1:0 (SSP7), short bridge–short intervening sequence 18-8/8-7:1:0 (SSP8), and short bridge–long intervening sequence 18-8/20-7:1:0 (SSP9) were tested at Tm 60–66 °C, while all other RT-PCR conditions were maintained similar to the above (Table S1). Amplification of mutant target using SuperSelective primer with a symmetric bubble 18-8/8-7:1:0 (SSP8) was the least affected by changes in Tm, although the window of discrimination (ΔCt) between the Ct for the mutant and the Ct for the WT increased with increasing temperature (Figure S2a). Amplification of the mutant target using both SuperSelective primers forming an asymmetric bubble (18-20/8-7:1:0 and 18-8/20-7:1:0) was delayed according to the Tm of the assay. Only 18-20/8-7:1:0 suppressed the amplification of the WT pDNA template (at Tm 66 °C) but also caused a delay (5 cycles) in the amplification of the intended mutant target (Figure S2b).

2.4. Changes in Foot Design

In some of the potential SuperSelective primer designs tested previously, we introduced a mismatch in the foot sequence at the fourth position from the interrogating nucleotide, i.e., 18-14/14-6:1:0 (inh15_SSP14), 18-14/13-7:1:0 (inh15_SSP15), and 18-20/8-7:1:0 (inh15_SSP14- inh15_SSP16) to destabilize the binding of the foot sequence with the WT DNA. We also tested SuperSelective primers 18-14/14-3:2:1, 18-14/14-6:2:1, and 18-14/14-6:2:0 (inh15_SSP23 to inh15_25, respectively) by introducing a mismatch nucleotide immediately after the interrogating nucleotide. In theory, this design would weaken the binding of the foot sequence with its target, with a more pronounced effect on the WT target with two mismatches compared to the mutant with a single mismatch. We also investigated the effect of varying the location of the interrogating nucleotide in the foot sequence on the primer's ability to discriminate mutant templates from WT templates using primer designs 18-14/14-6:1:2, 18-14/13-7:1:2, 18-20/8-7:1:2, and 18-14/14-3:2:1

(inh15_SSP19 to inh15_SSP22, respectively). Results from the PCR assays indicate that the variations in foot sequence design aimed to prevent the amplification of WT DNA led to a 8–16 cycle delay in the amplification of the mutant target.

2.5. SuperSelective Reverse Primers

To decrease the binding of SuperSelective forward primers paired with conventional reverse primers to WT DNA, the SuperSelective forward primers were paired with SuperSelective reverse primers [31]. We designed SuperSelective reverse primers 18-14/14-6:1:0 (inh15_SSP10) and 18-14/13-7:1:0 (inh15_SSP11) with varying foot length (7/8 nt), 18-14/8-6:1:0 (inh15_SSP12) and 18-14/8-7:1:0 (inh15_SSP13) with varying foot length (7/8 nt) plus a shorter intervening sequence (8 bp), and 20-14/14-6:1:0 (inh15_SSP17) and Rev 20-14/13-7:1:0 (inh15_SSP18) with varying foot length (7/8 nt) and a longer anchor (20 nt). These primers were paired with SuperSelective primers 18-14/13-7:1:0 (inh15_SSP4) and 18-20/8-7:1:0 (inh15_SSP7) or their respective anchor sequences, which acted as conventional forward primers. The effect of foot length and intervening sequence on a reverse SuperSelective primer paired with a conventional forward primer was as shown above for a forward SuperSelective primer paired with a conventional reverse primer. We compared the Ct values for detecting 10^5 copies of mutant and 10^5 copies of WT with each of these SuperSelective primers paired either with a conventional second primer to their pairing with another SuperSelective primer. Results show that using two SuperSelective primers drastically increased the window of discrimination (ΔCt) for mutant and WT targets from <5 to 8–13, yet it did not completely suppress the amplification of WT templates.

Some of these variations in SuperSelective primer design were also tested on other mutations in the panel, i.e., *inhA* (-8T→A, -17G→T), *katG* (S315T, AGC→ACA, and AGC→ACC), and *rpoB* (D516V, GAC→GTC; H526D, CAC→GAC; H526Y, CAC→TAC; and S531L, TCG→TTG) (Table S2). All the tested SuperSelective primer designs and combinations amplified the unintended WT target or the suppression of WT DNA was accompanied by a delay in amplification of 10^5 copies of mutant and/or suppression of amplification of ten and one copy of mutant. We repeatedly observed an overlap of low copy number mutant (ten and one) and WT amplification peaks. It should be noted that the Sybr Green reagent used to monitor amplification in these RT-PCR assays is a DNA-intercalating dye that detects specific and non-specific amplification products. Therefore, to confirm the specificity of the amplified products tested by Sybr Green, a subset of SuperSelective primers were tested for amplification using molecular beacon probes.

2.6. Validation of RT-PCR Assays for Mutant Detection with Molecular Beacon Probes

Amplification of WT DNA can lead to a false positive sample detection in RT-PCR assays, particularly for clinical samples with low copies of mutant DNA and high copies of WT DNA. To confirm whether the amplification products (from WT DNA) in our SuperSelective PCR assays were the intended amplification products and not the amplification of non-specific products, such as primer-dimers, we designed and tested amplification-product-specific molecular beacon probes in the PCR assays [32]. Since non-hybridized molecular beacons are virtually non-fluorescent and hybridized molecular beacons fluoresce brightly in their characteristic color [32,33], the amount of amplicon present during the annealing phase of each amplification cycle is automatically measured by the spectrofluorometric thermal cycler in which the PCR assays are carried out.

2.7. Targeting Closely Localized Mutations

Of the panel of mutations selected for this study, both *katG* mutations AGC→ACA and AGC→ACC are present at the same codon (315). The anchor and foot sequences of the SuperSelective primers designed to detect these mutations overlapped with each other. Similarly, *inhA* mutations -8T→A, -15C→T, and -17G→T are co-localized in the promoter region and *rpoB* mutations D516V (GAC→GTC), H526D (CAC→GAC), H526Y (CAC→TAC), and S531L (TCG→TTG) in the RIF-resistance-determining region (RRDR) with overlapping regions of the primer. The amplification products of both *katG* mutations, three *inhA* promoter mutations, and four *rpoB* mutations also had overlapping nucleotide sequences. These regions were exploited to design the molecular beacon probes to detect mutations. Thus, we could detect both *katG* mutations using the same/single probe. Similarly, we designed one probe for the identification of three mutations in the *inhA* promoter region, i.e., -8T→A, -15C→T, -17G→T, and one probe for all the *rpoB* mutations D516V (GAC→GTC), H526D (CAC→GAC), H526Y (CAC→TAC), and S531L (TCG→TTG). Thus, three molecular beacon probes were utilized to detect nine mutations. We designed the following probes, which included a unique sequence at both ends of the probe (in italics), to form the stem region and labeled the probe with FAM at the 5′ end and a quencher BHQ-1 at the 3′ end.

katG: 5′ FAM-CGCTCG GACGAACACCCCGACGAAAT CGAGCG-BHQ-1 3′
inhA: 5′ FAM-CGCTCG CCGGGCCGAAATCGGTATGT CGAGCG-BHQ-1 3′
rpoB: 5′ FAM-CGCTCG GAATTGGCTCAGCTGGCTGG CGAGCG-BHQ-1 3′

2.8. Use of Asymmetric PCR Conditions

We conducted three separate RT-PCR assays to optimize the molecular beacon probe concentrations for *katG*, *inhA*, and *rpoB*. The concentration of each probe was optimized for at least one mutation from the group for which it was specific. Thus, the molecular beacon probe *katG* was optimized for the detection of *katG* S315T (AGC→ACA), the *inhA* probe for *inhA* promoter mutation -8T→A, and the *rpoB* probe for *rpoB* mutation D516V (GAC→GTC). RT-PCR assays were carried out using 10^5 copies of linearized mutant plasmid in three sets of reactions with variable concentrations of the probe, i.e., 0.1, 0.125, and 0.25 μM. We used 0.1 μM each of forward and reverse primers in these assays. Fluorescence intensity with the 0.25 μM probe was the highest as expected (>2-fold), while there was no significant difference between relative fluorescence units (RFU) obtained with 0.1 and 0.125 μM probes. Figure S3 demonstrates an example of the increase in fluorescence signal intensity with increasing concentration of *inhA* probe with 10^5 copies of *inhA* -8T→A mutants, using SuperSelective primer 18-14/14-6:1:0.

In order to avoid probe crosstalk in multiplex reactions with a very high amount of probe, we tested asymmetric primer concentrations to enhance fluorescence signal intensity in the reactions to control the amount of molecular beacon probe. RT-PCR assays were performed in two sets with SuperSelective primer and conventional reverse primer in the ratios 1:5 and 1:10 on 10^5 copies of mutant target DNA using a 0.1 μM probe. SuperSelective primer and conventional reverse primer in the ratio 1:1 (symmetric PCR, 0.1 μM each of forward and reverse primer) were used as a control for these reactions. We observed a 2.5-3-fold rise in fluorescence signal intensity under both tested asymmetric PCR conditions, although there was no significant difference between primer ratios 1:5 and 1:10. Figure S4 shows the amplification of *rpoB* D516V GAC→GTC with SuperSelective primer 18-14/13-7:1:0 paired with a conventional reverse primer in 1:1, 1:5, and 1:10 ratios on 10^5 copies of mutant target DNA. These experiments demonstrate the advantage of using non-symmetric primer concentrations in RT-PCR assays over symmetric primer concentrations.

We extended our observations from the above experiments to other mutations included in this study, *katG* S315T AGC→ACC, *inhA* promoter -15C→T and -17G→T, *rpoB* H526D (CAC→GAC), H526Y (CAC→TAC), and S531L (TCG→TTG). All the subsequent monoplex RT-PCR assays with molecular beacon probes were performed under asymmetric PCR conditions utilizing 0.1 μM and 0.5 μM of forward and reverse primer, respectively, and 0.1 μM of the probe to maximize the fluorescence signal intensity. We further optimized the individual probe concentrations for the multiplexed assays (discussed later).

2.9. Selective Amplification of the Mutant from Mixed DNA

To detect each of the nine mutations in our selected panel of target mutant genes and to determine their selectivity, one to five SuperSelective primers were tested. RT-PCR experiments were performed using linearized mutant and WT pDNA templates in the ratios 10^5-1:10^5 (Mutant:WT) and 10^5-1:10^4 (Mutant:WT). Each primer was tested at least twice on duplicate samples. Low copy number mutants (ten and one) and WT samples were tested on three to eight replicates.

RT-PCR assay with the molecular beacon probes confirmed that the amplification products initiated from WT DNA templates (with 10^5 copies of WT DNA) resulted from specific amplification. Under the tested assay conditions, SuperSelective primers (Table 1) successfully suppressed the amplification of 10^4 copies of WT DNA while amplifying the mutant to one copy. Primers were tested for up to 60 RT-PCR cycles to check for the appearance of late amplification signals (after Ct 50) from WT. Using the selected SuperSelective primers (Table 1), we could detect all nine mutations in our panel with absolute specificity. We could detect down to one copy of mutant within 50 cycles of the RT-PCR, with detection of WT (up to 10^4 copies) suppressed until 60 cycles. These observations confirmed that SuperSelective primers are highly specific for amplifying the target mutant DNA (specificity 100%) while suppressing the amplification of up to 10^4 copies of WT DNA.

Table 1. Limit of detection (LoD) of mutant DNA from clinical isolates.

Mutation	SuperSelective Forward Primer	Conventional Reverse Primer	LoD *	No. of gDNA Samples Tested
katG (AGC→ACC)	20-14/14-5:1:0	*katG*_rev2	1	7
katG (AGC→ACA)	20-14/14-5:2:0	*katG*_rev2	1	2
inhA -8T→A	20-14/14-6:1:0	*inhA*_rev1	1	2
inhA -15C→T	18-14/13-7:1:0	*inhA*_rev1	1	4
inhA -17G→T	20-14/13-7:1:0	*inhA*_rev2	1	1
D516V (GAC→GTC)	20-14/13-7:1:0	*rpoB*_rev1	1	3
H526D (CAC→GAC)	20-14/13-7:1:0	*rpoB*_rev1	1	2
H526Y (CAC→TAC)	20-14/13-7:1:0	*rpoB*_rev1	1	2
S531L (TCG→TTG)	20-14/13-7:1:0	*rpoB*_rev1	1	4

* LoD: The lowest number of detectable copies of mutant DNA in a background of 10^4 copies of WT DNA.

2.10. Highly Sensitive Method of Mutant Detection in a Background of Susceptible M. tuberculosis DNA

The parameters of SuperSelective primer design and RT-PCR assay conditions were optimized initially using linearized plasmids containing the mutation of interest. Plasmids, which are limited in their total length, generate a low probability of SuperSelective primers to find non-specific binding sites, which will increase many-fold in a sample containing the genome of *M. tuberculosis*. To validate our SuperSelective primer designs and RT-PCR assay conditions, we tested selected SuperSelective forward primer and conventional reverse primer combinations (Table 1) on several *M. tuberculosis* gDNA samples containing the mutations of interest. We tested these conditions on 15 mutant gDNA samples containing the panel of mutations, as shown in Table 1. Of these, ten gDNA samples were MDR with *katG* S315T/*inhA*-15/*inhA*-17 mutation and one of the four RRDR mutations. (Table 1).

We tested each of the selected SuperSelective primer and RT-PCR assay conditions on 10^5 to one copy of mutant gDNA in a background of 10^4 copies of WT *M. tuberculosis* H37Rv gDNA. We tested two replicates for high copy numbers (10^5, 10^4, 10^3, 10^2, 50) of mutant, eight replicates for the low copy numbers (25, 10, 1) of mutant, and at least four replicates of 10^4 copies of WT. Consistent with our findings in the plasmid-based assays (Table 1), we were able to amplify up to one copy of mutant DNA (in at least two out of eight replicates) for all the nine mutations in *katG*, *inhA*, and *rpoB*. Ten copies of mutant DNA were amplified in at least four out of eight replicates and twenty-five copies in at least six out of eight replicates, while eight out of eight replicates were positively detected for fifty copies of mutant DNA or more. Figure 1a shows the amplification peaks obtained in RT-PCR assays containing 10^5, 10^4, 10^3, 10^2, fifty, twenty-five, ten, one, and zero copies of *katG* S315T AGC > ACC mutant gDNA sample in a background of 10^4 copies of WT *M. tuberculosis* H37Rv DNA, utilizing the SuperSelective primer 20-14/14-5:1:0 combined with a conventional reverse primer and using a FAM-labeled molecular beacon probe for the detection.

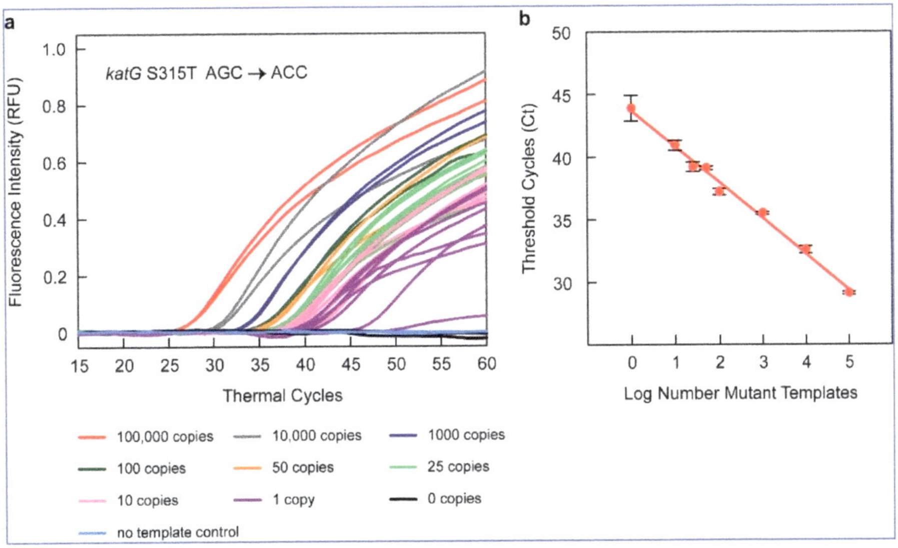

Figure 1. SuperSelective primer-based detection of *katG* S315T AGC > ACC mutant sequences in the presence of abundant WT sequences. (**a**) Results of RT-PCR assays to detect *katG* S315T AGC > ACC mutations with a FAM-labeled molecular beacon probe utilizing the SuperSelective primer 20-14/14-5:1:0 with gDNA as a template. The figure shows amplification of 10^5 (red), 10^4 (grey), 10^3 (blue), 10^2 (dark green), 50 (orange), 25 (light green), 10 (pink), 1 (violet) copies of mutant DNA template in a background of 10^4 copies of WT DNA template; 10^4 copies of WT (black) and nuclease-free water (yellow) were used as control. (**b**) The threshold cycle measured for each reaction that contained mutant templates is plotted as a function of the logarithm of the number of mutant templates initially present in each reaction.

Thus, our SuperSelective primer-based RT-PCR assays could detect one in ten thousand copies of mutant DNA for all nine gene mutations included in our panel, confirming the ultra-high sensitivity (0.01%) of these SuperSelective primers for the detection of INH and RIF resistance in *M. tuberculosis.*

2.11. Quantifiable Mutant Detection

The clinical goal of these RT-PCR assays that utilize SuperSelective primers is to measure the relative abundance of mutant DNA fragments relevant to INH and RIF resistance in TB in the context of the amount of WT *M. tuberculosis* DNA present in the sample. In our RT-PCR assays, control reactions that contained no template DNA did not produce any false amplicons, such as primer-dimers, despite the longer length of the SuperSelective primers. The reaction containing 10^4 WT templates and no mutant templates was suppressed to such an extent that it did not produce a significant number of amplicons until after 50 cycles of amplification had been carried out (Figure 1a).

To examine the potential of SuperSelective primer pairs for mutant quantification, we compared the mean Ct values (in an RT-PCR assay) with each SuperSelective primer to the mutant copy number tested. The mean Ct values were inversely linearly proportional to the logarithm of the number of copies of mutant target DNA (Figure 1b). This inverse linear relationship between the logarithm of the number of mutant targets initially present in a sample and the Ct value observed for that sample is the hallmark of quantitative exponential amplification assays. Thus, RT-PCR assays employing SuperSelective primer pairs could be utilized to determine the actual amount of mutant target DNA even in a high background (10^4 copies) of WT DNA, based only on Ct values. These results demonstrate the potential of SuperSelective primer in RT-PCR assays to not only detect but also quantify the amount of rare mutant DNA in the presence of abundant WT *M. tuberculosis* DNA, providing a novel and powerful tool for detecting and quantifying heteroresistance in TB patients.

2.12. Multiplexed Assay

The mutations causing INH and RIF resistance in *M. tuberculosis* are localized within a compact region of katG gene, *inhA* promoter, and RRDR of the rpoB gene. For mutant sequences occurring in separate genes, multiplex real-time PCR assays can be designed relatively easily to quantitate the amount of each mutant gene in the same sample. SuperSelective primers for these mutations would have different anchor and foot sequences. The amplicons generated from each mutant gene will possess a unique sequence that can be easily distinguished from the sequences of amplicons generated from the other mutant genes, making it easy to distinguish these amplicons from each other through the use of differently colored molecular beacon probes. Thus, we decided to multiplex the RT-PCR assays to detect mutations present on different drug resistance genes.

Amongst the drug-resistant TB cases, *katG* S315T mutation is the most common mutation and *inhA* mutation (-15C→T) is the second most frequently observed mutation associated with INH resistance [34], while *rpoB* mutation S531L (TCG→TTG) is the most widespread mutation responsible for high-level resistance to RIF [35]. Considering the importance of these three mutations in MDR-TB cases, we designed a prototype multiplex assay incorporating SuperSelective primers to simultaneously detect katG gene S315T AGC→ACA, *inhA* -15 C→T, and S531L TCG→TTG.

2.13. Optimization of Assay Components

For the simultaneous detection of katG gene S315T AGC→ACA, *inhA* -15 C→T, and S531L TCG→TTG, we utilized SuperSelective primers having LoD of one copy of mutant in the corresponding monoplex assays (Table 1) and tested the efficiency and specificity of these primers in a multiplexed environment. To ensure that the selected primer pairs do not have any cross-reactivity with each other, we tested these three primer pairs in various possible combinations. For example, we tested the *katG* S315T AGC→ACA and *inhA* -15 C→T primer pair in a duplex RT-PCR assay on 10^5 copies of *katG* S315T AGC→ACA mutation containing linearized pDNA to study the effect of *inhA* -15 C→T primer pair addition on the amplification threshold of *katG* S315T AGC→ACA mutant. A monoplex RT-PCR assay with a *katG* S315T AGC→ACA primer pair and 10^5 copies of *katG* S315T mutation containing linearized pDNA was used as a control. Sybr Green was used to monitor the amplification of these reactions. We also studied the effect of *katG* S315T AGC→ACA primer pair addition on the amplification threshold of *inhA* -15 C→T by using *inhA* -15 C→T mutation containing linearized pDNA.

Similarly, *katG-rpoB*, *inhA-rpoB*, and *katG-inhA-rpoB* primer pairs were also tested. The *rpoB* primer pair addition delayed the amplification threshold of both *katG* S315T AGC→ACA and *inhA* -15 C→T mutants in the duplex assays, but the Ct values in the three-plex assays were unaffected. Thus, these three primer pairs were utilized to further optimize the multiplex RT-PCR assay. For these three-plex RT-PCR assays, we used three molecular beacon probes specific for amplicons generated by SuperSelective primers corresponding to *katG*, *inhA*, and *rpoB*. Each probe was labeled with a different fluorophore to detect signals for each mutation in a different optical channel.

katG: 5′ FAM-CGCTCG GACGAACACCCCGACGAAAT CGAGCG-BHQ-1 3′
inhA: 5′ Cal Fluor Red 610-CGCTCG CCGGGCCGAAATCGGTATGT CGAGCG-BHQ-2 3′
rpoB: 5′ Quasar 670-CGCTCG GAATTGGCTCAGCTGGCTGG CGAGCG-BHQ-2 3′

Probe concentrations were re-optimized for multiplex RT-PCR assays. Asymmetric RT-PCR assays were carried out on a Bio-Rad CFX96 Real-Time System in a final volume of 25 μL with 1X PCR buffer supplemented with 25 mM TMAC and 0.25% Tween20, 3 mM MgCl2, 0.05 U/μL Platinum Taq DNA polymerase, 0.25 mM dNTPs, each forward primer (*katG*, *inhA*, *rpoB*) 0.1 μM, each reverse primer (*katG*, *inhA*, *rpoB*) 0.5 μM, probe *katG* 0.25 μM, probe *inhA* 0.1 μM, and probe *rpoB* 0.25 μM. Amplification was performed for 60 cycles with Tm 60 °C for 20 s and extension at 72 °C for 20 s.

2.14. Selective Amplification

To ensure that each amplicon is only copied by its "correct" SuperSelective primer under the three-plex assay conditions, we carried out three sets of experiments, each with linearized plasmids containing *katG* S315T AGC→ACA, *inhA* -15 C→T, or S531L TCG→TTG mutation. We tested 10^5-1:10^4 copies of mutant:WT pDNA. Each condition was tested on at least two replicates.

In our first set of RT-PCR multiplex assays, with *katG* S315T AGC→ACA mutant-containing pDNA, we detected amplification signals only in the FAM channel (Figure 2a) down to one copy of mutant DNA in a background of 10^4 copies of WT DNA. No signals were detected in CAL Fluor Red 610 (CFR610) and Quasar 670 (Q670) channels. Similarly, in multiplex reactions with pDNA containing *inhA* -15 C→T and S531L TCG→TTG mutant, fluorescence was recorded only in the CFR610 (Figure 2b) and Q670 (Figure 2c) channels, respectively, while no amplification signal was observed in the non-specific channels. These results confirmed that the SuperSelective primers detect only specific mutants even in the presence of abundant WT DNA fragments. None of the primer pairs cross-reacted to amplify a non-specific target, and they completely suppressed the unintended WT target in the background, signifying a high specificity of SuperSelective primers even in a multiplexed environment.

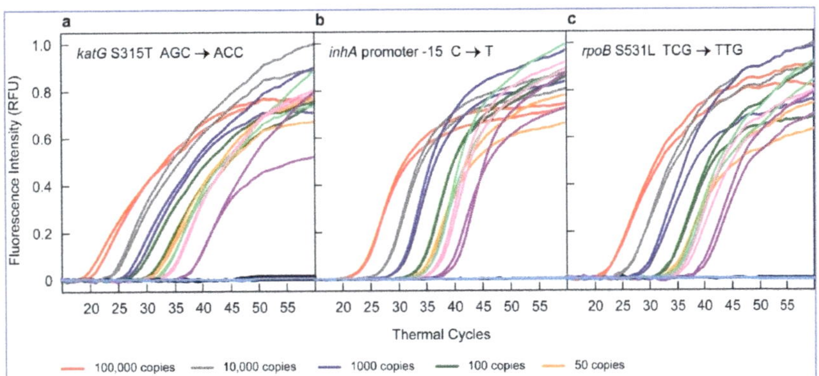

Figure 2. Results of multiplex RT-PCR assays to detect *katG* S315T AGC > ACC, *inhA* -15 C→T, and S531L TCG→TTG mutations utilizing the SuperSelective primers 20-14/14-5:1:0, 18-14/13-7:1:0, and 20-14/13-7:1:0, respectively, with pDNA used as a template. Molecular beacon probes labeled with FAM, CFR610, and Q670 were used to detect amplification signals. The figure illustrates amplification of 10^5 (red), 10^4 (grey), 10^3 (blue), 10^2 (dark green), 50 (orange), 25 (light green), 10 (pink), 1 (violet) copies of mutant DNA template in a background of 10^4 copies of WT DNA template; 10^4 copies of WT (black) and nuclease-free water (yellow) were used as control. The figure shows amplification peaks of (**a**) *katG* S315T AGC > ACC pDNA detected in FAM channel, (**b**) *inhA* -15 C→T pDNA detected in CFR610 channel, and (**c**) S531L TCG→TTG detected in Q670 channel. There was no amplification in non-specific channels; hence, the data has been excluded from these graphs.

2.15. Sensitivity on Clinical gDNA with Potential Sputum Carryover

To expand the utility of our optimized SuperSelective PCR to clinical applications, we tested 23 *M. tuberculosis* gDNA samples isolated from the sputum of pulmonary TB patients. The sputum samples were obtained from microbiologically, radiologically, and clinically confirmed TB cases (Table 2). The samples were single-blinded to the status of phenotypic or genotypic drug resistance in these *M. tuberculosis* gDNA samples.

Table 2. Detection of mutations in clinical samples using SuperSelective primers.

			katG (AGC > ACC)		*inhA* (-15C→T)		*rpoB* (S531L)	
Sample	**Age/Sex**	**Total DNA (*rrs* Copies)**	**Number of Mutant DNA Copies Detected by SuperSelective Primers**	**DNA Sequencing Result**	**Number of Mutant DNA Copies Detected by SuperSelective Primers**	**DNA Sequencing Result**	**Number of Mutant DNA Copies Detected by SuperSelective Primers**	**DNA Sequencing Result**
AVR1	36/M	2595	2328	AGC > ACC	40	WT	2167	S531L
AVR2	28/M	254	5	WT	38	WT	284	S531L
AVR4	39/F	498	679	-	0	-	0	-
AVR7	33/M	540	1086	AGC > ACC	0	WT	1332	S531L
AVR8	49/M	979	601	-	0	-	0	-
AVR9	37/F	2867	2039	AGC > ACC	0	WT	2827	S531L
AVR11	62/F	24,693	2	WT	0	WT	4	WT
AVR12	28/F	3294	2262	AGC > ACC	0	WT	3992	S531L
AVR13	24/F	32	33	AGC > ACC	0	WT	42	S531L
AVR14	70/F	129	36	AGC > ACC	0	WT	57	S531L
AVR16	26/M	2038	534	AGC > ACC	0	WT	1184	S531L
AVR18	27/F	2235	0	-	84	-	0	-
AVR19	42/M	2680	0	-	0	-	0	-
AVR23	26/F	2867	0	-	0	-	0	-
AVR24	19/F	2434	0	-	0	-	0	-
AVR25	38/F	238	350	-	0	-	0	-
AVR26	18/F	197	197	AGC > ACC	0	WT	208	S531L
AVR27	26/F	2857	0	-	0	-	0	-
AVR30	28/M	1617	3885	-	0	-	0	-
AVS32	18/M	855	0	-	0	-	0	-
AVS34	42/F	3758	0	-	0	-	0	-
AVS35	57/M	5559	0	WT	10	WT	2	WT
AVS38	35/F	280	38	-	0	-	0	-

We first assessed the total amount of DNA present in each of these 23 clinical gDNA samples by amplifying a reference *M. tuberculosis* WT sequence, 16S rRNA (rrs), present in the sample and unrelated to the mutations of interest. The generation of amplicons from this reference sequence served as a quality indicator control for the gDNA samples. The RT-PCR assays were performed using the conventional forward and reverse primer to amplify a region of rrs gene using Sybr Green to monitor amplification. The Ct value of these WT amplicons reflects the amount of DNA present in the sample. The number of copies of rrs was calculated using the slope and intercept of the standard curve, which was plotted using 10^5-1 copies of WT H37Rv gDNA. If the number of genomes turned out to be zero or lower than a predetermined value, the assay results would have been ignored due to there being too little DNA in the sample for the rare target mutations, if they exist, to be present.

This assay confirmed that all the clinical samples had detectable copies of the rrs, ranging from 33 to 25,000. Next, we performed RT-PCR using the mutation-specific SuperSelective primers (Table 1) to determine the number of mutant DNA copies present in each of these 23 clinical *M. tuberculosis* gDNA samples, with mutant gDNA used to prepare the standards at copy numbers ranging from 1 to 10^5. We first plotted the standard curve using mean Ct values versus the logarithm of the number of copies of mutant target DNA for each mutation (as shown in Figure 1b). Next, the clinical *M. tuberculosis* gDNA samples were subjected to RT-PCR with the SuperSelective primers, and the number of copies of each mutant gene was calculated based on the slope and intercept of the standard curve.

To validate our assessment of drug resistance detection based on SuperSelective primers, sections of katG, inhA, and rpoB genes containing the mutations of interest were sequenced in 11/23 samples (Table 2). Eight out of twenty-three samples were also tested for phenotypic drug susceptibility using pDST. Based on the RT-PCR assay using SuperSelective primers, six out of twenty-three samples were found to be WT, two of which were also confirmed by pDST. Using SuperSelective primers, six out of twenty-three samples were found to be INH-monoresistant, including five with *katG* S315T AGC→ACA mutation and one with *inhA* -15 C→T mutation. Of these, one of the *katG* mutant sample was confirmed to be INH-resistant by pDST. Using SuperSelective primers, we also identified 11/23 clinical gDNA samples as MDR with mutations conferring both INH and RIF resistance. Five of eleven of these samples had a complete agreement between the SuperSelective primer RT-PCR assay and sequencing, and three out of eleven tested by pDST were confirmed to be INH- and RIF-resistant (Table 2). In three of the eleven samples (AVR2, AVR11, and AVS35), there was a discordance between the SuperSelective primer RT-PCR assay and sequencing. The SuperSelective primer RT-PCR assay detected sample AVR2 as MDR, demonstrating five copies of *katG* S315T AGC→ACA and thirty-eight copies of *inhA* -15 C→T mutant DNA in a total sample of two hundred fifty-four gDNA copies. Neither of these INH mutations were detected by DNA sequencing; thus, this sample would be considered as RIF-monoresistant, rather than MDR. With the SuperSelective primer RT-PCR assay, sample AVR11 had two and four copies of *katG* S315T AGC→ACA and *rpoB* S531L TCG→TTG mutant DNA, respectively, in a total background of over twenty-four thousand copies of gDNA. This sample was detected as WT by sequencing, yet pDST results of this sample classified it as MDR. In another sample (AVS35), pDST and sequencing failed to detect drug resistance. However, SuperSelective primers identified ten and two copies of *inhA* -15 C→T and *rpoB* S531L TCG→TTG mutant DNA, respectively, with rrs copies being >5500 in this sample.

While targeted sequencing is known to detect heteroresistance with a limit of detection (LoD) of 10%, SuperSelective primers were able to detect specific mutants as low as two copies in the presence of abundant DNA fragments from the entire *M. tuberculosis* genome with exceptional sensitivity, due to the suppression of amplification of the abundant WT DNA fragments.

3. Discussion

We report the successful use of SuperSelective PCR primers for detecting RIF and INH heteroresistance in TB with a detection limit of 0.01% for the mutant DNA/WT DNA ratio. Our method shows a 10-fold improvement in detecting mutant gene copies in the presence of abundant WT genes of *M. tuberculosis* genome over the most sensitive methods reported to date, such as digital PCR [10].

We present nine SuperSelective primer designs which, under optimized RT-PCR assay conditions, can detect nine mutations commonly associated with RIF and INH resistance in TB disease, including *katG* (S315T, AGC→ACA, and AGC→ACC), *inhA* promoter (-8T→A, -15C→T, and -17G→T), and *rpoB* (D516V, GAC→GTC; H526D, CAC→GAC; H526Y, CAC→TAC; and S531L, TCG→TTG). After generating primers for mutations associated with resistance to INH and RIF, we performed clinical validation of the approach in *M. tuberculosis* DNA isolated from sputum samples obtained from TB patients, which included a mixture of drug-sensitive and drug-resistant cases. While our SuperSelective primer-based approach confirmed the presence of mutant DNA in patients with sequencing results demonstrating the mutation of interest in all instances, we also detected low copy numbers of mutant DNA among several patients with putative WT DNA by sequencing, confirming the presence of heteroresistant *M. tuberculosis* below the threshold that can be amplified by standard methods of DNA sequencing.

Heteroresistance, known as the detection of both the mutant and the wild-type strains, is considered to be the early stage in the development of drug-resistant TB and is reflective of the slow evolution of bacteria from a sensitive to resistant profile [36]. Heteroresistance was initially attributed to infections with multiple infecting *M. tuberculosis* strains (with mixed phenotypic resistance patterns); strains from different lung regions demonstrate different drug susceptibilities. In recent years, the implementation of molecular diagnostic techniques has led to an alternate conceptualization of heteroresistance, recognizing that infection by a single *M. tuberculosis* isolate may contain both mutant (drug-resistant) and wild-type (drug-susceptible) genomic DNA of the pathogen. Standard phenotypic culture-based drug resistance assays lead to the loss of these heteroresistant variants during the subculture process, interfering with our understanding of this phenomenon. The sensitivity of GenoType MTBDRplus and LPA Nipro for heteroresistance detection is about 5% [12,14]. While with more advanced techniques, such as digital PCR, the detection threshold can reach up to 0.1% mutant:WT [10], these assays require sophisticated equipment and/or are more expensive to implement in TB-endemic countries. In contrast, our method involves a carefully designed PCR primer strategy that can be used in a standard RT-PCR machine. However, as with all other primer- and probe-based amplification assays, and unlike sequencing-based methods, it is necessary to know the exact nature of the polymorphism to develop a SuperSelective primer-based RT-PCR assay.

Early identification of heteroresistant TB may support efforts to optimize TB therapy on an individual basis, for example, with adherence interventions or clinical decisions on drug selection, dosing, or therapeutic drug monitoring. SuperSelective primers, despite their extraordinary discriminatory ability (higher than conventional methods), are supremely easy to use. The only difference between these assays and conventional multiplex RT-PCR assays is the substitution of SuperSelective primers for conventional primers, while the sample preparation, amplification, and assessment are carried out in the same manner. Thus, RT-PCR assays with SuperSelective primers can be used as an accessory technique for resistance detection along with other methods of TB detection. Although the current study was performed using sputum samples from primary TB cases, utility of SuperSelective PCR primers can be extended to re-treatment cases, which usually have a higher probability of a low to moderate number of copies of mutant DNA in the sputum samples and are relatively easier to detect.

In summary, we report the successful development and application of "SuperSelective" primer-based RT-PCR to detect gene mutations in *M. tuberculosis* associated with heteroresistance in an abundant wild-type background. Improved diagnostic tools, such as the one described here, would enable better and improved therapeutic management of drug-sensitive and drug-resistant TB cases in general and heteroresistant TB cases in particular.

4. Materials and Methods

4.1. Selected Antibiotic-Drug-Resistant Mutations

SuperSelective primers were designed to amplify mutant *M. tuberculosis* DNA corresponding to INH and RIF resistance in a WT background. For INH resistance, SuperSelective primers were designed to target *katG* (S315T, AGC→ACA, and AGC→ACC) and the promoter region of *inhA* (-8T→A, -15C→T, and -17G→T) [25–27]. For RIF resistance, the primers were designed to focus on the most common mutations in the RRDR of *rpoB* (D516V, GAC→GTC; H526D, CAC→GAC; H526Y, CAC→TAC; and S531L, TCG→TTG) [37].

4.2. Structure and Functionality of SuperSelective Primers

The SuperSelective PCR primers were designed to have a minimal probability of initiating synthesis on WT sequences, even when the only difference between the mutant and the WT targets was a single-nucleotide polymorphism. The SuperSelective primer designs were adapted from our previous reports [25,38]. Briefly, SuperSelective primers used in the present study consist of the 3 sequence segments in order from 5′ to 3′: (i) an anchor sequence complementary to the target DNA that is sufficiently long to ensure strong hybridization to its DNA target fragment under PCR annealing conditions; (ii) a distinctive bridge sequence, which is non-complementary to the corresponding intervening sequence in the DNA target fragment; and (iii) a short 3′-foot sequence that is entirely complementary to the corresponding sequence in the mutant DNA target fragments, but is discrepant to the corresponding sequence in the related WT DNA fragments.

In a PCR assay, the function of a SuperSelective primer is to maintain a delicate balance between the anchor sequence (long 5′ segment), which promotes its efficient binding to a gene of interest, and the foot sequence (short 3′ segment), which restricts binding to the subsequence that includes the target mutation. Complementarity between the "interrogating nucleotide" in the foot to the corresponding nucleotide in the mutant target sequence makes the foot and mutant target perfectly compatible and leads to priming of the synthesis of an amplicon, while the mismatch at the corresponding nucleotide in the WT target sequence inhibits the amplification of the WT target. DNA-DNA hybrids formed by the anchor sequence with target DNA and the short foot sequence with target DNA are separated from each other by a single-stranded bubble formed by the non-complimentary bridge sequence (which should not form any secondary structures) and the intervening sequence in the DNA target fragment. This bubble effectively separates the primer's polymerization initiation from the primer's target recognition. Figure 3A shows an example of a SuperSelective primer bound to its mutant target sequence. This particular primer was designed to selectively amplify DNA fragments containing *katG* S315T (AGC→ACA) single-nucleotide polymorphism, which is commonly associated with INH resistance in TB patients.

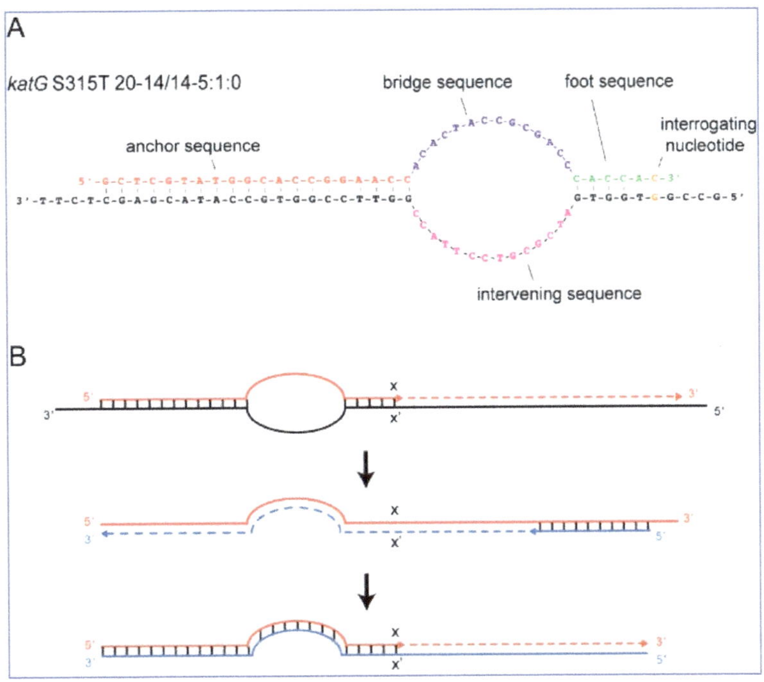

Figure 3. Structure and principle of operation of SuperSelective primers. (**A**) Structure of a SuperSelec-

tive primer for detecting *katG* S315T (AGC→ACA) mutant sequences in the presence of *katG* WT sequences. (**B**) Principle of operation of SuperSelective primers. The selective step occurs only when a SuperSelective primer hybridizes to a DNA (−) template fragment present in the sample. Due to the small size of the foot sequence, the probability of initiation of a (+) amplicon is significantly greater if the target sequence of the foot in the (−) template fragment is a completely complementary mutant (x) sequence than if the target sequence of the foot in the (−) template fragment is a mismatched WT sequence. Suppose (+) amplicon synthesis does occur. In that case, the resulting (+) amplicon serves as a template for a conventional reverse primer and is efficiently copied during the next thermal cycle, generating a (−) amplicon in which the complement of the unique bridge sequence that was present in the SuperSelective primer is substituted for the intervening sequence that was present in the original (−) template fragment. As a result, in subsequent thermal cycles, the entire SuperSelective primer sequence is complementary to the (−) amplicon strands, and exponential amplification occurs efficiently and can be followed in real time.

The nomenclature for primer labeling was as we described earlier [25]. For example, a SuperSelective primer 20-14/14-5:1:0 (shown in Figure 3A) indicates that the anchor sequence is 20 nucleotides long, the bridge sequence is 14 nucleotides long (across from an intervening sequence in the template that is 14 nucleotides long), and the foot sequence is 6 nucleotides long, with one interrogating nucleotide located at the last position from the primer's 3′ end.

In an RT-PCR assay with a SuperSelective primer, the selective step occurs when a SuperSelective primer (designed against DNA (−) strand template) is bound to the original DNA sample being analyzed. The foot sequence of a SuperSelective primer initiates the synthesis of an amplicon, and as the reaction proceeds, the entire sequence of the SuperSelective primer (including the "artificial" bridge sequence) is incorporated into that (+) amplicon. In the successive thermal cycles, the resulting amplicons are amplified efficiently in the normal manner, with the entire SuperSelective primer sequence serving as a long conventional primer that is completely complementary to the (−) amplicon, which includes the complement of the primer's bridge sequence in place of the intervening sequence that was present in the original template (Figure 3B).

4.3. Primers, Molecular Beacons, and PCR Reagents

In all assays, a SuperSelective forward primer was paired with a conventional reverse primer unless mentioned otherwise. SuperSelective primer sequences were assessed using the Mfold web server [39] and the OligoAnalyzer computer program (Integrated DNA Technologies, Coralville, IA, USA) to ensure these sequences are unlikely to form internal secondary structures like hairpin loops, self-dimers, or heterodimers with conventional reverse primers. Conventional and SuperSelective primers were purchased from Integrated DNA Technologies. Molecular beacon probes were purchased from LGC Biosearch Technologies (Petaluma, CA, USA). Platinum Taq DNA Polymerase, deoxynucleotide triphosphates (dNTPs), and Sybr Green were purchased from Thermo Fisher Scientific, Inc. (Waltham, MA, USA). Nuclease-free water was purchased from Ambion, Inc. (Austin, TX, USA).

4.4. Target DNA

Complete WT and mutant (AGC→ACA) katG genes of *M. tuberculosis* were cloned into the vector pcDNA3.1 [40] and used as the source of pDNA for assays. For the optimization of assays to detect AGC→ACC mutation, plasmids (pUC) containing a 584 bp subsequence of katG gene (WT and AGC→ACC) were synthesized (Integrated DNA Technologies). To detect *inhA* promoter mutations, plasmids (pUC) containing a 387 bp region covering mutations (-8T→A, -15C→T, and -17G→T) and the corresponding WT were synthesized (Integrated DNA Technologies). To detect *rpoB* mutations, a 488 bp subsequence of mutant genes containing mutations D516V (GAC→GTC), *rpoB* H526D (CAC→GAC), H526Y (CAC→TAC), *rpoB* S531L (TCG→TTG), and WT were synthesized (Integrated DNA

Technologies). The selected *M. tuberculosis* genes targeting INH and RIF resistance are commonly associated with MDR-TB cases [37,41–43].

The plasmids containing mutations of interest were linearized by restriction digestion for 120 min at 37 °C in 50 µL containing 10 units of restriction endonuclease *ClaI* with appropriate buffer supplied by the manufacturer (New England Biolabs, Ipswich, MA, USA). The reaction mix was incubated for 20 min at 65 °C to inactivate the enzyme, and the concentration of each linearized plasmid was determined in a NanoDrop instrument (Thermo Fisher Scientific, Inc., Waltham, MA, USA). The plasmids were diluted in nuclease-free water to create stock solutions containing known quantities of linearized target plasmids/µL.

The genomic DNA (gDNA) of INH- and RIF-resistant *M. tuberculosis* was provided by the Kreiswirth laboratory (CDI, Hackensack Meridian Health, Nutley, NJ, USA). The gDNA sequences were verified by DNA sequence analysis (Integrated DNA Technologies). WT H37Rv DNA was obtained from BEI resources (NIAID, NIH, Washington, DC, USA).

4.5. Preparation of DNA Samples

Template DNA for RT-PCR assays was prepared using pDNA/gDNA. A calculated number of mutant DNA copies was mixed with a calculated number of WT DNA copies to create varying ratios of mutant DNA against a WT background. Serial dilutions of mutant DNA (10^5-1 copies) were mixed with 10^5 or 10^4 copies of WT DNA. Tested Mutant:WT ratios were 10^5:10^5, 10^4:10^5, 10^3:10^5, 10^2:10^5, 50:10^5, 25:10^5, 10:10^5, 1:10^5, 0:10^5 and 10^5:10^4, 10^4:10^4, 10^3:10^4, 10^2:10^4, 50:10^4, 25:10^4, 10:10^4, 1:10^4, 0:10^4. Mutant DNA copies 10^5, 10^4, 10^3, 10^2, 50, 25, 10 and 1 were used as control.

4.6. Assay Composition and Testing Procedure

Monoplex RT-PCR assays were performed with a SuperSelective forward primer and a conventional reverse primer on an AriaMx Real-Time PCR System (Agilent Technologies, Santa Clara, CA, USA). All amplifications were carried out in 0.2 mL clear tubes (Agilent Technologies, Santa Clara, CA, USA) in a final volume of 25 µL, containing 1X PCR buffer supplemented with 25 mM tetramethylammonium chloride (TMAC) (Sigma-Aldrich, St. Louis, MO, USA), 0.25% Tween20 (Sigma-Aldrich, St. Louis, MO, USA), 3 mM MgCl2, 0.05 U/µL Platinum Taq DNA polymerase (Thermo Fisher Scientific, Inc., Waltham, MA, USA), 0.25 mM dNTP mix, 0.1 µM each of forward and reverse primer, and 0.5X Sybr Green (Thermo Fisher Scientific, Inc., Waltham, MA, USA) to monitor template amplification. The reaction mixtures were incubated for 2 min at 95 °C to activate the Platinum Taq DNA polymerase, followed by 50–65 cycles of DNA denaturation at 95 °C for 15 s, primer annealing at 60 °C for 20 s, and extension at 72 °C for 20 s. Fluorescence was monitored and measured during the annealing step.

For the detection of mutants using molecular beacon probes in monoplex assays, asymmetric PCR conditions were optimized (Figures S3 and S4) to maximize the intensity of fluorescence signals. RT-PCR assay conditions were the same as described above, except that the forward and reverse primers were used at a 1:5 ratio, with final concentrations of 0.1 µM and 0.5 µM, respectively. This assay used 0.1 µM of molecular beacon probe instead of Sybr Green to monitor amplification.

Multiplex assays with molecular beacon probes were carried out in 0.2 mL white polypropylene PCR tubes (USA Scientific, Ocala, FL, USA) in a Bio-Rad CFX96 Real-Time System (Bio-Rad Laboratories, Hercules, CA, USA). RT-PCR assay conditions were the same as those used in the monoplex assays, except that 3 primer pairs and 3 probes were used (*katG*, *inhA*, *rpoB*) for multiplexing. The final concentration of each forward primer (*katG*, *inhA*, *rpoB*) was 0.1 µM, each reverse primer (*katG*, *inhA*, *rpoB*) 0.5 µM, probe *katG* 0.25 µM, probe *inhA* 0.1 µM, and probe *rpoB* 0.25 µM.

The RT-PCR assays were initially optimized on linearized plasmids containing a subsequence of the gene containing respective mutations as template DNA because of the availability of a large amount of material. The assay results were then confirmed by testing the optimized assay conditions on *M. tuberculosis* gDNA. RT-PCR conditions were tested for mutant DNA template at copy numbers 10^5-1 in a background of $10^5/10^4$ copies of WT DNA template. Mutant DNA at the copy numbers 10^5-1 with no background of WT DNA was used as an assay control. Nuclease-free water was used as a no template control. Experiments were performed on duplicate samples and repeated at least 3 times.

4.7. Clinical Samples

Twenty-three *M. tuberculosis* DNA samples were isolated from archived sputum samples of anonymized TB patients with unknown drug resistance. These de-identified and anonymized sputum specimens were thrashed samples collected as part of routine TB diagnosis screening. Hence, no informed consent was required to extract bacterial DNA from the specimen, and the archived samples were accessed exclusively to rationally generate beneficial and scientifically valid assays to improve patient health. To isolate the gDNA, the sputum samples were first decontaminated and homogenized by using NALC-NaOH, then 2 mL of sputum was taken in an Oakridge tube, and an equal amount of 1% N-Acetyl-L-cysteine, 4% sodium hydroxide, and 2.9% sodium citrate was added. The suspension was incubated at 37 °C for 15 min and added with 45 mL of phosphate buffer, followed by centrifugation at 5000 rpm for 15 min. The obtained pellet was re-suspended in 1 mL of nuclease-free water. For DNA extraction, 1 mL of the decontaminated sample was spun again and re-suspended in 50 µL of Genolyse Lysis buffer (Hain Lifesciences GmbH, Hardwiesentrabe, Nehren, Germany), followed by heat-killing at 95 °C for 5 min to lyse the bacterial cells. Subsequently, the Genolyse Neutralizing buffer (50 µL) was added to the suspension, and the resulting DNA in the supernatant was pelleted by centrifugation at 13,000 rpm for 5 min.

Supplementary Materials: The following supporting information can be downloaded at: https://www.mdpi.com/article/10.3390/ijms232415752/s1

Author Contributions: Conceptualization, funding, and study design: S.A.E.M., C.V. and S.S. Literature search, methodology, and figures: A.N., S.A.E.M., V.C. and M.V.-B. Data collection, analysis, interpretation: A.N., S.A.E.M., N.K., E.S., V.C., B.K., M.V.-B., C.V. and S.S. Writing: A.N., S.A.E.M., C.V. and S.S. Review and editing: A.N., S.A.E.M., N.K., E.S., B.K., M.V.-B., C.V. and S.S. All authors have read and agreed to the published version of the manuscript.

Funding: This study was funded by an NIAID/NIH grant (R21AI146820 to CV, SAEM, and SS). SAEM is partly supported by an NCI/NIH grant (R01CA22729). The funders had no role in study design, data collection and analysis, decision to publish, or preparation of the manuscript.

Institutional Review Board Statement: All biocontainment level-3 (BSL3) work was performed as per the Rutgers University Institutional Biosafety Committee (IBC) approved protocols (approval# 13-499v800). Archived sputum samples of anonymized and de-identified tuberculosis patients were used in this study.

Informed Consent Statement: The sputum was initially collected as part of routine diagnostic screening for tuberculosis; hence, no informed consent was required to extract bacterial DNA from these thrashed specimens. The archived samples were accessed for this study to rationally and exclusively generate beneficial and scientifically valid assays to improve patient health.

Data Availability Statement: All data generated and analyzed for this study are included in this article. All materials used and described in the methods are commercially available.

Acknowledgments: The following reagent was obtained through BEI Resources, NIAID, NIH (Manassas, VA, USA): Genomic DNA from *Mycobacterium tuberculosis*, Strain H37Rv, NR-48669.

Conflicts of Interest: The authors have declared that no competing interest exist.

References

1. World Health Organization. *Global Tuberculosis Report*; WHO: Geneva, Switzerland, 2021; Volume 57.
2. Gygli, S.M.; Borrell, S.; Trauner, A.; Gagneux, S. Antimicrobial resistance in *Mycobacterium tuberculosis*: Mechanistic and evolutionary perspectives. *FEMS Microbiol. Rev.* **2017**, *41*, 354–373. [CrossRef] [PubMed]

3. Law, S.; Piatek, A.S.; Vincent, C.; Oxlade, O.; Menzies, D. Emergence of drug resistance in patients with tuberculosis cared for by the Indian health-care system: A dynamic modelling study. *Lancet Public Health* **2017**, *2*, e47–e55. [CrossRef]
4. Seaworth, B.J.; Griffith, D.E. Therapy of Multidrug-Resistant and Extensively Drug-Resistant Tuberculosis. *Microbiol. Spectr.* **2017**, *5*, 1–28. [CrossRef] [PubMed]
5. Schwoebel, V.; Trebucq, A.; Kashongwe, Z.; Bakayoko, A.S.; Kuaban, C.; Noeske, J.; Harouna, S.H.; Souleymane, M.B.; Piubello, A.; Ciza, F.; et al. Outcomes of a nine-month regimen for rifampicin-resistant tuberculosis up to 24 months after treatment completion in nine African countries. *eClinicalMedicine* **2020**, *20*, 100268. [CrossRef] [PubMed]
6. Dheda, K.; Barry, C.E., 3rd; Maartens, G. Tuberculosis. *Lancet* **2016**, *387*, 1211–1226. [CrossRef] [PubMed]
7. Palaci, M.; Dietze, R.; Hadad, D.J.; Ribeiro, F.K.; Peres, R.L.; Vinhas, S.A.; Maciel, E.L.; do Valle Dettoni, V.; Horter, L.; Boom, W.H.; et al. Cavitary disease and quantitative sputum bacillary load in cases of pulmonary tuberculosis. *J. Clin. Microbiol.* **2007**, *45*, 4064–4066. [CrossRef]
8. Chen, L.; Zhang, J.; Zhang, H. Heteroresistance of *Mycobacterium tuberculosis* strains may be associated more strongly with poor treatment outcomes than within-host heterogeneity of *M. tuberculosis* infection. *J. Infect. Dis.* **2016**, *214*, 1286–1287. [CrossRef]
9. Eilertson, B.; Maruri, F.; Blackman, A.; Herrera, M.; Samuels, D.C.; Sterling, T.R. High proportion of heteroresistance in *gyrA* and *gyrB* in fluoroquinolone-resistant *Mycobacterium tuberculosis* clinical isolates. *Antimicrob. Agents Chemother.* **2014**, *58*, 3270–3275. [CrossRef]
10. Pholwat, S.; Stroup, S.; Foongladda, S.; Houpt, E. Digital PCR to detect and quantify heteroresistance in drug resistant *Mycobacterium tuberculosis*. *PLoS ONE* **2013**, *8*, e57238. [CrossRef]
11. Rinder, H.; Mieskes, K.T.; Loscher, T. Heteroresistance in *Mycobacterium tuberculosis*. *Int. J. Tuberc. Lung Dis.* **2001**, *5*, 339–345.
12. Ng, K.C.S.; Supply, P.; Cobelens, F.G.J.; Gaudin, C.; Gonzalez-Martin, J.; de Jong, B.C.; Rigouts, L. How well Do Routine Molecular Diagnostics Detect Rifampin Heteroresistance in *Mycobacterium tuberculosis*? *J. Clin. Microbiol.* **2019**, *57*, e00717-19. [CrossRef] [PubMed]
13. Tolani, M.P.; D'Souza, D.T.; Mistry, N.F. Drug resistance mutations and heteroresistance detected using the GenoType MTBDR*plus*assay and their implication for treatment outcomes in patients from Mumbai, India. *BMC Infect. Dis.* **2012**, *12*, 9. [CrossRef] [PubMed]
14. Folkvardsen, D.B.; Thomsen, V.O.; Rigouts, L.; Rasmussen, E.M.; Bang, D.; Bernaerts, G.; Werngren, J.; Toro, J.C.; Hoffner, S.; Hillemann, D.; et al. Rifampin heteroresistance in *Mycobacterium tuberculosis* cultures as detected by phenotypic and genotypic drug susceptibility test methods. *J. Clin. Microbiol.* **2013**, *51*, 4220–4222. [CrossRef] [PubMed]
15. Maningi, N.E.; Daum, L.T.; Rodriguez, J.D.; Mphahlele, M.; Peters, R.P.; Fischer, G.W.; Chambers, J.P.; Fourie, P.B. Improved Detection by Next-Generation Sequencing of Pyrazinamide Resistance in *Mycobacterium tuberculosis* Isolates. *J. Clin. Microbiol.* **2015**, *53*, 3779–3783. [CrossRef]
16. Zhang, Z.; Wang, Y.; Pang, Y.; Liu, C. Comparison of different drug susceptibility test methods to detect rifampin heteroresistance in *Mycobacterium tuberculosis*. *Antimicrob. Agents Chemother.* **2014**, *58*, 5632–5635. [CrossRef]
17. Ye, M.; Yuan, W.; Molaeipour, L.; Azizian, K.; Ahmadi, A.; Kouhsari, E. Antibiotic heteroresistance in *Mycobacterium tuberculosis* isolates: A systematic review and meta-analysis. *Ann. Clin. Microbiol. Antimicrob.* **2021**, *20*, 73. [CrossRef]
18. Cohen, T.; Chindelevitch, L.; Misra, R.; Kempner, M.E.; Galea, J.; Moodley, P.; Wilson, D. Within-Host Heterogeneity of *Mycobacterium tuberculosis* Infection Is Associated with Poor Early Treatment Response: A Prospective Cohort Study. *J. Infect. Dis.* **2016**, *213*, 1796–1799. [CrossRef]
19. Shin, S.S.; Modongo, C.; Zetola, N.M. The impact of mixed infections on the interpretation of molecular epidemiology studies of tuberculosis. *Int. J. Tuberc. Lung Dis.* **2016**, *20*, 423–424. [CrossRef]
20. Zetola, N.M.; Shin, S.S.; Tumedi, K.A.; Moeti, K.; Ncube, R.; Nicol, M.; Collman, R.G.; Klausner, J.D.; Modongo, C. Mixed *Mycobacterium tuberculosis* complex infections and false-negative results for rifampin resistance by GeneXpert MTB/RIF are associated with poor clinical outcomes. *J. Clin. Microbiol.* **2014**, *52*, 2422–2429. [CrossRef]
21. Mendez, M.P.; Landon, M.E.; McCloud, M.K.; Davidson, P.; Christensen, P.J. Co-infection with pansensitive and multidrug-resistant strains of *Mycobacterium tuberculosis*. *Emerg. Infect. Dis.* **2009**, *15*, 578–580. [CrossRef]
22. Hingley-Wilson, S.M.; Casey, R.; Connell, D.; Bremang, S.; Evans, J.T.; Hawkey, P.M.; Smith, G.E.; Jepson, A.; Philip, S.; Kon, O.M.; et al. Undetected multidrug-resistant tuberculosis amplified by first-line therapy in mixed infection. *Emerg. Infect. Dis.* **2013**, *19*, 1138–1141. [CrossRef] [PubMed]
23. Xu, Y.; Wu, J.; Liao, S.; Sun, Z. Treating tuberculosis with high doses of anti-TB drugs: Mechanisms and outcomes. *Ann. Clin. Microbiol. Antimicrob.* **2017**, *16*, 67. [CrossRef] [PubMed]
24. Devaleenal Daniel, B.; Ramachandran, G.; Swaminathan, S. The challenges of pharmacokinetic variability of first-line anti-TB drugs. *Expert Rev. Clin. Pharmacol.* **2017**, *10*, 47–58. [CrossRef]
25. Vargas, D.Y.; Kramer, F.R.; Tyagi, S.; Marras, S.A. Multiplex Real-Time PCR Assays that Measure the Abundance of Extremely Rare Mutations Associated with Cancer. *PLoS ONE* **2016**, *11*, e0156546. [CrossRef] [PubMed]
26. Newton, C.R.; Graham, A.; Heptinstall, L.E.; Powell, S.J.; Summers, C.; Kalsheker, N.; Smith, J.C.; Markham, A.F. Analysis of any point mutation in DNA. The amplification refractory mutation system (ARMS). *Nucleic Acids Res.* **1989**, *17*, 2503–2516. [CrossRef]
27. Chun, J.Y.; Kim, K.J.; Hwang, I.T.; Kim, Y.J.; Lee, D.H.; Lee, I.K.; Kim, J.K. Dual priming oligonucleotide system for the multiplex detection of respiratory viruses and SNP genotyping of CYP2C19 gene. *Nucleic Acids Res.* **2007**, *35*, e40. [CrossRef]

28. Hazbon, M.H.; Alland, D. Hairpin primers for simplified single-nucleotide polymorphism analysis of *Mycobacterium tuberculosis* and other organisms. *J. Clin. Microbiol.* **2004**, *42*, 1236–1242. [CrossRef] [PubMed]
29. Tabrizi, S.N.; Tan, L.Y.; Walker, S.; Twin, J.; Poljak, M.; Bradshaw, C.S.; Fairley, C.K.; Bissessor, M.; Mokany, E.; Todd, A.V.; et al. Multiplex Assay for Simultaneous Detection of *Mycoplasma genitalium* and Macrolide Resistance Using PlexZyme and PlexPrime Technology. *PLoS ONE* **2016**, *11*, e0156740. [CrossRef]
30. Cole, S.T.; Brosch, R.; Parkhill, J.; Garnier, T.; Churcher, C.; Harris, D.; Gordon, S.V.; Eiglmeier, K.; Gas, S.; Barry, C.E., 3rd; et al. Deciphering the biology of *Mycobacterium tuberculosis* from the complete genome sequence. *Nature* **1998**, *393*, 537–544. [CrossRef]
31. Kramer, F.R.; Vargas, D.Y. SuperSelective primer pairs for sensitive detection of rare somatic mutations. *Sci. Rep.* **2021**, *11*, 22384. [CrossRef]
32. Tyagi, S.; Kramer, F.R. Molecular beacons: Probes that fluoresce upon hybridization. *Nat. Biotechnol.* **1996**, *14*, 303–308. [CrossRef] [PubMed]
33. Tyagi, S.; Bratu, D.P.; Kramer, F.R. Multicolor molecular beacons for allele discrimination. *Nat. Biotechnol.* **1998**, *16*, 49–53. [CrossRef] [PubMed]
34. Seifert, M.; Catanzaro, D.; Catanzaro, A.; Rodwell, T.C. Genetic mutations associated with isoniazid resistance in *Mycobacterium tuberculosis*: A systematic review. *PLoS ONE* **2015**, *10*, e0119628. [CrossRef] [PubMed]
35. Somoskovi, A.; Parsons, L.M.; Salfinger, M. The molecular basis of resistance to isoniazid, rifampin, and pyrazinamide in *Mycobacterium tuberculosis*. *Respir. Res.* **2001**, *2*, 164–168. [CrossRef] [PubMed]
36. Hofmann-Thiel, S.; van Ingen, J.; Feldmann, K.; Turaev, L.; Uzakova, G.T.; Murmusaeva, G.; van Soolingen, D.; Hoffmann, H. Mechanisms of heteroresistance to isoniazid and rifampin of *Mycobacterium tuberculosis* in Tashkent, Uzbekistan. *Eur. Respir. J.* **2009**, *33*, 368–374. [CrossRef]
37. Rosales-Klintz, S.; Jureen, P.; Zalutskayae, A.; Skrahina, A.; Xu, B.; Hu, Y.; Pineda-Garcia, L.; Merza, M.A.; Muntean, I.; Bwanga, F.; et al. Drug resistance-related mutations in multidrug-resistant *Mycobacterium tuberculosis* isolates from diverse geographical regions. *Int. J. Mycobacteriol.* **2012**, *1*, 124–130. [CrossRef]
38. Vargas, D.Y.; Marras, S.A.E.; Tyagi, S.; Kramer, F.R. Suppression of Wild-Type Amplification by Selectivity Enhancing Agents in PCR Assays that Utilize SuperSelective Primers for the Detection of Rare Somatic Mutations. *J. Mol. Diagn.* **2018**, *20*, 415–427. [CrossRef]
39. Zuker, M. Mfold web server for nucleic acid folding and hybridization prediction. *Nucleic Acids Res.* **2003**, *31*, 3406–3415. [CrossRef]
40. Weeks, S.D.; Drinker, M.; Loll, P.J. Ligation independent cloning vectors for expression of SUMO fusions. *Protein Expr. Purif.* **2007**, *53*, 40–50. [CrossRef]
41. Pym, A.S.; Saint-Joanis, B.; Cole, S.T. Effect of katG mutations on the virulence of *Mycobacterium tuberculosis* and the implication for transmission in humans. *Infect. Immun.* **2002**, *70*, 4955–4960. [CrossRef]
42. Torres, J.N.; Paul, L.V.; Rodwell, T.C.; Victor, T.C.; Amallraja, A.M.; Elghraoui, A.; Goodmanson, A.P.; Ramirez-Busby, S.M.; Chawla, A.; Zadorozhny, V.; et al. Novel *katG* mutations causing isoniazid resistance in clinical *M. tuberculosis* isolates. *Emerg. Microbes Infect.* **2015**, *4*, e42. [CrossRef] [PubMed]
43. Yu, S.; Girotto, S.; Lee, C.; Magliozzo, R.S. Reduced affinity for Isoniazid in the S315T mutant of *Mycobacterium tuberculosis* KatG is a key factor in antibiotic resistance. *J. Biol. Chem.* **2003**, *278*, 14769–14775. [CrossRef] [PubMed]

International Journal of
Molecular Sciences

Article

A Novel Ag@AgCl Nanoparticle Synthesized by Arctic Marine Bacterium: Characterization, Activity and Mechanism

Shuang Li [1,2,†], Hui Zhang [2,†], Bailin Cong [2,3,*], Peiqing He [2], Wenqi Liu [3] and Shenghao Liu [2,3]

1 College of Chemistry and Chemical Engineering, Ocean University of China, Qingdao 266100, China
2 First Institute of Oceanography, Ministry of Natural Resources, Qingdao 266061, China
3 School of Advanced Manufacturing, Fuzhou University, Fuzhou 350108, China
* Correspondence: biolin@fio.org.cn
† These authors contributed equally to this work.

Abstract: An additive- and pollution-free method for the preparation of biogenic silver and silver chloride nanoparticles (Ag@AgCl NPs) was developed from the bacteria *Shewanella* sp. Arc9-LZ, which was isolated from the deep sea of the Arctic Ocean. The optimal synthesizing conditions were explored, including light, pH, Ag^+ concentration and time. The nanoparticles were studied by means of ultraviolet-visible (UV-Vis) spectrophotometry, energy dispersive spectrometry (EDS), X-ray diffraction (XRD) and inductively coupled plasma optical emission spectrometers (ICP-OES). The transmission electron microscope (TEM) showed that the nanoparticles were spherical and well dispersed, with particle sizes less than 20.00 nm. With Ag@AgCl nanoparticles, the kinetic rate constants for congo red (CR) and rhodamine B (RhB) dye degradation were 2.74×10^{-1} min^{-1} and 7.78×10^{-1} min^{-1}, respectively. The maximum decolourization efficiencies of CR and RhB were 93.36% and 99.52%, respectively. Ag@AgCl nanoparticles also showed high antibacterial activities against the Gram-positive and Gram-negative bacteria. The Fourier transform infrared spectroscopy (FTIR) spectrum indicated that the O-H, N-H and -COO- groups in the supernatant of Arc9-LZ might participate in the reduction, stabilization and capping of nanoparticles. We mapped the schematic diagram on possible mechanisms for synthesizing Ag@AgCl NPs.

Keywords: Ag@AgCl nanoparticles; synthesis mechanisms; *Shewanella* sp.; dye reduction; antibacterial activity

Citation: Li, S.; Zhang, H.; Cong, B.; He, P.; Liu, W.; Liu, S. A Novel Ag@AgCl Nanoparticle Synthesized by Arctic Marine Bacterium: Characterization, Activity and Mechanism. *Int. J. Mol. Sci.* **2022**, *23*, 15558. https://doi.org/10.3390/ijms232415558

Academic Editor: Helena Felgueiras

Received: 13 November 2022
Accepted: 1 December 2022
Published: 8 December 2022

1. Introduction

Since they display a high surface-to-volume ratio and exhibit improved properties compared with bulk materials, metal nanoparticles (MNPs) have been extensively applied in catalysis, biological labelling, optoelectronics, surface-enhanced Raman scattering detection, etc [1,2]. Among all MNP-based goods available on the market, silver nanoparticles (AgNPs) show the highest degree of commercialization, accounting for approximately 55.4% [3]. Currently, AgNPs have been widely used for making catalysts for the degradation of environmental pollutants [4], developing new antimicrobial [5], antiviral [6] and anticancer drugs [7], and manufacturing packing materials [8].

Due to the COVID-19 pandemic, it is reasonable to find an agent with long-term sterilization properties and broad-spectrum antimicrobial activity. AgNPs have strong antibacterial activity against a wide range of pathogens and low cytotoxicity towards mammalian cells [9,10]. Notably, the combination of Ag^+ ions and Cl radicals will further enhance their antibacterial effects [11]. Highly reactive chlorine-free radicals from silver and silver chloride nanoparticles (Ag@AgCl NPs) could efficiently inactivate pathogenic bacteria by attacking the bacterial cell wall and disrupting cellular metabolism [11,12]. However, most studies focus on the antibacterial activities of AgNPs rather than Ag@AgCl NPs.

Conventional physiochemical methods for MNP synthesis involve the use of toxic solvents, harsh chemicals and complicated, expensive techniques, limiting their use in biomedical and clinical fields [13]. Biological nanoparticles are often water soluble and biocompatible, which are essential properties for many applications [14], especially medical applications, such as coatings for bone prostheses, surgical devices, silver-impregnated catheters, infusion systems and dental composites [3]. To date, AgNPs have been synthesized via biological systems including bacteria, fungi, yeast, viruses, algae and plants [15–17]. However, Ag@AgCl NPs are mainly synthesized via chemical methods, such as the ion-exchange process [18], vapour diffusion strategy [19], precipitation method [20] and solvothermal method [21]. The preparation of Ag@AgCl NPs via plant extract is emerging as a new approach, including Saccharum officinarum juice [13], *Aquilaria agallocha* leaf juice [22], *Salvia officinalis* leaf extract [23], *Elaeagnus angustifolia* leaf extract [24], beetroot juice [25] and *Allium sativum* leaf extract [26]. To the best of our knowledge, *Staphylococcus pasteuri* sp. nov., ZAR1, which is isolated from the Zarshouran gold mine, is the only bacteria reported to have the ability to synthesize Ag@AgCl NPs [27]. So, the preparation of Ag@AgCl NPs via bacteria has great significance and broad implications.

As a well-known dissimilatory reducing bacteria, the *Shewanella* genus plays crucial roles in biogeochemical cycles, nanomaterials formation, microbial fuel cells and bioremediation owing to its ability to use a wide range of terminal electron acceptors [28–30]. *Shewanella* microbial fuel cells boosted the charge-extraction efficiency substantially by introducing transmembrane and outer-membrane silver nanoparticles [29]. Remarkably, despite the large number of metal nanoparticles synthesized by the *Shewanella* genus, there are very limited reports on Ag@AgCl NPs.

The concrete mechanisms of AgNPs and Ag@AgCl NPs remain unknown. Previous studies suggested that plasmid-mediated silver resistance rendered bacteria capable of accumulating the silver intracellular and silver resistance gene homologues including *silE*, *silP* and *silS* [31,32]. The reductases, reducing agents, amino acids and peptides produced by microorganisms might participate in metal reduction as well as the capping of nanoparticles to a narrow size range [33,34]. However, for extracellular metal ion reduction, microorganisms face the dilemma of how to transfer electrons derived from central metabolisms onto extracellular electron acceptors [35]. The *Shewanella* species appear to be specially adapted for the reduction of extracellular electron acceptors, and the electron transfer mechanisms involve c-type cytochromes, extracellular electron shuttles and direct interspecies electron transfer [35]. Thus, it is significant to investigate the mechanism of nanomaterial synthesis by microorganisms from the perspective of electron transfer.

Here, we synthesized biogenic Ag@AgCl NPs via the cell-free supernatant of *Shewanella* sp. Arc9-LZ in darkness. The bacteria were isolated from the marine sediments of the Arctic Ocean (158°01′12″W; 84°28′38″N). The synthesis conditions of Ag@AgCl were optimized for silver nitrate concentrations, pH and duration. The nanoparticles were studied by means of ultraviolet-visible (UV-Vis) spectrophotometry, transmission electron microscopes (TEM), energy dispersive spectrometry (EDS), X-ray diffraction (XRD) and inductively coupled plasma optical emission spectrometers (ICP-OES). The degradation kinetic rate constants for congo red (CR) and rhodamine B (RhB) dye were calculated, and the antimicrobial activity against Gram-negative (*Pseudomonas aeruginosa* and *Escherichia coli*) and Gram-positive bacteria (*Bacillus subtilis* and *Staphylococcus aureus*) were investigated. Combined with genome annotation, we mapped the electron transfer and synthesis mechanism of Ag@AgCl NPs in *Shewanella* sp. Arc9-LZ.

2. Results and Discussion

2.1. Identification of the Bacterium

The scanning electron microscope (SEM) results showed that *Shewanella* sp. Arc9-LZ is rod-shaped, lacks flagella, and is 1.2–1.5 μm in length and 0.25–0.40 μm in width (Figure 1A,B). The 16S rRNA gene sequence of this strain was blasted against the National Centre for Biotechnology Information (NCBI) and showed 99.50% similarity with *Shewanella livingstonensis* LMG 19866 (MK131328.1). The phylogenetic trees based on the 16S rRNA gene sequence also indicated that strain Arc9-LZ was clustered with the genus *Shewanella* and showed the highest similarity with *Shewanella livingstonensis* LMG 19866 (Figure 1C). This strain has been deposited in the China General MicroBiological Culture Collection Centre (CGMCC) with accession number CGMCC 1.18550.

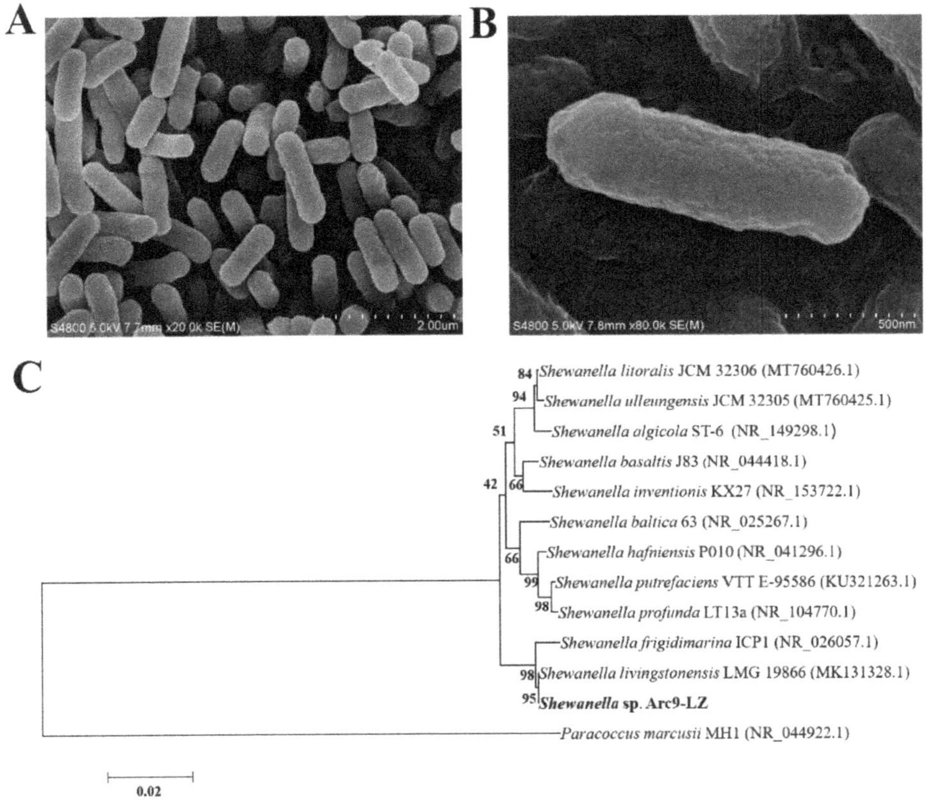

Figure 1. SEM results of the stain *Shewanella* sp. Arc9-LZ (**A**,**B**). Phylogenetic tree based on 16S rRNA gene sequences, Bootstrap ≥ 1000. GenBank accession numbers are indicated in parentheses (**C**).

The biological nanoparticles synthesised from the genus *Shewanella* are summarized in Table 1. All the genera mentioned were isolated from warm and tropical regions, including *Shewanella* sp. PV-4 (isolated from deep-sea, Hydrothermal Naha vent, HI, USA) [36], *Shewanella* sp. HN-41 (isolated from tidal flats, Haenam, the Republic of Korea) [37,38], *Shewanella algae bangaramma* (isolated from the coast, Pudhumadam, India) [39], *Shewanella algae* ATCC 51181 (isolated from bottom sediments, Great Bay estuary, New Hampshire) [40–42], *Shewanella oneidensis* KR-12 (isolated from Ke-Ya River, Hsinchu, Taiwan) [43] and *Shewanella oneidensis* MR-1 (isolated from Oneida Lake, NY, USA) [28,44–50]. To the best of our knowledge, no research shows that genus *Shewanella* isolated from cold environments has the ability to reduce metal ions, and Ag@AgCl NPs are a novel material compared to the metal nanoparticle synthesis by *Shewanella* that was previously investigated. It gives clues for understanding the biogeochemical cycles of silver in mid to high ocean latitudes, especially in polar regions.

Table 1. The summary of biological nanoparticle synthesizing from the genus *Shewanella*.

Bacteria	Size (nm)	Shape	Type of Nanoparticle	Location	Reference
S. sp. HN-41	140–221	Spherical	Se	Intracellular	[37]
	10–30	Short rod	Magnetite, siderite	Intracellular	[38]
S. sp. PV-4	-	-	Magnetite	Extracellular	[36]
	>35	-	magnetite	Intracellular	[51]
S. algae bangaramma	5–30	Spherical	Ag	Intracellular	[39]
S. oneidensis KR-12	2–8	Spherical	Pb	Intracellular	[43]
	5	-	Pt	Intracellular	[41]
	10–20	Spherical, triangle	Au	Extracellular	[40]
	40–50	Rectangular, rhombic, hexagonal	Fe_3O_4	Extracellular	[44]
S. algae ATCC 51181	2–50	Spherical	Au	Extracellular	[45]
	-	Spherical	Pd	Intracellular	[46]
	nano-sized	Needle, hexagonal	Jarosite	Extracellular	[47]
	30–43	Pseudohexagonal, irregular, rhombohedral	Fe_2O_3	Intracellular	[28]
	17–23	Spherical	Ag and Au	Extracellular	[48]
			V	Intracellular	[49]
	nano-sized	Amorphous round	Manganese oxide	Intracellular	[30,50]

2.2. Synthesis of Biogenic Ag@AgCl Nanoparticles

Surface plasmon resonance (SPR) excitation in the collective oscillation of free conduction electrons is provoked by an interacting electromagnetic field, leading to colour changes [52]. Thus, AgNP formation could be estimated visually by the observed colour change from light yellow to reddish brown (Figure 2) [53]. The presence of AgNPs rather than AgCl enabled the absorption of light in the visible region, with an absorption peak of approximately 410 nm [13]. This solution presented fine homogeneity, and no precipitation was detected, suggesting that the nanoparticles were stable and well dispersed [54]. No obvious biogenic AgNP formation was detected in the control (only $AgNO_3$ and culture medium with $AgNO_3$) (Figure 2).

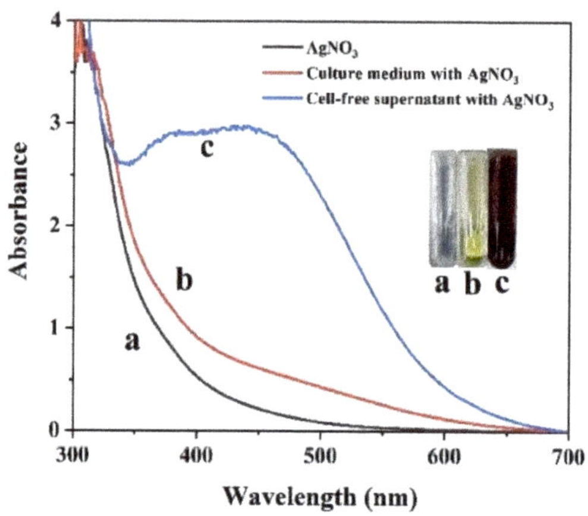

Figure 2. UV/vis spectral analysis of $AgNO_3$ (**a**), culture medium with $AgNO_3$ (**b**) and the culture supernatant of *Shewanella* sp. Arc9-LZ with $AgNO_3$ (**c**) over the wavelength range of 300–700 nm after 48 h of reaction.

2.3. Characterization of Biogenic Ag@AgCl Nanoparticles

After freeze-drying, brown-red Ag@AgCl nanoparticle powder was obtained (Figure 3A). On the basis of SPR, optical absorption peaks at 3 keV targeted the presence of metallic silver nanoparticles [55] (Figure 3B). The EDS results illustrated the presence of silver and chlorine elements, accounting for 6.53% and 2.08% of the total Ag@AgCl NPs, respectively (Figure 3B). The strong signals targeting C and O might indicate the presence of proteins acting as capping material on the nanoparticle's surface [56]. The XRD pattern of the biogenic Ag@AgCl nanoparticles is shown in Figure 3C. It exhibits peaks at $2\theta = 32.24, 46.13, 54.88, 57.52, 67.2, 76.8$, corresponding well to the (200), (220), (311), (222), (400) and (420) planes of AgCl and matching the JCPDS file 31-1238 for solid AgCl. This spectrum also shows peaks at $2\theta = 38.30, 44.01, 64.24$, and 77.48, which can be assigned to the (111), (200), (220) and (311) planes, corresponding well to the face-centred cubic structure of metallic silver, matching the JCPDS file 65-2871 for cubic Ag.

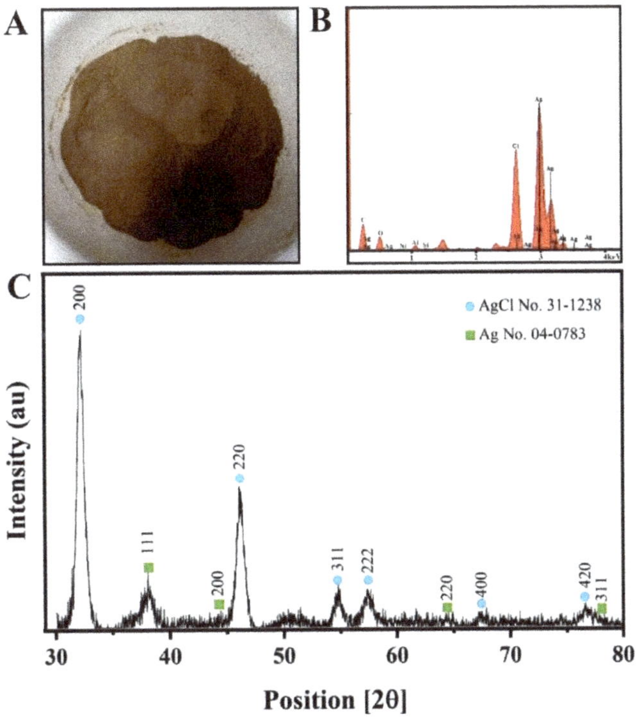

Figure 3. Powder of nanoparticles (**A**), EDS spectrum (**B**) and XRD spectrum (**C**) of biogenic Ag@AgCl nanoparticles.

The morphology, size distribution and dispersibility of biogenic nanoparticles are shown in Figure 4. The Ag@AgCl NPs were spherical and ellipsoidal with beneficial dispersibility (Figure 4A). The size ranged from 4 nm to 20 nm, and most nanoparticles concentrated in the size of 8–16 nm, which was in accordance with the normal distribution (Figure 4B). There are limited reports on the biogenic Ag@AgCl NPs, so we summarized the morphology and size of AgNPs synthesized by bacteria (Table 2). Most biogenic nanomaterials from other reports were spherical. The nanoparticles in this study have larger specific surface areas, which may improve the loading of the surfaces or enable a greater release of ions into the solutions [1]. For example, small nanoparticles display larger surface areas than large particles, leading to higher antimicrobial and catalytic activity [57]. The nanoparticles will have great significance when they are uniform in size and shape and are well dispersed [13]. The results demonstrated that *Shewanella* sp. Arc9-LZ cell-free supernatant provided natural capping to synthesize nanoparticles and prevent the aggregation of nanoparticles [34].

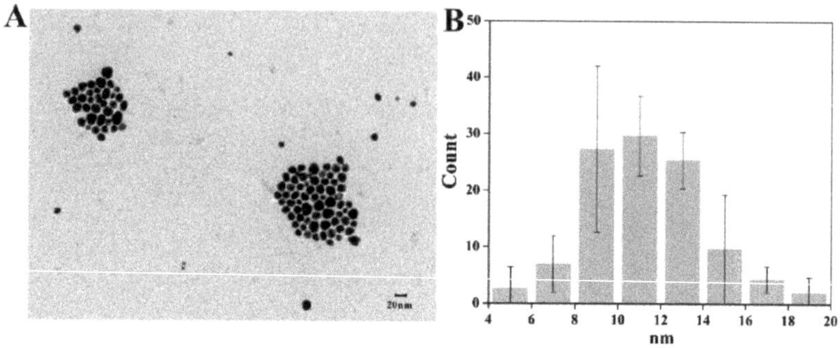

Figure 4. TEM image (**A**) and size distribution (**B**) of Ag@AgCl NPs.

Table 2. The summary of AgNP synthesis from the bacteria.

Stains	Location	Morphology	Size (nm)	Reference
Acinetobacter calcoaceticus PUCM 1005	Extracellular	Spherical	4–40 nm	[58]
Pseudomonas stutzeri AG259	Intracellular	Subglobose	5–25 nm	[59]
Bacillus subtilis MSBN 17	Intracellular	Spherical	60 nm	[60]
Ureibacillus thermosphaericus	Extracellular	Spherical	10–100 nm	[61]
Pseudomonas aeruginosa ATCC 27853	Extracellular	Spherical	33–300 nm	[62]
Streptacidiphilus durhamensis HGG16n	Extracellular	Spherical	8–48 nm	[63]
Paracoccus sp. Arc7-R13	Extracellular	Subglobose	2–25 nm	[64]
Staphylococcus epidermidis ATCC 12228	Intracellular	Spherical, oval, short rod, triangle	10–100 nm	[65]
Klebsiella pneumoniae	Extracellular	Spherical	5–32 nm	[66]
Aeromonas sp. THG-FG1.2	Extracellular	Spherical	8–16 nm	[67]
Actinomycetes MRS- 1	Extracellular	Spherical	4.7–18.8 nm	[68]
Fusarium oxysporum	Extracellular	Spherical	40 nm	[69]
Trichoderma spp.	Intracellular	Spherical	14–25 nm	[70]

2.4. Influence of Synthesizing Conditions on Ag@AgCl NPs

The pH, Ag^+ concentration and reaction time are important parameters affecting morphology, diameter and dispersity [71,72]. Therefore, we investigated the influence of pH, Ag^+ concentration and reaction time on biogenic Ag@AgCl NP production (Figure 5). The reaction system has the highest absorbance at 410 nm and the deepest reddish brown under neutral conditions, followed by alkaline and acidic conditions (Figure 5A,B). There was a flocculent precipitate existing in the solution under acidic conditions (pH = 3.0 and pH = 5.0), which may be attributed to the protein denaturation. We concluded that the optimal pH for microbial synthesis of nanoparticles was related to the pH of the microbial environment. As shown in Figure 5C,D, the producing number of AgNP nanoparticles positively correlated with the $AgNO_3$ concentration (0–8 mmol/L) added to the cell-free supernatant of Shewanella sp. Arc9-LZ ($R^2 = 0.98$, $p < 0.01$). Abundant reducing and capping agents exist in the Arc9-LZ cell-free supernatant. The synthesis efficiency of the Ag nanoparticles increases with increasing Ag^+ concentration, which may be attributed to the increased probability of the reducing agent colliding with silver ions. As shown in Figure 5E, the SPR intensity at approximately 410 nm increased steadily as a function of reaction time, without an obvious band shift that targets the increase in particle size [73]. Therefore, our results preliminarily indicated that the amount of biogenic Ag@AgCl nanoparticles grew over time ($R^2 = 0.98$, $p < 0.01$) and that the size was stable from 1 d to 9 d (Figure 5E,F).

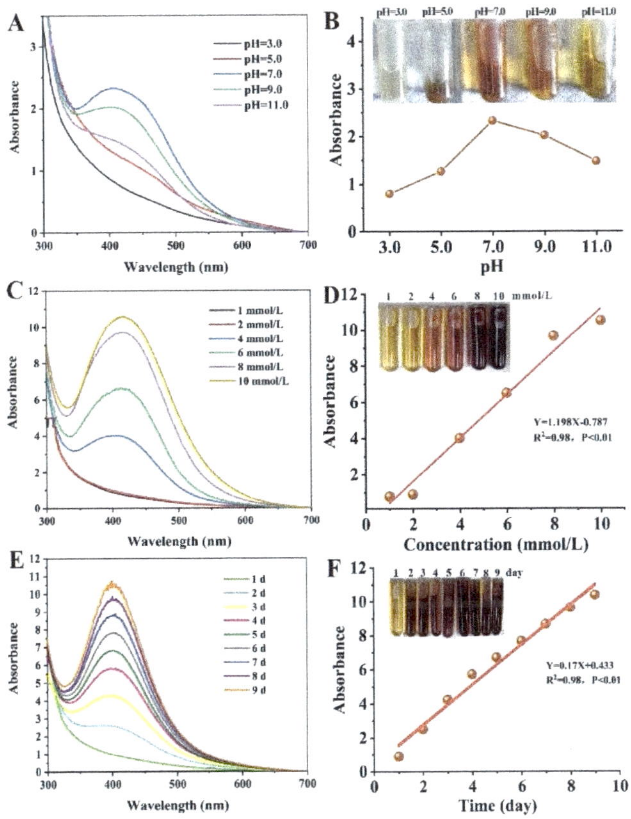

Figure 5. Influence of pH, Ag^+ concentration and reaction time on nanoparticle synthesis: Uv-vis spectra of reaction system with different pH (**A**), Ag^+ concentration (**C**) and reaction time (**E**). Digital images and linear graphs of reaction mixtures with different pH (**B**), Ag^+ concentration (**D**) and reaction time (**F**).

2.5. Catalytic Activity of the Biogenic Ag@AgCl Nanoparticles

Dyes and dyestuffs are widely used within the food, pharmaceutical, cosmetic, textile and leather industries [74]. Over 7×10^5 tons of synthetic dyes are produced per year, and approximately 10–15% is discharged into the environment [75]. Both azo and rhodamine B dyes can cause direct destruction of aquatic wildlife and are mutagenic and carcinogenic to humans [76]. Newer and stricter legislation has been established in many countries to enhance enforcement concerning wastewater discharge [75]. Thus, there is a growing need to develop eco-friendly methods to remove dyes from wastewater. Capping agents and stabilizers might lead to the production of small nanoparticles. However, steric hindrance on the surface of the nanoparticle might affect its activities [77]. Hence, it is necessary to verify the catalytic activity of our biogenic Ag@AgCl nanoparticles.

Figure 6A demonstrates the UV-Vis absorption spectra and colour changes of the CR solution treated with $NaBH_4$ in the presence of biogenic Ag@AgCl nanoparticles over a 12 min period. Before degradation, the bright red CR solution showed a strong absorption peak of 497 nm. Upon the addition of 0.025 mg/L (ultimate concentration) biogenic

Ag@AgCl nanoparticles into the reaction system, the colour of the solution changed from bright red to colourless within 10 min. Meanwhile, the absorption peak of 497 nm significantly decreased over time. Figure 6C shows the UV-Vis absorption spectra of the CR solution treated with $NaBH_4$ in the absence of the Ag@AgCl nanoparticles. It is obvious that there is no change in the colour or the maximum absorption peak of the CR solution. Thus, the degradation of CR by $NaBH_4$ is limited in the absence of the Ag@AgCl nanoparticles. The pseudo-first-order linear relation of $\ln(A_t/A_0)$ versus the reaction time of the degradation of CR is depicted in Figure 6E. According to A_t and A_0, the kinetic rate constant (k) of the catalytic reaction in the presence of the biogenic Ag@AgCl nanoparticles is 2.74×10^{-1} min^{-1}. Figure 6F shows the decolourization efficiency of the CR solution with the Ag@AgCl nanoparticles. The maximum decolourization efficiency of CR by the Ag@AgCl nanoparticles was 93.36% (10 min), which was much higher than the natural decolourization efficiency (9.51%, 8 min) in the absence of Ag@AgCl nanoparticles. We, therefore, suggested that the biogenic Ag@AgCl nanoparticles synthesized by *Shewanella* sp. Arc9-LZ exhibited excellent catalytic activity.

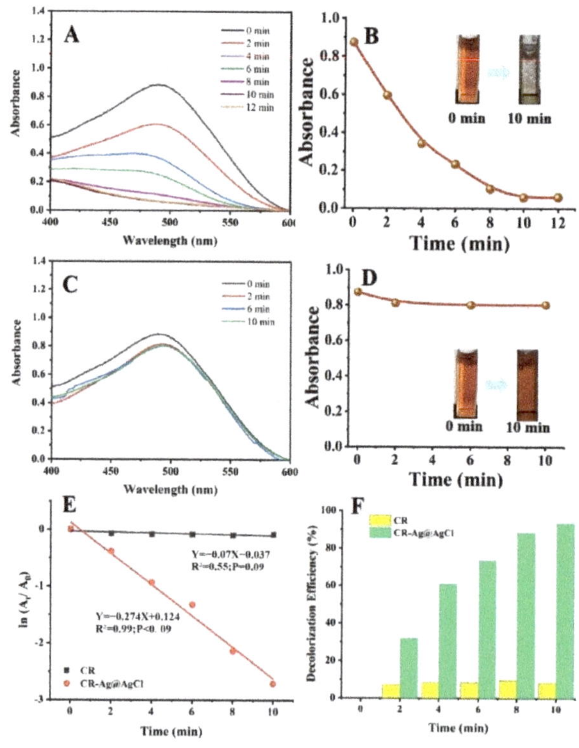

Figure 6. The reduction of CR in aqueous solution recorded every 2 min: the Uv-vis spectral of CR reaction with Ag@AgCl nanoparticles (**A**) and without (**C**); Digital images of CR reaction with Ag@AgCl nanoparticles (**B**) and without (**D**); $\ln(A_t/A_0)$ versus reaction time for CR reduction (**E**); Decolourization efficiency versus reaction time for CR reduction (**F**).

The RhB dye will absorb light at 554 nm in the visible region, showing a bright rose-red colour. After the addition of the biogenic Ag@AgCl nanoparticles to the RhB dye solution, the absorbance intensity at 554 nm showed a sharp decline within 6 min (Figure 7A). In contrast, the absorbance intensity of the dye solution lacking Ag@AgCl nanoparticles remained stable as the reaction time proceeded to 22 min (Figure 7C). Figure 7E indicates that after the addition of $NaBH_4$, the kinetic rate constant (k) for RhB degradation with Ag@AgCl nanoparticle treatment was 7.78×10^{-1} min^{-1}, which was higher than that of the control (1.38×10^{-1} min^{-1}). Additionally, the RhB decolourization efficiency with Ag@AgCl nanoparticle treatment at 6 min was 99.52%, higher than the control of 74.12% (6 min) and 93.91% (22 min) (Figure 7F). It was obvious that the biogenic Ag@AgCl nanoparticles could accelerate the reduction rate of RhB.

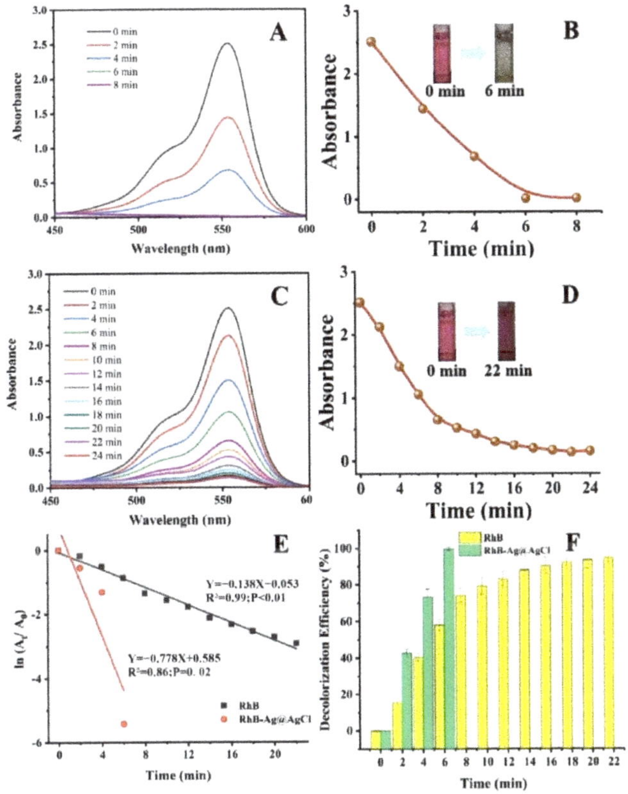

Figure 7. The reduction of RhB in aqueous solution recorded every 2 min: the Uv-vis spectral of RhB reaction with Ag@AgCl nanoparticles (**A**) and without (**C**). Digital images of RhB reaction with Ag@AgCl nanoparticles (**B**) and without (**D**). $\ln(A_t/A_0)$ versus reaction time for RhB reduction (**E**). Decolourization efficiency versus reaction time for RhB reduction (**F**).

The colour fading of the dyes might be attributed to efficient particle-mediated electron transfer from the BH_4^- ion to the dye resulting in the break of azo bonds [78]. With a large surface-to-volume ratio, the nanoparticles expose atoms on the surface as potential catalytic sites [79] and act as a substrate for the electron transfer reaction [80]. They might also indirectly increase the collision probability between the dyes and $NaBH_4$ through physical factors [81]. Therefore, the small biogenic Ag@AgCl NPs synthesized by *Shewanella.* sp. Arc9-LZ enables good catalysis.

2.6. Antibacterial Activity

The nanoparticles' antimicrobial ability is not only influenced by the compounds of the material, but it is also related to the material's size. Smaller nanoparticles have stronger antibacterial effects with larger total surface area per unit volume [82]. In this study, Ag@AgCl nanoparticles showed obvious antimicrobial activity against Gram-positive (Bacillus subtilis ATCC6633 and Staphylococcus aureus ATCC6538) and Gram-negative (Pseudomonas aeruginosa PAO1 and Escherichia coli CGMCC1.2340) bacteria. Compared to the control, the treatments with the Ag@AgCl nanoparticles presented an obvious inhibition zone when the concentrations of nanoparticles reached 20 μg/mL. With increasing Ag@AgCl nanoparticle concentration, the diameter of the inhibition zone tended to increase (Figure 8A,B). Among the strains, Gram-negative Pseudomonas aeruginosa PAO1 had the greatest response to the Ag@AgCl nanoparticles, which can be attributed to the structure of its cell wall.

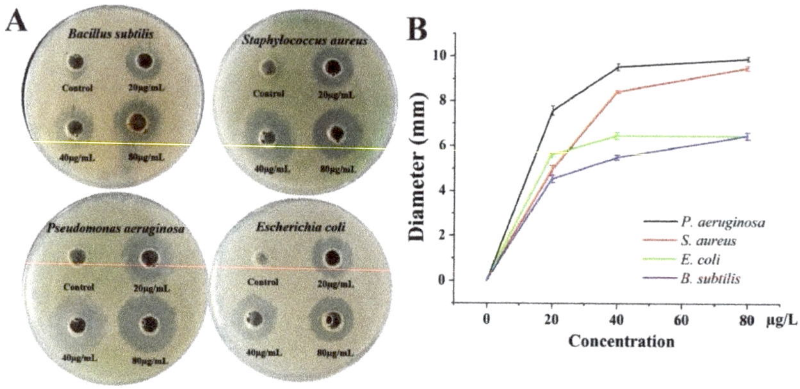

Figure 8. Antimicrobial activities of biogenic Ag@AgCl nanoparticles against *S. aureus*, *P. aeruginosa*, *E. coli* and *B. subtilis*: digital images (**A**) and inhibitory zone diameter (**B**).

Ag@AgCl nanoparticles may be related to the release of silver ions, which adhere to the membrane surface, disturb its normal function, cause protein denaturation, affect the respiratory chain and cause irreversible DNA damage, eventually leading to microbial death [9,12]. Tamboli and Lee [83] believed that the changes and damage to the membranes engendered by the Ag@AgCl nanoparticles caused a significant increase in permeability, leaving bacterial cells incapable of properly regulating transport through the plasma membrane and destroying the double-stranded DNA structure, resulting in cell death. Studies have found that the antibacterial effects of silver nanoparticles increased with decreasing particle size [84]. Particles with higher specific surface areas dissolve faster than those with smaller surface areas, which is why smaller particles have higher effective Ag^+ concentrations and better antibacterial effects [82].

2.7. The Mechanism of Nanoparticle Formation

Duan et al. [33] believed that proteins, amino acids, organic acids and secondary metabolites are related to reducing, capping and stabilizing nanoparticle formation. According to the FTIR spectrum of the cell-free supernatant and biogenic Ag@AgCl solution, the functional groups responsible for the reduction of Ag^+ were tentatively explored (Figure 9). During the reduction, three bands at 1403.21, 1453.76 and 1652.95 cm^{-1} weakened or disappeared, and two major peaks at 3422.05 and 668.90 cm^{-1} appeared. The stretching vibrations of the O–H and N–H groups are located at 3422.05 cm^{-1} [85]. The bands observed at 1403.21 cm^{-1} and 668.90 cm^{-1} can be assigned to the O–H stretching vibrations of the carboxylate and N–H deformation vibrations of the amine. The band at 1453.76 represents the stretching vibrations of –COO– groups of amino acids with free carboxylate groups [86]. The peak of 1650 cm^{-1} is attributed to carbonyl stretching and is a typical indicator of amide linkages [87]. The FTIR spectra showed that amino acids, proteins and organic molecules with amide linkages in the supernatant of Arc9-LZ might participate in the reduction, stabilization and capping of the biogenic Ag@AgCl nanoparticles.

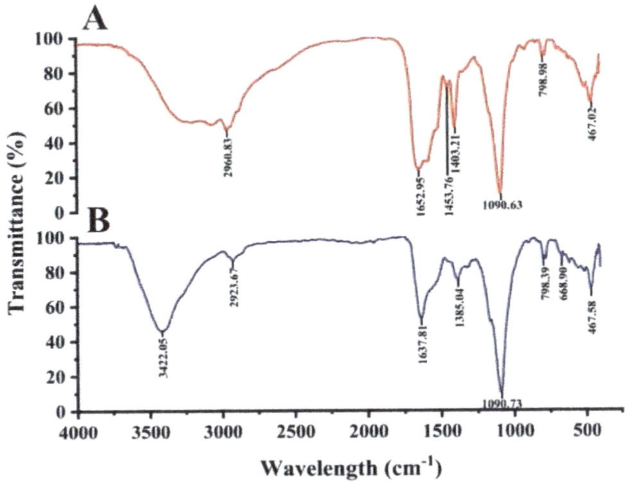

Figure 9. FTIR spectra recorded from microbially synthesized Ag@AgCl nanoparticles: cell-free supernatant (**A**) and biogenic Ag@AgCl nanoparticle solution (**B**).

To speculate the mechanism of Ag@AgCl NPs synthesis, we summarized the genes, enzymes, proteins and small molecules that may be involved in the synthesis of Ag@AgCl NPs in Table 3. The complete genome sequence of *Shewanella* sp. Arc9-LZ has been submitted to the GenBank database under accession number CP048031 [88]. The whole genome of *Shewanella* sp. Arc9-LZ was annotated by databases of Kyoto Encyclopedia of Genes and Genomes (KEGG), Cluster of Orthologous Groups of proteins (COG), Non-Redundant Protein Database (NR), Transporter Classification Database (TCDB), Swiss-Protand Database and Carbohydrate-Active enZYmes Database (CAZy) [88]. We concluded that the mechanism of Ag@AgCl NP synthesizing by *Shewanella* sp. Arc9-LZ is not plasmid-mediated silver resistance but extracellular electron shuttles. The schematic diagram of mechanisms is shown in Figure 10, which needs further evidence.

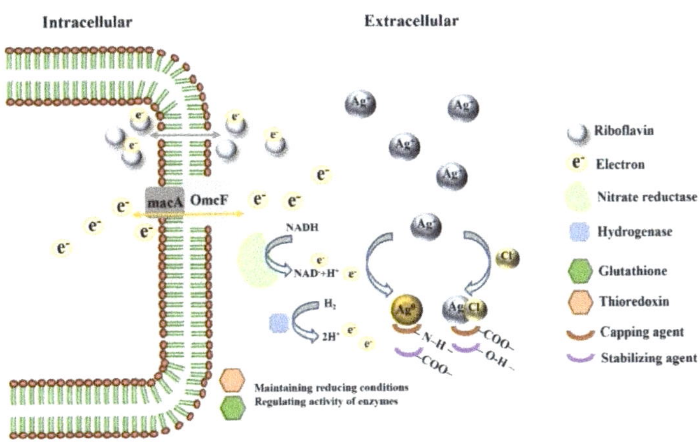

Figure 10. Schematic diagram on mechanisms of synthesizing Ag@AgCl NPs.

Table 3. Microbial substances that may be involved in the synthesis of silver nanomaterials.

Substances	Functions	Reference
Silver-binding gene homologue: *silE*	Encoding a periplasmic silver-binding protein that presents histidine sites for silver ion binding.	[32]
Three major gene homologues of *silE*, *silP*, and *silS*	Participation in silver resistance	[31]
Periplasmic c-type cytochrome (MacA) and outer membrane c-type cytochrome (OmcF)	Reduction of Ag^+ to Ag^0	[89]
NADH-dependent enzymes, especially nitrate reductase	Roles in AgNP synthesis	[90]
Reducing enzymes belonging to class of nitrogenase and hydrogenase	Reduce silver ions to nano-silver	[91]
Cellular nitrogenase	Enzyme concentration dictating the size of AgNPs	[91]
NfsA, an oxygen-insensitive nitroreductase	Reducing $AgNO_3$ to AgNPs	[66]
Spore-associated enzymes, like glucose oxidase, alkaline phosphatase, laccase and catalase	Generating the reducing cofactors and stimulating the biogenesis of AgNPs	[92]
Glutathione and thioredoxin systems	Maintaining the reducing conditions indirectly and regulating the activity of enzymes	[93]
Riboflavin	Acting as a soluble redox shuttle to mediate metal reduction	[94]

3. Materials and Methods

3.1. Materials

The strains *Bacillus subtilis* ATCC6633, *Staphylococcus aureus* ATCC6538, *Pseudomonas aeruginosa* PAO1 and *Escherichia coli* CGMCC1.2340 were used for assaying antibacterial activities and were stored at −80 °C. The dyes CR and RhB were purchased from Sangon Biotech (Shanghai, China) and Aladdin (Shanghai, China), respectively. $NaBH_4$ was supplied by Sinopharm Chemical Reagent Co., Ltd. (Shanghai, China). AgNO3 was obtained from the Shanghai Chemical Reagent Factory (Shanghai, China). The yeast extract, tryptone and agar powder were all purchased from Solarbio Ltd. (Beijing, China).

3.2. Isolation and Identification of the Strain *Shewanella* sp. Arc9-LZ

Shewanella sp. Arc9–LZ bacteria were isolated from the marine sediments of the Arctic Ocean (158°01′12″W; 84°28′38″N) collected during the 9th Chinese National Arctic Expedition in 2018. Firstly, *Shewanella* sp. Arc9-LZ was activated in the marine ZoBell 2216E medium (peptone, 5 g/L; yeast extract, 1 g/L; natural seawater, 1 L) at 150 r/min. After incubation at 15 °C for 2 d, the culture was serially diluted and spread on 2216E medium agar plates (ZoBell 2216E medium with 15 g/L agar powder) to isolate the single clone.

Based on the 16S rRNA sequence alignment, the strain was identified by PCR with the Bact27F and Univ1492R primers. PCR amplicons of 16S rRNA genes were sequenced by Sangon Biotech (Shanghai, China). Sequences (1500 bp fragments) were analysed using CodonCode Aligner software. Additionally, 16S rRNA gene sequences were aligned with the closest matches available in GenBank and EzTaxon server 2.1 with the Clustal W function of BioEdit software (7.1.3.0). Phylogenetic trees were constructed with the Molecular Evolutionary Genetics Analysis software (MEGA version 4.0) using the neighbour-joining method, and 1000 bootstraps were performed to assign confidence levels to the tree nodes.

The DNA of this stain was extracted with a Bacteria DNA kit (TIANamp, DP302, Tiangen Biotech, Beijing, China). Gene Pools were constructed on the Pacbio platform with an SMRT bell TM Template kit (version 1.0) and Illumina PE150 platform with NEBNext®Ultra™ DNA Library Prep Kit for Illumina (NEB, Ipswich, Ipswich, MA, USA), respectively. The sequence was analysed by PacBio Sequeland Illumina NovaSeq PE150 for different libraries [88].

3.3. Biosynthesis of the Silver Nanoparticles in the Dark

Shewanella sp. Arc9–LZ was cultured in the YP medium (peptone, 10 g/L; yeast extract, 5 g/L; ultra-pure water, 1 L) and incubated at 150 r/min and 15 °C for 48 h. To obtain 100 mL of cell-free extracts, the fermented liquid was centrifuged at 12,000 r/min for 15 min before being filtered through a 0.22 μm syringe filter. Then, biogenic Ag@AgCl nanoparticles were synthesized by mixing the cell-free supernatant and $AgNO_3$ at 150 r/min and 35 °C in the dark. The cell-free supernatant without $AgNO_3$ and the liquid media with $AgNO_3$ were kept under the same conditions and set as controls.

3.4. Characterization of Biogenic Ag@AgCl Nanoparticles

To characterize biogenic Ag@AgCl nanoparticles, samples were monitored using ultraviolet-visible (UV-Vis) spectrophotometry (Shimadzu Model UV 2550, China) in the range of 300–700 nm at a resolution of 1 nm. The biogenic Ag@AgCl nanoparticles were designed under different pH conditions (pH = 3.0, 5.0, 7.0, 9.0, 11.0), final concentrations of $AgNO_3$ (1 mmol/L, 2 mmol/L, 4 mmol/L, 8 mmol/L and 10 mmol/L) and lengths of time (1 d, 2 d, 3 d, 4 d, 5 d, 6 d, 7 d, 8 d and 9 d). The morphology and size of the produced Ag@AgCl nanoparticles were discerned by means of transmission electron microscopy (TEM) (Hitachi HT7700, Tokyo, Japan). The presence and structure of the biogenic Ag@AgCl nanoparticles in the samples were determined by means of X-ray diffraction (XRD) (Bruker D8 Advance, Karlsruhe, Germany) and energy dispersive spectrometry (EDS) (IXRF, USA). Fourier transform infrared spectroscopy (FTIR) spectra were obtained on a Nicolet iN10 (Thermo Electron Scientific Instruments LLC, Madison, WI, USA) with wavelengths ranging from 500 to 4000 cm^{-1}. The concentration of biogenic Ag@AgCl nanoparticles was analysed by means of inductively coupled plasma optical emission spectrometry (ICP-OES) (Perkin-Elmer Optima, Waltham, MA, USA).

3.5. Catalytic Activity for Azo Dyes and RhB

The catalytic activity of biogenic Ag@AgCl nanoparticles for dye reduction was carried out as follows. Firstly, 10 mL of 100 mg/L dyes (CR and RhB), 10 mL of 50 mmol/L $NaBH_4$ and 30 mL of ultrapure water were mixed in a conical flask at 40 °C in the dark. Two reaction setups were employed at the same time. In the first setup, biogenic Ag@AgCl nanoparticles were present in the reaction system at a final concentration of 0.025 mg/L. In the second setup, an equal volume of ultrapure water instead of biogenic Ag@AgCl nanoparticles was put into the reaction system as a control. At pre-determined time intervals, a UV-Vis spectrophotometer recorded the reduction of CR and RhB in the ranges of 400–600 nm and 450–600 nm, respectively. Considering that the amount of added $NaBH_4$ was much larger than that of the dyes, the catalytic reduction should follow pseudo-first-order kinetics, which can be expressed as $\ln(A_t/A_0) = -kt$ [72]. In this equation, At represents the dye absorbance at time t, A_0 represents the initial dye absorbance and slope k represents the apparent reduction rate constant [52].

3.6. Antibacterial Activity

The antibacterial activity of biogenic Ag@AgCl nanoparticles was based on the standard agar-well diffusion method with some modifications [95]. The tested strains were cultured in the LB medium (peptone, 10 g/L; yeast extract, 5 g/L; NaCl, 10 g/L, ultra-pure water, 1 L) and spread uniformly on LB agar plates with cotton swabs at a final concentration of 1×10^5–1×10^6 CFU/mL. A sterile cork borer was used to punch 4 circular holes of 8 mm diameter into the plates, and 100 μL of various concentrations of Ag@AgCl (0, 20, 40, 80 μg/mL) was added to the holes. The negative control in the antibacterial study was the LB medium with the same volume. The plates were incubated at 37 °C for 12 h for the observation and calculation of the inhibition zone.

4. Conclusions

Ag@AgCl NPs, which are a novel material, were synthesized by the strain *Shewanella* sp. Arc9-LZ, which was isolated from the Arctic Ocean. The methods are eco-friendly (no additional chemical reductant, low energy consumption and waste emission), simple (mild reaction conditions, one-pot process), repeatable, sustainable and renewable. With a small size ($\leq$20 nm), spherical shape and beneficial dispersity, Ag@AgCl nanoparticles exhibited excellent catalysis and antibacterial ability application prospects. The kinetic rate constants (k) for CR and RhB degradation with Ag@AgCl nanoparticles were 2.74×10^{-1} min^{-1} and 7.78×10^{-1} min^{-1}, and the maximum decolourization efficiency of CR and RhB were 93.36% and 99.52%, respectively. Moreover, Ag@AgCl nanoparticles showed high antibacterial ability against the Gram-positive and Gram-negative bacteria investigated. In addition to excellent material and good applications, this method might give clues for the further development of the synthesis of new biomaterials, utilization of abundant microbial resources from the Arctic, explanation of the origin of deep-sea metal resources, and understanding of life and the biogeochemical silver cycle in aqueous environments. The schematic diagram of mechanisms for the synthesis of Ag@AgCl NPs was mapped.

Author Contributions: S.L. (Shuang Li) conducted the treatment, analysed the data and wrote the manuscript. H.Z. conducted the treatment and analysed the data. B.C. and P.H. conceptualized original research. H.Z. and W.L. conducted some of the experiments. S.L. (Shenghao Liu) supervised and revised the manuscript. All authors have read and agreed to the published version of the manuscript.

Funding: This work was financially supported by the GEF-FAO Project: Demonstration of Estuarine Biodiversity Conservation, Restoration and Protected Area Networking in China (Grant No: GCP/CPR/045/GFF), the National Key R&D Programs of China (Grant No: 2018YFC1406701), Basic Scientific Fund for National Public Research Institutes of China (Grant No: 2018Q03) and Chinese National Natural Science Foundation (Grant No: 41006102).

Institutional Review Board Statement: Not applicable.

Informed Consent Statement: Not applicable.

Data Availability Statement: The *Shewanella* sp. Arc9-LZ has been deposited in the China General MicroBiological Culture Collection Centre (CGMCC) with accession number CGMCC 1.18550. The complete genome sequence of this strain has been submitted to the GenBank database under accession number CP048031.

Conflicts of Interest: The authors declare no conflict of interest.

References

1. Perni, S.; Hakala, V.; Prokopovich, P. Biogenic synthesis of antimicrobial silver nanoparticles capped with l-cysteine. *Colloids Surf. A Physicochem. Eng. Asp.* **2014**, *460*, 219–224. [CrossRef]
2. Yin, Y.; Li, Z.-Y.; Zhong, Z.; Gates, B.; Xia, Y.; Venkateswaran, S. Synthesis and characterization of stable aqueous dispersions of silver nanoparticles through the Tollens process. *J. Mater. Chem.* **2002**, *12*, 522–527. [CrossRef]
3. Agnihotri, S.; Mukherji, S.; Mukherji, S. Immobilized silver nanoparticles enhance contact killing and show highest efficacy: Elucidation of the mechanism of bactericidal action of silver. *Nanoscale* **2013**, *5*, 7328–7340. [CrossRef]
4. Yu, J.; Zhang, L.; Qian, J.; Zhu, Z.; Ni, S.; Liu, G.; Xu, X. In situ exsolution of silver nanoparticles on $AgTaO_3$-$SrTiO_3$ solid solutions as efficient plasmonic photocatalysts for water splitting. *Appl. Catal. B-Environ.* **2019**, *256*, 117818. [CrossRef]
5. Iqtedar, M.; Aslam, M.; Akhyar, M.; Shehzaad, A.; Abdullah, R.; Kaleem, A. Extracellular biosynthesis, characterization, optimization of silver nanoparticles (AgNPs) using *Bacillus mojavensis* BTCB15 and its antimicrobial activity against multidrug resistant pathogens. *Prep. Biochem. Biotechnol.* **2019**, *49*, 136–142. [CrossRef]
6. Szymanska, E.; Orlowski, P.; Winnicka, K.; Tomaszewska, E.; Baska, P.; Celichowski, G.; Grobelny, J.; Basa, A.; Krzyzowska, M. Multifunctional Tannic Acid/Silver Nanoparticle-Based Mucoadhesive Hydrogel for Improved Local Treatment of HSV Infection: In Vitro and In Vivo Studies. *Int. J. Mol. Sci.* **2018**, *19*, 387. [CrossRef]
7. Govindaraju, K.; Krishnamoorthy, K.; Alsagaby, S.A.; Singaravelu, G.; Premanathan, M. Green synthesis of silver nanoparticles for selective toxicity towards cancer cells. *IET Nanobiotechnol.* **2015**, *9*, 325–330. [CrossRef]
8. Huang, Y.; Chen, S.; Bing, X.; Gao, C.; Wang, T.; Yuan, B. Nanosilver Migrated into Food-Simulating Solutions from Commercially Available Food Fresh Containers. *Packag. Technol. Sci.* **2011**, *24*, 291–297. [CrossRef]

9. Mahmoudi, M.; Serpooshan, V. Silver-Coated Engineered Magnetic Nanoparticles Are Promising for the Success in the Fight against Antibacterial Resistance Threat. *ACS Nano* **2012**, *6*, 2656–2664. [CrossRef]
10. Kumar, A.; Vemula, P.K.; Ajayan, P.M.; John, G. Silver-nanoparticle-embedded antimicrobial paints based on vegetable oil. *Nat. Mater.* **2008**, *7*, 236–241. [CrossRef] [PubMed]
11. Thangudu, S.; Kulkarni, S.S.; Vankayala, R.; Chiang, C.S.; Hwang, K.C. Photosensitized reactive chlorine species-mediated therapeutic destruction of drug-resistant bacteria using plasmonic core-shell Ag@AgCl nanocubes as an external nanomedicine. *Nanoscale* **2020**, *12*, 12970–12984. [CrossRef] [PubMed]
12. Morones, J.R.; Elechiguerra, J.L.; Camacho, A.; Holt, K.; Kouri, J.B.; Ramirez, J.T.; Yacaman, M.J. The bactericidal effect of silver nanoparticles. *Nanotechnology* **2005**, *16*, 2346–2353. [CrossRef]
13. Kulkarni, A.A.; Bhanage, B.M. Ag@AgCl Nanomaterial Synthesis Using Sugar Cane Juice and Its Application in Degradation of Azo Dyes. *ACS Sustain. Chem. Eng.* **2014**, *2*, 1007–1013. [CrossRef]
14. Krumov, N.; Perner-Nochta, I.; Oder, S.; Gotchev, V.; Angelov, A.; Posten, C. Production of Inorganic Nanoparticles by Microorganisms. *Chem. Eng. Technol.* **2009**, *32*, 1026–1035. [CrossRef]
15. Zhao, C.; Xu, S.; Feng, C.; Liu, Y.; Yang, Y.; Wang, Y.; Xiao, Y.; Song, Z.; Liu, Q.; Li, J. Characterization and differential expression of three GnRH forms during reproductive development in cultured turbot *Schophthalmus maximus*. *J. Oceanol. Limnol.* **2018**, *36*, 1360–1373. [CrossRef]
16. Apte, M.; Sambre, D.; Gaikawad, S.; Joshi, S.; Bankar, A.; Kumar, A.R.; Zinjarde, S. Psychrotrophic yeast *Yarrowia lipolytica* NCYC 789 mediates the synthesis of antimicrobial silver nanoparticles via cell-associated melanin. *AMB Express* **2013**, *3*, 32. [CrossRef]
17. Mohapatra, B.; Kuriakose, S.; Mohapatra, S. Rapid green synthesis of silver nanoparticles and nanorods using *Piper nigrum* extract. *J. Alloys Compd.* **2015**, *637*, 119–126. [CrossRef]
18. Wang, P.; Huang, B.; Lou, Z.; Zhang, X.; Qin, X.; Dai, Y.; Zheng, Z.; Wang, X. Synthesis of highly efficient Ag@AgCl plasmonic photocatalysts with various structures. *Chemistry* **2010**, *16*, 538–544. [CrossRef]
19. Xu, Z.; Han, L.; Hu, P.; Dong, S. Facile synthesis of small Ag@AgCl nanoparticles via a vapor diffusion strategy and their highly efficient visible-light-driven photocatalytic performance. *Catal. Sci. Technol.* **2014**, *4*, 3615–3619. [CrossRef]
20. Song, L.; Zhao, X.; Zhang, S. Preparation and photocatalytic activity of carbon dot/Ag/AgCl. *Appl. Organomet. Chem.* **2019**, *33*, e5019. [CrossRef]
21. Ge, J.; Chen, Y.; Xu, J.; Liu, Y.; Zhang, L.; Zha, F. Fabrication of Ag@AgCl with Enhanced Plasmonic Photocatalysis Performance via a Deep Eutectic Solvent. *Aust. J. Chem.* **2019**, *72*, 200–205. [CrossRef]
22. Devi, T.B.; Begum, S.; Ahmaruzzaman, M. Photo-catalytic activity of Plasmonic Ag@AgCl nanoparticles (synthesized via a green route) for the effective degradation of Victoria Blue B from aqueous phase. *J. Photochem. Photobiol. B* **2016**, *160*, 260–270. [CrossRef] [PubMed]
23. Albeladi, S.S.R.; Malik, M.A.; Al-thabaiti, S.A. Facile biofabrication of silver nanoparticles using *Salvia officinalis* leaf extract and its catalytic activity towards Congo red dye degradation. *J. Mater. Res. Technol.-JMR&T* **2020**, *9*, 10031–10044. [CrossRef]
24. Rashidi, M.; Islami, M.R. Green synthesis of Ag@AgCl/*Elaeagnus angustifolia* seed nanocomposite using *Elaeagnus angustifolia* leaves: An amazing nanophotocatalyst with highly photocatalytic activity under sunlight irradiation. *Environ. Sci. Pollut. Res. Int.* **2020**, *27*, 21455–21467. [CrossRef] [PubMed]
25. Kou, J.; Varma, R.S. Beet juice-induced green fabrication of plasmonic AgCl/Ag nanoparticles. *ChemSusChem* **2012**, *5*, 2435–2441. [CrossRef]
26. Boonupara, T.; Kajitvichyanukul, P. Facile synthesis of plasmonic Ag/AgCl nanoparticles with aqueous garlic extract (*Allium sativum* L.) for visible-light triggered antibacterial activity. *Mater. Lett.* **2020**, *277*, 128362. [CrossRef]
27. Fakher, S.N.; Kashi, F.J. Microbial Synthesized Ag/AgCl Nanoparticles Using *Staphylococcus pasteuri* sp. nov., ZAR1: Antimutagenicity, Antimicrobial Agent. *J. Inorg. Organomet. Polym. Mater.* **2021**, *31*, 1688–1703. [CrossRef]
28. Bose, S.; Hochella, M.F.; Gorby, Y.A.; Kennedy, D.W.; McCready, D.E.; Madden, A.S.; Lower, B.H. Bioreduction of hematite nanoparticles by the dissimilatory iron reducing bacterium *Shewanella oneidensis* MR-1. *Geochim. Cosmochim. Acta* **2009**, *73*, 962–976. [CrossRef]
29. Cao, B.; Zhao, Z.; Peng, L.; Shiu, H.-Y.; Ding, M.; Song, F.; Guan, X.; Lee, C.K.; Huang, J.; Zhu, D.; et al. Silver nanoparticles boost charge-extraction efficiency in *Shewanella* microbial fuel cells. *Science* **2021**, *373*, 1336–1340. [CrossRef]
30. Wright, M.H.; Farooqui, S.M.; White, A.R.; Greene, A.C.; Voordouw, G. Production of Manganese Oxide Nanoparticles by *Shewanella* Species. *Appl. Environ. Microbiol.* **2016**, *82*, 5402–5409. [CrossRef]
31. Parikh, R.Y.; Singh, S.; Prasad, B.L.V.; Patole, M.S.; Sastry, M.; Shouche, Y.S. Extracellular synthesis of crystalline silver nanoparticles and molecular evidence of silver resistance from *Morganella* sp.: Towards understanding biochemical synthesis mechanism. *Chembiochem* **2008**, *9*, 1415–1422. [CrossRef]
32. Singh, R.; Shedbalkar, U.U.; Wadhwani, S.A.; Chopade, B.A. Bacteriagenic silver nanoparticles: Synthesis, mechanism, and applications. *Appl. Microbiol. Biotechnol.* **2015**, *99*, 4579–4593. [CrossRef] [PubMed]
33. Duan, H.; Wang, D.; Li, Y. Green chemistry for nanoparticle synthesis. *Chem. Soc. Rev.* **2015**, *44*, 5778–5792. [CrossRef]
34. Singh, P.; Kim, Y.J.; Zhang, D.; Yang, D.C. Biological Synthesis of Nanoparticles from Plants and Microorganisms. *Trends Biotechnol.* **2016**, *34*, 588–599. [CrossRef] [PubMed]
35. Lovley, D.R.; Holmes, D.E.; Nevin, K.P. Dissimilatory Fe(III) and Mn(IV) Reduction. In *Advances in Microbial Physiology*; Academic Press: Cambridge, MA, USA, 2004; pp. 219–286.

36. Moon, J.W.; Roh, Y.; Lauf, R.J.; Vali, H.; Yeary, L.W.; Phelps, T.J. Microbial preparation of metal-substituted magnetite nanoparticles. *J. Microbiol. Methods* **2007**, *70*, 150–158. [CrossRef] [PubMed]
37. Lee, J.H.; Han, J.; Choi, H.; Hur, H.G. Effects of temperature and dissolved oxygen on Se(IV) removal and Se(0) precipitation by *Shewanella* sp HN-41. *Chemosphere* **2007**, *68*, 1898–1905. [CrossRef]
38. Lee, J.H.; Roh, Y.; Kim, K.W.; Hur, H.G. Organic acid-dependent iron mineral formation by a newly isolated iron-reducing bacterium, *Shewanella* sp HN-41. *Geomicrobiol. J.* **2007**, *24*, 31–41. [CrossRef]
39. Babu, M.Y.; Devi, V.J.; Ramakritinan, C.M.; Umarani, R.; Taredahalli, N.; Kumaraguru, A.K. Application of Biosynthesized Silver Nanoparticles in Agricultural and Marine Pest Control. *Curr. Nanosci.* **2014**, *10*, 374–381. [CrossRef]
40. Konishi, Y.; Tsukiyama, T.; Tachimi, T.; Saitoh, N.; Nomura, T.; Nagamine, S. Microbial deposition of gold nanoparticles by the metal-reducing bacterium *Shewanella algae*. *Electrochim. Acta* **2007**, *53*, 186–192. [CrossRef]
41. Konishi, Y.; Ohno, K.; Saitoh, N.; Nomura, T.; Nagamine, S.; Hishida, H.; Takahashi, Y.; Uruga, T. Bioreductive deposition of platinum nanoparticles on the bacterium *Shewanella algae*. *J. Biotechnol.* **2007**, *128*, 648–653. [CrossRef] [PubMed]
42. Caccavo, F.; Blakemore, R.P.; Lovley, D.R. A Hydrogen-Oxidizing, Fe(III)-Reducing Microorganism from the Great Bay Estuary, New-Hampshire. *Appl. Environ. Microbiol.* **1992**, *58*, 3211–3216. [CrossRef] [PubMed]
43. Liu, C.L.; Yen, J.H. Characterization of lead nanoparticles formed by *Shewanella* sp KR-12. *J. Nanopart. Res.* **2016**, *18*, 30. [CrossRef]
44. Perez-Gonzalez, T.; Jimenez-Lopez, C.; Neal, A.L.; Rull-Perez, F.; Rodriguez-Navarro, A.; Fernandez-Vivas, A.; Ianez-Pareja, E. Magnetite biomineralization induced by *Shewanella oneidensis*. *Geochim. Cosmochim. Acta* **2010**, *74*, 967–979. [CrossRef]
45. Suresh, A.K.; Pelletier, D.A.; Wang, W.; Broich, M.L.; Moon, J.W.; Gu, B.; Allison, D.P.; Joy, D.C.; Phelps, T.J.; Doktycz, M.J. Biofabrication of discrete spherical gold nanoparticles using the metal-reducing bacterium *Shewanella oneidensis*. *Acta Biomater.* **2011**, *7*, 2148–2152. [CrossRef] [PubMed]
46. De Windt, W.; Aelterman, P.; Verstraete, W. Bioreductive deposition of palladium (0) nanoparticles on *Shewanella oneidensis* with catalytic activity towards reductive dechlorination of polychlorinated biphenyls. *Environ. Microbiol.* **2005**, *7*, 314–325. [CrossRef] [PubMed]
47. Ouyang, B.J.; Lu, X.C.; Liu, H.; Li, J.; Zhu, T.T.; Zhu, X.Y.; Lu, J.J.; Wang, R.C. Reduction of jarosite by *Shewanella oneidensis* MR-1 and secondary mineralization. *Geochim. Cosmochim. Acta* **2014**, *124*, 54–71. [CrossRef]
48. Ramasamy, M.; Lee, J.H.; Lee, J. Potent antimicrobial and antibiofilm activities of bacteriogenically synthesized gold-silver nanoparticles against pathogenic bacteria and their physiochemical characterizations. *J. Biomater. Appl.* **2016**, *31*, 366–378. [CrossRef]
49. Myers, J.M.; Antholine, W.E.; Myers, C.R. Vanadium(V) reduction by Shewanella oneidensis MR-1 requires menaquinone and cytochromes from the cytoplasmic and outer membranes 2. *Appl. Environ. Microbiol.* **2004**, *70*, 1405–1412. [CrossRef]
50. Nealson, K.H.; Scott, J. *Ecophysiology of the Genus Shewanella*; Springer: New York, NY, USA, 2006.
51. Roh, Y.; Gao, H.C.; Vali, H.; Kennedy, D.W.; Yang, Z.K.; Gao, W.M.; Dohnalkova, A.C.; Stapleton, R.D.; Moon, J.W.; Phelps, T.J.; et al. Metal reduction and iron biomineralization by a psychrotolerant Fe(III)-reducing bacterium, *Shewanella* sp strain PV-4. *Appl. Environ. Microbiol.* **2006**, *72*, 3236–3244. [CrossRef]
52. Veisi, H.; Azizi, S.; Mohammadi, P. Green synthesis of the silver nanoparticles mediated by *Thymbra spicata* extract and its application as a heterogeneous and recyclable nanocatalyst for catalytic reduction of a variety of dyes in water. *J. Clean. Prod.* **2018**, *170*, 1536–1543. [CrossRef]
53. Eustis, S.; El-Sayed, M.A. Why gold nanoparticles are more precious than pretty gold: Noble metal surface plasmon resonance and its enhancement of the radiative and nonradiative properties of nanocrystals of different shapes. *Chem. Soc. Rev.* **2006**, *35*, 209–217. [CrossRef] [PubMed]
54. Kariuki, V.M.; Yazgan, I.; Akgul, A.; Kowal, A.; Parlinska, M.; Sadik, O.A. Synthesis and catalytic, antimicrobial and cytotoxicity evaluation of gold and silver nanoparticles using biodegradable, II-conjugated polyamic acid. *Environ. Sci.-Nano* **2015**, *2*, 518–527. [CrossRef]
55. Magudapathy, P.; Gangopadhyay, P.; Panigrahi, B.K.; Nair, K.G.M.; Dhara, S. Electrical transport studies of Ag nanoclusters embedded in glass matrix. *Phys. B* **2001**, *299*, 142–146. [CrossRef]
56. Jain, N.; Bhargava, A.; Majumdar, S.; Tarafdar, J.C.; Panwar, J. Extracellular biosynthesis and characterization of silver nanoparticles using *Aspergillus flavus* NJP08: A mechanism perspective. *Nanoscale* **2011**, *3*, 635–641. [CrossRef]
57. Kulthong, K.; Srisung, S.; Boonpavanitchakul, K.; Kangwansupamonkon, W.; Maniratanachote, R. Determination of silver nanoparticle release from antibacterial fabrics into artificial sweat. *Part. Fibre Toxicol.* **2010**, *7*, 8. [CrossRef]
58. Gaidhani, S.; Singh, R.; Singh, D.; Patel, U.; Shevade, K.; Yeshvekar, R.; Chopade, B.A. Biofilm disruption activity of silver nanoparticles synthesized by *Acinetobacter calcoaceticus* PUCM 1005. *Mater. Lett.* **2013**, *108*, 324–327. [CrossRef]
59. Otaqsara, S.M.T. Biosynthesis of quasi-spherical Ag nanoparticle by *Pseudomonas aeruginosa* as a bioreducing agent. *Eur. Phys. J.-Appl. Phys.* **2011**, *56*, 4. [CrossRef]
60. Sathiyanarayanan, G.; Kiran, G.S.; Selvin, J. Synthesis of silver nanoparticles by polysaccharide bioflocculant produced from marine *Bacillus subtilis* MSBN17. *Colloids Surf. B-Biointerfaces* **2013**, *102*, 13–20. [CrossRef]
61. Juibari, M.M.; Abbasalizadeh, S.; Jouzani, G.S.; Noruzi, M. Intensified biosynthesis of silver nanoparticles using a native extremophilic *Ureibacillus thermosphaericus* strain. *Mater. Lett.* **2011**, *65*, 1014–1017. [CrossRef]
62. Peiris, M.K.; Gunasekara, C.P.; Jayaweera, P.M.; Arachchi, N.D.; Fernando, N. Biosynthesized silver nanoparticles: Are they effective antimicrobials? *Mem. Inst. Oswaldo Cruz* **2017**, *112*, 537–543. [CrossRef]

63. Buszewski, B.; Railean-Plugaru, V.; Pomastowski, P.; Rafinska, K.; Szultka-Mlynska, M.; Golinska, P.; Wypij, M.; Laskowski, D.; Dahm, H. Antimicrobial activity of biosilver nanoparticles produced by a novel *Streptacidiphilus durhamensis* strain. *J. Microbiol. Immunol. Infect.* **2018**, *51*, 45–54. [CrossRef] [PubMed]
64. Zhang, Z.; Li, S.; Gu, X.; Li, J.; Lin, X. Biosynthesis, characterization and antibacterial activity of silver nanoparticles by the Arctic anti-oxidative bacterium *Paracoccus* sp. Arc7-R13. *Artif. Cells Nanomed. Biotechnol.* **2019**, *47*, 1488–1495. [CrossRef]
65. Amin, Z.R.; Khashyarmanesh, Z.; Bazzaz, B.S.F. Different behavior of *Staphylococcus epidermidis* in intracellular biosynthesis of silver and cadmium sulfide nanoparticles: More stability and lower toxicity of extracted nanoparticles. *World J. Microbiol. Biotechnol.* **2016**, *32*, 140. [CrossRef]
66. Shahverdi, A.R.; Fakhimi, A.; Shahverdi, H.R.; Minaian, S. Synthesis and effect of silver nanoparticles on the antibacterial activity of different antibiotics against *Staphylococcus aureus* and *Escherichia coli*. *Nanomed. Nanotechnol. Biol. Med.* **2007**, *3*, 168–171. [CrossRef] [PubMed]
67. Singh, H.; Du, J.; Yi, T.H. Biosynthesis of silver nanoparticles using *Aeromonas* sp THG-FG1.2 and its antibacterial activity against pathogenic microbes. *Artif. Cells Nanomed. Biotechnol.* **2017**, *45*, 584–590. [CrossRef] [PubMed]
68. Rajegaonkar, P.S.; Deshpande, B.A.; More, M.S.; Waghmare, S.S.; Sangawe, V.V.; Inamdar, A.; Shirsat, M.D.; Adhapure, N.N. Catalytic reduction of p-nitrophenol and methylene blue by microbiologically synthesized silver nanoparticles. *Mater. Sci. Eng. C-Mater. Biol. Appl.* **2018**, *93*, 623–629. [CrossRef]
69. Gupta, K.; Chundawat, T.S.; Malek, N. Antibacterial, Antifungal, Photocatalytic Activities and Seed Germination Effect of Mycosynthesized Silver Nanoparticles using *Fusarium oxysporum*. *Biointerface Res. Appl. Chem.* **2021**, *11*, 12082–12091. [CrossRef]
70. Ramos, M.M.; Dos, S.M.E.; da, S.S.I.; Lima, A.L.; de Oliveira, F.R.; de Freitas, C.M.; Fernandes, C.P.; de Carvalho, J.C.T.; Ferreira, I.M. Silver nanoparticle from whole cells of the fungi *Trichoderma* spp. isolated from Brazilian Amazon. *Biotechnol. Lett.* **2020**, *42*, 833–843. [CrossRef] [PubMed]
71. Gilroy, K.D.; Ruditskiy, A.; Peng, H.-C.; Qin, D.; Xia, Y. Bimetallic Nanocrystals: Syntheses, Properties, and Applications. *Chem. Rev.* **2016**, *116*, 10414–10472. [CrossRef]
72. Shen, W.; Qu, Y.; Pei, X.; Li, S.; You, S.; Wang, J.; Zhang, Z.; Zhou, J. Catalytic reduction of 4-nitrophenol using gold nanoparticles biosynthesized by cell-free extracts of *Aspergillus* sp. WL-Au. *J. Hazard. Mater.* **2017**, *321*, 299–306. [CrossRef]
73. Singaravelan, R.; Alwar, S.B.S. Electrochemical synthesis, characterisation and phytogenic properties of silver nanoparticles. *Appl. Nanosci.* **2015**, *5*, 983–991. [CrossRef]
74. Mcmullan, G.; Meehan, C.; Conneely, A.; Kirby, N.; Robinson, T.; Nigam, P.; Banat, I.M.; Marchant, R.; Smyth, W.F. Microbial decolourisation and degradation of textile dyes. *Appl. Microbiol. Biotechnol.* **2001**, *56*, 81–87. [CrossRef] [PubMed]
75. Papić, S.; Koprivanac, N.; Božić, A.L.; Meteš, A. Removal of some reactive dyes from synthetic wastewater by combined Al(III) coagulation/carbon adsorption process. *Dyes Pigment.* **2004**, *62*, 291–298. [CrossRef]
76. Wang, T.Y.; Liu, S.X.; Mao, W.; Bai, Y.C.; Chiang, K.; Shah, K.; Paz-Ferreiro, J. Novel Bi_2WO_6 loaded N-biochar composites with enhanced photocatalytic degradation of rhodamine B and Cr(VI). *J. Hazard. Mater.* **2020**, *389*, 121827. [CrossRef]
77. Liu, X.; Li, L.; Ye, M.; Xue, Y.; Chen, S. Polyaniline: Poly(sodium 4-styrenesulfonate)-stabilized gold nanoparticles as efficient, versatile catalysts. *Nanoscale* **2014**, *6*, 5223–5229. [CrossRef] [PubMed]
78. Jana, N.R.; Sau, T.K.; Pal, T. Growing small silver particle as redox catalyst. *J. Phys. Chem. B* **1999**, *103*, 115–121. [CrossRef]
79. Zhang, H.; Li, X.; Chen, G. Ionic liquid-facilitated synthesis and catalytic activity of highly dispersed Ag nanoclusters supported on TiO_2. *J. Mater. Chem.* **2009**, *19*, 8223–8231. [CrossRef]
80. Vidhu, V.K.; Philip, D. Catalytic degradation of organic dyes using biosynthesized silver nanoparticles. *Micron* **2014**, *56*, 54–62. [CrossRef]
81. Ghosh, S.K.; Kundu, S.; Mandal, M.; Pal, T. Silver and gold nanocluster catalyzed reduction of methylene blue by arsine in a micellar medium. *Langmuir* **2002**, *18*, 8756–8760. [CrossRef]
82. Sotiriou, G.A.; Pratsinis, S.E. Engineering nanosilver as an antibacterial, biosensor and bioimaging material. *Curr. Opin. Chem. Eng.* **2011**, *1*, 3–10. [CrossRef] [PubMed]
83. Tamboli, D.P.; Lee, D.S. Mechanistic antimicrobial approach of extracellularly synthesized silver nanoparticles against gram positive and gram negative bacteria. *J. Hazard. Mater.* **2013**, *260*, 878–884. [CrossRef] [PubMed]
84. Helmlinger, J.; Sengstock, C.; Gross-Heitfeld, C.; Mayer, C.; Schildhauer, T.A.; Koeller, M.; Epple, M. Silver nanoparticles with different size and shape: Equal cytotoxicity, but different antibacterial effects. *RSC Adv.* **2016**, *6*, 18490–18501. [CrossRef]
85. Emmanuel, R.; Karuppiah, C.; Chen, S.-M.; Palanisamy, S.; Padmavathy, S.; Prakash, P. Green synthesis of gold nanoparticles for trace level detection of a hazardous pollutant (nitrobenzene) causing *Methemoglobinaemia*. *J. Hazard. Mater.* **2014**, *279*, 117–124. [CrossRef]
86. Shankar, S.S.; Ahmad, A.; Sastry, M. Geranium leaf assisted biosynthesis of silver nanoparticles. *Biotechnol. Prog.* **2003**, *19*, 1627–1631. [CrossRef]
87. Sharma, N.; Pinnaka, A.K.; Raje, M.; Fnu, A.; Bhattacharyya, M.S.; Choudhury, A.R. Exploitation of marine bacteria for production of gold nanoparticles. *Microb. Cell Fact.* **2012**, *11*, 86. [CrossRef] [PubMed]
88. Li, S.; Niu, Y.; Chen, H.; He, P. Complete genome sequence of an Arctic Ocean bacterium *Shewanella* sp. Arc9-LZ with capacity of synthesizing silver nanoparticles in darkness. *Mar. Genomics* **2021**, *56*, 100808. [CrossRef] [PubMed]
89. Law, N.; Ansari, S.; Livens, F.R.; Renshaw, J.C.; Lloyd, J.R. Formation of nanoscale elemental silver particles via enzymatic reduction by *Geobacter sulfurreducens*. *Appl. Environ. Microbiol.* **2008**, *74*, 7090–7093. [CrossRef]

90. Kalimuthu, K.; Suresh Babu, R.; Venkataraman, D.; Bilal, M.; Gurunathan, S. Biosynthesis of silver nanocrystals by *Bacillus licheniformis*. *Colloids Surf. B Biointerfaces* **2008**, *65*, 150–153. [CrossRef]
91. Brayner, R.; Barberousse, H.; Hernadi, M.; Djedjat, C.; Yepremian, C.; Coradin, T.; Livage, J.; Fievet, F.; Coute, A. Cyanobacteria as Bioreactors for the synthesis of Au, Ag, Pd, and Pt nanoparticles via an enzyme-mediated route. *J. Nanosci. Nanotechnol.* **2007**, *7*, 2696–2708. [CrossRef] [PubMed]
92. Hosseini-Abari, A.; Emtiazi, G.; Ghasemi, S.M. Development of an eco-friendly approach for biogenesis of silver nanoparticles using spores of *Bacillus athrophaeus*. *World J. Microbiol. Biotechnol.* **2013**, *29*, 2359–2364. [CrossRef] [PubMed]
93. Jha, A.K.; Prasad, K. Biosynthesis of metal and oxide nanoparticles using *Lactobacilli* from yoghurt and probiotic spore tablets. *Biotechnol. J.* **2010**, *5*, 285–291. [CrossRef] [PubMed]
94. Marsili, E.; Baron, D.B.; Shikhare, I.D.; Coursolle, D.; Gralnick, J.A.; Bond, D.R. *Shewanella* Secretes flavins that mediate extracellular electron transfer. *Proc. Natl. Acad. Sci. USA* **2008**, *105*, 3968–3973. [CrossRef] [PubMed]
95. Nathan, P.; Law, E.J.; Murphy, D.F.; MacMillan, B.G. A laboratory method for selection of topical antimicrobial agents to treat infected burn wounds. *Burns* **1978**, *4*, 177–187. [CrossRef]

International Journal of
Molecular Sciences

MDPI

Article

Preharvest Application of Commercial Products Based on Chitosan, Phosphoric Acid Plus Micronutrients, and Orange Essential Oil on Postharvest Quality and Gray Mold Infections of Strawberry

Razieh Rajestary [1,†], Panayiota Xylia [2,†], Antonios Chrysargyris [2], Gianfranco Romanazzi [1] and Nikolaos Tzortzakis [2,*]

[1] Department of Agricultural Food and Environmental Sciences, Marche Polytechnic University, Via Brecce Bianche, 60131 Ancona, Italy

[2] Department of Agricultural Sciences, Biotechnology and Food Science, Cyprus University of Technology, Limassol 3603, Cyprus

* Correspondence: nikolaos.tzortzakis@cut.ac.cy

† These authors contributed equally to this work.

Abstract: Strawberry is a perishable fruit with a limited shelf life after harvest due to deterioration of quality and the development of gray mold, Rhizopus rot and other minor diseases. In this study, the effectiveness of commercial compounds based on chitosan, phosphoric acid plus micronutrients, and sweet orange essential oil (EO) in reducing decay and optimizing the quality of strawberries was analyzed. The plant canopy of a greenhouse crop was sprayed once and strawberry fruit were harvested three days later. Gray mold infections were evaluated after chilled storage for seven days at 4 ± 0.5 °C followed by five days shelf life. The qualitative parameters were recorded at harvest (initial day) and after three days of storage at room temperature (RT, 20 °C) or after cold storage and shelf life (CS, 4 °C). The application of sweet orange EO increased the antioxidant and flavonoid content at harvest, while a decrease was reported following three days of storage at RT. At the same time, increased ethylene production and weight loss were observed during CS three days after harvesting. Chitosan treatment maintained the harvest fruit quality and was effective in the control of postharvest decay. Our results suggest that the investigated natural compounds could improve strawberry quality after harvest. Since chitosan performed best in terms of maintaining quality and reducing postharvest decay, it could be considered as a good substitute for chemical-synthetic fungicides for the preservation of strawberry postharvest gray mold.

Keywords: chitosan; fruit quality parameters; sweet orange essential oil; phosphoric acid; preharvest treatments

Citation: Rajestary, R.; Xylia, P.; Chrysargyris, A.; Romanazzi, G.; Tzortzakis, N. Preharvest Application of Commercial Products Based on Chitosan, Phosphoric Acid Plus Micronutrients, and Orange Essential Oil on Postharvest Quality and Gray Mold Infections of Strawberry. *Int. J. Mol. Sci.* **2022**, *23*, 15472. https://doi.org/10.3390/ijms232415472

Academic Editor: Helena Felgueiras

Received: 24 October 2022
Accepted: 3 December 2022
Published: 7 December 2022

1. Introduction

Consumers adore strawberries (*Fragaria* × *ananassa* Duch.) for their distinct flavor, taste, and health benefits. Strawberries include a lot of bioactive components, including vitamins, amino acids, polyphenols, natural antioxidants, and anthocyanins [1]. A diet rich in antioxidants can prevent and minimize the risk of cardiovascular disease, cancer, and diabetes [2]. However, strawberry has a limited shelf life after harvest because of its susceptibility to fungal pathogens [3].

In the field, strawberry fruit can be infected with a fungus that can spread to nearby fruit during the postharvest period [4,5]. The primary fungal disease associated with strawberries while they are in storage is *Botrytis cinerea*, which is the cause of gray mold. Until now, the first line of defense against gray mold in strawberries in conventional agriculture has been the use of synthetic fungicides and preservation chemicals. However, when their use became more restricted due to legislation, pathogen resistance, and consumer concern over food safety and sustainability, research began to focus on alternate methods. One of the most promising alternatives for preventing fruit infections and delaying harvested product senescence in this context was determined to be the usage of natural bioactive substances [5–10].

Among the natural products investigated for their suitability for pre- and postharvest applications for the preservation of fresh commodities, edible coatings seem to be gaining more and more attention [11]. The increasing interest in these compounds derives from their ability to form a protective membrane that can serve as a barrier, controlling and/or decreasing the respiration and transpiration of the produce, thereby minimizing moisture loss [12]. In addition, edible coatings of various compositions present antimicrobial properties (antibacterial and antifungal), resulting in a decrease of postharvest decay [13]. In addition, plant-derived extracts, including essential oils (EOs), have been assessed for pre- and postharvest applications on fresh produce due to the antibacterial and antifungal properties that they possess [8,9,14–21].

The use of edible coatings with highly active antimicrobial ingredients, such as chitosan or EOs with multicomponent compositions, was recommended for reducing the potential for pathogen development [22–27]. Several studies have assessed the antifungal effectiveness of various EOs or chitosan on *B. cinerea*, examining the abilities of these compounds to stimulate plant defenses or form a film on fruit [28–31]. Moreover, exposure to EOs and chitosan has been shown to enhance the preservation of strawberry fruit quality, exhibiting increased concentrations of phenolics, anthocyanins, and flavonoids [15,32,33].

The objectives of the present study were to: (i) assess the efficacy of natural substances in preventing strawberry fruit postharvest degradation and gray mold development via preharvest greenhouse application of commercial products based on chitosan, phosphoric acid plus micronutrients and sweet orange EO; and (ii) to examine the impacts of these treatments on fruit qualitative parameters.

2. Results

2.1. Decay Evaluation

Chitosan was found to be the most effective compound to reduce gray mold (Figure S1) after 5 d shelf life, with a McKinney's index of 36.19% and a 35% reduction in comparison to the control (Table 1). No significant difference was observed in the McKinney index between fruit treated with PA+MN and sweet orange (*Citrus* × *aurantium* L.) EO and the control at 5 d shelf life, although a tendency toward decay reduction was observed.

Table 1. McKinney index of gray mold recorded on strawberries treated with commercial natural compound formulations. The fruit were kept for 7 d at 4 ± 1 °C, followed by 5 d shelf life at 20 °C and 95–98% RH. Different letters indicate that the values are significantly different according to a Tukey's test ($p \leq 0.05$).

Treatment	McKinney's Index (%)
Control	55.71 ± 15.48 a
Chitosan	36.19 ± 9.11 b
Phosphoric acid plus micronutrients (PA+MN)	49.05 ± 13.57 ab
Sweet orange EO	49.52 ± 6.51 ab

2.2. Fruit Quality

2.2.1. Effects at Harvest Time

Preharvest use of chitosan, PA+MN, and sweet orange EO affected fruit quality parameters on the initial day of the quality experiment (three days after spraying and prior storage), as presented in Table 2. A reduction in fresh strawberry weight with chitosan compared to the non-treated (control) fruits was observed on the day of harvesting. The application of chitosan resulted in significantly lower titratable acidity (TA) (0.64%)

in comparison to sweet orange EO treated and untreated (control) fruit (0.93 and 0.93%, respectively). Additionally, the ascorbic acid (AA) content of strawberries was found to be decreased with all three applied products compared to the non-treated fruits (control). According to the results from the ferric reducing antioxidant power (FRAP) assay, the antioxidant activity in strawberry fruit treated with sweet orange EO was higher than in the control (5.14 and 3.53 g kg^{-1} trolox, respectively) on Day 0. In contrast, the 2,2′-azino-bis(3-ethylbenzothiazoline-6-sulphonic acid (ABTS) assay showed that fruit treated with PA+MN and sweet orange EO had decreased antioxidants (7.19 and 7.46 g kg^{-1} trolox, respectively) compared to chitosan treatment and control (10.51 and 10.34 g kg^{-1} trolox, respectively). Among phenolic compounds, the flavonoid content in fruit treated with sweet orange EO (1.25 g kg^{-1} rutin) and PA+MN (1.07 g kg^{-1} rutin) was found to be increased compared to the control (0.66 g kg^{-1} rutin). Moreover, chitosan resulted in significantly decreased flavonoids compared to sweet orange EO treatment (0.80 and 1.25 g kg^{-1} rutin, respectively). Interestingly, the other examined parameters did not yield significantly different results between the applied treatments and the control (Table 2).

Table 2. Effect of foliar spray of chitosan, phosphoric acid plus micronutrients (PA+MN), and sweet orange EO on strawberry fruit quality after harvest. Values (n = 6 for color measurements; n = 3 for others quality attributes) in rows followed by the same letter are not significantly different; $p \leq 0.05$.

Quality Attributes	Control	Chitosan	PA+NM	Sweet Orange EO
Fresh weight (g)	19.89 ± 1.16 a	10.75 ± 2.94 b	14.50 ± 2.37 ab	17.44 ± 3.79 ab
Color L^*	35.98 ± 1.55 a	39.01 ± 4.80 a	39.66 ± 3.75 a	38.4 ± 3.43 a
Color a^*	34.45 ± 1.89 a	34.46 ± 4.08 a	32.00 ± 4.38 a	31.07 ± 3.21 a
Color b^*	19.12 ± 2.74 a	33.70 ± 19.45 a	28.14 ± 17.21 a	22.00 ± 3.10 a
Chroma	39.48 ± 2.62 a	50.65 ± 17.09 a	44.24 ± 14.78 a	38.15 ± 3.91 a
Hue	0.50 ± 0.05 a	0.67 ± 0.18 a	0.65 ± 0.17 a	0.61 ± 0.06 a
WI	24.72 ± 1.62 a	19.15 ± 13.45 a	24.02 ± 10.03 a	27.23 ± 1.37 a
Color index	51.76 ± 7.79 a	38.12 ± 14.93 a	36.19 ± 12.02 a	38.23 ± 7.14 a
Respiration (mL kg^{-1} h^{-1} CO_2)	26.75 ± 8.22 a	32.84 ± 6.80 a	29.31 ± 10.88 a	35.69 ± 15.84 a
Ethylene (mL kg^{-1} h^{-1})	19.40 ± 11.98 a	26.79 ± 16.63 a	37.55 ± 24.42 a	23.10 ± 19.22 a
Texture (N)	1.74 ± 0.60 a	2.11 ± 1.06 a	2.74 ± 0.68 a	2.88 ± 1.17 a
TSS (%)	8.06 ± 0.93 a	6.26 ± 1.97 a	7.90 ± 1.28 a	9.03 ± 1.60 a
TA (%)	0.93 ± 0.12 a	0.64 ± 0.09 b	0.81 ± 0.02 ab	0.93 ± 0.07 a
Ascorbic acid (g kg^{-1} AA)	1.40 ± 0.21 a	0.89 ± 0.13 b	0.79 ± 0.20 b	0.82 ± 0.14 b
Anthocyanins (g kg^{-1} cyn-3-glu)	0.03 ± 0.01 a	0.04 ± 0.03 a	0.02 ± 0.01 a	0.03 ± 0.01 a
Phenol (g kg^{-1} GAE)	2.56 ± 0.47 a	2.66 ± 0.12 a	2.92 ± 0.21 a	3.37 ± 0.48 a
FRAP (g kg^{-1} trolox)	3.53 ± 0.79 b	3.89 ± 0.97 ab	3.78 ± 0.50 ab	5.14 ± 0.65 a
ABTS (g kg^{-1} trolox)	10.34 ± 1.67 a	10.51 ± 1.03 a	7.18 ± 0.46 b	7.46 ± 0.66 b
Flavonoids (g kg^{-1} rutin)	0.66 ± 0.59 c	0.80 ± 0.20 bc	1.07 ± 0.69 ab	1.25 ± 0.21 a

2.2.2. Effects after Storage

The effects of the applied preharvest treatments on the quality attributes of strawberry fruit after storage for three days under different storage conditions (RT-20 °C and CS-4 °C) are presented in Figures 1–4. After three days of storage at RT (20 °C), ethylene production in strawberry fruit treated with sweet orange EO increased (170 μL kg^{-1} h^{-1} ethylene) compared to control (non-treated fruits) (33.15 μL kg^{-1} h^{-1} ethylene) (Figure 1C). Total soluble solids (TSS) in fruit treated with PA+MN were lower (14.11%) than in all the other applied treatments (Figure 3B). The application of sweet orange EO was found to decrease the antioxidant content of strawberry fruit (FRAP: 2.26 g kg^{-1} trolox) in comparison to all other treatments, whereas the ABTS assay showed that PA+MN was able to increase fruit antioxidant content (6.92 g kg^{-1} trolox), as opposed to sweet orange EO (4.29 g kg^{-1} trolox), during three days of storage at 20 °C (Figure 4D,E). Interestingly, the other examined parameters did not yield significantly different results between the applied treatments and the control.

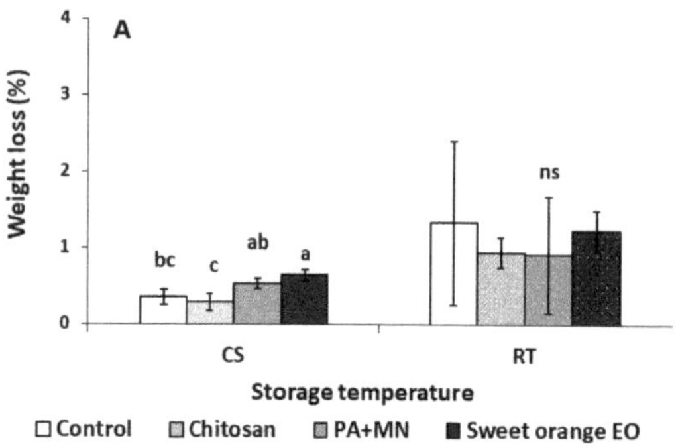

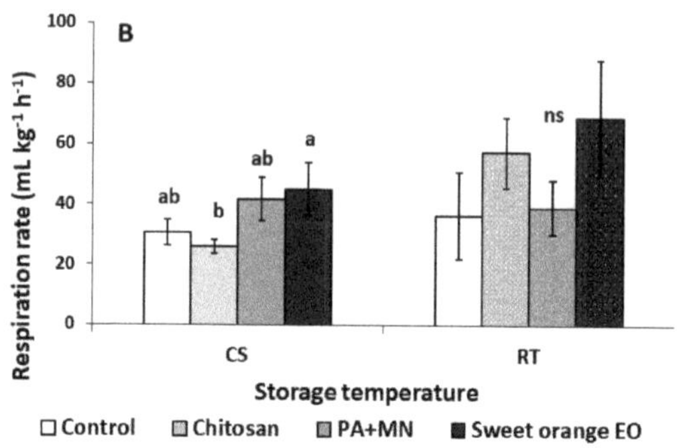

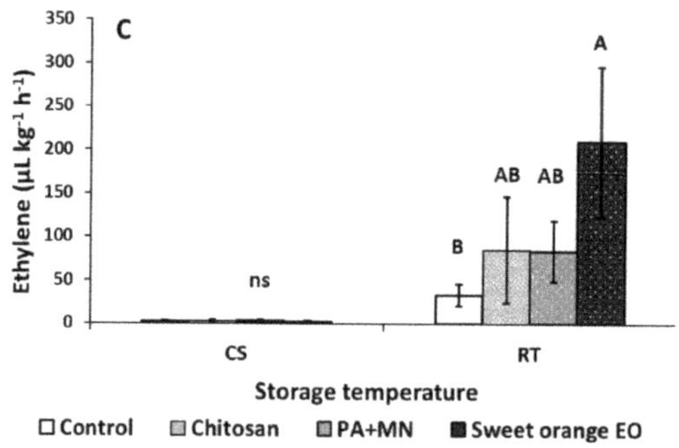

Figure 1. Impact of chitosan, phosphoric acid plus micronutrients (PA+MN), and sweet orange EO treatment on (**A**) weight loss (%), (**B**) respiration rate, and (**C**) ethylene production of strawberry fruit after three days of storage at RT (20 °C) and CS (4 °C). Error bars represent the standard deviation of the mean of three/six replicates. Treatments indicated with different letters for each storage condition differed significantly (Tukey's honestly significant difference, at $p \leq 0.05$), ns: not significant.

As illustrated in Figure 1A, after three days of CS (4 °C), the weight loss in fruit treated with sweet orange EO was higher (0.65%) compared to the control and chitosan treated group (0.36 and 0.30%, respectively). In addition, the application of PA+MN increased the weight loss (0.54%) in comparison to chitosan (0.30%). As shown in Figure 1B, sweet orange EO application yielded an increase of fruit respiration rates compared to chitosan after storage for three days at CS (45.00 and 25.78 mL kg^{-1} h^{-1} CO_2, respectively). The application of PA+MN was found to increase the L^* color value (38.22) of strawberry fruit, as opposed to sweet orange EO (34.60), after three days of storage at CS (Figure 2A). The b^* color value increased with the application of chitosan (26.91) compared to PA+MN treatment and non-treated fruit (21.75 and 20.18, respectively) (Figure 2C). Chitosan was found to increase the chroma value of strawberry fruit (45.93) after three days of storage at 4 °C in comparison to PA+MN (38.62) (Figure 2C). The whiteness index (WI) increased with the application of PA+MN (27.06), as opposed to chitosan and sweet orange EO (21.63 and 22.38, respectively) (Figure 2C). Regarding the TA in strawberry fruit, treatment with chitosan resulted in an increased value (0.83%) compared to control and sweet orange EO (0.65 and 0.64%, respectively) (Figure 3C). No differences were reported between the treatments and the control for all other tested parameters (Figures 2–4).

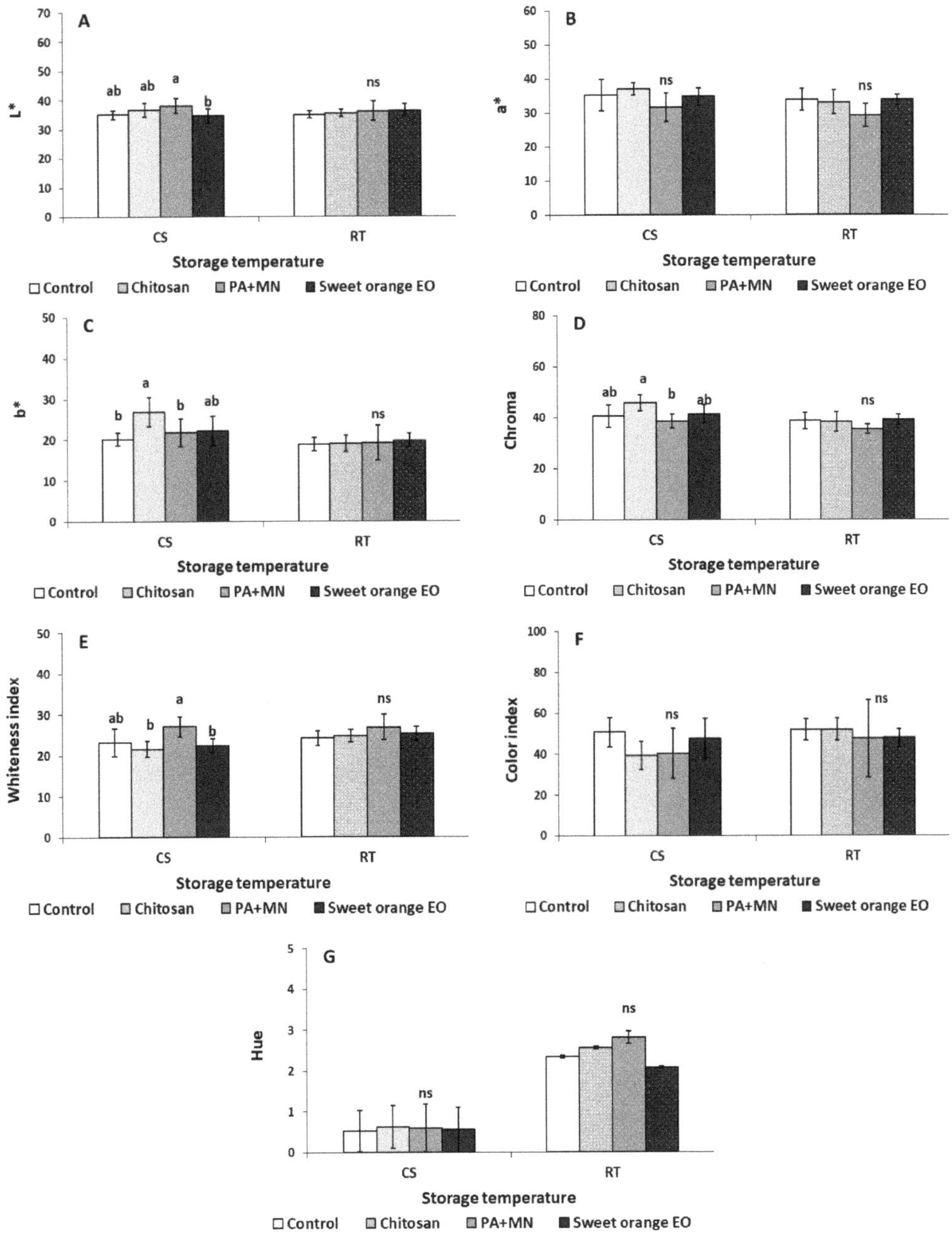

Figure 2. Effect of chitosan, phosphoric acid plus micronutrients (PA+MN), and sweet orange EO treatment and three days of storage at RT (20 °C) and CS (4 °C) on strawberry color parameters: (**A**) L^* color, (**B**) a^* color, (**C**) b^* color, (**D**) Chroma value, (**E**) Whiteness index, (**F**) Color index, and (**G**) Hue angle. Error bars represent the standard deviation of the mean of three/six replicates. Treatments indicated with different letters for each storage condition differed significantly (Tukey's honestly significant difference, at $p \leq 0.05$); ns: not significant.

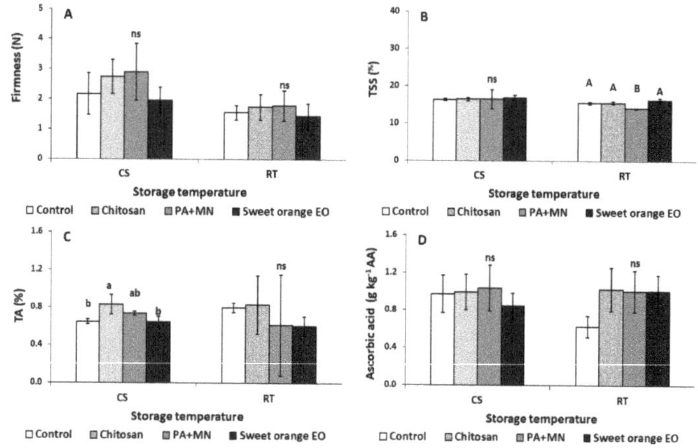

Figure 3. Effect of chitosan, phosphoric acid plus micronutrients (PA+MN), and sweet orange EO treatment on (**A**) firmness, (**B**) total soluble solids (TSS), (**C**) total acidity (TA), and (**D**) ascorbic acid (AA) of strawberry fruit after three days of storage at RT (20 °C) and CS (4 °C). Error bars represent the standard deviation of the mean of (three/six fruit) replicates. Treatments indicated with different letters for each storage condition differed significantly (Tukey's honestly significant difference, at $p \leq 0.05$); ns: not significant.

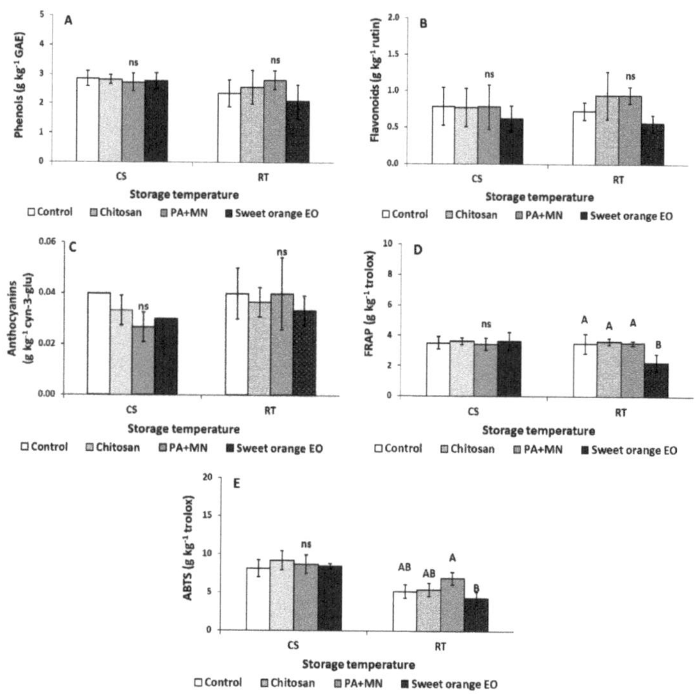

Figure 4. Effect of chitosan, phosphoric acid plus micronutrients (PA+MN), and sweet orange EO treatment and three days of storage at RT (20 °C) and CS (4 °C) on (**A**) total phenols, (**B**) total flavonoids, (**C**) anthocyanins, and (**D**,**E**) antioxidants (FRAP, ABTS). Error bars represent the standard deviation of the mean of (three/six fruit) replicates. Treatments indicated with different letters for each storage condition differed significantly (Tukey's test significant difference, at $p \leq 0.05$); ns: not significant.

3. Discussion

The pre- and postharvest use of many natural compounds (including chitosan, phosphoric acid plus micronutrients, and essential oils) has been investigated in an effort to decrease the use of chemical fungicides and control the development of postharvest diseases [9,16,33–35]. In the present study, preharvest application of chitosan was found to significantly reduce the development of gray mold on strawberry fruit cv. 'Festival' during shelf life. This compound interferes with *B. cinerea*, slowing gray mold infection, since this natural biopolymer possess film-forming, antimicrobial, and eliciting properties [36,37]. Chitosan has been previously shown to be responsible for decreasing the spore germination or lysis of *B. cinerea* mycelium following preharvest application [38,39]. For example, a previous study demonstrated that the use of chitosan (1% w/v Chito Plant) among other natural compounds significantly reduced postharvest decay of sweet cherries in vivo and, at the same time, reduced the growth of decay causing fungi in vitro [34]. However, in the present study, differently from chitosan, PA+MN and sweet orange EO were found to be ineffective in controlling gray mold development on strawberry during storage conditions, even though they have the potential to express eliciting (PA+MN) and antimicrobial (sweet orange EO) activities toward gray mold [34]. These results can be ascribed to the ability of *B. cinerea* to survive at a latent stage in plant tissues and to its ability to develop at cold storage temperatures [40].

Regarding how preharvest treatments affect strawberry fruit quality before harvest and after three days of storage under various conditions (RT-20 °C and CS-4 °C), sweet orange EO treatment was found to have a greater impact on strawberry characteristics on the first day of application and during storage at 20 and 4 °C, compared to the other applied treatments. Significant weight loss (moisture loss) of strawberry fruit was observed with CS three days after storage following sweet orange EO and PA+MN application, in contrast to chitosan application. This might be due to the difference in composition of the applied products and especially the presence of chitosan. It is well known that chitosan can form a semi-permeable coating around fresh produce; this allows gaseous exchanges and minimizes fruit weight loss [33]. With these observations, the assumption that the application of chitosan can result in lower postharvest weight loss can be made.

The influence of chitosan on the control of weight loss was a vital factor with a great impact on the appearance of strawberry fruit, attributed to the formation of an additional layer by the natural compounds coating the surface stomata. This resulted in a reduction of transpiration against oxygen, carbon dioxide, and ethylene levels and dehydration, thereby limiting weight loss [41–43]. These results revealed that the firmness stability (as a symbol of high quality, related to pectin and other cell wall components) was maintained. Additionally, the presence of anthocyanin in fruit epidermis and cortex related to the red pigment did not change after coating with all compounds [44,45]. Our work confirmed that edible coatings served as barriers, preventing moisture loss, water transfer, and weight loss. This barrier also decreased the ability of the strawberries to absorb oxygen, which, in turn, delayed the rates of respiration and the associated loss of weight from the fruit's surface. This finding was in agreement with the result of other studies [46,47]. Interestingly, after three days of CS, sweet orange EO was found to increase the respiration rate of strawberry fruits compared to chitosan treatment. Strawberries, as perishable fruit, tend to present a high respiration rate and are affected by storage conditions (i.e., temperature and duration) [16]. Sweet orange EO was reported to have fungicidal and insecticidal properties [48]. A previous study showed a decrease in the respiration rate of strawberry fruits following the postharvest application of 2% citrus (lemon, orange and mandarin) EOs during storage at 2 °C [16]. The findings in our study are not in accordance with that study; this might be attributed to the lower EO concentration applied (0.08% vs. 2%), the use of cold storage, and the time of application (pre- vs. postharvest), as well as the volatile nature of EOs, that could have evaporated prior to harvesting and/or during storage. Moreover, sweet orange EO, if used at higher concentrations on strawberry canopies, may induce phytotoxic effects.

Strawberries are non-climacteric fruit that can produce ethylene in limited amounts [49]. After storage at RT (20 °C) for three days, fruit treated with sweet orange EO presented increased ethylene production in comparison to non-treated fruit (control). Ethylene is elaborated in a number of processes besides ripening, including responses to pathogens and wounding, senescence of leaves, and the responses of various abiotic and biotic stress factors [50,51]. This result could be connected with the molecular characteristics of sweet orange EO, as volatile components might affect fruit quality parameters [16,52].

One of the key characteristics that influences consumer purchasing decisions is the bright red color of strawberry fruit. In the current study, the preharvest application of chitosan following three days of storage at 4 °C resulted in an increased chroma value and b^* color value compared to PA+MN, indicating a darker red color in the fruit. According to a previous study, strawberry fruits treated with chitosan (1%) prior to harvest were able to retain their dark red color [33]. The results from the present work are in agreement with these observations; this could be correlated with the fact that no significant differences in total flavonoid content were reported, with flavonoids being the main pigments responsible for strawberry fruit's color (including anthocyanins). Chitosan is approved as a basic substance for plant protection, and several applications are ongoing, both in experimental and practical applications, with the potential to replace synthetic fungicides in some settings [53].

The application of chitosan resulted in decreased TA of strawberry fruit three days after preharvest application, compared to sweet orange EO and control, whereas after three days of storage at CS, the opposite effect was observed. Interestingly, a minor impact of strawberry quality was observed with chitosan-based compounds, except for an increase in TA, which is known to influence the microbiological stability of fruit and phenolic compound content [54,55]. This could be attributed to the combination of chitosan application and cold storage (4 °C), as, in a previous study, it was shown that a chitosan coating and storage at 10 °C for six days delayed the ripening, decay, and senescing of strawberry fruit [42]. Fruit becomes sweeter as a result of an increase in TSS and a decrease in TA throughout the ripening process. In a prior study, an eco-product made of rosemary and eucalyptus EOs, applied preharvest to tomatoes, yielded biostimulant properties, increasing the ripening metabolism of the fruit while lowering firmness and the phenol and antioxidant contents [20]. The same EO-based product was further applied postharvest to tomato fruits (dipping and vapor application); it was found that this product was able to maintain the fruit's quality attributes during 14 d of storage at 11 °C [21]. These previous observations, together with the results of the current investigation (the application of treatment based on sweet orange EO), suggest that EOs could be considered for the preservation of fresh fruits, with either pre- or postharvest application.

Strawberries are a good source of antioxidants; including polyphenols, vitamins (especially vitamin C), and flavonoids (i.e., anthocyanins), which makes them an important part of a diet that can minimize the oxidative stress related with chronic diseases [56]. In our study, the preharvest application of sweet orange EO was found to increase the antioxidant and flavonoid contents in strawberry fruits during harvesting time (three days after preharvest application), while following three days of storage at 20 °C (RT), a reduction in antioxidants was reported with the use of sweet orange EO. In a previous study, the postharvest application of citrus EOs to strawberry fruit resulted in an increase in antioxidants in the fruit, even after 4 d of storage at 2 °C [16]. The increase in antioxidants might be associated with the fact that EOs consist of a mixture of compounds that present antioxidant activities, while the loss of antioxidants during storage at RT might be connected to the evaporation of these EOs at higher storage temperatures, as well to storage duration.

Overall, the preharvest use of natural compounds such as chitosan and EOs on strawberries for the improvement of their postharvest quality and to prevent fruit decay is very encouraging. However, it is important to further investigate the type and time of application, the concentrations used, as well as the postharvest conditions during storage.

4. Materials and Methods

4.1. Plant Material and Growth Conditions

The cover material of strawberry plants (*Fragaria* × *ananassa* Duch, cv. 'Festival'), grown in a commercial greenhouse, consisted of transparent polyethylene sheets. Experiments were undertaken in Kolossi, Limassol (Cyprus). Common cultivation practices were applied during the study: plants were fertigated (Plant Prod 20-20-20; Plant Prod 18-9-27, related to the percentage of N-P-K in the form of N-P_2O_5-K_2O) using an irrigation system at the beginning of, and later in, the crop season. Additionally, azoxystrobin for *Sphaerotheca macularis* f. sp. *fragariae* and emamectin benzoate for worm and orange oil were applied once via spraying. The soil type was considered silt loam (6.75% clay, 39.21% sand and 53.98% silt) with 3.03% organic matter, a pH of 7.93, an electrical conductivity (EC) of 6.58 mS cm^{-1}, and a calcium carbonate content of 21.67%. A completely randomized design was set up and six-month-old strawberry plant were used for this experiment.

4.2. Preharvest Treatments

Three different commercial products based on chitosan (Chito Plant, ChiPro GmbH, Bremen, Germany), phosphoric acid plus micronutrients (PA+MN) (KiForce, Alba Milagro, Milano, Italy), and sweet orange EO (Prev-Am Plus, Nufarm, Milano, Italy) were tested, resulting in four treatments, namely, (i) control, (ii) chitosan (1%), (iii) phosphoric acid plus micronutrients (PA+MN) (1%), and (iv) sweet orange EO (0.08%). The preparation of the solutions involved dissolving them in distilled water and stirring for 60 min. The control was distilled water. For each preharvest treatment, 90 plants were selected. Freshly prepared solutions were applied once by spraying 1.5 L of each compound throughout the plants with a mechanical mist sprayer. Fruit was harvested when it reached the commercial ripening stage based on the typical color (red skin) and shape, three days after spraying. For each treatment, 200 strawberries were counted at the picking date. Small fruit with defects were discarded. Strawberries were transferred to the laboratory within 1 h of picking for further experiments. Treated strawberries without visual damage which were homogeneous in color and size were selected and placed into small containers (1 L capacity).

4.3. Decay Evaluation

For each treatment, 42 fruit were tested (six for each of the seven replicates). Strawberry fruits were cold stored (CS) (4 °C) for one week and then exposed to five days of shelf life at 20 ± 1 °C. Strawberries were examined every day for decay symptoms and fungal growth (i.e., *B. cinerea*), and the number of infected fruit (disease incidence, expressed as a percentage) was recorded. According to a five-degree empirical scale, disease severity was recorded as follows: 0—fruit that is healthy; 1—fruit that is surface-infected between 1% and 20%; 2—fruit surface-infected between 21% and 40%; 3—fruit surface-infected between 41% and 60%; 4—fruit surface-infected between 61% and 80%; and 5—fruit that is surface-infected and sporulating at ≥81% [40]. The infection index, also known as the McKinney index, was represented as a percentage of the highest possible level and incorporates both the incidence and severity of the deterioration [57].

4.4. Quality Evaluation

According to each foliar application, strawberries were randomly selected, labeled, and weighed for fruit quality assessments (three replications with six fruits). Fruit was examined at harvest (initial day) and following three days of storage at room temperature (20 ± 1 °C) (RT) or shelf life (4 ± 0.5 °C plus 7 d at RT) (CS) with 95% relative humidity (RH). The physical and chemical properties of the fruit that were measured (on the initial and last day) included weight loss, color, firmness, rates of respiration, rates of ethylene production, total soluble solids-TSS, titratable acidity-TA, ascorbic acid-AA, anthocyanin content, total phenolics content, total flavonoid content, and antioxidant capacity. To this end, two different protocols were used: ferric reducing antioxidant power-FRAP and 2,2′-azino-bis3-ethylbenzothiazoline-6-sulfonic acid-ABTS.

4.4.1. Weight Loss, Respiration Rate, and Ethylene Production

The weight difference of strawberries between the initial day and the third day was measured, and the results were expressed as weight loss percentage (%) = $[(m_0 - m_3)/m_0] \times 100$, where m_0 is the initial weight and m_3 is the weight after three days of storage.

The rates of respiration of the strawberries were analyzed by employing an electronic dual gas analyzer, (GCS 250 Analyzer, International Control Analyzer Ltd., Kent, UK); the results are expressed as mL of carbon dioxide (CO_2) produced per kg of strawberry fresh weight per hour (mL kg^{-1} h^{-1} CO_2). After being taken out of cold storage (4 °C), strawberries were placed on a lab bench for about one hour to warm to room temperature (20 °C). Fruit was then placed in a 1 L capacity container. A 1 mm diameter hole was drilled into the lead of the container and sealed with a tape to prevent gas flow. In order to stop gas from leaking from the packaging, gaseous samples were taken through septa with a syringe. Strawberries of known fresh weight and volume were considered for calculations of respiration rates, as follows: Respiration rate (mL kg^{-1} h^{-1} CO_2) = $(V_1 - V_2) \times \%CO_2 *10/W \times T$ where, V_1 = Volume of container (L), V_2 = Volume of fruit (L), W = Weight of fruit (kg), T = Time of enclosure (h).

The ethylene production of the strawberries was analyzed using an ethylene analyzer (ICA 56 Analyzer, International Control Analyzer Ltd., Kent, UK). Following respiration measurement, gaseous samples were tested for ethylene production. Ethylene rate was expressed as µL of ethylene produced per kg of strawberry fresh weight per hour (µL kg^{-1} h^{-1} ethylene).

4.4.2. Color Evaluation

A Konica Minolta colorimeter (Chroma Meter CR 400 Konica Minolta, Japan) was employed to measure the surface color of the samples on days 0 and 3 (two values in each replicate/three replicates in each treatment). Fruit color was assessed by measuring the values of L^* (lightness), a^* (greenness [−] to redness [+]) and b^* (blueness [−] to yellowness [+]). The chroma value (C), whiteness index (WI), color index (CI), and hue angle (h) were computed according to the following equations: $C = \left(a^{*2} + b^{*2}\right)^{1/2}$, $WI = 100 - \lfloor (100 - L^*)^2 + a^{*2} + b^{*2} \rfloor^{1/2}$, $C = (a^* \times 1000)/(L^* \times b^*)$ and $h = \tan^{-1}(a^*/b^*)$, respectively [58,59].

4.4.3. Firmness, Total Soluble Solids, Titratable Acidity, and Ascorbic Acid

Strawberry fruit firmness was measured by a texture analyzer (TA-XT plus, Stable Micro Systems, Surrey, UK) equipped with a 3 mm diameter cylinder stainless probe. The probe penetrating depth was 7 mm and the cross-head speed of the texture analyzer was 1 mm s^{-1}. The results were reported as the maximum peak force in Newton (N) using the Texture Exponent 32 program (Stable Micro Systems, Surrey, UK).

The total soluble solids content of storage strawberry fruit was estimated using a digital pocket refractometer (Atago, Tokyo, Japan). Fruit was homogenized with a blender for 2 min before being filtered through muslin cloth for squeezing fruit juice. A few drops of the extract were placed on the refractometer prism glass and data were recorded through direct reading. Results of TSS were expressed as percentage (%).

Titratable acidity was analyzed using Mettler Toledo Food and Beverage Titrator (Mettler Toledo DL22, Mettler-Toledo AG, Schwerzenbach, Switzerland), as described by Tzortzakis [60]. Homogenized strawberry juice (5 mL) was placed in a 50 mL volumetric flask and the desired volume was obtained with distilled water. The extract was then titrated with 0.1 N NaOH (pH 8.2). The results are reported as citric acid percentage (%).

The ascorbic acid content was calculated according to the 2,6-dichlorophenolindophenol titration method [61]. Briefly, 5 g strawberry was extracted with 50 mL of 3% metaphosphoric acid solution. The upper aqueous phase was then recovered by centrifugation at 3000× *g* for 15 min at 25 °C. An aliquot of 10 mL was titrated using 0.1% 2,6-dichlorophenolindophenol to obtain a pink color. Ascorbic acid concentration was computed and reported as g of AA per kg of fresh weight (g kg^{-1} AA).

4.4.4. Polyphenol, Flavonoid, Anthocyanin, and Antioxidant Activity

Strawberry fruit tissue (1 g) was milled with 10 mL of methanol 50% (*v*/*v*) for 30 s. The extract was then ultra-sonicated for 30 min. Next, samples were placed in a shaker for 1 h at 200 rpm. The extract was centrifuged (Sigma 3-18K, Sigma Laboratory Centrifuge, Osterode am Harz, Germany) at 4000× *g* for 15 min at 4 °C. The supernatant was recovered and used for the analysis of total phenol content, total flavonoid content, and total antioxidant capacity (FRAP, ABTS).

The content of polyphenols was determined by employing the Folin-Ciocalteu method at 755 nm, according to Marinou et al. [62]. Crude extract (0.05 mL) was added to dH_2O (final volume 1.625 mL) and mixed thoroughly with 0.125 mL of Folin-Ciocalteu reagent. The reaction was left to stand for 5 min, followed by the addition of 1.25 mL of 7% (*w*/*v*) Na_2CO_3 (final reaction volume 3 mL). The mixture was allowed to stand for a further 60 min at room temperature in the dark. The absorbance was measured at 755 nm using a spectrophotometer (Multiskan GO, Thermo Fisher Scientific Oy, Finland). The total phenol content was computed through a calibration curve of gallic acid, and the results were reported as g of gallic acid equivalents per kg of fresh weight (g kg^{-1} GAE).

The total flavonoids content of crude extract was determined using the aluminum chloride method [63]. The fruit extract (0.25 mL) was mixed with 1.525 mL of dH_2O and 0.075 mL of 5% (*w*/*v*) $NaNO_2$ and left to stand for 5 min. Then, 0.15 mL of $AlCl_3$ 10% (*w*/*v*) was added to the mixture and the reaction was incubated for 5 min. Afterwards, 0.5 mL of 4% (*w*/*v*) NaOH was added to the mixture (total volume achieved 2.5 mL). The absorbance was measured at 510 nm. The result was reported as g of rutin per kg of fresh weight (g kg^{-1} rutin), using a calibration curve of prepared rutin dilutions.

The ability of the extract to reduce Fe^{3+} to Fe^{2+} and the formation of the blue colored $[Fe\ (TPTZ)_2]^{2+}$ complex was determined based on the following (FRAP) method from Chrysargyris et al. [64]. The FRAP reaction solution was prepared by combining 0.04 mL sample with 0.3M CH_3COONa (pH = 3.6), 10 mM Tripyridil-s-triazine (TPTZ) and 20 mM $FeCl_3$. A standard curve was also prepared using known (±)-6-hydroxy-2,5,7,8-tetramethylchromane-2-carboxylic acid (trolox) concentrations. The absorbance was measured spectrophotometrically at 593 nm after incubation at 37 °C for 4 min; the results were reported as g of trolox per kg of fresh weight (g kg^{-1} trolox).

The antioxidant capacity of strawberry fruit in the reaction with the ABTS radical cation was assessed according to the method outlined by Wojdyło et al. [65]. Briefly, the procedure was performed by mixing the ABTS solution 7 mM with 0.01 mL of sample. The mixture was incubated for 6 min at RT. A standard curve was prepared using known trolox concentrations. The absorbance at 734 nm was measured spectrophotometrically and the results were reported as g of trolox per kg of fresh weight (g kg^{-1} trolox).

The pH-differential method was used to determine the total anthocyanin content, using two buffers systems [66]. Strawberry tissue (2 g) was homogenized with 15 mL of methanol: dH_2O:HCl (70:29:1, *v*/*v*/*v*). The homogenized extract was then centrifuged at 4000× *g* for 5 min at 4 °C. Next, the prepared extract was mixed with corresponding buffers (0.025 M KCl, pH = 1 and 0.4 M CH_3COONa, pH = 4.5). After 15 min, the absorbance of each solution was measured at 520 and 700 nm using a spectrophotometer and calculated as g of cyanidin 3-glucoside equivalents per kg of fresh weight (g kg^{-1} cyn-3-glu).

4.5. Statistical Analysis

Analysis of variance (ANOVA) was used to analyze the data with the SPSS software (IBM SPSS version 22, Armonk, New York, NY, USA). Mean comparisons were made by Tukey's multiple range test. Significance was defined at $p \leq 0.05$.

5. Conclusions

This research investigated the preharvest effectiveness of commercial products based on natural compounds, i.e., chitosan, phosphoric acid plus micronutrients, and sweet orange EO, on strawberry postharvest decay and physiological qualitative parameters. Our results showed the greenhouse preharvest application of chitosan was effective against gray mold during shelf life. At the same time, the application of this compound maintained excellent qualitative parameters (i.e., color, respiration rate and TA) in the fruit. Our research suggests that this compound may be one of the most important natural compounds in related research and could be used for further investigation at the commercial level.

Supplementary Materials: The following supporting information can be downloaded at: https://www.mdpi.com/article/10.3390/ijms232415472/s1, Figure S1: Infection of postharvest gray mold on strawberries.

Author Contributions: Conceptualization, G.R. and N.T.; methodology, G.R. and N.T.; software, P.X.; validation, A.C., P.X. and G.R.; formal analysis, P.X. and R.R.; investigation, R.R. and P.X.; resources, G.R. and N.T.; data curation, A.C. and P.X.; writing—original draft preparation, R.R. and P.X.; writing—review and editing, A.C., G.R. and N.T.; visualization, R.R. and G.R.; supervision, A.C., P.X., G.R. and N.T.; project administration, G.R. and N.T.; funding acquisition, N.T. All authors have read and agreed to the published version of the manuscript.

Funding: This research was funded by PRIMA StopMedWaste project, which is funded by PRIMA, a programme supported by the European Union with co-funding by the Funding Agencies of Italy and RIF–Cyprus.

Institutional Review Board Statement: Not applicable.

Data Availability Statement: Not applicable.

Conflicts of Interest: The authors declare no conflict of interest.

References

1. Afrin, S.; Gasparrini, M.; Forbes-Hernandez, T.Y.; Reboredo-Rodriguez, P.; Mezzetti, B.; Varela-López, A.; Giampieri, F.; Battino, M. Promising health benefits of the strawberry: A focus on clinical studies. *J. Agric. Food Chem.* **2016**, *64*, 4435–4449. [CrossRef] [PubMed]
2. Food and Agriculture Organization of the United Nations—FAO. *Promotion of Fruit and Vegetables for Health: Report of the Pacific Regional Workshop*; FAO: Rome, Italy, 2015; ISBN 978-92-5-108879-1.
3. Feliziani, E.; Romanazzi, G. Postharvest decay of strawberry fruit: Etiology, epidemiology, and disease management. *J. Berry Res.* **2016**, *6*, 47–63. [CrossRef]
4. Maas, J.L. *Compendium of Strawberry Diseases*, 2nd ed.; The American Phytopathological Society Press: St. Paul, MN, USA, 1998.
5. Ugolini, L.; Pagnotta, E.; Matteo, R.; Malaguti, L.; Di Francesco, A.; Lazzeri, L. Brassica meal-derived allyl-isothiocyanate postharvest application: Influence on strawberry nutraceutical and biochemical parameters. *J. Sci. Food Agric.* **2019**, *99*, 4235–4241. [CrossRef] [PubMed]
6. Hua, C.; Kai, K.; Wang, X.; Shi, W.; Zhang, D.; Liu, Y. Curcumin inhibits gray mold development in kiwifruit by targeting mitogen-activated protein kinase (MAPK) cascades in *Botrytis cinerea*. *Postharvest Biol. Technol.* **2019**, *151*, 152–159. [CrossRef]
7. Landi, L.; De Miccolis Angelini, R.M.; Pollastro, S.; Feliziani, E.; Faretra, F.; Romanazzi, G. Global transcriptome analysis and identification of differentially expressed genes in strawberry after preharvest application of benzothiadiazole and chitosan. *Front. Plant Sci.* **2017**, *8*, 235. [CrossRef]
8. Xylia, P.; Chrysargyris, A.; Ahmed, Z.F.R.; Tzortzakis, N. Application of rosemary and eucalyptus essential oils and their main component on the preservation of apple and pear fruits. *Horticulture* **2021**, *7*, 479. [CrossRef]
9. Tzortzakis, N.; Chrysargyris, A.; Sivakumar, D.; Loulakakis, K. Vapour or dipping applications of methyl jasmonate, vinegar and sage oil for pepper fruit sanitation towards grey mould. *Postharvest Biol. Technol.* **2016**, *118*, 120–127. [CrossRef]
10. Chrysargyris, A.; Nikou, A.; Tzortzakis, N. Effectiveness of *Aloe vera* gel coating for maintaining tomato fruit quality. *N. Z. J. Crops Hortic. Sci.* **2016**, *44*, 203–217. [CrossRef]
11. Romanazzi, G.; Moumni, M. Chitosan and other edible coatings to manage postharvest decay, extend shelf life, and reduce loss and waste of fresh fruit and vegetables. *Curr. Opin. Biotechnol.* **2022**, *78*, 102834. [CrossRef]
12. Fallik, E.; Ilic, Z. Pre- and postharvest treatments affecting flavor quality of fruits and vegetables. In *Preharvest Modulation of Postharvest Fruit and Vegetable Quality*; Siddiqui, M.W., Ed.; Academic Press: Cambridge, MA, USA, 2018; pp. 139–168; ISBN 9780128098080.
13. Palou, L.; Ali, A.; Fallik, E.; Romanazzi, G. GRAS, plant- and animal-derived compounds as alternatives to conventional fungicides for the control of postharvest diseases of fresh horticultural produce. *Postharvest Biol. Technol.* **2016**, *122*, 41–52. [CrossRef]
14. Kalemba, D.; Kunicka, A. Antibacterial and antifungal properties of essential oils. *Curr. Med. Chem.* **2003**, *10*, 813–829. [CrossRef] [PubMed]

15. Sivakumar, D.; Bautista-Baños, S. A review on the use of essential oils for postharvest decay control and maintenance of fruit quality during storage. *Crops Prot.* **2014**, *64*, 27–37. [CrossRef]
16. Shehata, S.A.; Abdeldaym, E.A.; Ali, M.R.; Mohamed, R.M.; Bob, R.I.; Abdelgawad, K.F. Effect of some citrus essential oils on post-harvest shelf life and physicochemical quality of strawberries during cold storage. *Agronomy* **2020**, *10*, 1466. [CrossRef]
17. Chrysargyris, A.; Rousos, C.; Xylia, P.; Tzortzakis, N. Vapour application of sage essential oil maintain tomato fruit quality in breaker and red ripening stages. *Plants* **2021**, *10*, 2645. [CrossRef]
18. Xylia, P.; Chrysargyris, A.; Botsaris, G.; Tzortzakis, N. Mint and pomegranate extracts/oils as antibacterial agents against *Escherichia coli* O157:H7 and *Listeria monocytogenes* on shredded carrots. *J. Food Saf.* **2018**, *38*, e12423. [CrossRef]
19. Xylia, P.; Chrysargyris, A.; Botsaris, G.; Tzortzakis, N. Potential application of spearmint and lavender essential oils for assuring endive quality and safety. *Crops Prot.* **2017**, *102*, 94–103. [CrossRef]
20. Chrysargyris, A.; Charalambous, S.; Xylia, P.; Litskas, V.; Stavrinides, M.; Tzortzakis, N. Assessing the biostimulant effects of a novel plant-based formulation on tomato crop. *Sustainability* **2020**, *12*, 8432. [CrossRef]
21. Xylia, P.; Ioannou, I.; Chrysargyris, A.; Stavrinides, M.C.; Tzortzakis, N. Quality attributes and storage of tomato fruits as affected by an eco-friendly, essential oil-based product. *Plants* **2021**, *10*, 1125. [CrossRef]
22. Basak, S.; Guha, P. Betel leaf (*Piper betle* L.) essential oil microemulsion: Characterization and antifungal activity on growth, and apparent lag time of *Aspergillus flavus* in tomato paste. *LWT Food Sci. Technol.* **2017**, *75*, 616–623. [CrossRef]
23. Chen, C.; Nie, Z.; Wan, C.; Chen, J. Preservation of Xinyu tangerines with an edible coating using *Ficus hirta* Vahl. fruits extract-incorporated chitosan. *Biomolecules* **2019**, *9*, 46. [CrossRef]
24. Grande-Tovar, C.D.; Chaves-Lopez, C.; Serio, A.; Rossi, C.; Paparella, A. Chitosan coatings enriched with essential oils: Effects on fungi involve in fruit decay and mechanisms of action. *Trends Food Sci. Technol.* **2018**, *78*, 61–71. [CrossRef]
25. Rehman, R.; Hanif, M.A.; Mushtaq, Z.; Al-Sadi, A.M. Biosynthesis of essential oils in aromatic plants: A review. *Food Rev. Int.* **2016**, *32*, 117–160. [CrossRef]
26. Falguera, V.; Quintero, J.P.; Jiménez, A.; Muñoz, J.A.; Ibarz, A. Edible films and coatings: Structures, active functions and trends in their use. *Trends Food Sci. Technol.* **2011**, *22*, 292–303. [CrossRef]
27. Romanazzi, G.; Smilanick, J.L.; Feliziani, E.; Droby, S. Integrated management of postharvest gray mold on fruit crops. *Postharvest Biol. Technol.* **2016**, *113*, 69–76. [CrossRef]
28. Abbey, J.A.; Percival, D.; Abbey, L.; Asiedu, S.K.; Prithiviraj, B.; Schilder, A. Biofungicides as alternative to synthetic fungicide control of grey mould (*Botrytis cinerea*)-prospects and challenges. *Biocontrol Sci. Technol.* **2019**, *29*, 241–262. [CrossRef]
29. Romanazzi, G.; Feliziani, E.; Sivakumar, D. Chitosan, a biopolymer with triple action on postharvest decay of fruit and vegetables: Eliciting, antimicrobial and film-forming properties. *Front. Microbiol.* **2018**, *9*, 2745. [CrossRef]
30. Shi, Z.; Yang, H.; Jiao, J.; Wang, F.; Lu, Y.; Deng, J. Effects of graft copolymer of chitosan and salicylic acid on reducing rot of postharvest fruit and retarding cell wall degradation in grapefruit during storage. *Food Chem.* **2019**, *283*, 92–100. [CrossRef]
31. Feliziani, E.; Romanazzi, G. Preharvest application of synthetic fungicides and alternative treatments to control postharvest decay of fruit. *Stewart Postharvest Rev.* **2013**, *9*, 1–6. [CrossRef]
32. Wang, S.Y.; Gao, H. Effect of chitosan-based edible coating on antioxidants, antioxidant enzyme system, and postharvest fruit quality of strawberries (*Fragaria* x *ananassa* Duch.). *LWT Food Sci. Technol.* **2013**, *52*, 71–79. [CrossRef]
33. Feliziani, E.; Landi, L.; Romanazzi, G. Preharvest treatments with chitosan and other alternatives to conventional fungicides to control postharvest decay of strawberry. *Carbohydr. Polym.* **2015**, *132*, 111–117. [CrossRef]
34. Feliziani, E.; Santini, M.; Landi, L.; Romanazzi, G. Pre- and postharvest treatment with alternatives to synthetic fungicides to control postharvest decay of sweet cherry. *Postharvest Biol. Technol.* **2013**, *78*, 133–138. [CrossRef]
35. Hendges, C.; Stangarlin, J.R.; Zamban, V.C.; de Mascaro, M.H.N.; Carmelo, D.B. Antifungal activity and control of the early blight in tomato through tea tree essential oil. *Crops Prot.* **2021**, *148*, 105728. [CrossRef]
36. de Oliveira, A.M.; Enio, N.; de Oliveira, J. Antifungal and dilmogenic properties of micro- and nanostructures of chitosan and its derivatives. In *Nanomycotoxicology: Treating Mycotoxins in the Nano Way*; Mahendra, R., Kamel, A.A.-E., Eds.; Academic Press: Cambridge, MA, USA, 2020; pp. 397–421.
37. Rajestary, R.; Landi, L.; Romanazzi, G. Chitosan and postharvest decay of fresh fruit: Meta-analysis of disease control and antimicrobial and eliciting activities. *Compr. Rev. Food Sci. Food Saf.* **2021**, *20*, 563–582. [CrossRef] [PubMed]
38. Muñoz, Z.; Moret, A. Sensitivity of *Botrytis cinerea* to chitosan and acibenzolar-S-methyl. *Pest Manag. Sci.* **2010**, *66*, 974–979. [CrossRef]
39. Saavedra, G.M.; Sanfuentes, E.; Figueroa, P.M.; Figueroa, C.R. Independent preharvest applications of methyl jasmonate and chitosan elicit differential upregulation of defense-related genes with reduced incidence of gray mold decay during postharvest storage of *Fragaria chiloensis* fruit. *Int. J. Mol. Sci.* **2017**, *18*, 1420. [CrossRef] [PubMed]
40. Romanazzi, G.; Feliziani, E.; Santini, M.; Landi, L. Effectiveness of postharvest treatment with chitosan and other resistance inducers in the control of storage decay of strawberry. *Postharvest Biol. Technol.* **2013**, *75*, 24–27. [CrossRef]
41. Maqbool, M.; Ali, A.; Alderson, P.G.; Zahid, N.; Siddiqui, Y. Effect of a novel edible composite coating based on gum arabic and chitosan on biochemical and physiological responses of banana fruits during cold storage. *J. Agric. Food Chem.* **2011**, *59*, 5474–5482. [CrossRef]

42. Hernández-Muñoz, P.; Almenar, E.; Del Valle, V.; Velez, D.; Gavara, R. Effect of chitosan coating combined with postharvest calcium treatment on strawberry (*Fragaria* × *ananassa*) quality during refrigerated storage. *Food Chem.* **2008**, *110*, 428–435. [CrossRef]
43. Guerreiro, A.C.; Gago, C.M.L.; Faleiro, M.L.; Miguel, M.G.C.; Antunes, M.D.C. The use of polysaccharide-based edible coatings enriched with essential oils to improve shelf-life of strawberries. *Postharvest Biol. Technol.* **2015**, *110*, 51–60. [CrossRef]
44. Rodrigo, M.J.; Zacarias, L. Effect of postharvest ethylene treatment on carotenoid accumulation and the expression of carotenoid biosynthetic genes in the flavedo of orange (*Citrus sinensis* L. Osbeck) fruit. *Postharvest Biol. Technol.* **2007**, *43*, 14–22. [CrossRef]
45. Rico, D.; Barcenilla, B.; Meabe, A.; González, C.; Martín-Diana, A.B. Mechanical properties and quality parameters of chitosan-edible algae (*Palmaria palmata*) on ready-to-eat strawberries. *J. Sci. Food Agric.* **2019**, *99*, 2910–2921. [CrossRef] [PubMed]
46. Arowora, K.A.; Williams, J.; Adetunji, C.O.; Afolayan, S.S.; Olaleye, O.O.; Ogundele, B.A.; Fawole, O.B.; Adetunji, J.B. Effects of *Aloe vera* coatings on quality characteristics of oranges stored under cold storage. *Greener J. Agric. Sci.* **2013**, *3*, 039–047. [CrossRef]
47. Yuan, X.; Amarnath Praphakar, R.; Munusamy, M.A.; Alarfaj, A.A.; Suresh Kumar, S.; Rajan, M. Mucoadhesive guargum hydrogel inter-connected chitosan-g-polycaprolactone micelles for rifampicin delivery. *Carbohydr. Polym.* **2019**, *206*, 1–10. [CrossRef] [PubMed]
48. Tacoli, F.; Bell, V.A.; Cargnus, E.; Pavan, F. Insecticidal activity of natural products against vineyard mealybugs (Hemiptera: Pseudococcidae). *Crops Prot.* **2018**, *111*, 50–57. [CrossRef]
49. Trainotti, L.; Pavanello, A.; Casadoro, G. Different ethylene receptors show an increased expression during the ripening of strawberries: Does such an increment imply a role for ethylene in the ripening of these non-climacteric fruits? *J. Exp. Bot.* **2005**, *56*, 2037–2046. [CrossRef]
50. Alexander, L.; Grierson, D. Ethylene biosynthesis and action in tomato: A model for climacteric fruit ripening. *J. Exp. Bot.* **2002**, *53*, 2039–2055. [CrossRef]
51. Jia, M.; Du, P.; Ding, N.; Zhang, Q.; Xing, S.; Wei, L.; Zhao, Y.; Mao, W.; Li, J.; Li, B.; et al. Two FERONIA-like receptor kinases regulate apple fruit ripening by modulating ethylene production. *Front. Plant Sci.* **2017**, *8*, 1406. [CrossRef]
52. Xylia, P.; Clark, A.; Chrysargyris, A.; Romanazzi, G.; Tzortzakis, N. Quality and safety attributes on shredded carrots by using *Origanum majorana* and ascorbic acid. *Postharvest Biol. Technol.* **2019**, *155*, 120–129. [CrossRef]
53. Romanazzi, G.; Orçonneau, Y.; Moumni, M.; Davillerd, Y.; Marchand, P.A. Basic substances, a sustainable tool to complement and eventually replace synthetic pesticides in the management of pre and postharvest diseases: Reviewed instructions for users. *Molecules* **2022**, *27*, 3484. [CrossRef]
54. Olawuyi, I.F.; Lee, W. Influence of chitosan coating and packaging materials on the quality characteristics of fresh-cut cucumber. *Korean J. Food Preserv.* **2019**, *26*, 371–380. [CrossRef]
55. Xu, D.; Deng, Y.; Han, T.; Jiang, L.; Xi, P.; Wang, Q.; Jiang, Z.; Gao, L. In vitro and in vivo effectiveness of phenolic compounds for the control of postharvest gray mold of table grapes. *Postharvest Biol. Technol.* **2018**, *139*, 106–114. [CrossRef]
56. Pineli, L.d.L.d.O.; Moretti, C.L.; dos Santos, M.S.; Campos, A.B.; Brasileiro, A.V.; Córdova, A.C.; Chiarello, M.D. Antioxidants and other chemical and physical characteristics of two strawberry cultivars at different ripeness stages. *J. Food Compos. Anal.* **2011**, *24*, 11–16. [CrossRef]
57. McKinney, H.H. Influence of soil temperature and moisture on infection of wheat seedlings by *Helminthosporium sativum*. *J. Agric. Res.* **1923**, *26*, 195.
58. Bolin, H.R.; Huxsoll, C.C. Effect of preparation procedures and storage parameters on quality retension of salad-cut lettuce. *J. Food Sci.* **1991**, *56*, 60–62. [CrossRef]
59. Goyeneche, R.; Agüero, M.V.; Roura, S.; Di Scala, K. Application of citric acid and mild heat shock to minimally processed sliced radish: Color evaluation. *Postharvest Biol. Technol.* **2014**, *93*, 106–113. [CrossRef]
60. Tzortzakis, N. Physiological and proteomic approaches to address the active role of *Botrytis cinerea* inoculation in tomato postharvest ripening. *Microorganisms* **2019**, *7*, 681. [CrossRef]
61. AOAC International. *Official Methods of Analysis*, 18th ed.; AOAC International: Gaithersburg, MD, USA, 2007.
62. Marinou, E.; Chrysargyris, A.; Tzortzakis, N. Use of sawdust, coco soil and pumice in hydroponically grown strawberry. *Plant Soil Environ.* **2013**, *59*, 452–459. [CrossRef]
63. Meyers, K.J.; Watkins, C.B.; Pritts, M.P.; Liu, R.H. Antioxidant and antiproliferative activities of strawberries. *J. Agric. Food Chem.* **2003**, *51*, 6887–6892. [CrossRef]
64. Chrysargyris, A.; Panayiotou, C.; Tzortzakis, N. Nitrogen and phosphorus levels affected plant growth, essential oil composition and antioxidant status of lavender plant (*Lavandula angustifolia* Mill.). *Ind. Crops Prod.* **2016**, *83*, 577–586. [CrossRef]
65. Wojdyło, A.; Oszmiański, J.; Czemerys, R. Antioxidant activity and phenolic compounds in 32 selected herbs. *Food Chem.* **2007**, *105*, 940–949. [CrossRef]
66. Lee, J.; Durst, R.W.; Wrolstad, R.E. Determination of total monomeric anthocyanin pigment content of fruit juices, beverages, natural colorants, and wines by the pH differential method: Collaborative study. *J. AOAC Int.* **2005**, *88*, 1269–1278. [CrossRef] [PubMed]

International Journal of
Molecular Sciences

MDPI

Review

The Potential of Antibiotics and Nanomaterial Combinations as Therapeutic Strategies in the Management of Multidrug-Resistant Infections: A Review

Oluwaseun Ola Adeniji [1,2,*], Nolonwabo Nontongana [1,2], Janet Chiyem Okoh [3] and Anthony Ifeanyi Okoh [1,2,4]

1 SAMRC Microbial Water Quality Monitoring Centre, University of Fort Hare, Alice 5700, South Africa
2 Applied and Environmental Microbiology Research Group (AEMREG), Department of Biochemistry and Microbiology, University of Fort Hare, Alice 5700, South Africa
3 Department of Medicine, Faculty of Health Sciences, University of Cape Town, Cape Town 7700, South Africa
4 Department of Environmental Health Sciences, College of Health Sciences, University of Sharjah, Sharjah P.O. Box 26666, United Arab Emirates
* Correspondence: seunola2012@gmail.com

Abstract: Antibiotic resistance has become a major public health concern around the world. This is exacerbated by the non-discovery of novel drugs, the development of resistance mechanisms in most of the clinical isolates of bacteria, as well as recurring infections, hindering disease treatment efficacy. In vitro data has shown that antibiotic combinations can be effective when microorganisms are resistant to individual drugs. Recently, advances in the direction of combination therapy for the treatment of multidrug-resistant (MDR) bacterial infections have embraced antibiotic combinations and the use of nanoparticles conjugated with antibiotics. Nanoparticles (NPs) can penetrate the cellular membrane of disease-causing organisms and obstruct essential molecular pathways, showing unique antibacterial mechanisms. Combined with the optimal drugs, NPs have established synergy and may assist in regulating the general threat of emergent bacterial resistance. This review comprises a general overview of antibiotic combinations strategies for the treatment of microbial infections. The potential of antibiotic combinations with NPs as new entrants in the antimicrobial therapy domain is discussed.

Keywords: multidrug resistance; nanoparticles; antibiotic combinations; infectious disease

Citation: Adeniji, O.O.; Nontongana, N.; Okoh, J.C.; Okoh, A.I. The Potential of Antibiotics and Nanomaterial Combinations as Therapeutic Strategies in the Management of Multidrug-Resistant Infections: A Review. *Int. J. Mol. Sci.* **2022**, *23*, 15038. https://doi.org/10.3390/ijms232315038

Academic Editor: Helena Felgueiras

Received: 21 October 2022
Accepted: 21 November 2022
Published: 30 November 2022

1. Introduction

Globally, the emergence of multidrug-resistant organisms (MDROs), which account for approximately 16% of hospital-acquired infections, has been bolstered by the indiscriminate use of drugs, self-medication, and exposure to infectious diseases [1–3]. It is becoming increasingly difficult for physicians to care for patients with pan-drug-resistant (PDR), extensively resistant (XDR), or MDR organisms [4]. Resistance against more than one class of antibiotics characterizes the MDR organisms, whereas PDR counterparts are resistant against entirely clinical practice-recognized antimicrobial agents, established on sensitivity tests in the laboratory [5]. Patients infected with antimicrobial-resistant bacteria (ARB) require long-term antimicrobial therapy, and the prospect of effective clinical treatment deteriorates [6,7]. Hwang et al. [8] in their study described a woman who developed an actinomycotic brain abscess 15 months after being treated for non-invasive nasopharyngeal actinomycosis, which recurred as an invasive form. Carbapenems have been considered to be the most effective broad-spectrum beta-lactam antibiotic treatment against MDR Gram-negative microorganisms [4]. Consequently, colistin and tigecycline are currently regarded as "last resort" drugs for the treatment of carbapenem-resistant microorganisms. Nevertheless, as the use of these two medications has increased, there have been more reports of tigecycline- or colistin-resistant organisms emerging in the last five years [4,9,10].

The pharmaceutical manufacturing industry's lack of investment in drug discovery because of the intrinsically low rate of profit for antibiotics versus medicines geared toward chronic illnesses exacerbated the problem challenge of increasing bacterial resistance [11]. This condition is so threatening that the World Health Organization has identified MDROs as one of the three most serious threats to human health [12,13], even though the Infectious Disease Society of America (IDSA) has issued a call for biomedical community action to address the risks associated with MDR pathogens [14]. Aside from the development of sophisticated strategies to fight MDR disease-causing organisms, several medical and paramedical disciplines will be negatively influenced [15]. As a result, other methods of regulating bacterial infectious diseases, particularly Gram-negative bacteria, are desperately needed. Also, it is imperative to develop scientific knowledge concerning antibacterial combinations and pharmacological treatments to safeguard human well-being; since the previous resolution to the spread of infections found in antibiotics is currently unexpectedly the major basis of antibiotic resistance [16]. Therefore, the primarily newly established therapies are dual antibacterial combinations with nanoparticles (NPs) [16,17]. This review aims to outline the effect of antibiotics and antibiotic combinations, the use of nanomaterials as alternate antibacterial agents in the treatment of severe bacterial infectious illnesses, nanomaterial types, synthesis methods, and the characterization, applications, and toxicity of NPs.

1.1. *Genesis of Combination Therapy*

The increase in the incidence of MDR microorganisms, especially MDR Gram-negative microorganisms, indicates that monotherapy is becoming increasingly insufficient and thus the usage of combined treatments is advocated [18]. For example, combined therapy treatments have for a long time been employed in treating patients infected with HIV [19], patients with prosthetic joint infectious disease due to *methicillin-sensitive Staphylococcus. aureus* (MSSA), and for the treatment of prosthetic heart valve infections due to coagulase-negative staphylococci [20,21]. More so, they are almost entirely used for the treatment of *Mycobacterium tuberculosis* infectious diseases and are also critical for the treatment of bacterial infections [22]. The combination of streptomycin with penicillin was reported in 1950 [23], and trimethoprim with sulfonamides in 1968 [24]. Both combinations improved efficiency and antimicrobial range. The combination of drugs to defeat the selection for resistance to single agents in the treatment of tuberculosis was documented in the early 1950s [25], whereas Noordeen [26] documented that the "advantages of antibiotic combinations for treatment of leprosy were detected in the 1960s". Several new alternative strategies to conventional therapy regimens like bacteriophage [27], antibacterial antibodies [28], antibiotics combinations [29], photothermal treatment [30], Molecular hybridization (Synthesis of new hybrid drug) [31,32], and nanomaterials [33] have been tested.

1.1.1. Antibiotic Combination Therapy

Antibiotic combination therapy is a treatment that requires the prescription and usage of multiple drugs at the same time to treat an infection [34] because the combined drugs can better defeat bacterial resistance than when the drugs are used individually [34]. The combination could work either by:

(i). Targeting diverse pathways as it has been "for the combination of isoniazid, rifampicin, ethambutol, and pyrazinamide for use in the treatment of tuberculosis" [35,36]. This combination comprises two or more drugs that individually target various facets of the infection, as many ailments have been discovered to be multicomponent or multifactorial. Also, drugs A and B's mechanisms are very much liable to be impartial in their exact activities. Nevertheless, they could have interactive impacts on patient status's general upturn [37].

(ii). Inhibiting different targets through a single pathway. The mechanism involves several drug combinations that aim at only one "disease component and usually a single cell type or even a single response pathway in the cell type, but with various targeted sites" [36]. This precise targeting reduced drug doses and toxicities, thus allowing more noteworthy results. For example, the drug cotrimoxazole (sulfamethoxazole + trimethoprim) utilizes two medicines functioning at various stages in a single pathway to achieve better inhibition than using either of those drugs alone [37]. "It is a combination of sulfamethoxazole, inhibits folic acid production by suppressing dihydropteroate synthetase, and trimethoprim, which functions at a later step in nucleotide production to subdue dihydrofolate reductase" [37].

(iii). Inhibiting the same target in different ways, for example, in cases regarding streptogramins and virginiamycin [35,36]. "The mode of action involves one drug that is useful on its own but is not sufficiently efficient or too toxic and a second drug that does not share the same activity as the first drug and may have no beneficial effect on its own, but that can improve the efficacy of the first drug by either pharmacokinetic/pharmacodynamic mechanism" [37].

1.1.2. The Pros and Cons of Antibiotic Combination Therapy

Antibiotic combination therapy is an extremely selective procedure, although it often comprises a beta-lactam with an aminoglycoside [34]. Whether or not this establishes a positive effect for infected patients has become controversial because antibiotic combination therapy has been discovered to work in some instances. Nevertheless, the pros and cons of antibiotic–antibiotic combination therapy can help choose the best circumstances for its application [34].

The Pros (Benefits) of Combination Antibiotic Therapy

(a) Synergy in action: The first and most important justification for the application of combined antibiotic treatment is to generate synergistic drug interactions [37]. Synergistic combinations of antibiotics can exterminate MDR maladies much more efficiently than antibiotics taken independently [37]. For example, the tigecycline and carbapenem combination has been used to efficiently treat carbapenemase-containing *Klebsiella pneumonia*-infected patients [34]. Moreso, the combination of ceftazidime with tobramycin has been demonstrated to be a potent therapy for patients with cystic fibrosis [35,38].

(b) Polymicrobial infections: Combination therapy effectively treats polymicrobial infections—diseases that comprise multiple bacterial pathogens that are usually detected in the pelvic region, intra-abdominal and urogenital tract A diseases. Combinations of ciprofloxacin with metronidazole drugs are the basis for this treatment [39].

(c) Improved uptake and sequential blockage: In the improvement of the uptake and inhibition of consecutive steps, combining treatments also helps. Beta-lactam combined with aminoglycoside drugs results in antibacterial synergy with increased uptake. This behavior is facilitated by way of beta-lactam bringing about cell membrane impairment, which enables the aminoglycoside movement into the bacteria cells, thus improving the bactericidal activity [18].

(d) Reduced toxic effects and diminished death rate: Coherent drug combination therapy reduces the concentration necessary for therapeutics and minimizes the dose associated with toxicity and reduced mortality rate. Though, there is no information "from clinical trials that prove beyond a reasonable doubt that combination therapy with different agents allows for a reduction of the drug dose sufficient to decrease dose-related toxicity" [40].

(e) Prevention of Drug Resistance: Different molecular targets of individual agents permit the investigation of the use of drug combination therapy and thus, widen the activity spectrum. With their broad ranges of action and multimodal activity, antimicrobial substances may impede the emergence of drug resistance [18].

(f) Empirical Treatment: In a situation where the nature of the infectious disease is not evident, i.e., undiagnosed infections, empiric antibiotic combinations help begin the treatment. Utilizing empiric antibiotic therapy with an agent to which bacteria is sensitive has been linked to a decrease in death rate and improvement in the results [41,42].

Adverse Effects of Combination Antibiotic Therapy

Although antibiotics combination therapy has a lot of advantages, there are some significant drawbacks, which are as follows:

(i). Antagonism: In contrast to synergistic drug interactions, some combinations might show antagonism, where one drug can either invalidate or diminish the effect of the other on organisms [38]. For example, the initiation of Beta-lactamase by one agent leaves the second agent unsuccessful in treating *Enterobacter*, *Serratia*, or *Pseudomonas* with combination therapy. Antagonism may make bactericidal agent bacteriostatic [18].

(ii). *Clostridium difficile* infection (CDI): This is another known unfavorable effect of antibiotic use. Any wide-spectrum antibiotic (aminoglycosides and beta-lactams) can cause an overgrowth of *C. difficile*. "Among these, fluoroquinolones were documented being an independent risk factor for CDI" [43,44].

(iii). Other effects of combination therapy include: drug toxicity such as nephrotoxicity and ototoxicity, fungal overgrowth, drug interactions, irrational drug use, and a rise in the cost of treatment [42].

2. Drug Interactions

The drug-to-drug interactions may be pharmacokinetic or pharmacodynamic. Pharmacokinetic interactions comprise drug absorption, metabolism, distribution, and excretion, whereas pharmacodynamic interactivities describe the relationship between drug concentration on the site of action and the effect on the body. It is essential to think about various antibiotics' reactions to discover the most effective drug combinations. The result of this combination can be "synergism ($C > a + b$), antagonism ($C < a + b$), additivity ($C = a + b$), autonomy ($C \sim a$ or $C \sim b$), depending on whether or not the combined effect of the drugs is more than, equal to, or smaller than, the result predicted by their activities" [43–46]; where 'a' is drug 1, 'b' is drug 2, and 'C' is the combination of drugs 1 & 2 [36,47]. The assessment of action can be performed in-silico by the Kirby–Bauer disc diffusion method, agar-well diffusion, or checkerboard assay [36]. Synergistic combinations are more efficient, while antagonistic combinations are less effective, and additivity drug combinations are as efficient at inhibiting microbial growth as each antibiotic's sum when used independently [48]. To quantify the interaction between the antibiotics being tested using standard checkerboard assay, "the fractional inhibitory concentration (FIC) index" value is estimated as shown below.

"The fractional inhibitory concentration (FIC) is the Minimum Inhibitory concentration (MIC) of drug A in the company of B divided by the MIC of A".

In contrast, the FIC index is the sum of the FIC of drugs A and B. Often, time-kill assays are used as a follow-up to confirm synergism. Synergistic combinations ought to decrease the colony-forming unit concentration with a factor of at least 2 $\log_{10}$ per milliliter [49].

FIC index = MIC of antibiotic in combination/MIC of antibiotics alone.

FIC of the two drugs = FIC index of drug A + FIC index of drug B.

An FIC index of <0.5 indicates synergism, >0.5–1 indicates additive effects, >1 to <2 indicates indifference, and ≥ 2 is considered to be antagonism [49,50].

2.1. Combination of Beta-Lactams with Aminoglycosides and Fluoroquinolones

Antimicrobial synergy has conventionally been observed with combinations of beta-lactams and aminoglycosides for diseases of Gram-negative microorganism origin. The combinations of a beta-lactam and an aminoglycoside enables various mechanisms of action of destroying microorganisms [18,48,51–53]. Beta-lactam facilitated the disruption of Gram-negative bacilli's cell walls and enabled aminoglycoside movement into the periplasmic space [54,55]. Though antibiotic synergy seems to be best proven for the combination of beta-lactam and aminoglycoside, similar information on synergistic effect has become apparent for "combinations of beta-lactams and aminoglycosides" [18,56–58].

2.1.1. Combination Therapy of Beta-Lactams with Lipopeptide and Glycopeptide

Combination therapy with daptomycin and beta-lactams has also been studied in vitro; comparable to in vitro vancomycin studies, many combination-therapy studies have shown synergy, particularly among those with some degree of daptomycin resistance [59–61]. Gritsenko et al. [62] published a case series demonstrating clearance of refractory MRSA bacteremia in five patients treated with vancomycin and ceftaroline. Likewise, following microbiological failure or relapse with more conventional regimens (vancomycin monotherapy, daptomycin monotherapy, and daptomycin plus gentamicin), Dhand and Sakoulas, [63] reported a 2011 case series that evaluated seven patients treated with daptomycin and nafcillin or oxacillin. All patients were bacteremic for at least 5 days before receiving daptomycin in combination with an anti-staphylococcal beta-lactam, and all patients had documented bacterial clearance within 24–48 h of receiving daptomycin in combination with an anti-staphylococcal beta-lactam [63].

2.1.2. Beta-Lactam/Beta-Lactamase Inhibitor Combinations

Several unique beta-lactam/beta-lactamase inhibitor combinations have recently been developed, which include ceftolozane/tazobactam, ceftazidime/avibactam, meropenem/vaborbactam, imipenem-cilastatin/relebactam, aztreonam/avibactam, cefepime/tazobactam, and ceftaroline/avibactam. The Food and Drug Administration has approved ceftolozane/tazobactam, ceftazidime/avibactam, meropenem/vaborbactam, and imipenem-cilastatin/relebactam for clinical use [64–69]. These new beta-lactam/beta-lactamase inhibitor combinations are effective against a wide range of bacteria, including the most common Gram-negative bacteria that cause complicated urinary tract infections and acute pyelonephritis. Furthermore, they have strong in vitro activity against a wide range of multidrug-resistant organisms [70–73].

2.1.3. Clinical Cases of Combination Therapy

In medicine, combination therapy is frequently used when monotherapy does not produce an adequate therapeutic response. Patients at Seoul St. Mary's Hospital who were under the age of 19 and had been diagnosed with *Pseudomonas aeruginosa* while having hematologic or oncologic comorbidities were studied in the retrospective observational study that was published by Kim et al. [74]. It was discovered that 36 different cases of P. aeruginosa infection, affecting 31 of the patients, had already occurred. The cohort's *P. aeruginosa* infections were found to be moderately resistant to piperacillin-tazobactam and cefepime, with only 67.6% and 88.9% susceptibility, compared to these isolates being highly susceptible to amikacin, colistin, and ciprofloxacin (100%, 100%, and 97.2%, respectively). The cohort's mean age was 9.5 ± 5.4. The "patients who did receive a combination therapy were treated with either piperacillin–tazobactam plus an aminoglycoside (16 (44.4%)), cefepime with an aminoglycoside (2 (5.6%)), or meropenem with an aminoglycoside (1 (2.8%)). The monotherapy treatments consisted of either meropenem (14 (38.9%)) or cefepime (3 (7.4%)". Overall, it was discovered that monotherapies had a mortality rate of 17 (or 58.8%), whereas combination therapies had a mortality rate of 4 (or 21%). Cefepime or cefepime plus aminoglycoside were the most effective therapies, both of which had a 0% fatality rate.

Likewise, according to Ceravolo et al. [75] reports, a seven-year-old Caucasian child with neuronopathic Gaucher disease who was homozygous for L444P mutations was treated with a combination of substrate reduction therapy and enzyme replacement therapy. He had been receiving enzyme replacement therapy since he was 18 months old, and concurrent miglustat treatment started when he was 30 months old. Dosing was increased over the course of a month in accordance with his body surface area. After starting his miglustat therapy, he suffered mild diarrhea, which become progressively less frequent and severe. His hematological parameters and plasma angiotensin-converting enzyme activity returned to normal, and his splenomegaly was reduced. Additionally, there was a significant and consistent drop in plasma chitotriosidase. The patient displayed no symptoms of neurological impairment after receiving combination therapy for five years.

2.2. *Nano-Materials Combinations with Antibacterial Drugs*

Adjuvants are molecules that, when combined with drugs, make an ineffectual medicine efficient. They have hardly any antibacterial properties of their own [76], but extend the lifetime of antibiotics, and impede resistance mechanisms [77,78]. However, they habitually produce some harmful results due to drug-drug interactivities [36]. Nanomaterials are materials having a minimum of one of their dimensions not exceeding 100 nm, and they are the key constituents of nanotechnology [36,79]. Nanoparticles (NPs) act like magic bullets with the right concentration to target the delivery of the drug at the proper place and at a suitable time [80]. This is beneficial over conventional systems as they can surmount drug resistance given their multi-performance nature, as microorganisms will not be able to utilize multiple gene mutations at the same time [36].

Antibacterial agents can be combined with NPs to overcome antibiotic resistance and enhance their efficacy. They can reduce the dose and toxicity of antibiotics to be taken [81]. And since NPs act on bacteria via multiple targets and/or mechanisms, it is extremely difficult for bacteria to develop resistance. In other words, the likelihood of simultaneous mutations required for resistance formation is extremely low. This is especially unlikely when NPs are combined with antimicrobials [82]. As a result, the use of NPs in combination with antibiotics is regarded as a method for preventing the development of bacterial resistance [82]. Antibiotics used in conjunction with NPs are more effective against Gram-positive and Gram-negative bacteria, as well as drug-resistant bacteria. Aabed and Mohammed [83] demonstrated synergistic effects of AgNPs in combination with bacitracin, ciprofloxacin, tetracycline, and cefixime against *P. aeruginosa*, *E. coli*, *S. aureus*, and *Candida albicans*. In another study, ZnO NPs combined with Vancomycin, and ampicillin had a synergistic effect against MDR *Enterococcus feacium* [84].

3. Effect of Nanomaterials on Bacteria

Some distinguishing characteristics of nanomaterials that make them a potential preference for antibiotics are discussed below.

- They can simply infiltrate the cell membrane of bacteria and potentially harm its composition, which eventually causes cell lysis [85].
- The modality of action of nanomaterials' antimicrobial effect is similar to the mechanism of antibiotics, such as cell membrane disruption, "reactive oxygen species (ROS)-facilitated oxidative stress, intracellular protein production inhibition, and leaking of intracellular components" [86].
- Several nanoparticles can serve as carriers for antibiotic drugs to distribute them efficiently to their action sites by reducing the drugs' likely negative results [33].
- The retaining power of NPs in the body far exceeds that of antibiotics, and this could be beneficial for lasting therapeutic results [87].
- Nanomaterials can be highly functional with respect to their target and their goal because they may be efficient against the cells of bacteria without actually being noxious to animal cells [33].

3.1. *Organic versus Inorganic Nanoparticles*

Nanoparticles (NPs) may be classified as organic or inorganic based on the desirable action. Organic NPs consist of "liposomes, polymeric NPs, polymeric micelles, and solid lipid NPs (SLNs)" utilized in main treatments. Their main benefits comprise biodegradability, low systemic toxicity, compatibility, and handling of either hydrophilic or hydrophobic drugs. Nevertheless, organic NPs have some limitations, which include low encapsulation efficiency, short shelf life, poor stability at elevated temperatures, and a lack of tolerance for severe processing environments [88]. Inorganic NPs harbor exceptional physicochemical attributes due to their high surface volume, which makes them an extremely promising therapy as an antibacterial by overcoming the detriments of antibiotics and bulk metals [89]. For the synthesis of metal-based NPs, transition metals are likely to be the most suitable because they have partially filled d-orbitals that confer better redox activity on them, a characteristic that enables nanoparticle aggregation [90].

3.2. Types of Inorganic Nanoparticles

The most tested metallic NPs are copper, silver, gold, titanium, aluminum, zinc, and iron [91,92]. Compared to antibiotics, the different antibacterial mechanisms of NPs depend on their unusual crystal morphologies (edges and corners), size, large surface, and reactive sites [93]. They are reported to show wide-spectrum antibacterial properties against mycobacteria, Gram-positive and -negative organisms, and fungi. Nevertheless, their antibacterial properties differ amid the various kinds of nanoparticles as along with the diverse organisms [94,95].

3.3. Silver Nanoparticles

Silver nanoparticles (AgNPs) have been reported as the most investigated nanomaterials for antimicrobial activity because of their broad range of actions against several microbes [96–98]. Silver compounds such as silver nitrate, metallic silver, and silver sulfadiazine are used for various medical uses including dental work, burns, disinfection of medical equipment wound treatment, and controlling bacterial contamination, among others [99]. Moreover, due to their multi-dimensional strategies to convey antimicrobial action, the possibility of resistance of bacteria to AgNPs is low [100].

Several researchers have proposed various antimicrobial actions for AgNPs, which include impairment of the external membrane of bacteria [101], interface with enzymes, and disintegration of the cellular elements [64], as shown in Figure 1 [98]. It was also reported that the antibacterial activity of AgNPs depends on their size and shape. In this respect, Lu et al. [102] affirmed that the antibacterial property and particle size of AgNPs do not correlate directly but inversely. The three distinct types of AgNPs with good antibacterial activity are spherical, rod-shaped, and truncated triangular AgNPs (Figure 2) [84,103]. The properties of these nanoparticles, which influence their activity towards bacterial pathogens, are presented in Table 1.

Table 1. Properties of nanoparticles that influence their activity toward bacterial pathogens.

Property	Functions	References
Particle Size	"Smaller Ag-NPs have larger specific surface areas, which result in higher antimicrobial activity". Because there is a higher chance of being in contact with, and transient on, the cellular membrane of the organism than with bigger nanoparticles.	[64,104,105]
Particle Shape	Ag-NPs with various forms can result in various microbial cell injury degrees through interacting with periplasmic enzymes. For example, cube-shaped silver nanoparticles demonstrate more potent antimicrobial effect than sphere-shaped and wire-shaped silver nanoparticles with comparable diameters, implying that the shape impact on antimicrobial activities is because of the surface area and facet reactivity.	[105–107]
Roughness	"As the roughness of Ag-NPs increases, the size and the surface area-to-mass ratio promotes the adsorption of bacterial proteins, which is followed by a reduction in bacterial adhesion".	[64]

Table 1. *Cont.*

Property	Functions	References
Environmental Condition	Temperature and pH cause considerable differences in antimicrobial effect. For example, the temperature of the environment has a strong impact on antimicrobial activity due to its impact on ROS generation rate. Likewise, "at low pH, the surfaces of the NPs were positively charged, which is beneficial to the interaction with the negatively charged groups of the bacterial cell barrier, inducing strong multivalent electrostatic regulation".	[64,108]
Doping Modification	Prevents the assemblage of gaps and permits their dispersion in aqueous conditions or other hydrophilic media.	[109]
Zeta Potential	AgNPs with a positive surface charge are susceptible to the fact that they are absorbed on bacterial surfaces, in contrast to their counterparts with negative charge.	[110]

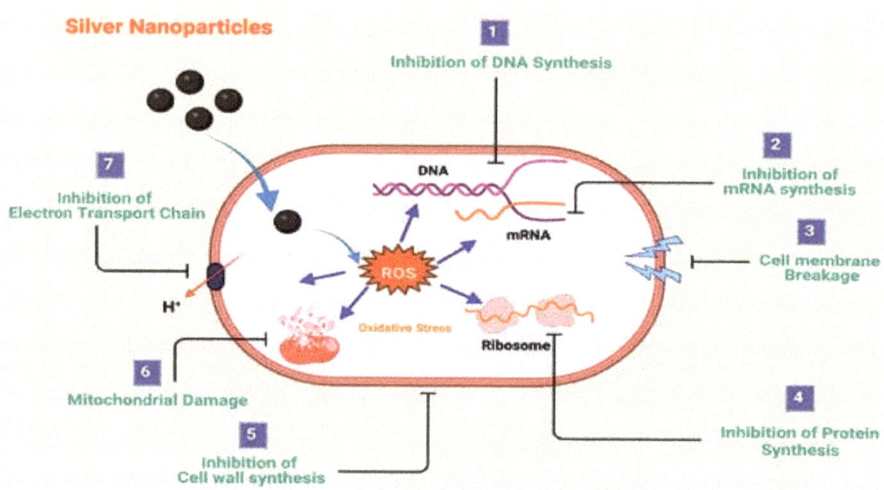

Figure 1. Antibacterial mechanism of silver nanoparticles. Source: Jain et al. [98] with permission to reuse the figure under a CC BY open access license.

Figure 2. Scanning electron micrograph of silver nanoparticles (Adeniji et al. [84]).

3.4. Zinc Oxide Nanoparticles (ZnO NPs)

NPs of ZnO origin are well known to efficiently hamper the development of a broad range of susceptible and non-susceptible microorganisms, rising as optimistic candidates to combat antibacterial resistance [111,112]. Iron oxide (Fe_3O_4), copper oxide, and zinc oxide (ZnO) have antibacterial effects that enable their application in medical care [113]. "Regarding the intrinsic photocatalytic property of metal oxides, they generate ROS and become potent agents against bacteria [114,115]. Gelabert et al. [116] and Nagvenkar et al. [117] reported that the antibacterial mechanism of the NPs is linked partially to the dissolution of metal ions and the formation of ROS" (Figure 3). "ZnO releases Zn^{2+} in a liquid medium and is adsorbed at the surface of bacteria, where it interacts with functional groups in proteins and nucleic acids, obstructing enzyme activity and the normal physiological processes [118]. However, some authors demonstrated that Zn ions have little antimicrobial activity, implying that dissolution of Zn^{2+} might not be the main mechanism of action" [119,120].

Zinc oxide NPs can be produced through chemical and physical processes, such as chemical vapor deposition, solvothermal, sol-gel hydrothermal, laser exposer, and spray pyrolysis [121,122]. Zinc oxide has been found useful in photocatalysis, light-emitting diodes, UV filtration, solar cells, piezo-electric transducers, memory devices, and photodetectors [121,123]. ZnO is also used "in the structure of electrochemical sensors and biosensors" [124,125], the food industry [126], wastewater treatment [127], sunscreens [128], composites [129], dental cement, drug delivery system, and in cancer treatment [130]. Thus, Zinc oxide NPs might be utilized as prospective antimicrobials against several pathogenic organisms [131].

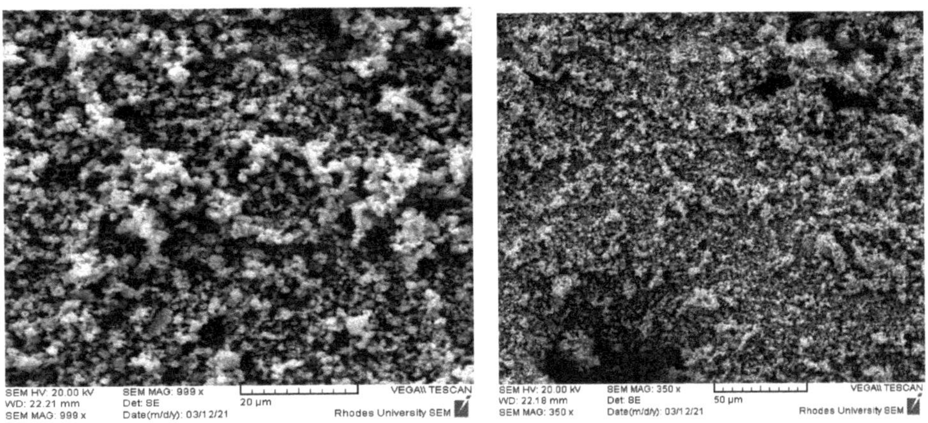

Figure 3. Scanning electron micrograph of Zinc oxide nanoparticles (Adeniji et al. [84]).

3.5. Other Nanoparticles

Several other NPs have been documented, including Si, SiO_2 [132], MgO [133], CaO [134], Al_2O_3 [135], and bismuth [136]. Yamamoto et al. [137], "reported that the generation of superoxide on their surface was the major reaction mechanism of antibacterial action by CaO and MgO". It was also reported that the "antibacterial mechanism of Al_2O_3 is dependent on the interaction between NPs and bacterial cell membranes" [138]. Metal-derived NPs and their application are shown in Figure 4 [139].

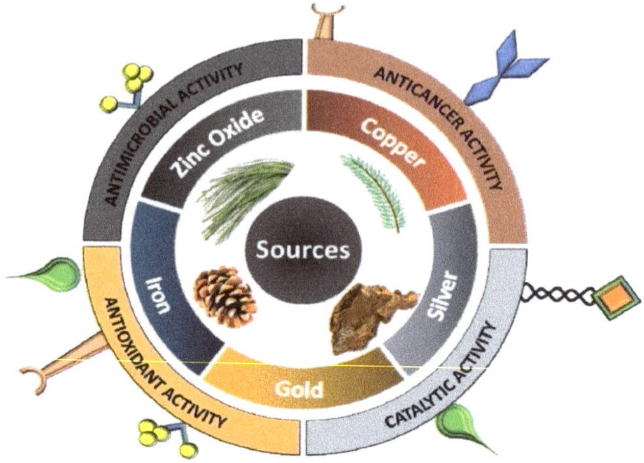

Figure 4. Metal-derived NPs and their application. Source: Bhardwaj et al. [139] with permission to reuse the figure under a CC BY open access license.

4. Mode of Action of Nanomaterials

NPs have an antimicrobial property that can subjugate common resistant mechanisms such as reduced "cell permeability, alteration of target sites, enzyme inactivation, and increased efflux through overexpression of efflux pumps to scarper from the antibacterial effect of antibiotics" [139]. The antimicrobial action of nanoparticles against MDR organisms and biofilms is based on several features: their high surface area in connection with bacteria via electrostatic attraction, hydrophobic interactions, or van der Waals forces on the NPs size and stability; and how concentrated the drug is [140,141]. The interplay of nanoparticles with bacteria usually activates oxidative stress actions, inhibition of the enzyme, deactivation of protein, and changes in gene expression Typical mechanisms of antibacterial action are associated with metal ion release, oxidative stress, and non-oxidative actions, as shown in Figure 5 [85,142].

Due to electrostatic interactions, there is an attraction between the bacteria cell walls, which have negative charges, and the surfaces of the NPs, which are positively charged. Alternately, metal-based NPs with positive charges build a powerful bond with membranes following the intrusion of cell walls and, subsequently, enhance their permeability [143]. Moreso, NPs are also able to discharge metal ions from the extracellular space, which are proficient in penetrating the cell and disturbing genetic activities [143]. Metal ions of NPs inside the cell can induce the making of ROS. "Oxidative stress instigated by ROS is among the utmost critical mechanisms aiding the antimicrobial effect of nanoparticles [144,145]. ROS are raw by-products of cellular oxidative metabolism with substantial essential functions in the modulation of cell survival and death, cell signaling, and differentiation" [146] The oxidative stress produced leads to glutathione oxidation, hence defeating the antioxidant protective mechanism of the microorganisms against ROS. The metal ions could then move freely within the cellular structure disrupting cell functions [143].

Membrane infiltration can also be accomplished by "interactions with surface lipids [147]. The curative application of NPs is improved by their capability to bestow physical protection on the mechanisms of microbial resistance [148]. Ansari et al. [138] documented that the buildup of NPs in the bacterial cell wall results in an irregularly shaped pit, and perforation and interferes with metabolic activities. Research carried out by Joost et al. [149] reported that therapy with TiO_2 NPs amplified the bacterial cell volume, bringing about membrane leakage. Furthermore, NPs bonded with antibiotics exhibit co-adjuvant "effects against bacteria, interdict biofilm formation, and are employed to fight MDROs [87,120]. "While inadequate membrane transport confines the efficacy of several drugs [150], drug-loaded NPs' vehicles can move into host cells using endocytosis, enabling their intracellular entry" [85].

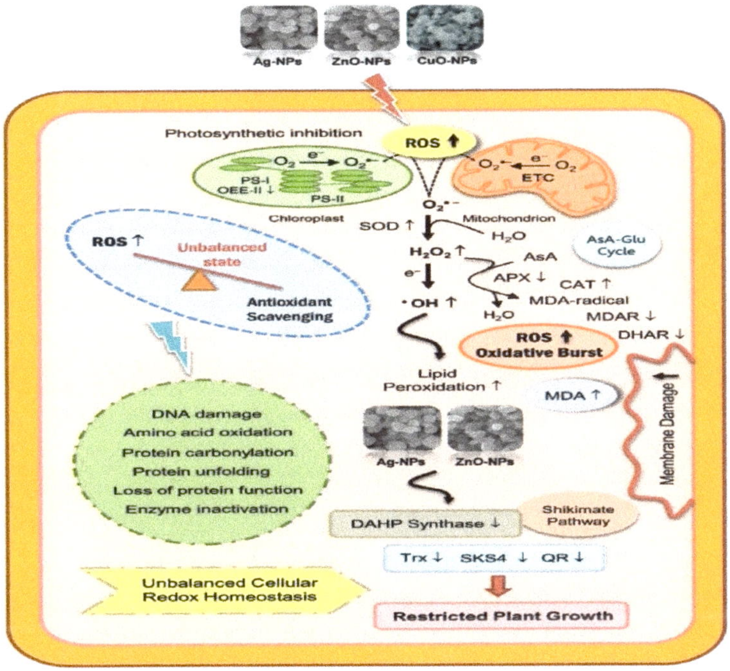

Figure 5. Mode of action of nanoparticles. Adapted from Hossain et al. [131] with permission to reuse the figure under a CC BY open access license.

4.1. *Advantages of Nanomaterials in Combating MDR Pathogens*

Nanomaterials might be efficaciously modified to have an antibacterial property with no noxious side effects based on their unique physicochemical properties [87]. In the same vein, they can be distributed in appropriate and affordable means with reduced administration frequency by numerous paths [151]. "Synergistic antibacterial activity, improved solubility, and suspension of drugs are additional advantages of nanomaterials. Besides their excellent antibacterial properties, nanomaterials can be used as carriers for the delivery of antimicrobial moieties to regions of poor absorption in the body" [33]. The benefits of nanomaterials as antibacterial drug delivery vehicles are listed below.

- As a result of the inadequate membrane transport process of some drugs, their impact on intracellular disease-causing organisms is constricted. Meanwhile, the controllable size of the NPs aids in designing targeted antibiotics [143,152].
- The time of drug retention in the blood can be enhanced as a result of using NPs as an antimicrobial drug delivery vehicle [153].
- The solubility of nanomaterials in the bloodstream is allowed by the surface chemistry of NPs [33].
- Opsonization is an additional biological impediment where the physicochemical properties of nanomaterials have been well applied for the efficient delivery of antibacterial drugs to the site of action. It also allows a high measure at the site of infection site and therefore reduces the harmful effects [154,155]. For example, vancomycin is an effective Gram-positive bacteria drug but can be harmful to the kidney and ear [85]. In that regard, "Qi et al. [156] documented that the vancomycin-modified mesoporous

silica NPs can be aimed at a particular Gram-positive disease-causing microorganism and selectively exterminate them over macrophage-like cells" [130].

- Antibiotics can be protected from damage to the chemical reaction and resistance against targeted bacteria by nanomaterials. It has been proven by many researchers that several NPs can "overcome the traditional efflux mechanism of bacteria cells that often obstruct the uptake of antibiotics by the cells" [157]. For example, "Liu et al. [158] reported that the dendrimers could impede P-glycoprotein-mediated efflux in the gastrointestinal tract".

4.2. *Antibacterial Effect of Nanomaterials-Antibiotics Combination*

Functionalization of NPs with antibiotics has the potential to fight against resistant bacteria because NPs can take antimicrobial agents to the sites of infection and lessen the toxicity and dosage of drugs [155]. Combining nanomaterials with antibiotics has been widely reported as being potent in combating bacterial resistance [127]. Panacek et al. [159] and Scandorieiro et al. [160] documented the synergistic antimicrobial efficacy of AgNPs and antibiotics against *Staphylococcus aureus*, beta-lactamase- or carbapenemase-producing *Escherichia coli*, *Pseudomonas. Aeruginosa*, and *Acinetobacter baumannii* strains at very small concentrations via the infiltration of the bacterial cell membrane and the intrusion with essential molecular pathways, creating distinctive antimicrobial mechanisms [161].

To ascertain the interactive impact of antibiotics and nanoparticles, the fold increment in the diameter of the inhibition zone of each antibiotic after combination with nanoparticles is determined according to Sindhu et al. [162].

$$\textit{The fold increase} = (b^2 - a^2)/a^2$$

"where; (a) is the inhibition zone of antibiotic alone and (b) is the inhibition zone of antibiotic plus nanoparticles".

The nanomaterial–antibiotics blend has efficaciously mired the growth of MDR microorganisms according to various reports. Silver nanoparticles with ciprofloxacin [161], vancomycin [163], and clotrimazole [164] have effectively prevented the growth of Vancomycin-resistant *Enterococci* (VRE) and Methicillin-resistant *Staphylococcus aureus* (MRSA) species. Gold NPs with vancomycin [165] or ampicillin [166], and ZnO NPs along with ciprofloxacin [167], similarly displayed antibacterial effects against MRSA and MDR *A. baumannii*, respectively. The efficiency of drugs combined with nanoparticles was the same in Gram-positive and -negative microorganisms, in contrast to the complexity of exterminating MDROs with antibiotics only [155]. Thus, further studies are needed on the blend of nanomaterials with antibiotics, as this could offer an unimaginable breakthrough in the treatment of diseases caused by pathogenic MDR organisms.

5. Cytotoxicity of Nanomaterials

The concerns over the use of NPs are local and systemic noxious problems, plus harmful effects on helpful bacteria in humans [87,168]. Kandi and Kandi [169] reported that nanoparticles and their toxic disintegration products could bring about hemolysis and interfere with blood clotting pathways. Meanwhile, the mechanism of toxic problems remains unclear, but noticeably, large-sized nanoparticles have a higher tendency of endangering human health [170].

AgNPs are reported as the most powerful nano weapon in the fight against bacterial diseases [103,171]. However, resistance to AgNPs is now increasingly reported because of the genetic modification of bacteria [172]. The deposition of silver nanoparticles in the liver, lungs, spleen and other organs is responsible for the potential harm and dysfunction of such body organs and the extreme reduction of their effectiveness [155]. The oxidative harm of CuO NPs and damage to DNA caused by zinc oxide nanoparticles or TiO_2 NPs have limited their usage [155]. The buildup of metallic NPs in the tissues can lead to long-term toxicity, like hepatotoxicity, or nephrotoxicity, among others [173–175]. Nevertheless, some in vivo investigations have indicated there are no obvious life-threatening poisonous effects connected to nanoparticles [174–176]. Reports about the biocompatibility of various nanomaterials in cases wherein toxicity depends on the size, concentration, and time of treatment are numerous [177]. For example, Naskar et al. [178] stated that the Ag–ZnO–graphene nanocomposite is not lethal at low amounts but poisonous when the levels are higher. Nevertheless, nanoparticles have become known as the alternative antibacterial strategy for the fight against biofilms and the treatment of serious bacterial infectious diseases [179].

Approaches to Addressing Nanomaterials Toxicity Dilemma

Nanoparticles have long been regarded as a potential solution to the rising resistance to common antibiotics and the emergence of multidrug-resistant bacteria. Despite the enormous potency of these nanoparticles, reports on bacterial resistance to nanoparticles are starting to emerge. Its prevalent clinical application raises the prospect of resistance to these potential biomolecules [180,181]. Some studies have shown that AgNPs have cytotoxic effects in various cell lines based on size, shape, concentration, or capping agent [182–184]. As a result, combinations with other antimicrobial treatments have been suggested to enhance antimicrobial properties while decreasing AgNP cytotoxicity. Many devices have already been employed to combat nanomaterial toxicity. The best tactic is to cap the nanoparticle with a biocompatible polymer such as *polyethylene glycol* (PEG) or chitosan. PEG is a synthetic polymer that is well-suited for biomedical applications such as bioconjugation, drug delivery, and bio-sensing [185]. Reports have shown that PEG capping NPs reduces the noxiousness of NPs and enhances biological compatibility [177,186]. Likewise, these polymers have their own antibacterial properties. Hence, the capping of nanoparticles with *polyethylene glycol* or chitosan not only improves their biological compatibility but the capping materials also exterminate bacterial cells collectively with nanoparticles [186,187] Abdalla et al. [188], for example, tested polyvinyl alcohol (PV) and chitosan (C) as capping agents in AgNPs against clinical isolates of *Staphylococcus epidermis*, *Staphylococcus aureus*, *Klebsiella. pneumoniae*, and *Escherichia coli*. Surprisingly, this combination had a beneficial antibacterial effect and inhibited biofilm formation in all isolates tested. Chia et al. [189] also described the use of a silica coating to reduce the toxicity of ZnO NPs. Doping is also an efficient approach to lessen the perniciousness of nanomaterials. In this regard, Xia et al. [190] stated that Fe-doped ZnO lessened toxicity in rodent lungs and zebrafish embryos by reducing the dissolution of zinc oxide nanoparticles. In addition, Limayem et al. [191] investigated the antibacterial activity of chitosan NPs, ZnO alone, and a combination of chitosan and ZnO against MDR and wild-type strains. On MDR *E. coli* ZnO combined with chitosan demonstrated a synergistic effect as well as with MDR *Enterococcus faecium*. Later, Mehta et al. [192] tested biofilm activity against MDR *Enterococcus faecium* in a lipid micelle with the previously described composite (ZnO-chitosan composite). Surprisingly, the results showed a 50% reduction in bacterial biofilm size when compared to chitosan and ZnO alone. Still, it should be stated that a well-organized review is required before any clinical application of nanomaterials as drugs for antimicrobial activity.

6. Limitations and Strengths of the Application of Nanoparticles

Although NPs have the potential to treat infectious diseases, several challenges remain for their clinically effective translation, including further evaluation of their interactions with cells, tissues, and organs; optimal dose; identification of appropriate administration routes; and toxicity upon acute and long-term exposure [120,193]. It is critical to consider nanoparticle dosage. Drugs that are beneficial at low doses (concentration) may be toxic at high doses. Most studies report varying concentrations, and the number of cells exposed is rarely reported. Because nanoparticles have such promising potential, one important goal of the nanomaterials research community is to synthesize nanoparticles or nanoparticles that can conjugate very effectively at low doses (concentration). Non-toxic biological materials that can increase the potency of nanoparticles without increasing the concentration that may be toxic to biological systems should be studied. The combination of different nanoparticles can also aid in dosage reduction. Individual nanoparticles are also less effective than nanocomposites. As a result, more emphasis should be placed on their formulation [194]. Furthermore, creating nanoparticles that can bind to proteins, polysaccharides, or small bioactive compounds could be important in enhancing their antibacterial properties. The combination of nanoparticles and antibiotics significantly reduces the number of antibiotics that must be administered. This helps to reduce the toxicity of several antibiotics as well as the acquisition of resistance. Combination therapy will set the stage for nanoparticles to be used as adjuncts to the existing antibiotics, assisting in the reduction of resistance associated with the majority of bacteria.

7. Conclusions and Recommendations

MDR pathogens are becoming an emergent public health crisis, making many healthcare-related diseases demanding to treat with existing antibiotics. The use of nanomaterials provides a possible approach to controlling diseases caused by these pathogens. Nanoparticles demonstrating antibacterial properties can target numerous biomolecules and potentially lessen or eradicate the evolution of MDR. At the same time, the transformation of NPs to medical use necessitates suitable techniques for the preparation of NPs and comprehensive knowledge of the physicochemical distinctive features, in silico and in vivo effects, biodistribution pharmacokinetics, and pharmacodynamics of nanoparticles. Precise blends of NPs and antibiotics can help inhibit the occurrence of resistance or drive resistant bacteria back toward drug sensitivity. Furthermore, the clinical data on the nanomaterial-based antibacterial property are scanty, so there is a need for more in-depth studies, including in vivo findings towards a successful translation of nanomaterials to medical applications in attacking MDR organisms.

Author Contributions: O.O.A.: Conception and design of the study, manuscript preparation. N.N.: manuscript review, curation and proofreading. J.C.O.: manuscript review, curation and proofreading. A.I.O.: Conception and design of the study, manuscript review, proofreading and grant owner. All authors have read and agreed to the published version of the manuscript.

Funding: The authors appreciate the South African Medical Research Council for funding this research.

Institutional Review Board Statement: Not applicable.

Informed Consent Statement: Not applicable.

Data Availability Statement: Not applicable.

Conflicts of Interest: The authors declare no conflict of interest.

References

1. Ibrahim, M.E.; Bilal, N.E.; Hamid, M.E. Increased multi-drug resistant *Escherichia coli* from hospitals in Khartoum state, Sudan. *Afr. Health Sci.* **2012**, *12*, 368–375. [CrossRef] [PubMed]
2. Pendleton, J.N.; Gorman, S.P.; Gilmore, B.F. Clinical relevance of the ESKAPE pathogens. *Expert Rev. Anti-Infect. Ther.* **2013**, *11*, 297–308. [CrossRef] [PubMed]
3. Mulani, M.S.; Kamble, E.; Kumkar, S.N.; Tawre, M.S.; Pardesi, K.R. Emerging Strategies to Combat ESKAPE Pathogens in the Era of Antimicrobial Resistance: A Review. *Front. Microbiol.* **2019**, *10*, 539. [CrossRef] [PubMed]
4. Wang, C.-H.; Hsieh, Y.-H.; Powers, Z.M.; Kao, C.-Y. Defeating Antibiotic-Resistant Bacteria: Exploring Alternative Therapies for a Post-Antibiotic Era. *Int. J. Mol. Sci.* **2020**, *21*, 1061. [CrossRef] [PubMed]
5. Magiorakos, A.-P.; Srinivasan, A.; Carey, R.B.; Carmeli, Y.; Falagas, M.E.; Giske, C.G.; Harbarth, S.; Hindler, J.F.; Kahlmeter, G.; Olsson-Liljequist, B.; et al. Multidrug-resistant, extensively drug-resistant and pandrug-resistant bacteria: An international expert proposal for interim standard definitions for acquired resistance. *Clin. Microbiol. Infect.* **2012**, *18*, 268–281. [CrossRef] [PubMed]
6. Bonnet, V.; Dupont, H.; Glorion, S.; Aupée, M.; Kipnis, E.; Gérard, J.; Hanouz, J.; Fischer, M. Influence of bacterial resistance on mortality in intensive care units: A registry study from 2000 to 2013 (IICU Study). *J. Hosp. Infect.* **2019**, *102*, 317–324. [CrossRef]
7. Naylor, N.R.; Pouwels, K.B.; Hope, R.; Green, N.; Henderson, K.L.; Knight, G.M.; Atun, R.; Robotham, J.V.; Deeny, S.R. The health and cost burden of antibiotic-resistant and susceptible *Escherichia coli* bacteraemia in the English hospital setting: A national retro-spective cohort study. *PLoS ONE* **2019**, *14*, e0221944. [CrossRef]

8. Hwang, C.S.; Lee, H.; Hong, M.P.; Kim, J.H.; Kim, K.S. Brain abscess caused by chronic invasive actinomycosis in the nasopharynx: A case report and literature review. *Medicine* **2018**, *97*, e0406. [CrossRef]
9. Sekyere, J.O.; Govinden, U.; Bester, L.; Essack, S. Colistin and tigecycline resistance in carbapenemase-producing Gram-negative bacteria: Emerging resistance mechanisms and detection methods. *J. Appl. Microbiol.* **2016**, *121*, 601–617. [CrossRef]
10. Wang, Q.; Zhang, P.; Zhao, D.; Jiang, Y.; Zhao, F.; Wang, Y.; Li, X.; Du, X.; Yu, Y. Emergence of tigecycline resistance in *Escherichia coli* co-producing MCR-1 and NDM-5 during tigecycline salvage treatment. *Infect. Drug Resist.* **2018**, *11*, 2241–2248. [CrossRef]
11. Spellberg, B.; Miller, L.G.; Kuo, M.N.; Bradley, J.; Scheld, W.M.; Edwards, J.E. Societal Costs Versus Savings from Wild-Card Patent Extension Legislation to Spur Critically Needed Antibiotic Development. *Infection* **2007**, *35*, 167–174. [CrossRef] [PubMed]
12. Bassetti, M.; Ginocchio, F.; Mikulska, M. New Treatment Options against Gram-Negative Organisms. In *Annual Update in Intensive Care and Emergency Medicine*; Springer: Berlin/Heidelberg, Germany, 2011; pp. 501–515.
13. World Health Organization. WHO Priority Pathogens List for R&D of New Antibiotics. 2017. Available online: http://www.who.int/mediacentre/news/releases/2017/bacteria-antibiotics-needed/en/ (accessed on 27 February 2017).
14. Spellberg, B.; Guidos, R.; Gilbert, D.; Bradley, J.; Boucher, H.W.; Scheld, W.M.; Bartlett, J.G.; Edwards, J., Jr.; Infectious Diseases Society of America. The epidemic of antibiotic-resistant infections: A call to action for the medical community from the Infectious Diseases Society of America. *Clin. Infect. Dis.* **2008**, *46*, 155–164. [CrossRef] [PubMed]
15. Worthington, R.J.; Melander, C. Combination approaches to combat multidrug-resistant bacteria. *Trends Biotechnol.* **2013**, *31*, 177–184. [CrossRef]
16. De la Fuente-Salcido, N.M.; López-De la Cruz, D.; Andrade, A.A. Bacterial Resistance Challenged by Binary Antimicrobial Combinations. In *Antimicrobial Research: Novel Bioknowledge and Educational Programs*; Méndez-Vilas, A., Ed.; Bioprospecting and Bioprocessing; Autonomous University of Coahuila: Saltillo, Mexico, 2017.
17. Stoimenov, P.K.; Klinger, R.L.; Marchin, G.L.; Klabunde, K.J. Metal Oxide Nanoparticles as Bactericidal Agents. *Langmuir* **2002**, *18*, 6679–6686. [CrossRef]
18. Tamma, P.D.; Cosgrove, S.E.; Maragakis, L.L. Combination Therapy for Treatment of Infections with Gram-Negative Bacteria. *Clin. Microbiol. Rev.* **2012**, *25*, 450–470. [CrossRef]
19. Maenza, J.; Flexner, C. Combination antiretroviral therapy for HIV infection. *Am. Fam. Physician* **1998**, *57*, 2789–2798.
20. Forrest, G.N.; Tamura, K. Rifampin Combination Therapy for Nonmycobacterial Infections. *Clin. Microbiol. Rev.* **2010**, *23*, 14–34. [CrossRef] [PubMed]
21. Habib, G.; Lancellotti, P.; Antunes, M.J.; Bongiorni, M.G.; Casalta, J.P.; Del Zotti, F.; Dulgheru, R.; El Khoury, G.; Erba, P.A.; Iung, B.; et al. 2015 ESC guidelines for the management of infective endocarditis: The task force for the management of infective endocarditis of the European Society of Cardiology (ESC) endorsed by: European Association for Cardio-Thoracic Surgery (EACTS), the European Association of Nuclear Medicine (EANM). *Eur. Heart J.* **2015**, *36*, 3075–3128.
22. Iseman, M.D. Extensively Drug-Resistant *Mycobacterium tuberculosis*: Charles Darwin Would Understand. *Clin. Infect. Dis.* **2007**, *45*, 1415–1416. [CrossRef]
23. Jawetz, E.; Gunnison, J.B.; Bruff, J.B.; Coleman, V.R. Studies on Antibiotic Synergism and Antagonism: Synergism among Seven Antibiotics against Various Bacteria in vitro. *J. Bacteriol.* **1952**, *64*, 29–39. [CrossRef]
24. Bushby, S.R.M.; Hitchings, G.H. Trimethoprim, a Sulphonamide Potentiator. *Br. J. Pharmacol. Chemother.* **1968**, *33*, 72–90. [CrossRef] [PubMed]
25. Kerantzas, C.A.; Jacobs, W.R., Jr. Origins of Combination Therapy for Tuberculosis: Lessons for Future Antimicrobial development and application. *MBio* **2017**, *8*, e01586-16. [CrossRef] [PubMed]
26. Noordeen, S.K. History of chemotherapy of leprosy. *Clin. Dermatol.* **2016**, *34*, 32–36. [CrossRef] [PubMed]
27. Qadir, M.I.; Chauhdary, Z. Antibacterial Activity of Novel Strains of Bacteriophages: An Experimental Approach. *Crit. Rev. Eukaryot. Gene Expr.* **2018**, *28*, 1–12. [CrossRef] [PubMed]
28. Pelfrene, E.; Mura, M.; Sanches, A.C.; Cavaleri, M. Monoclonal antibodies as anti-infective products: A promising future? *Clin. Microbiol. Infect.* **2019**, *25*, 60–64. [CrossRef] [PubMed]
29. Tyers, M.; Wright, G.D. Drug combinations: A strategy to extend the life of antibiotics in the 21st century. *Nat. Rev. Microbiol.* **2019**, *17*, 141–155. [CrossRef]
30. Xu, J.-W.; Yao, K.; Xu, Z.-K. Nanomaterials with a photothermal effect for antibacterial activities: An overview. *Nanoscale* **2019**, *11*, 8680–8691. [CrossRef]
31. Viegas-Junior, C.; Danuello, A.; da Silva Bolzani, V.; Barreiro, E.J.; Fraga, C.A.M. Molecular Hybridization: A Useful Tool in the Design of New Drug Prototypes. *Curr. Med. Chem.* **2007**, *14*, 1829–1852. [CrossRef]
32. Lipeeva, A.V.; Zakharov, D.O.; Burova, L.G.; Frolova, T.S.; Baev, D.S.; Shirokikh, I.V.; Evstropov, A.N.; Sinitsyna, O.I.; Tolsikova, T.G.; Shults, E.E. Design, Synthesis and Antibacterial Activity of Coumarin-1,2,3-triazole Hybrids Obtained from Natural Furocoumarin Peucedanin. *Molecules* **2019**, *24*, 2126. [CrossRef]
33. Gupta, A.; Mumtaz, S.; Li, C.-H.; Hussain, I.; Rotello, V.M. Combatting antibiotic-resistant bacteria using nanomaterials. *Chem. Soc. Rev.* **2019**, *48*, 415–427. [CrossRef]
34. Ahmed, A.; Azim, A.; Gurjar, M.; Baronia, A.K. Current concepts in combination antibiotic therapy for critically ill patients. *Indian J. Crit. Care Med.* **2014**, *18*, 310–314. [PubMed]
35. Fischbach, M.A. Combination therapies for combating antimicrobial resistance. *Curr. Opin. Microbiol.* **2011**, *14*, 519–523. [CrossRef] [PubMed]

36. Ruddaraju, L.K.; Pammi, S.V.N.; Guntuku, G.S.; Padavala, V.S.; Kolapalli, V.R.M. A review on anti-bacterials to combat resistance: From ancient era of plants and metals to present and future perspectives of green nano technological combinations. *Asian J. Pharm. Sci.* **2020**, *15*, 42–59. [CrossRef] [PubMed]
37. Toews, M.L.; Bylund, D.B. Pharmacologic principles for combination therapy. *Proc. Am. Thorac. Soc.* **2005**, *2*, 282–289. [CrossRef]
38. Rybak, M.J.; McGrath, B.J. Combination Antimicrobial Therapy for Bacterial Infections. *Drugs* **1996**, *52*, 390–405. [CrossRef]
39. Caballero, J.; Rello, J. Combination antibiotic therapy for community-acquired pneumonia. *Ann. Intensive Care* **2011**, *1*, 48. [CrossRef] [PubMed]
40. Weiss, K.; Low, D.; Cortes, L.; Beaupre, A.; Gauthier, R.; Gregoire, P.; Legare, M.; Neoveu, F.; Thibert, D.; Tremblay, C. Clinical Characteristics at Initial Presentation and Impact of Dual Therapy on the Outcome of Bacteremic *Streptococcus pneumoniae* Pneumonia in Adults. *Can. Respir. J.* **2004**, *11*, 589–593. [CrossRef]
41. Kumar, A.; Zarychanski, R.; Light, B.; Parrillo, J.; Maki, D.; Simon, D.; LaPorta, D.; Lapinsky, S.; Ellis, P.; Mirzanejad, Y.; et al. Early combination antibiotic therapy yields improved survival compared with monotherapy in septic shock: A propensity-matched analysis. *Crit. Care Med.* **2010**, *38*, 1773–1785. [CrossRef]
42. Mehta, K.C.; Dargad, R.R.; Borade, D.M.; Swami, O.C. Burden of antibiotic resistance in common infectious diseases: Role of antibiotic combination therapy. *J. Clin. Diagn. Res.* **2014**, *8*, ME05–ME08. [CrossRef] [PubMed]
43. Pépin, J.; Saheb, N.; Coulombe, M.-A.; Alary, M.-E.; Corriveau, M.-P.; Authier, S.; Leblanc, M.; Rivard, G.; Bettez, M.; Primeau, V.; et al. Emergence of Fluoroquinolones as the Predominant Risk Factor for *Clostridium difficile*-Associated Diarrhea: A Cohort Study during an Epidemic in Quebec. *Clin. Infect. Dis.* **2005**, *41*, 1254–1260. [CrossRef]
44. Sandora, T.J.; Fung, M.; Flaherty, K.; Helsing, L.; Scanlon, P.; Potter-Bynoe, G.; Gidengil, C.A.; Lee, G.M. Epidemiology and Risk Factors for *Clostridium difficile* Infection in Children. *Pediatr. Infect. Dis. J.* **2011**, *30*, 580–584. [CrossRef] [PubMed]
45. Keith, C.T.; Borisy, A.A.; Stockwell, B.R. Multicomponent therapeutics for networked systems. *Nat. Rev. Drug Discov.* **2005**, *4*, 71–78. [CrossRef]
46. Loewe, S. The problem of synergism and antagonism of combined drugs. *Arzneimittelforschung* **1953**, *3*, 285–290. [PubMed]
47. Chait, R.; Craney, A.; Kishony, R. Antibiotic interactions that select against resistance. *Nature* **2007**, *446*, 668–671. [CrossRef] [PubMed]
48. Eliopoulos, G.M.; Eliopoulos, C.T. Antibiotic combinations: Should they be tested? *Clin. Microbiol. Rev.* **1988**, *1*, 139–156. [CrossRef] [PubMed]
49. Suárez-Cerda, J.; Espinoza-Gómez, H.; Alonso-Núñez, G.; Rivero, I.A.; Gochi-Ponce, Y.; Flores-López, L.Z. A green synthesis of copper nanoparticles using native cyclodextrins as stabilizing agents. *J. Saudi Chem. Soc.* **2017**, *21*, 341–348. [CrossRef]
50. Faleiro, M.; Miguel, M. Use of Essential Oils and Their Components against Multidrug-Resistant Bacteria. In *Fighting Multidrug Resistance with Herbal Extracts, Essential Oils and Their Components*; Academic Press: Cambridge, MA, USA, 2013; pp. 65–94.
51. American Thoracic Society and Infectious Diseases Society of America (ATS). Guidelines for the management of adults with hospital-acquired, ventilator-associated, and healthcare-associated pneumonia. *Am. J. Respir. Crit. Care Med.* **2005**, *171*, 388–416. [CrossRef]
52. Glew, R.H.; Pavuk, R.A. Early synergistic interaction between semisynthetic penicillins and aminoglycosidic aminocyclitols against Enterobacteriaceae. *Antimicrob. Agents Chemother.* **1983**, *23*, 902–906. [CrossRef]
53. Takahashi, K.; Kanno, H. Synergistic activities of combinations of beta-lactams, fosfomycin, and tobramycin against *Pseudomonas aeruginosa*. *Antimicrob. Agents Chemother.* **1984**, *26*, 789–791. [CrossRef]
54. Hancock, R.E.; Raffle, V.J.; Nicas, T.I. Involvement of the outer membrane in gentamicin and streptomycin uptake and killing in *Pseudomonas aeruginosa*. *Antimicrob. Agents Chemother.* **1981**, *19*, 777–785. [CrossRef]
55. Miller, M.H.; Feinstein, S.A.; Chow, R.T. Early effects of beta-lactams on aminoglycoside uptake, bactericidal rates, and turbidimetrically measured growth inhibition in *Pseudomonas aeruginosa*. *Antimicrob. Agents Chemother.* **1987**, *31*, 108–110. [CrossRef] [PubMed]
56. Fish, D.N.; Choi, M.K.; Jung, R. Synergic activity of cephalosporins plus fluoroquinolones against *Pseudomonas aeruginosa* with resistance to one or both drugs. *J. Antimicrob. Chemother.* **2002**, *50*, 1045–1049. [CrossRef] [PubMed]
57. Gradelski, E.; Kolek, B.; Bonner, D.P.; Valera, L.; Minassian, B.; Fung-Tomc, J. Activity of gatifloxacin and ciprofloxacin in combination with other antimicrobial agents. *Int. J. Antimicrob. Agents* **2001**, *17*, 103–107. [CrossRef]
58. Pankuch, G.A.; Lin, G.; Seifert, H.; Appelbaum, P.C. Activity of Meropenem with and without Ciprofloxacin and Colistin against *Pseudomonas aeruginosa* and *Acinetobacter baumannii*. *Antimicrob. Agents Chemother.* **2008**, *52*, 333–336. [CrossRef]
59. Davis, J.S.; van Hal, S.; Tong, S.Y. Combination Antibiotic Treatment of Serious Methicillin-Resistant *Staphylococcus aureus* Infections. In *Seminars in Respiratory and Critical Care Medicine*; Thieme Medical Publishers: New York, NY, USA, 2015; Volume 36, pp. 3–16.
60. Leonard, S.N.; Rolek, K.M. Evaluation of the combination of daptomycin and nafcillin against vancomycin-intermediate *Staphylococcus aureus*. *J. Antimicrob. Chemother.* **2013**, *68*, 644–647. [CrossRef] [PubMed]
61. Mehta, S.; Singh, C.; Plata, K.B.; Chanda, P.K.; Paul, A.; Riosa, S.; Rosato, R.R.; Rosato, A.E. β-Lactams Increase the Antibacterial Activity of Daptomycin against Clinical Methicillin-Resistant *Staphylococcus aureus* Strains and Prevent Selection of Daptomycin-Resistant Derivatives. *Antimicrob. Agents Chemother.* **2012**, *56*, 6192–6200. [CrossRef]
62. Gritsenko, D.; Fedorenko, M.; Ruhe, J.J.; Altshuler, J. Combination Therapy With Vancomycin and Ceftaroline for Refractory Methicillin-resistant *Staphylococcus aureus* Bacteremia: A Case Series. *Clin. Ther.* **2017**, *39*, 212–218. [CrossRef]

63. Dhand, A.; Sakoulas, G. Daptomycin in Combination With Other Antibiotics for the Treatment of Complicated Methicillin-Resistant *Staphylococcus aureus* Bacteremia. *Clin. Ther.* **2014**, *36*, 1303–1316. [CrossRef]
64. Van Duin, D.; Bonomo, R.A. Ceftazidime/avibactam and ceftolozane/tazobactam: Second-generation beta-Lactam/beta-lactamase inhibitor combinations. *Clin. Infect. Dis.* **2016**, *63*, 234–241. [CrossRef]
65. Jean, S.-S.; Gould, I.M.; Lee, W.-S.; Hsueh, P.-R.; International Society of Antimicrobial Chemotherapy (ISAC). New Drugs for Multidrug-Resistant Gram-Negative Organisms: Time for Stewardship. *Drugs* **2019**, *79*, 705–714. [CrossRef]
66. Cheng, I.-L.; Chen, Y.H.; Lai, C.-C.; Tang, H.-J. The use of ceftolozane-tazobactam in the treatment of complicated intra-abdominal infections and urinary tract infections—A meta-analysis of randomized controlled trials. *Int. J. Antimicrob. Agents* **2020**, *55*, 105858. [CrossRef] [PubMed]
67. Lai, C.-C.; Chen, C.-C.; Tang, H.-J. Meropenem-Vaborbactam in the Treatment of Acute Bacterial Infections. *J. Clin. Med.* **2019**, *8*, 1650. [CrossRef] [PubMed]
68. Tan, C.-K.; Lai, C.-C.; Chao, C.-M. Ceftazidime–Avibactam versus Meropenem for the Treatment of Complicated Intra-Abdominal Infections. *Antibiotics* **2019**, *8*, 255. [CrossRef] [PubMed]
69. Zhanel, G.G.; Lawrence, C.K.; Adam, H.; Schweizer, F.; Zelenitsky, S.; Zhanel, M.; Lagacé-Wiens, P.R.S.; Walkty, A.; Denisuik, A.; Golden, A.; et al. Imipenem-relebactam and meropenem-vaborbactam: Two novel carbapenem-beta-lactamase inhibitor combinations. *Drugs* **2018**, *78*, 65–98. [CrossRef] [PubMed]
70. Mirza, H.C.; Hortaç, E.; Koçak, A.A.; Demirkaya, M.H.; Yayla, B.; Güçlü, A.Ü.; Başustaoğlu, A. In vitro activity of ceftolozane-tazobactam and ceftazidime-avibactam against clinical Isolates of meropenem-non-susceptible *Pseudomonas aeruginosa*: A two-center study. *J. Glob. Antimicrob. Resist.* **2019**, *20*, 334–338. [CrossRef] [PubMed]
71. Spiliopoulou, I.; Kazmierczak, K.; Stone, G.G. In vitro activity of ceftazidime/avibactam against isolates of carbapenem-non-susceptible Enterobacteriaceae collected during the INFORM global surveillance programme (2015–17). *J. Antimicrob. Chemother.* **2019**, *75*, 384–391. [CrossRef]
72. Kaushik, A.; Ammerman, N.C.; Lee, J.; Martins, O.; Kreiswirth, B.N.; Lamichhane, G.; Parrish, N.M.; Nuermberger, E.L. In vitro activity of the new beta-lactamase inhibitors relebactam and vaborbactam in combination with beta-lactams against *Mycobacterium abscessus* complex clinical isolates. *Antimicrob. Agents Chemother.* **2019**, *63*, e02623-18.
73. Petty, L.A.; Henig, O.; Patel, T.S.; Pogue, J.M.; Kaye, K.S. Overview of meropenem-vaborbactam and newer antimicrobial agents for the treatment of carbapenem-resistant Enterobacteriaceae. *Infect. Drug Resist.* **2018**, *11*, 1461–1472. [CrossRef]
74. Kim, Y.J.; Jun, Y.H.; Kim, Y.R.; Park, K.G.; Park, Y.J.; Kang, J.Y.; Kim, S.I. Risk factors for mortality in patients with *Pseudomonas aeruginosa* bacteremia; Retrospective study of impact of combination antimicrobial therapy. *BMC Infect. Dis.* **2014**, *14*, 161. [CrossRef]
75. Ceravolo, F.; Grisolia, M.; Sestito, S.; Falvo, F.; Moricca, M.T.; Concolino, D. Combination therapy in a patient with chronic neu-ronopathic Gaucher disease: A case report. *J. Med. Case Rep.* **2017**, *11*, 19. [CrossRef]
76. González-Bello, C. Antibiotic adjuvants—A strategy to unlock bacterial resistance to antibiotics. *Bioorganic Med. Chem. Lett.* **2017**, *27*, 4221–4228. [CrossRef] [PubMed]
77. Kalan, L.; Wright, G.D. Antibiotic adjuvants: Multicomponent anti-infective strategies. *Expert Rev. Mol. Med.* **2011**, *13*, e5. [CrossRef]
78. Bernal, P.; Molina-Santiago, C.; Daddaoua, A.; Llamas, M.A. Antibiotic adjuvants: Identification and clinical use. *Microb. Biotechnol.* **2013**, *6*, 445–449. [CrossRef]
79. Food and Drug Administration; US Department of Health and Human Services. *Guidance for Industry Considering Whether an FDA-Regulated Product Involves the Application of Nanotechnology*; Food and Drug Administration: Silver Spring, MD, USA, 2014.
80. Plank, C. Silence the target. *Nat. Nanotechnol.* **2009**, *4*, 544–545. [CrossRef]
81. Hutchings, M.I.; Truman, A.W.; Wilkinson, B. Antibiotics: Past, present and future. *Curr. Opin. Microbiol.* **2019**, *51*, 72–80. [CrossRef] [PubMed]
82. Zhao, Y.; Jiang, X. Multiple strategies to activate gold nanoparticles as antibiotics. *Nanoscale* **2013**, *5*, 8340–8350. [CrossRef]
83. Aabed, K.; Mohammed, A.E. Synergistic and Antagonistic Effects of Biogenic Silver Nanoparticles in Combination With Antibiotics Against Some Pathogenic Microbes. *Front. Bioeng. Biotechnol.* **2021**, *9*, 652362. [CrossRef] [PubMed]
84. Adeniji, O.O.; Ojemaye, M.O.; Okoh, A.I. Antibacterial Activity of Metallic Nanoparticles against Multidrug-Resistant Pathogens Isolated from Environmental Samples: Nanoparticles/Antibiotic Combination Therapy and Cytotoxicity Study. *ACS Appl. Bio Mater.* **2022**, *5*, 4814–4826. [CrossRef] [PubMed]
85. Yin, I.X.; Zhang, J.; Zhao, I.S.; Mei, M.L.; Li, Q.; Chu, C.H. The Antibacterial Mechanism of Silver Nanoparticles and Its Application in Dentistry. *Int. J. Nanomed.* **2020**, *15*, 2555–2562. [CrossRef]
86. Raghunath, A.; Perumal, E. Metal oxide nanoparticles as antimicrobial agents: A promise for the future. *Int. J. Antimicrob. Agents* **2017**, *49*, 137–152. [CrossRef]
87. Zhang, L.; Pornpattananangkul, D.; Hu, C.-M.; Huang, C.-M. Development of Nanoparticles for Antimicrobial Drug Delivery. *Curr. Med. Chem.* **2010**, *17*, 585–594. [CrossRef]
88. Bayda, S.; Hadla, M.; Palazzolo, S.; Riello, P.; Corona, G.; Toffoli, G.; Rizzolio, F. Inorganic Nanoparticles for Cancer Therapy: A Transition from Lab to Clinic. *Curr. Med. Chem.* **2018**, *25*, 4269–4303. [CrossRef]
89. Pelgrift, R.Y.; Friedman, A.J. Nanotechnology as a therapeutic tool to combat microbial resistance. *Adv. Drug Deliv. Rev.* **2013**, *65*, 1803–1815. [CrossRef]

90. Watt, J.; Cheong, S.; Tilley, R.D. How to control the shape of metal nanostructures in organic solution phase synthesis for plasmonics and catalysis. *Nano Today* **2013**, *8*, 198–215. [CrossRef]
91. Aderibigbe, B.A. Metal-Based Nanoparticles for the Treatment of Infectious Diseases. *Molecules* **2017**, *22*, 1370. [CrossRef] [PubMed]
92. Webster, T.J.; Seil, J.T. Antimicrobial applications of nanotechnology: Methods and literature. *Int. J. Nanomed.* **2012**, *7*, 2767–2781. [CrossRef]
93. Slavin, Y.N.; Asnis, J.; Häfeli, U.O.; Bach, H. Metal nanoparticles: Understanding the mechanisms behind antibacterial activity. *J. Nanobiotechnol.* **2017**, *15*, 65. [CrossRef] [PubMed]
94. Lesniak, A.; Salvati, A.; Santos-Martinez, M.J.; Radomski, M.W.; Dawson, K.A.; Åberg, C. Nanoparticle Adhesion to the Cell Membrane and Its Effect on Nanoparticle Uptake Efficiency. *J. Am. Chem. Soc.* **2013**, *135*, 8–1444. [CrossRef] [PubMed]
95. Tang, S.; Zheng, J. Antibacterial Activity of Silver Nanoparticles: Structural Effects. *Adv. Healthc. Mater.* **2018**, *7*, 1701503. [CrossRef] [PubMed]
96. Sotiriou, G.A.; Pratsinis, S.E. Antibacterial Activity of Nanosilver Ions and Particles. *Environ. Sci. Technol.* **2010**, *44*, 5649–5654. [CrossRef]
97. Naskar, A.; Khan, H.; Sarkar, R.; Kumar, S.; Halder, D.; Jana, S. Anti-biofilm activity and food packaging application of room temperature solution process based polyethylene glycol capped Ag-ZnO-graphene nanocomposite. *Mater. Sci. Eng. C* **2018**, *91*, 743–753. [CrossRef]
98. Jain, A.S.; Pawar, P.S.; Sarkar, A.; Junnuthula, V.; Dyawanapelly, S. Bionanofactories for green synthesis of silver nanoparticles: Toward antimicrobial applications. *Int. J. Mol. Sci.* **2021**, *22*, 11993. [CrossRef]
99. Politano, A.D.; Campbell, K.T.; Rosenberger, L.H.; Sawyer, R.G. Use of Silver in the Prevention and Treatment of Infections: Silver Review. *Surg. Infect.* **2013**, *14*, 8–20. [CrossRef] [PubMed]
100. Qing, Y.; Chen, L.; Li, R.; Liu, G.; Zhang, Y.; Tang, X.; Wang, J.; Liu, H.; Qin, Y. Potential antibacterial mechanism of silver nanoparticles and the optimization of orthopedic implantswa by advanced modification technologies. *Int. J. Nanomed.* **2018**, *13*, 3311–3327. [CrossRef]
101. Bondarenko, O.M.; Sihtmäe, M.; Kuzmičiova, J.; Ragelienė, L.; Kahru, A.; Daugelavičius, R. Plasma membrane is the target of rapid antibacterial action of silver nanoparticles in *Escherichia coli* and *Pseudomonas aeruginosa*. *Int. J. Nanomed.* **2018**, *13*, 6779–6790. [CrossRef] [PubMed]
102. Lu, Z.; Rong, K.; Li, J.; Yang, H.; Chen, R. Size-dependent antibacterial activities of silver nanoparticles against oral an-aerobic pathogenic bacteria. *J. Mater. Sci. Mater. Med.* **2013**, *24*, 1465–1471. [CrossRef] [PubMed]
103. Durán, N.; Durán, M.; de Jesus, M.B.; Seabra, A.B.; Fávaro, W.J.; Nakazato, G. Silver nanoparticles: A new view on mechanistic aspects on antimicrobial activity. *Nanomed. Nanotechnol. Biol. Med.* **2016**, *12*, 789–799. [CrossRef]
104. Poulose, S.; Panda, T.; Nair, P.P.; Theodore, T. Biosynthesis of silver nanoparticles. *J. Nanosci. Nanotechnol.* **2014**, *14*, 2038–2049. [CrossRef]
105. Prasannakumar, J.B.; Vidya, Y.S.; Anantharaju, K.S.; Ramgopal, G.; Nagabhushana, H.; Sharma, S.C.; Prasad, B.D.; Prashantha, S.C.; Basavaraj, R.B.; Rajanaik, H.; et al. Bio-mediated route for the synthesis of shape tunable Y_2O_3: Tb^{3+} nano-particles: Photoluminescence and antibacterial properties. *Spectrochim. Acta Part A Mol. Biomol. Spectrosc.* **2015**, *151*, 131–140. [CrossRef]
106. Van Dong, P.; Ha, C.H.; Binh, L.T.; Kasbohm, J. Chemical synthesis and antibacterial activity of novel-shaped silver nanoparticles. *Int. Nano Lett.* **2012**, *2*, 9. [CrossRef]
107. Actis, L.; Srinivasan, A.; Lopez-Ribot, J.L.; Ramasubramanian, A.K.; Ong, J.L. Effect of silver nanoparticle geometry on methicillin susceptible and resistant *Staphylococcus aureus*, and osteoblast viability. *J. Mater. Sci. Mater. Med.* **2015**, *26*, 215. [CrossRef]
108. Radovic-Moreno, A.F.; Lu, T.K.; Puscasu, V.A.; Yoon, C.J.; Langer, R.; Farokhzad, O.C. Surface Charge-Switching Polymeric Nanoparticles for Bacterial Cell Wall-Targeted Delivery of Antibiotics. *ACS Nano* **2012**, *6*, 4279–4287. [CrossRef]
109. Hartmann, N.B.; Jensen, K.A.; Baun, A.; Rasmussen, K.; Rauscher, H.; Tantra, R.; Cupi, D.; Gilliland, D.; Pianella, F.; Riego Sintes, J.M. Techniques and protocols for dispersing nanoparticle powders in aqueous media—Is there a rationale for har-monization? *J. Toxicol. Environ.* **2015**, *18*, 299–326.
110. Halder, S.; Yadav, K.K.; Sarkar, R.; Mukherjee, S.; Saha, P.; Haldar, S.; Karmakar, S.; Sen, T. Alteration of Zeta potential and membrane permeability in bacteria: A study with cationic agents. *SpringerPlus* **2015**, *4*, 672. [CrossRef]
111. Reshma, V.G.; Syama, S.; Sruthi, S.; Reshma, S.C.; Remya, N.S.; Mohanan, P.V. Engineered nanoparticles with Antimicrobial property. *Curr. Drug Metab.* **2018**, *18*, 1040–1054. [CrossRef]
112. Kotov, N.A.; VanEpps, J.S. Antibacterial Metal Oxide Nanoparticles: Challenges in Interpreting the Literature. *Curr. Pharm. Des.* **2018**, *24*, 896–903.
113. Sinha, R.; Karan, R.; Sinha, A.; Khare, S.K. Interaction and nanotoxic effect of ZnO and Ag nanoparticles on mesophilic and halophilic bacterial cells. *Bioresour. Technol.* **2011**, *102*, 1516–1520. [CrossRef] [PubMed]
114. Tong, G.; Du, F.; Wu, W.; Wu, R.; Liu, F.; Liang, Y. Enhanced reactive oxygen species (ROS) yields and antibacterial activity of spongy ZnO/ZnFe2O4 hybrid micro-hexahedra selectively synthesized through a versatile glucose-engineered co-precipitation/annealing process. *J. Mater. Chem. B* **2013**, *1*, 2647–2657. [CrossRef] [PubMed]
115. Singh, K.; Panghal, M.; Kadyan, S.; Chaudhary, U.; Yadav, J.P. Green silver nanoparticles of *Phyllanthus amarus*: As an antibacterial agent against multi drug resistant clinical isolates of *Pseudomonas aeruginosa*. *J. Nanobiotechnol.* **2014**, *12*, 40. [CrossRef]

116. Gelabert, A.; Sivry, Y.; Gobbi, P.; Mansouri-Guilani, N.; Menguy, N.; Brayner, R.; Siron, V.; Benedetti, M.F.; Ferrari, R. Testing nanoeffect onto model bacteria: Impact of speciation and genotypes. *Nanotoxicology* **2016**, *10*, 216–225. [CrossRef] [PubMed]
117. Nagvenkar, A.P.; Deokar, A.; Perelshtein, I.; Gedanken, A. A one-step sonochemical synthesis of stable ZnO–PVA nano-colloid as a potential biocidal agent. *J. Mater. Chem. B* **2016**, *4*, 2124–2132. [CrossRef] [PubMed]
118. Yu, J.; Zhang, W.; Li, Y.; Wang, G.; Yang, L.; Jin, J.; Chen, Q.; Huang, M. Synthesis, characterization, antimicrobial activity and mechanism of a novel hydroxyapatite whisker/nano zinc oxide biomaterial. *Biomed. Mater.* **2014**, *10*, 015001. [CrossRef] [PubMed]
119. Aydin Sevinç, B.; Hanley, L. Antibacterial activity of dental composites containing zinc oxide nanoparticles. *J. Biomed. Mater. Res. Part B Appl. Biomater.* **2010**, *94B*, 22–31. [CrossRef] [PubMed]
120. Baptista, P.V.; McCusker, M.P.; Carvalho, A.; Ferreira, D.A.; Mohan, N.M.; Martins, M.; Fernandes, A.R. Nano-strategies to fight multidrug resistant bacteria "A Battle of the Titans". *Front. Microbiol.* **2018**, *9*, 1441. [CrossRef]
121. Miri, A.; Mahdinejad, N.; Ebrahimy, O.; Khatami, M.; Sarani, M. Zinc oxide nanoparticles: Biosynthesis, characterization, antifungal and cytotoxic activity. *Mater. Sci. Eng. C* **2019**, *104*, 109981. [CrossRef]
122. Zheng, Y.; Huang, Y.; Shi, H.; Fu, L. Green biosynthesis of ZnO nanoparticles by *Plectranthus amboinicus* leaf extract and their application for electrochemical determination of norfloxacin. *Inorg. Nano-Met. Chem.* **2019**, *49*, 277–282. [CrossRef]
123. Shetti, N.P.; Bukkitgar, S.D.; Reddy, K.R.; Reddy, C.V.; Aminabhavi, T.M. ZnO-based nanostructured electrodes for electrochemical sensors and biosensors in biomedical applications. *Biosens. Bioelectron.* **2019**, *141*, 111417. [CrossRef]
124. Hatami, Z.; Ragheb, E.; Jalali, F.; Tabrizi, M.A.; Shamsipur, M. Zinc oxide-gold nanocomposite as a proper platform for label-free DNA biosensor. *Bioelectrochemistry* **2020**, *133*, 107458. [CrossRef] [PubMed]
125. Kumar, S.; Boro, J.C.; Ray, D.; Mukherjee, A.; Dutta, J. Bionanocomposite films of agar incorporated with ZnO nanoparticles as an active packaging material for shelf life extension of green grape. *Heliyon* **2019**, *5*, e01867. [CrossRef]
126. Seo, K.H.; Markus, J.; Soshnikova, V.; Oh, K.H.; Anandapadmanaban, G.; Perez, Z.E.J.; Mathiyalagan, R.; Kim, Y.J.; Yang, D.C. Facile and green synthesis of zinc oxide particles by *Stevia Rebaudiana* and its in vitro photocatalytic activity. *Inorg. Nano-Met. Chem.* **2019**, *49*, 1–6.
127. Vicente, A.; Sohm, B.; Flayac, J.; Rousselle, P.; Bauda, P.; Pagnout, C. Toxicity mechanisms of ZnO UV-filters used in sunscreens toward the model cyanobacteria *Synechococcus elongatus* PCC 7942. *Environ. Sci. Pollut. Res.* **2019**, *26*, 22450–22463. [CrossRef] [PubMed]
128. Dodero, A.; Alloisio, M.; Vicini, S.; Castellano, M. Preparation of composite alginate-based electrospun membranes loaded with ZnO nanoparticles. *Carbohydr. Polym.* **2020**, *227*, 115371. [CrossRef] [PubMed]
129. Akbarian, M.; Mahjoub, S.; Elahi, S.M.; Zabihi, E.; Tashakkorian, H. Green synthesis, formulation and biological evaluation of a novel ZnO nanocarrier loaded with paclitaxel as drug delivery system on MCF-7 cell line. *Colloids Surf. B Biointerfaces* **2020**, *186*, 110686. [CrossRef] [PubMed]
130. Naskar, A.; Kim, K.-S. Nanomaterials as Delivery Vehicles and Components of New Strategies to Combat Bacterial Infections: Advantages and Limitations. *Microorganisms* **2019**, *7*, 356. [CrossRef]
131. Dizaj, S.M.; Lotfipour, F.; Barzegar-Jalali, M.; Zarrintan, M.H.; Adibkia, K. Antimicrobial activity of the metals and metal oxide nanoparticles. *Mater. Sci. Eng. C* **2014**, *44*, 278–284. [CrossRef]
132. Rajasekar, K.; Dinesh, A.; Durka, M.; Muthukumaravel, K. Facile synthesis and in vitro biological screening of Pd@ SiO2 core–shell nanoparticles. *J. Nanosci. Nanotechnol.* **2019**, *19*, 3536–3543. [CrossRef]
133. Nguyen, N.-Y.T.; Grelling, N.; Wetteland, C.L.; Rosario, R.; Liu, H.N. Antimicrobial Activities and Mechanisms of Magnesium Oxide Nanoparticles (nMgO) against Pathogenic Bacteria, Yeasts, and Biofilms. *Sci. Rep.* **2018**, *8*, 16260. [CrossRef]
134. Ijaz, U.; Bhatti, I.A.; Mirza, S.; Ashar, A. Characterization and evaluation of antibacterial activity of plant mediated calcium oxide (CaO) nanoparticles by employing *Mentha pipertia* extract. *Mater. Res. Express* **2017**, *4*, 105402. [CrossRef]
135. Manikandan, V.; Jayanthi, P.; Priyadharsan, A.; Vijayaprathap, E.; Anbarasan, P.M.; Velmurugan, P. Green synthesis of pH-responsive Al_2O_3 nanoparticles: Application to rapid removal of nitrate ions with enhanced antibacterial activity. *J. Photochem. Photobiol. A Chem.* **2019**, *371*, 205–215. [CrossRef]
136. Cui, Y.; Zhao, Y.; Tian, Y.; Zhang, W.; Lü, X.; Jiang, X. The molecular mechanism of action of bactericidal gold nanoparticles on *Escherichia coli*. *Biomaterials* **2012**, *33*, 2327–2333. [CrossRef]
137. Yamamoto, O.; Ohira, T.; Alvarez, K.; Fukuda, M. Antibacterial characteristics of CaCO3–MgO composites. *Mater. Sci. Eng. B* **2010**, *173*, 208–212. [CrossRef]
138. Ansari, M.A.; Khan, H.M.; Khan, A.A.; Cameotra, S.S.; Saquib, Q.; Musarrat, J. Interaction of Al_2O_3 nanoparticles with *Escherichia coli* and their cell envelope biomolecules. *J. Appl. Microbiol.* **2014**, *116*, 772–783. [CrossRef]
139. Mulvey, M.R.; Simor, A.E. Antimicrobial resistance in hospitals: How concerned should we be? *CMAJ* **2009**, *180*, 408–415. [CrossRef]
140. Chen, C.-W.; Hsu, C.-Y.; Lai, S.-M.; Syu, W.-J.; Wang, T.-Y.; Lai, P.-S. Metal nanobullets for multidrug resistant bacteria and biofilms. *Adv. Drug Deliv. Rev.* **2014**, *78*, 88–104. [CrossRef] [PubMed]
141. Li, H.; Chen, Q.; Zhao, J.; Urmila, K. Enhancing the antimicrobial activity of natural extraction using the synthetic ultrasmall metal nanoparticles. *Sci. Rep.* **2015**, *5*, 11033. [CrossRef] [PubMed]
142. Hossain, Z.; Yasmeen, F.; Komatsu, S. Nanoparticles: Synthesis, morphophysiological effects, and proteomic responses of crop plants. *Int. J. Mol. Sci.* **2020**, *21*, 3056. [CrossRef] [PubMed]

143. Stensberg, M.C.; Wei, Q.S.; McLamore, E.S.; Porterfield, D.M.; Wei, A.; Sepúlveda, M.S. Toxicological studies on silver nanoparticles: Challenges and opportunities in assessment, monitoring and imaging. *Nanomedicine* **2011**, *6*, 879–898. [CrossRef] [PubMed]
144. Dwivedi, S.; Wahab, R.; Khan, F.; Mishra, Y.K.; Musarrat, J.; Al-Khedhairy, A.A. Reactive Oxygen Species Mediated Bacterial Biofilm Inhibition via Zinc Oxide Nanoparticles and Their Statistical Determination. *PLoS ONE* **2014**, *9*, e111289. [CrossRef]
145. Rudramurthy, G.R.; Swamy, M.K.; Sinniah, U.R.; Ghasemzadeh, A. Nanoparticles: Alternatives Against Drug-Resistant Pathogenic Microbes. *Molecules* **2016**, *21*, 836. [CrossRef] [PubMed]
146. Sánchez-López, E.; Gomes, D.; Esteruelas, G.; Bonilla, L.; Lopez-Machado, A.L.; Galindo, R.; Cano, A.; Espina, M.; Ettcheto, M.; Camins, A.; et al. Metal-Based Nanoparticles as Antimicrobial Agents: An Overview. *Nanomaterials* **2020**, *10*, 292. [CrossRef]
147. Huang, Y.; Yu, F.; Park, Y.-S.; Wang, J.; Shin, M.-C.; Chung, H.S.; Yang, V.C. Co-administration of protein drugs with gold nanoparticles to enable percutaneous delivery. *Biomaterials* **2010**, *31*, 9086–9091. [CrossRef] [PubMed]
148. Huh, A.J.; Kwon, Y.J. "Nanoantibiotics": A new paradigm for treating infectious diseases using nanomaterials in the antibiotics resistant era. *J. Control. Release* **2011**, *156*, 128–145. [CrossRef] [PubMed]
149. Joost, U.; Juganson, K.; Visnapuu, M.; Mortimer, M.; Kahru, A.; Nõmmiste, E.; Joost, U.; Kisand, V.; Ivask, A. Photocatalytic antibacterial activity of nano-TiO_2 (anatase)-based thin films: Effects on *Escherichia coli* cells and fatty acids. *J. Photochem. Photobiol. B Biol.* **2015**, *142*, 178–185. [CrossRef] [PubMed]
150. Andrade, F.; Rafael, D.; Videira, M.; Ferreira, D.; Sosnik, A.; Sarmento, B. Nanotechnology and pulmonary delivery to overcome resistance in infectious diseases. *Adv. Drug Deliv. Rev.* **2013**, *65*, 1816–1827. [CrossRef] [PubMed]
151. Weir, E.; Lawlor, A.; Whelan, A.; Regan, F. The use of nanoparticles in anti-microbial materials and their characterization. *Analyst* **2008**, *133*, 835–845. [CrossRef] [PubMed]
152. Xie, S.; Yang, F.; Tao, Y.; Chen, D.; Qu, W.; Huang, L.; Liu, Z.; Pan, Y.; Yuan, Z. Enhanced intracellular delivery and antibacterial efficacy of enrofloxacin-loaded docosanoic acid solid lipid nanoparticles against intracellular Salmonella. *Sci. Rep.* **2017**, *7*, 41104. [CrossRef]
153. Chu, L.; Gao, H.; Cheng, T.; Zhang, Y.; Liu, J.; Huang, F.; Yang, C.; Shi, L.; Liu, J. A charge-adaptive nanosystem for prolonged and enhanced in vivo antibiotic delivery. *Chem. Commun.* **2016**, *52*, 6265–6268. [CrossRef]
154. Xiong, M.-H.; Li, Y.-J.; Bao, Y.; Yang, X.-Z.; Hu, B.; Wang, J. Bacteria-Responsive Multifunctional Nanogel for Targeted Antibiotic Delivery. *Adv. Mater.* **2012**, *24*, 6175–6180. [CrossRef]
155. Hemeg, H.A. Nanomaterials for alternative antibacterial therapy. *Int. J. Nanomed.* **2017**, *12*, 8211–8225. [CrossRef]
156. Qi, G.; Li, L.; Yu, F.; Wang, H. Vancomycin-Modified Mesoporous Silica Nanoparticles for Selective Recognition and Killing of Pathogenic Gram-Positive Bacteria Over Macrophage-Like Cells. *ACS Appl. Mater. Interfaces* **2013**, *5*, 10874–10881. [CrossRef]
157. Mühling, M.; Bradford, A.; Readman, J.W.; Somerfield, P.; Handy, R.D. An investigation into the effects of silver nanoparticles on antibiotic resistance of naturally occurring bacteria in an estuarine sediment. *Mar. Environ. Res.* **2009**, *68*, 278–283. [CrossRef] [PubMed]
158. Liu, Y.; Tee, J.K.; Chiu, G.N.C. Dendrimers in oral drug delivery application: Current explorations, toxicity issues and strategies for improvement. *Curr. Pharm. Des.* **2015**, *21*, 2629–2642. [CrossRef] [PubMed]
159. Panáček, A.; Smékalová, M.; Kilianová, M.; Prucek, R.; Bogdanová, K.; Večeřová, R.; Kolář, M.; Havrdová, M.; Płaza, G.A.; Chojniak, J.; et al. Strong and nonspecific synergistic antibacterial efficiency of antibiotics combined with silver nanoparticles at very low concentrations showing no cytotoxic effect. *Molecules* **2016**, *21*, 26. [CrossRef] [PubMed]
160. Scandorieiro, S.; De Camargo, L.C.; Lancheros, C.A.C.; Yamada-Ogatta, S.F.; Nakamura, C.V.; de Oliveira, A.G.; Andrade, C.G.T.J.; Durán, N.; Nakazato, G.; Kobayashi, R.K.T. Synergistic and Additive Effect of Oregano Essential Oil and Biological Silver Nanoparticles against Multidrug-Resistant Bacterial Strains. *Front. Microbiol.* **2016**, *7*, 760. [CrossRef] [PubMed]
161. Naqvi, S.Z.H.; Kiran, U.; Ali, M.I.; Jamal, A.; Hameed, A.; Ahmed, S.; Ali, N. Combined efficacy of biologically synthesized silver nanoparticles and different antibiotics against multidrug-resistant bacteria. *Int. J. Nanomed.* **2013**, *8*, 3187–3195. [CrossRef] [PubMed]
162. Sindhu, P.D.; Mukherjee, A.; Chandrasekaran, N. Synergistic effect of biogenic silver nanocolloid in combination with antibiotics: A potent therapeutic agent. *Int. J. Pharm. Pharm. Sci.* **2013**, *51*, 292–295.
163. Esmaeillou, M.; Zarrini, G.; Rezaee, M.A.; Mojarrad, J.S.; Bahadori, A. Vancomycin Capped with Silver Nanoparticles as an Antibacterial Agent against Multi-Drug Resistance Bacteria. *Adv. Pharm. Bull.* **2017**, *7*, 479–483. [CrossRef]
164. Kalhapure, R.S.; Sonawane, S.J.; Sikwal, D.R.; Jadhav, M.; Rambharose, S.; Mocktar, C.; Govender, T. Solid lipid nanoparticles of clotrimazole silver complex: An efficient nano antibacterial against *Staphylococcus aureus* and MRSA. *Colloids Surf. B Biointerfaces* **2015**, *136*, 651–658. [CrossRef]
165. Hur, Y.E.; Park, Y. Vancomycin-Functionalized Gold and Silver Nanoparticles as an Antibacterial Nanoplatform Against Methicillin-Resistant *Staphylococcus aureus*. *J. Nanosci. Nanotechnol.* **2016**, *16*, 6393–6399. [CrossRef]
166. Brown, A.N.; Smith, K.; Samuels, T.A.; Lu, J.; Obare, S.O.; Scott, M.E. Nanoparticles Functionalized with Ampicillin Destroy Multiple-Antibiotic-Resistant Isolates of *Pseudomonas aeruginosa* and *Enterobacter aerogenes* and Methicillin-Resistant *Staphylococcus aureus*. *Appl. Environ. Microbiol.* **2012**, *78*, 2768–2774. [CrossRef]
167. Ghasemi, F.; Jalal, R. Antimicrobial action of zinc oxide nanoparticles in combination with ciprofloxacin and ceftazidime against multidrug-resistant *Acinetobacter baumannii*. *J. Glob. Antimicrob. Resist.* **2016**, *6*, 118–122. [CrossRef] [PubMed]

168. Khan, M.F.; Ansari, A.H.; Hameedullah, M.; Ahmad, E.; Husain, F.M.; Zia, Q.; Baig, U.; Zaheer, M.R.; Alam, M.M.; Khan, A.M.; et al. Sol-gel synthesis of thorn-like ZnO nanoparticles endorsing mechanical stirring effect and their anti-microbial activities: Potential role as nano-antibiotics. *Sci. Rep.* **2016**, *6*, 27689. [CrossRef] [PubMed]
169. Kandi, V.; Kandi, S. Antimicrobial properties of nanomolecules: Potential candidates as antibiotics in the era of multi-drug resistance. *Epidemiol. Health* **2015**, *37*, e2015020. [CrossRef] [PubMed]
170. Dos Santos, C.A.; Seckler, M.M.; Ingle, A.P.; Gupta, I.; Galdiero, S.; Galdiero, M.; Gade, A.; Rai, M. Silver Nanoparticles: Therapeutical Uses, Toxicity, and Safety Issues. *J. Pharm. Sci.* **2014**, *103*, 1931–1944. [CrossRef] [PubMed]
171. Dalal, C.; Saha, A.; Jana, N.R. Nanoparticle Multivalency Directed Shifting of Cellular Uptake Mechanism. *J. Phys. Chem. C* **2016**, *120*, 6778–6786. [CrossRef]
172. Graves, J.L., Jr.; Tajkarimi, M.; Cunningham, Q.; Campbell, A.; Nonga, H.; Harrison, S.H.; Barrick, J.E. Rapid evolution of silver nanoparticle resistance in *Escherichia coli*. *Front. Genet.* **2015**, *6*, 42. [CrossRef]
173. Duncan, R.; Gaspar, R. Nanomedicine(s) under the Microscope. *Mol. Pharm.* **2011**, *8*, 2101–2141. [CrossRef]
174. Wei, L.; Lu, J.; Xu, H.; Patel, A.; Chen, Z.S.; Chen, G. Silver nanoparticles: Synthesis, properties, and therapeutic applications. *Drug Discov. Today Technol.* **2015**, *20*, 595–601. [CrossRef]
175. Zazo, H.; Colino, C.I.; Lanao, J.M. Current applications of nanoparticles in infectious diseases. *J. Control. Release* **2016**, *224*, 86–102.
176. Sengupta, J.; Ghosh, S.; Datta, P.; Gomes, A.; Gomes, A. Physiologically Important Metal Nanoparticles and Their Toxicity. *J. Nanosci. Nanotechnol.* **2014**, *14*, 990–1006. [CrossRef]
177. Dobrucka, R.; Szymanski, M.; Przekop, R. The study of toxicity effects of biosynthesized silver nanoparticles using *Veronica officinalis* extract. *Int. J. Environ. Sci. Technol.* **2019**, *16*, 8517–8526. [CrossRef]
178. Naskar, A.; Bera, S.; Bhattacharya, R.; Saha, P.; Roy, S.S.; Sen, T.; Jana, S. Synthesis, characterization and antibacterial activity of Ag incorporated ZnO–graphene nanocomposites. *RSC Adv.* **2016**, *6*, 88751–88761. [CrossRef]
179. Rai, M.K.; Deshmukh, S.D.; Ingle, A.P.; Gade, A.K. Silver nanoparticles: The powerful nanoweapon against multidrug-resistant bacteria. *J. Appl. Microbiol.* **2012**, *112*, 841–852. [CrossRef] [PubMed]
180. Barros, C.H.N.; Fulaz, S.; Stanisic, D.; Tasic, L. Biogenic Nanosilver against Multidrug-Resistant Bacteria (MDRB). *Antibiotics* **2018**, *7*, 69. [CrossRef]
181. Finley, P.J.; Norton, R.; Austin, C.; Mitchell, A.; Zank, S.; Durham, P. Unprecedented Silver Resistance in Clinically Isolated Enterobacteriaceae: Major Implications for Burn and Wound Management. *Antimicrob. Agents Chemother.* **2015**, *59*, 4734–4741. [CrossRef]
182. Gaillet, S.; Rouanet, J.-M. Silver nanoparticles: Their potential toxic effects after oral exposure and underlying mechanisms—A review. *Food Chem. Toxicol.* **2015**, *77*, 58–63. [CrossRef]
183. Akter, M.; Sikder, M.T.; Rahman, M.M.; Ullah, A.K.M.A.; Hossain, K.F.B.; Banik, S.; Hosokawa, T.; Saito, T.; Kurasaki, M. A systematic review on silver nanoparticles-induced cytotoxicity: Physicochemical properties and perspectives. *J. Adv. Res.* **2018**, *9*, 1–16. [CrossRef]
184. Liao, C.; Li, Y.; Tjong, S.C. Bactericidal and Cytotoxic Properties of Silver Nanoparticles. *Int. J. Mol. Sci.* **2019**, *20*, 449. [CrossRef]
185. Suk, J.S.; Xu, Q.; Kim, N.; Hanes, J.; Ensign, L.M. PEGylation as a strategy for improving nanoparticle-based drug and gene delivery. *Adv. Drug Deliv. Rev.* **2016**, *99 Pt A*, 28–51. [CrossRef]
186. Cinteza, L.O.; Scomoroscenco, C.; Voicu, S.N.; Nistor, C.L.; Nitu, S.G.; Trica, B.; Jecu, M.-L.; Petcu, C. Chitosan-Stabilized Ag Nanoparticles with Superior Biocompatibility and Their Synergistic Antibacterial Effect in Mixtures with Essential Oils. *Nanomaterials* **2018**, *8*, 826. [CrossRef]
187. Meshram, J.V.; Koli, V.B.; Kumbhar, S.G.; Borde, L.C.; Phadatare, M.R.; Pawar, S.H. Structural, spectroscopic and anti-microbial inspection of PEG capped ZnO nanoparticles for biomedical applications. *Mater. Res. Express* **2018**, *5*, 045016. [CrossRef]
188. Abdallah, O.M.; El-Baghdady, K.Z.; Khalil, M.M.H.; El Borhamy, M.I.; Meligi, G.A. Antibacterial, antibiofilm and cytotoxic activities of biogenic polyvinyl alcohol-silver and chitosan-silver nanocomposites. *J. Polym. Res.* **2020**, *27*, 74. [CrossRef]
189. Chia, S.L.; Leong, D.T. Reducing ZnO nanoparticles toxicity through silica coating. *Heliyon* **2016**, *2*, e00177. [CrossRef] [PubMed]
190. Xia, T.; Zhao, Y.; Sager, T.; George, S.; Pokhrel, S.; Li, N.; Schoenfeld, D.; Meng, H.; Lin, S.; Wang, X.; et al. Decreased Dissolution of ZnO by Iron Doping Yields Nanoparticles with Reduced Toxicity in the Rodent Lung and Zebrafish Embryos. *ACS Nano* **2011**, *5*, 1223–1235. [CrossRef]
191. Limayem, A.; Micciche, A.; Haller, E.; Zhang, C.; Mohapatra, S. Nanotherapeutics for mutating multi-drug resistant fecal bacteria. *J. Nanosci. Nanotechnol.* **2015**, *1*, 100–106.
192. Mehta, M.; Allen-Gipson, D.; Mohapatra, S.; Kindy, M.; Limayem, A. Study on the therapeutic index and synergistic effect of Chitosan-zinc oxide nanomicellar composites for drug-resistant bacterial biofilm inhibition. *Int. J. Pharm.* **2019**, *565*, 472–480. [CrossRef]
193. Sandhiya, S.; Dkhar, S.A.; Surendiran, A. Emerging trends of nanomedicine—An overview. *Fundam. Clin. Pharmacol.* **2009**, *23*, 263–269. [CrossRef]
194. Mba, I.E.; Nweze, E.I. Nanoparticles as therapeutic options for treating multidrug-resistant bacteria: Research progress, challenges, and prospects. *World J. Microbiol. Biotechnol.* **2021**, *37*, 108.

International Journal of
Molecular Sciences

MDPI

Article

Insights into the Biocompatibility and Biological Potential of a Chitosan Nanoencapsulated Textile Dye

Eduardo M. Costa *, Sara Silva, Freni K. Tavaria and Manuela Pintado

Universidade Católica Portuguesa, CBQF—Centro de Biotecnologia e Química Fina—Laboratório Associado, Escola Superior de Biotecnologia, Rua Diogo Botelho 1327, 4169-005 Porto, Portugal
* Correspondence: emcosta@ucp.pt; Tel.: +351-22-558-00-00

Abstract: Traditionally synthetic textile dyes are hazardous and toxic compounds devoid of any biological activity. As nanoencapsulation of yellow everzol textile dye with chitosan has been shown to produce biocompatible nanoparticles which were still capable of dyeing textiles, this work aims to further characterize the biocompatibility of yellow everzol nanoparticles (NPs) and to ascertain if the produced nanoencapsulated dyes possess any biological activity against various skin pathogens in vitro assays and in a cell infection model. The results showed that the NPs had no deleterious effects on the HaCat cells' metabolism and cell wall, contrary to the high toxicity of the dye. The biological activity evaluation showed that NPs had a significant antimicrobial activity, with low MICs (0.5–2 mg/mL) and MBCs (1–3 mg/mL) being registered. Additionally, NPs inhibited biofilm formation of all tested microorganisms (inhibitions between 30 and 87%) and biofilm quorum sensing. Lastly, the dye NPs were effective in managing MRSA infection of HaCat cells as they significantly reduced intracellular and extracellular bacterial counts.

Keywords: chitosan nanoparticles; yellow everzol textile dyes; biocompatibility; HaCat cells; antimicrobial activity; skin pathogens; cell infection assays

Citation: Costa, E.M.; Silva, S.; Tavaria, F.K.; Pintado, M. Insights into the Biocompatibility and Biological Potential of a Chitosan Nanoencapsulated Textile Dye. *Int. J. Mol. Sci.* **2022**, *23*, 14234. https://doi.org/10.3390/ijms232214234

Academic Editor: Helena Felgueiras

Received: 13 October 2022
Accepted: 15 November 2022
Published: 17 November 2022

1. Introduction

Synthetic textile dyes have been widely regarded as being hazardous for human health, with carcinogenic and mutagenic effects, and possible toxic effects ranging from contact dermatitis and skin allergies to tumours and heart problems [1–9]. Additionally, with the advent of textile functionalization, the need for biologically active molecules has risen and while some natural textile dyes have been reported as possessing some degree of antimicrobial activity, traditional synthetic dyes do not possess such properties and thus an additional antimicrobial coating to be applied to the textile [10–12]. Nowadays, alternatives are being sought to improve synthetic dye's biocompatibility and possibly grant them some biological potential. Among these, alternative nanotechnological approaches have risen to the forefront.

Chitosan (polycationic polysaccharide with high biocompatibility and biological activity)-based nanoparticles (NPs) are one the best candidates for the improvement of synthetic textile dyes as they have proven interactions with dyes [13–15]. They also have proven technological viability in the textile industry as they are already used as finishing agents capable of providing antimicrobial activity [16], enhancing the breaking strength, shrink-proofing, adding wrinkle-resistance properties [17,18] and even enhancing textile dyeing [19]. Furthermore, chitosan NPs high loading capacity and ability to limit the compound's interaction with the external environment leads to reduced compound-mediated toxicity [20].

Considering that a previous work [15] showed that dye nanoencapsulation with chitosan was a viable alternative for the development of novel textile dyeing methodology, while simultaneously producing a nanoencapsulated textile dye which was non-cytotoxic towards human skin cells. A question arose: would it be possible that the nanoencapsulated yellow everzol (NDYE) may have retained the biological potential of the encapsulation molecule (chitosan)? Considering this, we hypothesized that the developed NDYE may be a novel biologically active textile dyeing solution, with relevant antimicrobial and antibiofilm activity against skin pathogens and a capacity to prevent skin cells infection. To prove so, the nanoencapsulated dyes biocompatibility was further characterized, and the biological potential of the yellow everzol NPs against skin pathogens was evaluated through sessile (minimal inhibitory concentration (MIC) and minimal bacterial concentration (MBC) determinations) and planktonic (inhibition of biofilm formation and quorum sensing signalling) assays and through an ex vivo HaCat infection assay.

2. Results

2.1. Size and Zeta Potential Determination

Our results showed that we obtained a well-dispersed suspension of yellow everzol-loaded NPs (Figure 1). The NDYE prepared and analyzed showed an average size of 319.9 ± 0.81 nm with a polydispersity index (PDI) of 0.310. The zeta potential was 54.2 ± 2.14 mV, suggesting that the encapsulation procedure did not affect the positive zeta potential naturally found in chitosan–TPP nanoparticles. In comparison with the free dye and void NPs, it is possible to see that NDYE encapsulation had no statistically significant ($p > 0.05$) effects upon the nanoparticles in comparison with the void NPs but had significant ($p < 0.05$) differences relative to the free dye, especially at the zeta potential level as the free dye had a negative average particle charge.

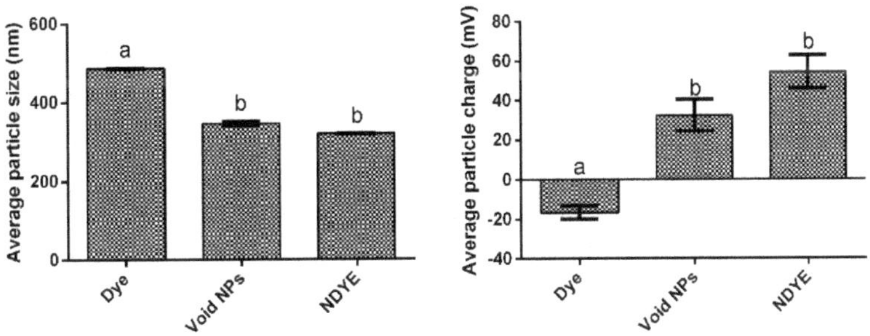

Figure 1. Average particle size (nm) and average particle charge for the free dye, void nanoparticles (NPs) and nanoencapsulated yellow everzol dye (NDYE). Different letters represent the statistically significant ($p < 0.05$) differences found between conditions assayed.

2.2. Biocompatibility Assays

As can be seen from Figure 2 both the void and the loaded NPs tested had no cytotoxic effects upon HaCat cells metabolism at the various concentrations tested.

When considering the activity of the yellow everzol dye on its own the data obtained showed a very high HaCat metabolism inhibition percentage, even at the lowest concentration tested. These values were statistically significantly ($p < 0.05$) higher than those registered for both the NPs loaded with dye and the void NPs. Interestingly enough, barring the higher concentration tested, statistically significant ($p < 0.05$) differences were found between the void and the loaded nanoparticles, with the latter presenting metabolism promotions in all concentrations tested.

When considering the lactate dehydrogenase (LDH) results obtained (Figure 3) it is possible to see that both NPs (void and NDYE) had no cytotoxic effects on HaCat cell walls.

On the other hand, yellow everzol dye on its own showed cytotoxic effects towards the HaCat cells in all tested conditions, with dye toxicity values varying from ca. 10 to 40%. Contrary to what was seen in the metabolism inhibition assay, no statistically significant ($p > 0.05$) differences were observed between NDYE and the void NPs while clear statistically significant ($p < 0.05$) differences were observed between both NPs and yellow everzol dye LDH cytotoxicity levels.

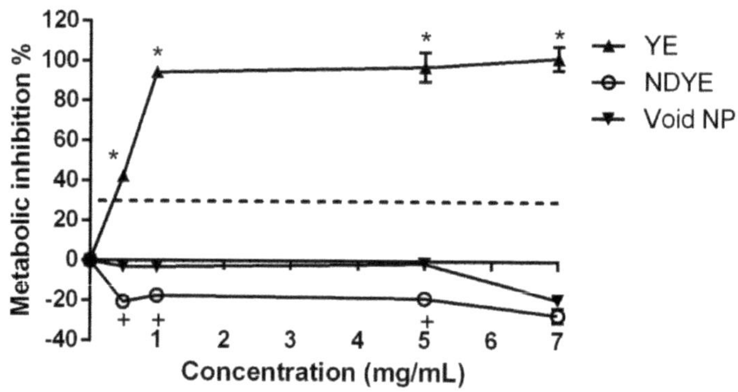

Figure 2. Evaluation of the effect of NDYE (○), void NPs (▼) and yellow everzol (YE) (▲) upon HaCat metabolism, at the different concentrations tested. The dotted line represents the 30% cytotoxicity limit as defined by the ISO 10993-5:2009 standard [21]. * Represents the statistically significant differences ($p < 0.05$) found between the textile dye and the tested nanoparticles. +Represents the statistically significant differences ($p < 0.05$) found between the void and the loaded nanoparticles.

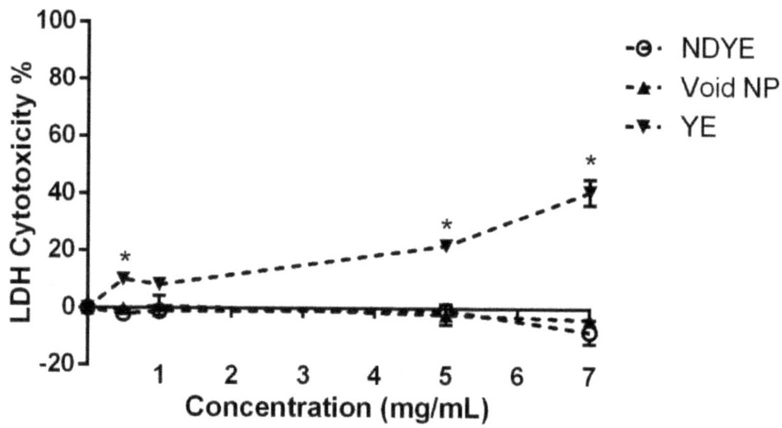

Figure 3. Evaluation of the effect of NDYE (○), void NPs (▼) and yellow everzol (▲) upon HaCat cell wall integrity, at the different concentrations tested. * Represents the statistically significant differences ($p < 0.05$) found between the textile dye and the tested nanoparticles.

2.3. MIC and MBC Determination

The MIC and MBC results obtained for NDYE can be seen in Table 1.

Overall, it stands to notice the lack of MBCs registered for *Pseudomonas aeruginosa* and the low MICs obtained for MRSE and *A. baumannii*. When considering only the MIC results obtained, it's interesting to see that the average MIC was 1 mg/mL and that no clear patterns of susceptibility were present. When considering antimicrobial resistance as a differentiating factor, antibiotic-resistant strains presented MICs that were either equal (MRSA), superior (VRSA) or inferior (MRSE) to the ones obtained for the sensitive strains (MSSA and *S. epidermidis*). Similarly, when considering the microorganism's cell wall as a distinguishing factor, the MICs values obtained showed no discernible sensibility patterns that could be observed. Likewise, for the MBC values, no discernible patterns of susceptibility were ascertainable. In fact, excluding the previously referred lack of MBCs registered for *P. aeruginosa*, the other outlier registered was the low MBC observed for MSSA. Furthermore, it also important to reference the low average MBC (2.28 mg/mL) registered and the small difference (1 to 2.28 mg/mL) registered between the average MBC and the average MIC.

Table 1. MIC and MBC values obtained for NDYE against the target microorganisms. All results are given in mg/mL of NDYE. nd—not determined.

Microorganism	MIC	MBC
Methicillin-sensitive *Staphylococcus aureus*	1	1
Methicillin-resistant *Staphylococcus aureus*	1	3
Vancomycin-resistant *Staphylococcus aureus*	2	3
Staphylococcus epidermidis	2	3
Methicillin-resistant *Staphylococcus epidermidis*	0.5	3
Pseudomonas aeruginosa	1	Nd
Acinetobacter baumannii	0.5	3

2.4. Biofilm Formation Inhibition

As can be seen from Figure 4, the NDYE was capable of having inhibitory activity upon the biofilm biomass formation of the targeted microorganisms.

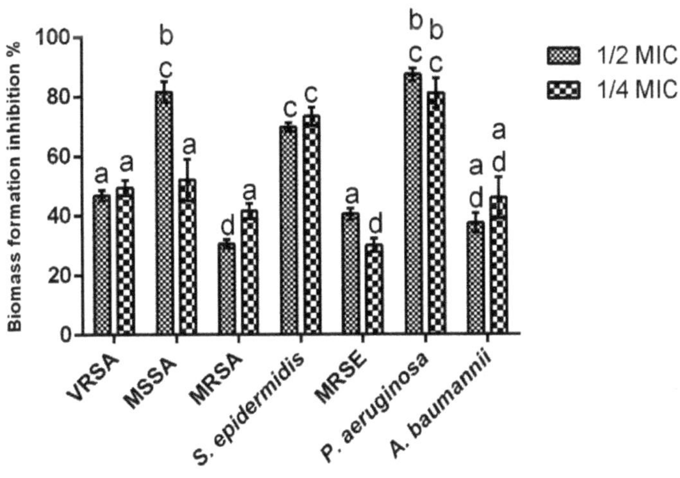

Figure 4. Evaluation of sub-MIC concentrations (1/2 and 1/4 of the MICs) of NDYE upon the selected microorganism's biofilm formation. Values obtained are given in percentage of biofilm formation inhibition. Different letters represent statistically significant differences found ($p < 0.05$).

Overall, all of the inhibition percentages registered were above 30%, with the average inhibition value registered being 54.83% and the highest inhibition obtained being 87.38%. Additionally, it bears notice that for four of the seven microorganisms assayed, the NDYE concentration was not a significant factor, as no statistically significant ($p > 0.05$) differences were found between the tested concentrations. For the cases where statistically significant ($p < 0.05$) differences were found, MRSA inhibition values were a particular case, as the highest inhibition values were registered at the lowest concentration tested. On a closer look, it is interesting to see that for *S. aureus*, the highest sensitivity was registered for

MSSA, followed by VRSA and MRSA with inhibition percentages registered for the first being 35 and 40% higher, respectively. A similar pattern of susceptibility was observed for *S. epidermidis* where the antibiotic resistance strain (MRSE) was significantly ($p < 0.05$) less susceptible to NDYE antibiofilm activity. When analyzing the results obtained for the gram-negative strains tested the high inhibition values obtained for *P. aeruginosa* stand out, especially in comparison with the values registered for *A. baumannii*. Interestingly, *A. baumannii* presented a susceptibility pattern similar to MRSA, as the highest biofilm inhibition percentage was obtained at the lowest concentration tested.

2.5. Quorum Sensing Inhibition

The quorum sensing inhibition results obtained (Figure 5) show that NDYE inhibited violacein production in the *Chromobacterium violaceum* (*C. violaceum*) reporter system. In fact, for NDYE concentrations between 7 and 3 mg/mL, no statistically significant ($p > 0.05$) differences were found and violacein inhibition percentages varied between 65 and 72%. For concentrations of 2 mg/mL or lower a significant ($p < 0.05$) drop-in activity was observed with violacein inhibition percentages reaching negative values at the lowest concentrations tested.

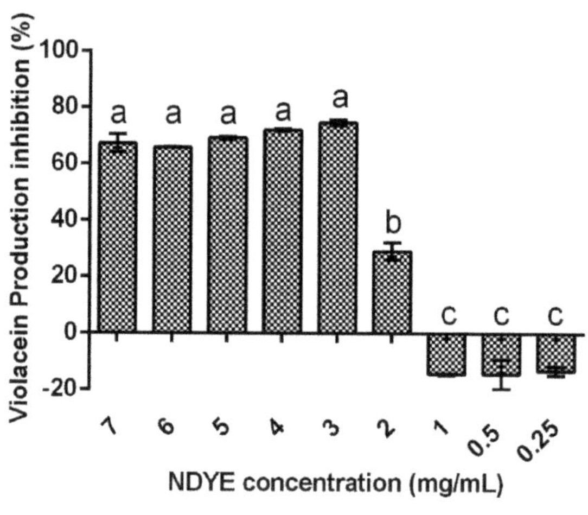

Figure 5. Effect of the tested NDYE concentrations upon the *C. violaceum* quorum sensing reporter system. Results are given in terms of violacein production inhibition percentages. Different letters represent statistically significant differences found ($p < 0.05$).

2.6. Cell Infection Assays

The cell infection results obtained are depicted in Figure 6. As can be seen, statistically significant differences ($p < 0.05$) were found both for the extracellular and intracellular bacterial counts.

When considering only the extracellular counts, NDYE reduced MRSA-viable counts by ca. 1.2 and 1.5 logs of colony forming units (CFUs) at the 24 h mark with no statistically significant ($p > 0.05$) differences being found between the MIC and the MBC. On the other hand, for the intracellular counts, while both concentrations still produced statistically significant ($p < 0.05$) viable counts reductions, the one obtained for the MBC was significantly ($p < 0.05$) higher than the one registered for the MIC. Additionally, it merits to notice the reduced activity that the MIC registered in the intracellular conditions as it only produced a viable count reduction of ca. 0.6 log of CFU versus the 1.2 log of CFU registered in the extracellular settings.

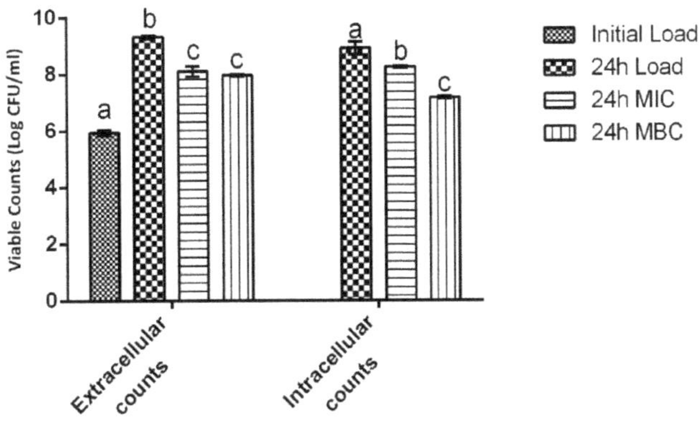

Figure 6. Effect of NDYE, at MIC and MBC concentraions, upon MRSA infection of HaCat cells. Results are given in terms of total bacterial counts. Different letters represent statistically significant differences found ($p < 0.05$) between the tested conditions.

3. Discussion

Synthetic textile dyes have been widely regarded as being hazardous to human health, with carcinogenic effects and high cellular cytotoxicity being reported [1–5]. The produced NDYEs presented a relatively small size, low PDI and positive particle charge, all abiding to the intrinsic known characteristics of chitosan NPs [15]. Furthermore, these nanoparticles have been previously shown to be not cytotoxicity towards HaCat cells; these results were further corroborated by the ones registered in this work, as both the XTT and LDH dose–response assays performed here showed that NDYEs did not affect HaCat cell line metabolism and cellular integrity. These results further corroborate the hypothesis advanced by Costa, Silva [15]; dyes entrapment in NPs will block them from interacting with the surrounding environment and thus prevent them from having cytotoxic effects on the HaCat cells.

While some natural textile dyes have been reported as possessing some degree of antimicrobial activity [12], traditional synthetic dyes, like the one used in this work, do not possess such properties and usually either require a modification or an additional application of an antimicrobial coating to the textile [10,11]. Chitosan NPs, on the other hand, have already been established as viable antimicrobial agents as shown in previous works [16,22–26]. However, and considering that, as shown previously [15], the yellow everzol encapsulation process occurs through an interaction of the dyes sulfuric moieties with chitosan amino groups, and it could stand to reason that this could originate loaded chitosan NPs with reduced or non-existent biological activity. As can be seen from the results obtained, this was not the case, as the yellow everzol-loaded NPs were capable of inhibiting the studied microorganisms in all of the different assays performed. Previous works have shown that encapsulating bioactive compounds in chitosan NPs, usually, leads to the formation of compounds with added antimicrobial activity as nanoencapsulation permits compounds to be internalized into cells and thus potentiate their activity [25]. As no previous works have explored the biological potential of nanoencapsulated textile dyes, no direct comparisons were possible. Nevertheless, comparisons, particularly against void chitosan NPs, can still be drawn.

When regarding *S. aureus* and his multidrug-resistant strains (MRSA and VRSA) the existing body of work is twofold. While on the one hand, chitosan NPs have been unanimously shown to be capable of inhibiting the planktonic growth of MSSA [16,23,24,27] (as seen in this work), on the other hand, for MRSA and VRSA of previous works were not as unanimous, with chitosan NPs being described as either capable [22,24,26]) or incapable [28–30] of inhibiting MRSA and VRSA. For *S. epidermidis* and MRSE, the obtained results do not find any support in the existing body of work. In fact, the values here reported are drastically different from those reported in previous works, as these have shown that chitosan NPs had either limited activity, as Silva, Silva [29] reported for *S. epidermidis* a MIC of 16 mg/mL, or possessed inhibitory but no bactericidal activity, as reported by Costa, Silva [24] for MRSE. Similarly, the results obtained for *P. aeruginosa* are somewhat contrary to those previously reported. While in previous works [15,31–33] chitosan NPs have been described as being effective inhibitors of *P. aeruginosa* growth, the same was not observed here, as no MBC value was registered for the yellow everzol NPs against *P. aeruginosa*. Lastly, for *A. baumannii* the MIC and MBC values obtained are similar to those previously reported by Costa, Silva [26], with the lack of any additional works curtailing any further comparisons. The discrepancies observed between this work and the previous can be justified by the existing literature. The first possible explanation may be related to the particle's zeta potential, a factor described as being fundamental in NPs antimicrobial activity. As microbial surfaces are negatively charged, NPs with higher zeta potential (as the ones obtained in this work) will have a stronger interaction with the bacteria and thus stronger antimicrobial activity [24,29,34]. Another possible explanation could be related to the NPs quantum size effect, as per this effect, higher antimicrobial activity is associated with smaller particles size, and the particles used in this work are engorged, thus larger, due to the encapsulation of textile dye [25]. Lastly, the last factor affecting the NPs activity may be the microorganisms themselves, because while NPs have been described as possessing no significant differences in sensibility between gram-positive or gram-negative bacteria [35], other microorganism intrinsic factors, such as hydrophilicity and charge density on the bacterial surface, may be at play and thus influence the NPs activity [36]. Additionally, regardless of all of the previous hypotheses, it must not be overlooked that these comparisons are being drawn between void NPs and NPs with altered surface structure (as previously shown by Costa, Silva [15]) due to the yellow everzol entrapment procedure and as such it is possible that any other number of unknown factors may be influencing the observed activity.

Biofilm formation is a crucial process in microbial infection as the majority of infections are attributed to biofilm formation and growth [37,38]. Furthermore, due to their large surface area and hydrophilic porous structure, textiles can retain water, oxygen and nutrients thus creating a propitious environment for microorganism attachment and growth. This attachment is of great concern as textile-embedded bacteria may double every 20–30 min, meaning that one single bacteria cell can increase to 1×10^6 cells in just 7 h [11,39]. As previously established, due to the lack of biological activity of synthetic textile dyes, the obtained results have to be compared to ones previously reported for void NPs. Concerning the activity of the yellow everzol NPs over the staphylococci tested, while previous works are in line with the results obtained here, as void chitosan NPs are reported as being capable of inhibiting MSSA, MRSA and *S. epidermidis* biofilm formation, the same cannot be said regarding MRSE as chitosan NPs are depicted as being incapable of inhibiting MRSE biofilm formation [22,24,40]. Considering that chitosan NPs biological activity is known to be strongly related to its zeta potential [29], and that previously Costa, Silva [24] have hypothesized that MRSE strong production of exopolysaccharides (EPS) during biofilm growth may hamper an NPs capacity to defuse into biofilms structure and interfere with membranes integrity, it is possible that the high zeta potential registered in this work (+54.2 mV to +17.3 mV) may be key in disrupting the biofilms EPS structure and allow NPs to interact with the MRSE cells surface. Regarding *P. aeruginosa* and *A. baumannii*, the obtained results are similar to those previously reported by Costa, Silva [26] for void chitosan NPs, with the higher zeta potential of the particles used in this work (+54.2 mV to +27.1 mV) explaining the higher inhibition percentages obtained here.

Quorum sensing is the way bacteria communicate with each other. By using small, diffusible signals that permit them to modulate various facets of microbial metabolism, such as biofilm formation, the QS molecules and systems present alternative targets for antimicrobial activity [41,42]. The values here registered fall in line with those registered in previous works (76.21%) for chitosan NPs inhibition of violacein production [42].

Methicillin-resistant *S. aureus's* capacity to invade human host cells is a fundamental part of its infectious process and its capacity to adhere to and infect HaCat cells has been described as being extremely high [43–45]. Works exploiting chitosan NPs', void or loaded, capacity to inhibit bacterial adhesion/infection are few and while none deal directly with textile dye-loaded NPs inhibition of MRSA in HaCat cells, some comparisons are still possible. When concerning chitosan NPs inhibition of *S. aureus* infection of human cells, the results obtained here find purchase in the existing work as Mu, Niu [46] comes in line with the activity here observed as it showed that chitosan conjugated with streptomycin eliminated bacterial intracellular counts in RAW264.7 macrophages. Additionally, when considering chitosan NPs' capacity to inhibit microorganisms in cellular systems, Mu, Niu [46] in the same work showed that these particles were also capable of eliminating *Listeria monocytogenes* and *Salmonella typhimurium* intracellular counts and that Elbi, Nimal [47] reported that chitosan NPs were capable of reducing, also in RAW264.7 cells, *Salmonella Paratyphi* A intracellular survival. Lastly, as a previous work has shown [26], void chitosan NPs could effectively reduce and inhibit *A. baumannii*'s infection of HaCat cells, thus lending credence to the results here reported.

Overall, the results obtained showed that the nanoencapsulation of textile dyes mitigates their known toxic potential as yellow everzol NPs were completely biocompatible with HaCat cells. Furthermore, yellow everzol NPs were biologically active as they were capable of inhibiting the planktonic and sessile growth of the tested microorganisms and had inhibitory activity in the *C. violaceum* quorum sensing reporter system. Additionally, the NPs were active in an ex vivo setting, as they reduced MRSA extracellular and intracellular counts in HaCat cells. Furthermore, these results open the possibility for the development of bioactive textile dyes and may allow for the elimination of the need for post-dying textile functionalization.

4. Materials and Methods

4.1. Sources of Chemicals and Solutions Preparation

Low-molecular-weight chitosan (LMW) was obtained from Sigma-Aldrich (St. Louis, MO, USA) and presented a DD between 75 and 85% and an MW of 107 kDa. Sodium Tripolyphosphate (TPP) was purchased from Sigma–Aldrich Chemical Co. Ltd. Reactive (yellow everzol) textile dye was kindly donated by Aquitex S.A. and was prepared at 10 mg/mL using ultra-pure water (Millipore SIM FILTER, Burlington, MA, USA) and stirred until complete dissolution.

4.2. Microorganisms and Cellular Line

Pseudomanas aeruginosa, VRSA, MSSA and MRSE strains were obtained from the American Type Culture Collection (ATCC 700699, ATCC10145, ATCC 25,923 and ATCC 51,625, respectively; Manassas, VA, USA). *A. baumannii* and MRSA strains were obtained from the culture collection of the Göteburg University (CCUG; Gothenburg, Sweden) (CCUG 61,012 and CCUG 60578). *S. epidermidis* was obtained from the Belgian Co-ordinated Collections of Microorganisms (Belgium) (LMG 10474; Gent, Belgium).

For antimicrobial testing inoculums were prepared in Luria-Bertani Broth (LB; ThermoFisher Scientific, Waltham, MA, USA) and for biofilms assays in Tryptic Soy Broth (TSB, Biokar Diagnostics, Beauvais, France).

For cell-based assays human keratinocytes (HaCat) were obtained from Cell Line Services (Oppenheim, Denmark). For assays cells were cultured, at 37 °C in a humidified atmosphere of 95% air and 5% CO_2, using Dulbecco's Modified Eagle's Medium (DMEM) with 4.5 g/L glucose, L-glutamine without pyruvate (Lonza, Verviers, Belgium) containing 10% foetal bovine serum (FBS, Biowest, Nuaillé, France) and 1% (*v*/*v*) Penicillin-Streptomycin-Fungizone (Lonza, Verviers, Belgium).

4.3. Nanoencapsulated Dyes Production and Characterization

Yellow everzol nanoencapsulation was performed as previously described [15]. Briefly, LMW chitosan was dissolved at 2 mg/mL in acetic acid that was 1.75× more concentrated and the pH was adjusted to 5 with NaOH. TPP was used at chitosan to TPP relation of 7:1. The process began with the addition of 4 mL of chitosan, placed under stirring at 500 rpm, to which a mixture of dye and TPP (1 and 2 mL, respectively) were gently added dropwise, at room temperature. At the end of the process, the loaded NPs were frozen and freeze-dried for posterior use in the biological assays.

The prepared dyed loaded NPs physical properties were analyzed through dynamic light scattering (DLS) using a Malvern Instruments NanoZSP (Worcestershire, UK) with particle size (PS), polydispersity index (PDI) and zeta potential (ZP) being determined. All assays were performed using a disposable folded capillary cell (Malvern, Worcestershire, UK), with a 90° laser angle, at room temperature (25 °C). Size values were determined based on intensity distribution. All assays were performed in sextuplicate.

4.4. Biocompatibility Assays

4.4.1. XTT Assay

The produced nanoparticles' impact upon HaCat metabolism was performed as previously described using the XTT viable dye [15,48]. Briefly, HaCat cells were seeded at 1×10^5 cells/mL in the wells of a 96-well microplate and allowed to adhere. After 24 h, the media was removed, and the cells were washed with PBS. Following this, media with loaded and void NPs (7 mg/mL to 0.5 mg/mL) was added. Simultaneously, two controls were accessed: one with textile dyes at 7 mg/mL and the other with sterile water. After 24 h, 25 µL of XTT working solution were added to each well and the cells were incubated, in the dark, for 2 h. The optical density (OD) at 485 nm was then measured using a microplate reader (FLUOstar, OPTIMA, BMG Labtech, Ortenberg, Germany). All assays were performed in quintuplicate.

4.4.2. LDH Leakage Assay

Nanoparticles cytotoxicity was assessed by LDH leakage into the culture medium. Following the exposure of HaCat cells to void and loaded nanoparticles (7 mg/mL to 0.5 mg/mL), the activity of LDH in the medium was determined using the commercially available kit from ThermoScientific (Rockford, IL, USA) and using accordingly to the manufacturer's instructions. Briefly, cells were seeded at 2×10^4 cells/100 µL of media, allowed to adhere for 24 h and then exposed to the nanoparticles. After 24 h incubation cells were processed accordingly to the kits instructions and the results were given accordingly to the following equation:

$$\text{\% Cytotoxicity} = [(\text{Compound treated LDH activity} - \text{Spontaneous LDH activity})/(\text{Maximum LDH activity} - \text{Spontaneous LDH activity})] \times 100 \quad (1)$$

4.5. Antimicrobial Activity

Evaluation of the compounds' MIC was executed as previously described Silva, Costa [49]. Briefly, a 0.5 MacFarland inoculum of each microorganism was prepared and inoculated in Muller Hinton Broth (Biokar Diagnostics, Beauvais, France) with dye loaded NPs concentrations ranging from 0.1 to 7 mg/mL. Simultaneously, a negative (non-inoculated media with NPs at 0.1 mg/mL) and a positive control (inoculated media with microorganism only) were also evaluated. The MIC was determined by observing the lowest sample concentration at which no bacterial growth was visible. All assays were performed in triplicate. Determination of the MBC was performed as previously described by Costa, Silva [26]. Succinctly, MBCs were defined as the lowest sample concentration at which bacterial growth in agar plates was prevented and it was evaluated by inoculation of 100 µL aliquots of negative wells (absence of turbidity in MIC determination) in Plate

Count Agar (PCA, Biokar Diagnostics, Beauvais, France), using the plate spread technique. All assays were performed in quadruplicate.

4.6. Antibiofilm Activity

Dye-loaded NPs' capacity to inhibit biofilm formation was evaluated as previously by Costa, Silva [50]. Briefly, in a flat-bottom 96-well microplate, TSB supplemented with 1% (*w*/*v*) glucose and dye NPs at sub-MIC concentrations (1/2 and 1/4 of the MIC) (was inoculated at 2% (*v*/*v*) using an overnight inoculum and was incubated for 24 h at 37 °C. Afterwards, the contents of the wells were discarded and each well was washed to remove non-adhered cells. Quantification of biofilm formation was performed through determination of the biomass present using the crystal violet assay. All assays were performed in quadruplicate.

4.7. Quorum Sensing (QS) Inhibition Assay

Screening of the QS inhibitory activity of NDYE was carried out based on their ability to inhibit the production of the purple pigment violacein by *C. violaceum*. The assay was performed as described by Costa, Silva [51]. Briefly, in a 96-well microplate, wells were filled with 200 mL of the test solution with NDYE added at concentrations between 7 and 0.25 mg/mL. Simultaneously, a positive control, where chitosan was substituted with water, was also assayed. Following that, the plate was incubated at 37 °C for 24 h in a microplate reader (FLUOstar, OPTIMA, BGM Labtech) with OD being measured at 577 nm for violacein production and 660 for bacterial growth. All assays were performed in triplicate.

4.8. Cell Infection Assays

The dyes NPs' antimicrobial in a cellular system was evaluated via adaptation of the work previously published by Pati, Mehta [52]. Briefly, human keratinocytes were seeded at 1×10^5 cells/mL and allowed to adhere. After 24 h, the media was removed, cells were washed with PBS and then three conditions were assayed: (i) cell culture media with MRSA (1×10^6 CFU/mL) and NDYEs at MIC; (ii) cell culture media with MRSA and NDYEs at MBC; (iii) cell culture media with MRSA only (infection control). After 24 h, the extracellular and intracellular viable counts were determined through serial dilution and plating. For the extracellular counts, the supernatant from the wells was harvested while for the intracellular counts HaCat cells were lysed with triton, harvested with PBS and then processed for viable counts. For both conditions, viable counts were determined by the drop method as previously described by Costa, Silva [26]. Results were given for both assays as total viable counts (extracellular counts) and total bacteria by HaCat cell (intracellular counts). All assays were performed in quadruplicate.

4.9. Statistical Analysis

This work's statistical analysis was performed using IBM SPSS Statistics v21.0.0 (New York, NY, USA) software. Considering that data followed a normal distribution, a one-way ANOVA coupled with Turkey's post hoc test was performed with Turkey's post hoc test being used with differences being considered significant for *p*-values below 0.05. Graphical processing of the data was performed using Graphad Prism 6 software (San Diego, CA, USA).

Author Contributions: Conceptualization, S.S. and E.M.C.; methodology, S.S. and E.M.C.; validation, F.K.T. and M.P.; writing—original draft preparation, E.M.C.; writing—review and editing, E.M.C. and S.S.; supervision, F.K.T. and M.P.; funding acquisition, E.M.C. and M.P. All authors have read and agreed to the published version of the manuscript.

Funding: This research was funded by Fundação para a Ciência e Tecnologia, through grant number UIDB/50016/2020 and E.M. Costa's PhD fellowship (SFRH/BDE/103957/2014, and through QREN-ANI via project 17819.

Institutional Review Board Statement: Not applicable.

Informed Consent Statement: Not applicable.

Data Availability Statement: The data presented in this study are available on request from the corresponding author. The data are not publicly available due to confidentiality agreements.

Conflicts of Interest: The authors declare no conflict of interest.

References

1. Nayak, M.C.; Isloor, A.M.; Moslehyani, A.; Ismail, N.; Ismail, A. Fabrication of novel PPSU/ZSM-5 ultrafiltration hollow fiber membranes for separation of proteins and hazardous reactive dyes. *J. Taiwan Inst. Chem. Eng.* **2018**, *82*, 342–350. [CrossRef]
2. Alver, E.; Bulut, M.; Metin, A.; Çiftçi, H. One step effective removal of Congo Red in chitosan nanoparticles by encapsulation. *Spectrochim. Acta Part A: Mol. Biomol. Spectrosc.* **2017**, *171*, 132–138. [CrossRef] [PubMed]
3. Wasim, M.; Sagar, S.; Sabir, A.; Shafiq, M.; Jamil, T. Decoration of open pore network in Polyvinylidene fluoride/MWCNTs with chitosan for the removal of reactive orange 16 dye. *Carbohydr. Polym.* **2017**, *174*, 474–483. [CrossRef] [PubMed]
4. Chaudhari, A.U.; Paul, D.; Dhotre, D.; Kodam, K.M. Effective biotransformation and detoxification of anthraquinone dye reactive blue 4 by using aerobic bacterial granules. *Water Res.* **2017**, *122*, 603–613. [CrossRef]
5. Jha, P.; Jobby, R.; Desai, N. Remediation of textile azo dye acid red 114 by hairy roots of Ipomoea carnea Jacq. and assessment of degraded dye toxicity with human keratinocyte cell line. *J. Hazard. Mater.* **2016**, *311*, 158–167. [CrossRef]
6. Cao, C.; Xiao, L.; Chen, C.; Shi, X.; Cao, Q.; Gao, L. In situ preparation of magnetic Fe3O4/chitosan nanoparticles via a novel reduction–precipitation method and their application in adsorption of reactive azo dye. *Powder Technol.* **2014**, *260*, 90–97. [CrossRef]
7. Luo, S.; Van Ooij, W.J. Surface modification of textile fibers for improvement of adhesion to polymeric matrices: A review. *J. Adhes. Sci. Technol.* **2002**, *16*, 1715–1735. [CrossRef]
8. Shahidi, S.; Wiener, J.; Ghoranneviss, M. Surface Modification Methods for Improving the Dyeability of Textile Fabrics. In *Eco-Friendly Textile Dyeing and Finishing*; Gunay, M., Ed.; IntechOpen: London, UK, 2013. [CrossRef]
9. Shahid, M.; Mohammad, F. Mohammad. Green Chemistry Approaches to Develop Antimicrobial Textiles Based on Sustainable Biopolymers—A Review. *Ind. Eng. Chem. Res.* **2013**, *52*, 5245–5260. [CrossRef]
10. Kasiri, M.B.; Safapour, S. Natural dyes and antimicrobials for green treatment of textiles. *Environ. Chem. Lett.* **2013**, *12*, 1–13. [CrossRef]
11. Coradi, M.; Zanetti, M.; Valério, A.; de Oliveira, D.; da Silva, A.; Souza, S.M.D.A.G.U.D.; de Souza, A.A.U. Production of antimicrobial textiles by cotton fabric functionalization and pectinolytic enzyme immobilization. *Mater. Chem. Phys.* **2018**, *208*, 28–34. [CrossRef]
12. Singh, R.; Jain, A.; Panwar, S.; Gupta, D.; Khare, S.K. Antimicrobial activity of some natural dyes. *Dye. Pigment.* **2005**, *66*, 99–102. [CrossRef]
13. Zhou, L.; Jin, J.; Liu, Z.; Liang, X.; Shang, C. Adsorption of acid dyes from aqueous solutions by the ethylenediamine-modified magnetic chitosan nanoparticles. *J. Hazard. Mater.* **2011**, *185*, 1045–1052. [CrossRef] [PubMed]
14. Hu, Z.; Zhang, J.; Chan, W.; Szeto, Y. The sorption of acid dye onto chitosan nanoparticles. *Polymer* **2006**, *47*, 5838–5842. [CrossRef]
15. Costa, E.M.; Silva, S.; Veiga, M.; Baptista, P.; Tavaria, F.K.; Pintado, M.E. Textile dyes loaded chitosan nanoparticles: Characterization, biocompatibility and staining capacity. *Carbohydr. Polym.* **2020**, *251*, 117120. [CrossRef] [PubMed]
16. Ali, S.W.; Rajendran, S.; Joshi, M. Synthesis and characterization of chitosan and silver loaded chitosan nanoparticles for bioactive polyester. *Carbohydr. Polym.* **2011**, *83*, 438–446. [CrossRef]
17. Yang, H.-C.; Wang, W.-H.; Huang, K.-S.; Hon, M.-H. Preparation and application of nanochitosan to finishing treatment with anti-microbial and anti-shrinking properties. *Carbohydr. Polym.* **2010**, *79*, 176–179. [CrossRef]
18. Lu, Y.-H.; Chen, Y.-Y.; Lin, H.; Wang, C.; Yang, Z.-D. Preparation of chitosan nanoparticles and their application to Antheraea pernyi silk. *J. Appl. Polym. Sci.* **2010**, *117*, 3362–3369. [CrossRef]
19. Cheung, A.; Szeto, Y.S.; McKay, G. Enhancing the adsorption capacities of acid dyes by chitosan nano particles. *Bioresour. Technol.* **2009**, *100*, 1143–1148. [CrossRef]
20. Zeng, Z.W.; Wang, J.J.; Xiao, R.Z.; Xie, T.; Zhou, G.L.; Zhan, X.R.; Wang, S.L. Recent advances of chitosan nanoparticles as drug carriers. *Int. J. Nanomed.* **2011**, *6*, 765–774. [CrossRef]
21. *ISO 10993-5:2009*; Biological Evaluation of Medical Devices in Tests for In Vitro Cytotoxicity. International Organization for Standardization: Geneva, Switzerland, 2009; p. 34.
22. Darabpour, E.; Kashef, N.; Mashayekhan, S. Chitosan nanoparticles enhance the efficiency of methylene blue-mediated antimicrobial photodynamic inactivation of bacterial biofilms: An in vitro study. *Photodiagnosis Photodyn. Ther.* **2016**, *14*, 211–217. [CrossRef]
23. Madureira, A.R.; Pereira, A.; Castro, P.M.; Pintado, M. Production of antimicrobial chitosan nanoparticles against food pathogens. *J. Food Eng.* **2015**, *167*, 210–216. [CrossRef]

24. Costa, E.; Silva, S.; Vicente, S.; Neto, C.; Castro, P.; Veiga, M.; Madureira, R.; Tavaria, F.; Pintado, M. Chitosan nanoparticles as alternative anti-staphylococci agents: Bactericidal, antibiofilm and antiadhesive effects. *Mater. Sci. Eng. C* **2017**, *79*, 221–226. [CrossRef] [PubMed]
25. Perinelli, D.R.; Fagioli, L.; Campana, R.; Lam, J.K.W.; Baffone, W.; Palmieri, G.F.; Casettari, L.; Bonacucina, G. Chitosan-based nanosystems and their exploited antimicrobial activity. *Eur. J. Pharm. Sci.* **2018**, *117*, 8–20. [CrossRef] [PubMed]
26. Costa, E.M.; Silva, S.; Veiga, M.; Tavaria, F.; Pintado, M.M. Exploring chitosan nanoparticles as effective inhibitors of antibiotic resistant skin microorganisms–From in vitro to ex vitro testing. *Carbohydr. Polym.* **2018**, *201*, 340–346. [CrossRef]
27. Du, W.-L.; Niu, S.-S.; Xu, Y.-L.; Xu, Z.-R.; Fan, C.-L. Antibacterial activity of chitosan tripolyphosphate nanoparticles loaded with various metal ions. *Carbohydr. Polym.* **2009**, *75*, 385–389. [CrossRef]
28. Chakraborty, S.P.; Sahu, S.K.; Mahapatra, S.K.; Santra, S.; Bal, M.; Roy, S.; Pramanik, P. Nanoconjugated vancomycin: New opportunities for the development of anti-VRSA agents. *Nanotechnology* **2010**, *21*, 105103. [CrossRef] [PubMed]
29. Silva, N.C.; Silva, S.; Sarmento, B.; Pintado, M. Chitosan nanoparticles for daptomycin delivery in ocular treatment of bacterial endophthalmitis. *Drug Deliv.* **2015**, *22*, 885–893. [CrossRef]
30. Bilal, M.; Rasheed, T.; Iqbal, H.M.; Hu, H.; Wang, W.; Zhang, X. Macromolecular agents with antimicrobial potentialities: A drive to combat antimicrobial resistance. *Int. J. Biol. Macromol.* **2017**, *103* (Suppl. C), 554–574. [CrossRef] [PubMed]
31. Aleanizy, F.S.; Alqahtani, F.Y.; Shazly, G.; Alfaraj, R.; Alsarra, I.; Alshamsan, A.; Abdulhady, H.G. Measurement and evaluation of the effects of pH gradients on the antimicrobial and antivirulence activities of chitosan nanoparticles in Pseudomonas aeruginosa. *Saudi Pharm. J.* **2017**, *26*, 79–83. [CrossRef]
32. Freire, P.L.; Albuquerque, A.J.; Farias, I.A.; da Silva, T.G.; Aguiar, J.S.; Galembeck, A.; Flores, M.A.; Sampaio, F.C.; Stamford, T.C.M.; Rosenblatt, A. Antimicrobial and cytotoxicity evaluation of colloidal chitosan–silver nanoparticles–fluoride nanocomposites. *Int. J. Biol. Macromol.* **2016**, *93*, 896–903. [CrossRef]
33. El-Naggar, M.Y.; Gohar, Y.M.; Sorour, M.A.; Waheeb, M.G. Hydrogel Dressing with a Nano-Formula against Methicillin-Resistant Staphylococcus aureus and Pseudomonas aeruginosa Diabetic Foot Bacteria. *J. Microbiol. Biotechnol.* **2016**, *26*, 408–420. [CrossRef] [PubMed]
34. Ing, L.Y.; Zin, N.M.; Sarwar, A.; Katas, H. Antifungal Activity of Chitosan Nanoparticles and Correlation with Their Physical Properties. *Int. J. Biomater.* **2012**, *2012*, 632698. [CrossRef] [PubMed]
35. Ma, Z.; Garrido-Maestu, A.; Jeong, K.C. Application, mode of action, and in vivo activity of chitosan and its micro- and nanoparticles as antimicrobial agents: A review. *Carbohydr. Polym.* **2017**, *176*, 257–265. [CrossRef] [PubMed]
36. Yildirim-Aksoy, M.; Beck, B. Antimicrobial activity of chitosan and a chitosan oligomer against bacterial pathogens of warmwater fish. *J. Appl. Microbiol.* **2017**, *122*, 1570–1578. [CrossRef] [PubMed]
37. Chopra, S.; Harjai, K.; Chhibber, S. Antibiotic susceptibility of ica-positive and ica-negative MRSA in different phases of biofilm growth. *J. Antibiot.* **2014**, *68*, 15–22. [CrossRef] [PubMed]
38. Chopra, S.; Harjai, K.; Chhibber, S. Potential of sequential treatment with minocycline and S. aureus specific phage lysin in eradication of MRSA biofilms: An in vitro study. *Appl. Microbiol. Biotechnol.* **2015**, *99*, 3201–3210. [CrossRef]
39. Morais, D.S.; Guedes, R.M.; Lopes, M.A. Antimicrobial Approaches for Textiles: From Research to Market. *Materials* **2016**, *9*, 498. [CrossRef]
40. Tan, Y.; Ma, S.; Liu, C.; Yu, W.; Han, F. Enhancing the stability and antibiofilm activity of DspB by immobilization on carboxymethyl chitosan nanoparticles. *Microbiol. Res.* **2015**, *178*, 35–41. [CrossRef]
41. Holban, A.M.; Gestal, M.C.; Grumezescu, A.M. Control of biofilm-associated infections by signaling molecules and nanoparticles. *Int. J. Pharm.* **2016**, *510*, 409–418. [CrossRef]
42. Ilk, S.; Sağlam, N.; Özgen, M.; Korkusuz, F. Chitosan nanoparticles enhances the anti-quorum sensing activity of kaempferol. *Int. J. Biol. Macromol.* **2017**, *94*, 653–662. [CrossRef]
43. Ballhausen, B.; Jung, P.; Kriegeskorte, A.; Makgotlho, P.E.; Ruffing, U.; von Müller, L.; Köck, R.; Peters, G.; Herrmann, M.; Ziebuhr, W.; et al. LA-MRSA CC398 differ from classical community acquired-MRSA and hospital acquired-MRSA lineages: Functional analysis of infection and colonization processes. *Int. J. Med. Microbiol.* **2014**, *304*, 777–786. [CrossRef] [PubMed]
44. Bünter, J.P.; Seth-Smith, H.M.; Rüegg, S.; Heikinheimo, A.; Borel, N.; Johler, S. Wild type agr-negative livestock-associated MRSA exhibits high adhesive capacity to human and porcine cells. *Res. Microbiol.* **2017**, *168*, 130–138. [CrossRef]
45. Fraunholz, M.; Sinha, B. Intracellular staphylococcus aureus: Live-in and let die. *Front. Cell. Infect. Microbiol.* **2012**, *2*, 43. [CrossRef] [PubMed]
46. Mu, H.; Niu, H.; Wang, D.; Sun, F.; Sun, Y.; Duan, J. Chitosan conjugation enables intracellular bacteria susceptible to aminoglycoside antibiotic. *Glycobiology* **2016**, *26*, 1190–1197. [CrossRef] [PubMed]
47. Elbi, S.; Nimal, T.R.; Rajan, V.K.; Baranwal, G.; Biswas, R.; Jayakumar, R.; Sathianarayanan, S. Fucoidan coated ciprofloxacin loaded chitosan nanoparticles for the treatment of intracellular and biofilm infections of Salmonella. *Colloids Surf. B Biointerfaces* **2017**, *160* (Suppl. C), 40–47.
48. Costa, E.M.; Pereira, C.F.; Ribeiro, A.A.; Casanova, F.; Freixo, R.; Pintado, M.; Ramos, O.L. Characterization and Evaluation of Commercial Carboxymethyl Cellulose Potential as an Active Ingredient for Cosmetics. *Appl. Sci.* **2022**, *12*, 6560. [CrossRef]
49. Silva, S.; Costa, E.M.; Costa, M.R.; Pereira, M.F.; Pereira, J.O.; Soares, J.C.; Pintado, M.M. Aqueous extracts of Vaccinium corymbosum as inhibitors of Staphylococcus aureus. *Food Control* **2015**, *51*, 314–320. [CrossRef]

50. Costa, E.M.; Silva, S.; Madureira, A.R.; Cardelle-Cobas, A.; Tavaria, F.K.; Pintado, M.M. A comprehensive study into the impact of a chitosan mouthwash upon oral microorganism's biofilm formation in vitro. *Carbohydr. Polym.* **2014**, *101*, 1081–1086. [CrossRef]
51. Costa, E.M.; Silva, S.; Pina, C.; Tavaria, F.K.; Pintado, M. Antimicrobial Effect of Chitosan against Periodontal Pathogens Biofilms. *SOJ Microbiol. Infect. Dis.* **2014**, *2*, 1–6.
52. Pati, R.; Mehta, R.K.; Mohanty, S.; Padhi, A.; Sengupta, M.; Vaseeharan, B.; Goswami, C.; Sonawane, A. Topical application of zinc oxide nanoparticles reduces bacterial skin infection in mice and exhibits antibacterial activity by inducing oxidative stress response and cell membrane disintegration in macrophages. *Nanomed. Nanotechnol. Biol. Med.* **2014**, *10*, 1195–1208. [CrossRef]

International Journal of
Molecular Sciences

MDPI

Review

Nanoparticles for Antimicrobial Agents Delivery—An Up-to-Date Review

Doina-Antonia Mercan [1], Adelina-Gabriela Niculescu [1] and Alexandru Mihai Grumezescu [1,2,3,*]

[1] Department of Science and Engineering of Oxide Materials and Nanomaterials, Politehnica University of Bucharest, 011061 Bucharest, Romania
[2] Research Institute of the University of Bucharest—ICUB, University of Bucharest, 050657 Bucharest, Romania
[3] Academy of Romanian Scientists, Ilfov No. 3, 050044 Bucharest, Romania
* Correspondence: grumezescu@yahoo.com

Abstract: Infectious diseases constitute an increasing threat to public health and medical systems worldwide. Particularly, the emergence of multidrug-resistant pathogens has left the pharmaceutical arsenal unarmed to fight against such severe microbial infections. Thus, the context has called for a paradigm shift in managing bacterial, fungal, viral, and parasitic infections, leading to the collision of medicine with nanotechnology. As a result, renewed research interest has been noted in utilizing various nanoparticles as drug delivery vehicles, aiming to overcome the limitations of current treatment options. In more detail, numerous studies have loaded natural and synthetic antimicrobial agents into different inorganic, lipid, and polymeric-based nanomaterials and tested them against clinically relevant pathogens. In this respect, this paper reviews the most recently reported successfully fabricated nanoformulations that demonstrated a great potential against bacteria, fungi, viruses, and parasites of interest for human medicine.

Keywords: antimicrobial therapy; antimicrobial resistance; drug delivery systems; antimicrobials delivery; nanocarriers; antibacterial nanoformulations; antifungal nanoformulations; antiviral nanoformulations; antiparasitic nanoformulations

Citation: Mercan, D.-A.; Niculescu, A.-G.; Grumezescu, A.M. Nanoparticles for Antimicrobial Agents Delivery—An Up-to-Date Review. *Int. J. Mol. Sci.* **2022**, *23*, 13862. https://doi.org/10.3390/ijms232213862

Academic Editor: Helena Felgueiras

Received: 24 October 2022
Accepted: 8 November 2022
Published: 10 November 2022

1. Introduction

Even though humans and pathogens have always dynamically interacted, this relationship became unbalanced. Human activities have caused pathogenic microbes, such as bacteria, fungi, viruses, and parasites, to appear and spread at a progressively distressing rate, rendering infectious diseases a common and burdensome health issue worldwide [1–4]. Numerous antimicrobial agents can be employed to fight against infections, yet they often face several limitations [5–7]. Specifically, the efficacy of conventional antimicrobial drugs is affected by their poor oral bioavailability and stability, low water solubility, low transportation rate across cellular membranes, lack of targeting, and systemic adverse effects [8–11]. Another important drawback of traditional drug-based therapeutic strategies is the inappropriate and inadequate administration of antimicrobial agents that have contributed to the emergence of drug-resistant pathogens and the formation of well-organized microbial communities called biofilms [12–16].

Unfortunately, biofilms have become a highly frequent problem in the clinical environment as microorganisms may adhere to and colonize the surfaces of biomedical devices. Thus, in order to avoid the acquiring of nosocomial infections, the burden of biofilms must be especially considered when using implantable and indwelling medical devices (e.g., catheters, stents, heart valves, pacemakers, prosthetic joints and implants, voice prostheses, contact lenses, internal and external fixation devices) [14,16–23]. Generally, the management of device-associated infections involves prolonged inpatient stay, surgical intervention, and long-term postoperative antibiotic therapy, all adding to healthcare costs and low patient compliance [10,13,24].

Despite existing therapies and medicines, infectious diseases in general, and biofilms in particular, remain difficult-to-eradicate problems [3]. Therefore, research efforts must be put into developing antimicrobial strategies able to surpass current challenges. In this context, the advances in nanotechnology represent promising opportunities for designing novel antimicrobial systems. A variety of nanoparticles (NPs) can be employed in developing performant delivery vehicles for natural and synthetic medicines capable of enhancing the activity of carried freight, ensuring a sustained drug release, and offering a chance of biofilm penetration and internalization into pathogenic microorganisms [8,25,26]. NPs loaded with antimicrobial agents can be used as effective therapeutics administered on different routes, but they can also be incorporated further into biomaterials for modifying surface nanotopography or coating biomedical devices, intending to potentiate or induce anti-infective properties [9,11,27–31].

In this respect, the present paper aims to overview the most recently developed antimicrobial nanoformulations, fabricated mainly between 2018 and 2022, that showed promising results when tested against clinically relevant pathogens, emphasizing their utility and versatility. Even though the topic has been previously addressed by several works [32–36], this review proposes a comprehensive path, correlating materials, fabrication methods, and delivered antimicrobial agents with targeted microorganisms, focusing on the applicability of drug delivery systems in treating and preventing bacterial, fungal, viral, and parasitic infections, and updating the literature with the newest developments in the field.

2. Nanoparticles for Antimicrobial Applications

NPs represent a key component in developing innovative anti-infectious strategies merging therapeutics and new materials toward enhancing antimicrobial potential. Their advantageous intrinsic properties, such as the high specific surface area in relation to volume and increased particle surface energy, render these materials more reactive and effective than their bulk counterparts [37–39]. Moreover, their small size is very suitable for antimicrobial biological operations, allowing NPs to interact with biological systems at the molecular level, permitting the targeted delivery of drugs and genes, and ensuring passage through biological barriers [39–44].

Moreover, the efficacy of NPs as antimicrobial agents' carriers can be improved for specific goals (e.g., increased cellular uptake, selective recognition, non-cytotoxicity, better payload binding capacity) through various surface-functionalization approaches. Specifically, stimuli-responsive ligands or functional groups can be used for modifying the surface layer of NPs through different physical, chemical, or biological methods toward achieving optimal antimicrobial activity [31]. In this respect, two main stimulation approaches have been actively researched for delivery nanosystems: locally stimulated or externally stimulated (Figure 1). The first category assumes cargo release in response to chemical and biochemical stimuli at intracellular (e.g., enzymatic activities, hydrolysis, pH, etc.) or tissue level (i.e., specific microenvironmental changes associated with pathological conditions). In contrast, the second category of delivery vehicles supposes the activated targeting and sustained release under the influence of external stimuli, including magnetic fields, electric stimulation, ultrasound, light, and temperature [11,45].

Taking into account the inherent beneficial properties of NPs and the variety of surface engineering possibilities, numerous studies have developed a wide range of antimicrobial nanoformulations aiming to bring effective solutions against relevant infections (Figure 2).

In this respect, the following subsections emphasize the recent progress in the fabrication of nanosystems for the delivery of antibacterial, antifungal, antiviral, and antiparasitic agents.

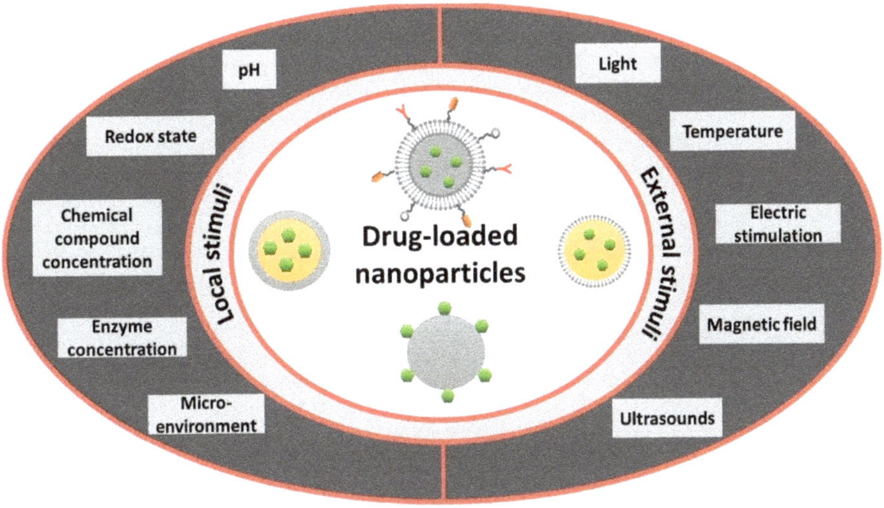

Figure 1. The main types of stimuli-responsive nanoparticles (NPs) for drug delivery.

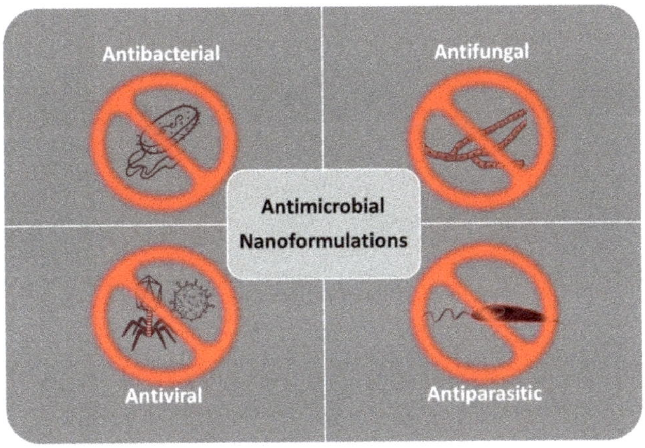

Figure 2. Antimicrobial nanoformulation possibilities.

2.1. Brief Overview of Nanoparticles Synthesis Methods

When discussing nanomaterials fabrication, two main approaches can be distinguished: top-down and bottom-up. The top-down approach implies starting from larger structures and reducing their size by means of mechanical force and the aid of finer and finer tools until reaching dimensions in the nano range [46,47]. These methods are preferred in industrial settings, as they can be easily scaled-up and produce fine particles with fine particle-producing capacity and reproducibility. Nonetheless, expensive equipment and intensive energy are required in such processes without guaranteeing control over particle growth and products' purity [46,48]. In opposition to top-down techniques, bottom-up processes assume the fabrication of nanoparticles through the growth and self-assembly of smaller components of atomic or molecular dimensions, conforming to a natural physical principle or an externally applied driving force [47]. Such methods are simple, rapid, energy-efficient, and cost-effective, being ideal options for laboratory-scale production of amorphous particles with reduced dimensions, narrow particle size distribution, increased solubility, and enhanced bioavailability. However, NPs obtained in this manner tend to agglomerate and might also present stability issues, while there are also several drawbacks associated with the fabrication processes (e.g., low yield, interbatch variability, scaling up challenges) [46,48].

A variety of physical, chemical, and biological fabrication methods are available for the synthesis of nanostructures (Table 1), including co-precipitation, hydrothermal synthesis, inert gas condensation, sputtering, microemulsion, microwave-assisted, laser ablation, sol-gel, ultrasound, spark discharge, template synthesis, and biological synthesis [39,46]. Depending on the chemical nature of the NPs, desired properties of the final product, and the cost-effectiveness of fabrication steps, one may prefer one method over the others. Some of the most employed techniques for the fabrication of magnetic NPs, polymeric NPs, and lipid-based NPs, have been gathered in Figures 3–5, respectively, offering a visual perspective on nanoconstructs' synthesis.

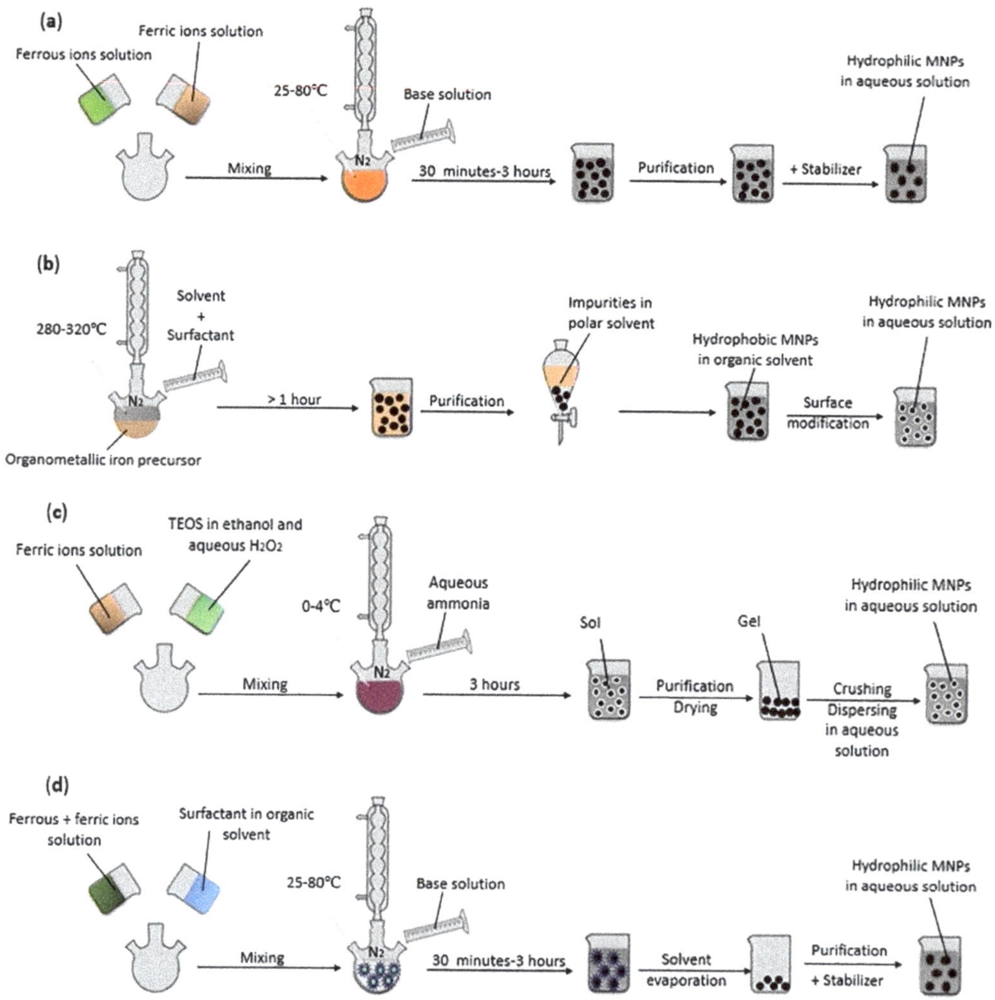

Figure 3. Schematic representation of commonly used synthesis methods to produce magnetite nanoparticles (MNPs): (**a**) co-precipitation; (**b**) thermal decomposition; (**c**) sol–gel; (**d**) microemulsion. Reprinted with permission from [49], © Elsevier, 2022.

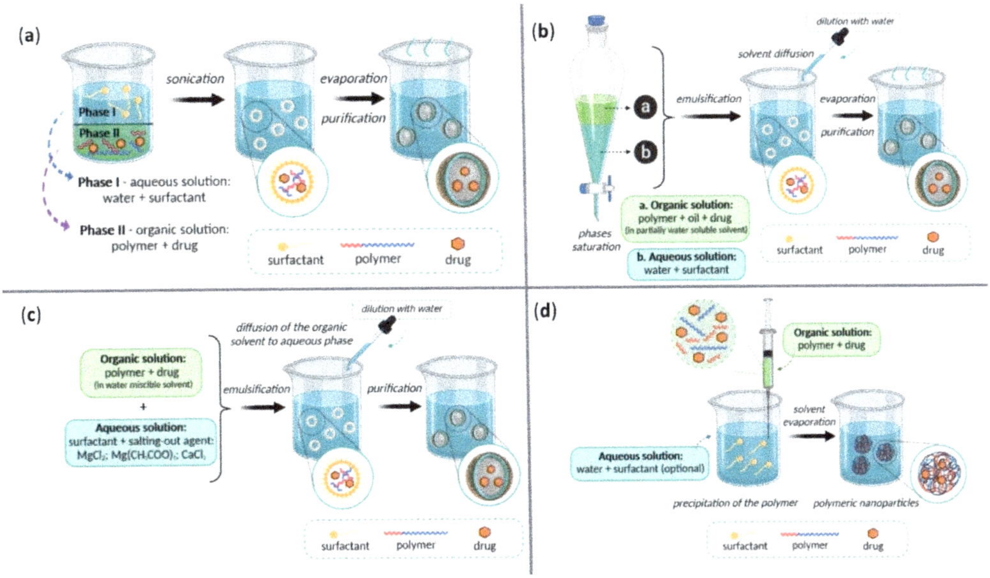

Figure 4. Schematic representation of commonly used synthesis methods to produce polymeric NPs: (**a**) solvent evaporation method; (**b**) emulsification/reverse salting-out method; (**c**) emulsification/solvent diffusion method; (**d**) nanoprecipitation method. Reprinted from an open-access source [50], adapted from [51].

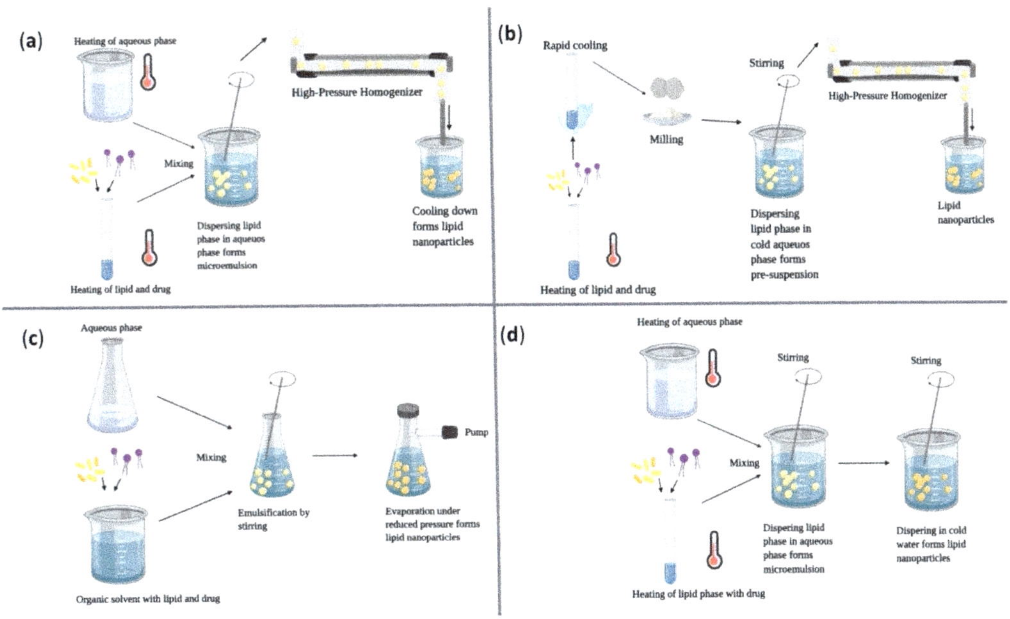

Figure 5. Schematic representation of commonly used synthesis methods to produce lipid-based NPs: (**a**) hot high-pressure homogenization method; (**b**) cold high-pressure homogenization method; (**c**) solvent evaporation method; (**d**) microemulsion method. Reprinted from an open-access source [52], adapted from [53].

Table 1. Classification of NP synthesis methods.

Synthesis Approach	Nature of Involved Processes	Examples of Techniques
Top-down approach	Physical methods	Ball milling, Laser ablation, Electron beam deposition, Sputtering, Aerosol spray
Bottom-up approach	Chemical methods	Co-precipitation, Thermal decomposition, Sol-gel, Microemulsion Sonochemical, Hydrothermal, Microwave assisted, Chemical reduction, Electrochemical, Solvothermal
	Biological methods	Bacteria-based, Plant-based

Moreover, for producing NPs with tailored structures, mixed approaches can be employed, generally involving a preprocessing step followed by a high-energy step. As each method has its advantages and limitations, thoughtful consideration is required when choosing the synthesis method so that the final nanostructures would have physicochemical stability, low polydispersity, reproducible size, high purity, and optimum morphology for antimicrobial drug delivery purposes [48].

2.2. *Antibacterial Nanoformulations*

Bacterial infections represent one of the biggest global health problems, remaining a significant cause of morbidity and mortality despite the numerous available antibiotics [54]. This is primarily due to the appearance of multi-drug resistant bacterial strains that cannot be effectively treated with conventional therapeutics [40,55–57]. Improper prescription of antimicrobial drugs and overuse and/or misuse of antibiotics has led to the current antimicrobial resistance growing crisis, enhancing microbial virulence and allowing bacteria to evade the host's immune response under the protection of a biofilm [40,56,58–61].

Therefore, a different approach had to be taken to enhance the antibacterial properties of existing drugs. Nanomaterials have become an attractive solution for transporting and releasing hydrophilic and lipophilic antibiotics or natural antimicrobial agents, as they can overpower bacterial resistance through several mechanisms. In more detail, NPs can ensure targeted delivery, allow passage through biological barriers, permeate and destroy the bacterial cell membrane, induce antimicrobial effects within cells, and impede biofilm formation [8,41,62–67].

Inorganic nanomaterials such as metal and metal oxide NPs have been of particular interest in creating antibacterial nanoformulations given their advantageous properties (e.g., low cost, long duration, safety, intrinsic antimicrobial activity) [59,68]. These materials act upon bacterial cells mainly through metal ions release, further increasing reactive oxygen species production and affecting bacterial metabolism. Nonetheless, repeated exposure can cause developing resistance even against these NPs. Thus, their antimicrobial efficacy is often enhanced through surface functionalization [69].

Some of the most commonly employed inorganic NPs for antibacterial applications are based on silver, iron oxide, zinc oxide, titanium oxide, magnesium oxide, and silica [69–71]. Drugs can be loaded into these systems either as coatings/shells on the NP surface [72], or they can be incorporated into the pores of the material [73]. Various synthetic antimicrobial agents have been used as cargos, including streptomycin [72], neomycin [72], vancomycin [74], cephalexin [75], and ciprofloxacin [76], leading to stronger effects against numerous pathogens. Specifically, antibiotic-loaded inorganic NPs have been tested with promising results against relevant bacterial strains, counting *Staphylococcus aureus* (*S. aureus*) [72,73,75,76], *Pseudomonas aeruginosa* (*P. aeruginosa*) [72], *Bacillus subtilis* (*B. subtilis*) [74], *Bacillus cereus* (*B. cereus*) [75], *Streptococcus* [74], *Escherichia coli* (*E. coli*) [73–75], and *Salmonella typhimurium* (*S. typhimurium*) [75].

Moreover, the nanosystems' antimicrobial properties and safety can be enhanced by adding biocompatible coatings [72,75,77]. In addition, surface functionalization of NPs can be performed to bypass triggering host defense mechanisms until reaching the site of infection and avoid potential adverse reactions or inhibition of NPs bioactivity [66,78–80].

Polymers represent another class of highly convenient materials for fabricating antimicrobial drug-delivery NPs. The main advantages of polymeric materials reside in their variety, versatility, and ease of functionalization. These characteristics render polymers suitable for improving drug solubility, delivering the cargo to the desired site, and targeting bacterial pathogens [50,67,81].

Recent studies have focused on developing vehicles from natural polymers, as they are generally recognized to possess superior biocompatibility to synthetic materials. The most commonly employed natural polymers for antibacterial agents encapsulation are polysaccharides, chitosan and alginate being the choice of numerous studies [55,82–93]. Nonetheless, synthetic polymers such as polylactic acid (PLA) [94,95], poly(lactic-co-glycolic acid (PLGA) [96,97], and polyvinylpyrrolidone (PVP) [98] have also attracted research interest as nanocarriers.

In what concerns the freight, a broad range of antimicrobial agents has been reported in the literature as suitable for polymer encapsulation. Synthetic drugs (e.g., levofloxacin [82], gentamicin [83,92], N′-((5-nitrofuran-2-yl)methylen)-2-benzhydrazide [86], rifampicin [55,94], ascorbic acid [55], doxycycline [88], rifaximin [89], ampicillin [91], teicoplanin [98], camptothecin [99], and vancomycin [100]) and natural antimicrobials (e.g., oregano oil [90,101], *Cinnamomum zeylanicum* (*C. zeylanicum*) essential oil [84], nettle essential oil [85], *Pistacia lentiscus* (*P. lentiscus*) L. var. *chia* essential oil [95], and red propolis extract [96]) have been successfully incorporated into nanosized polymeric materials.

The synergic properties of engineered polymeric NPs and carried antibacterial moieties have led to the obtaining of promising candidates for anti-infective therapeutics against *S. aureus* [55,82,83,85–87,90–92,94,96,97,99,100], *E. coli* [83–85,89,90,92,95,99], *P. aeruginosa* [82,89,90,92,93,96], *Erwinia carotovora* (*E. carotovora*) [84], *Pseudomonas fluorescens* (*P. fluorescens*) [84], *Enterococcus faecalis* (*E. faecalis*) [88,90], *Proteus mirabilis* (*P. mirabilis*) [88], *Bacillus haynesii* (*B. haynesii*) [89], *Streptococcus pyogenes* (*S. pyogenes*) [90], *Yersinia enterocolitica* (*Y. enerocolitica*) [90], *Listeria monocytogenes* (*L. monocytogenes*) [92], *B. subtilis* [95], *Streptococcus pneumoniae* (*S. pneumoniae*) [98], *Haemophilus influenzae* (*H. influenzae*) [98], and *Klebsiella pneumonia* (*K. pneumoniae)* [99].

To emphasize the variety and versatility of recently developed antibacterial nanoformulations, Table 2 summarizes several studies that fabricated promising drug delivery nanosystems for fighting against clinically relevant strains. In addition, Figure 6 provides a visual perspective over some of the discussed nanoconstructs.

Table 2. Examples of antibacterial nanoformulations.

NP Type	Fabrication Method	Physicochemical Properties	Antimicrobial Agent(s)	Targeted Pathogen(s)	Ref.
Magnetite NPs	Co-precipitation	Size range: ~2.8–~4.7 nm Shape: spherical	Streptomycin/ neomycin	*S. aureus*, *P. aeruginosa*	[72]
Magnetite NPs	Co-precipitation	Size range: 10–20 nm Average hydrodynamic diameter: 39.3 nm	Vancomycin	*B. subtilis*, *Streptococcus*, *E. coli*	[74]
Basil seed mucilage coated magnetite NPs	Co-precipitation	Mean size: 6 nm Specific surface area: 30.60 m^2g^{-1}	Cephalexin	*E. coli*, *S. typhimurium*, *S. aureus*, *B. cereus*	[75]

Table 2. *Cont.*

NP Type	Fabrication Method	Physicochemical Properties	Antimicrobial Agent(s)	Targeted Pathogen(s)	Ref.
Mesoporous iron oxide NPs	Co-precipitation	Average size: 78.34 ± 1.38 nm Zeta potential: −18.45 ± 1.89 mV Superficial area: 258.27 ± 7.51 m^2g^{-1}	Ciprofloxacin	*S. aureus*	[76]
Silanized mesoporous iron oxide NPs	Co-precipitation	Average size: 86.32 ± 2.0 nm Zeta potential: +5.76 ± 0.65 mV Superficial area: 186.27 ± 6.68 m^2g^{-1}	Ciprofloxacin	*S. aureus*	[76]
CTAB-loaded mesoporous silica NPs	Hydrothermal method	Size range: ~100–110 nm Shape: quasi-spherical	Silver NPs	*E. coli, S. aureus*	[73]
Chitosan NPs	Ionic gelation	Mean size: ranging from 161.90 ± 3.32 nm to 283.97 ± 4.21 nm Zeta potential: ranging from +30.43 ± 1.08 to +21.87 ± 1.87 mV	Levofloxacin	*P. aeruginosa, S. aureus*	[82]
Chitosan NPs	Ionotropic gelation	Average size: 135.2 ± 3.24 nm Shape: spherical	Gentamicin	*E. coli, S. aureus*	[83]
Chitosan NPs	Ionic gelation	Size range: 20–80 nm Hydrodynamic diameter: 141.4–181.6 nm Zeta potential: ranging from +49.9 to +38.7 mV Shape: spherical	*C. zeylanicum* essential oil	*E. coli, E. carotovora, P. fluorescens*	[84]
Chitosan NPs	Two-stage emulsion-ionic gelation method	Mean size: ranging from 208.3 ± 44.5 to 369.4 ± 48.1 nm Zeta potential: ranging from +30.1 ± 2.3 to +14.46 ± 0.9 mV	Nettle essential oil	*E. coli, S. aureus*	[85]
Polysorbate 20 micelles loaded in chitosan NPs	Ionic gelation	Average size: 321 nm Zeta potential: +37 mV Shape: spherical	N′-((5-nitrofuran-2-yl)methylen)-2-benzhydrazide	Multidrug-resistant *S. aureus*	[86]
Alginate-chitosan NPs	Ionic gelation	Average hydrodynamic diameter: 380 ± 15 nm Zeta potential: −28.5 ± 0.03	Rifampicin and ascorbic acid	MSSA, MRSA	[55]
Alginate-chitosan NPs	Calcium ion-induced pre-gelation of alginate core and further complexation with chitosan	Average hydrodynamic diameter: 276.5 ± 42 nm Zeta potential: −25 mV	LysMR-5	*S. aureus*	[87]
Chitosan-alginate NPs	Ionotropic gelation	Average size: 61.9 nm	Doxycycline	*E. faecalis, P. mirabilis*	[88]
Chitosan-alginate core-shell NPs	Precipitation/coacervation method	Size range: 700–1150 nm Zeta potential: −16.61 mV	Rifaximin	*E. coli, P. aeruginosa, Bacillus haynesii*	[89]
Chitosan-alginate NPs	Emulsification and consequent electrostatic gelation	Average size: 320 nm Zeta potential: −25 mV	Oregano oil	MSSA, MRSA, *E. faecalis, S. pyogenes, E. coli, P. aeruginosa, Y. enterocolitica*	[90]
Chitosan-polyanion NPs	Ionic gelation and polyelectrolyte complexation assisted by high-intensity sonication	Average size: ranging from 130.7 to 249.2 nm Zeta potential: ranging from +39.5 to +49.2 mV	Ampicillin	*S. aureus*	[91]
Phosphatidylcholine-chitosan liposome NPs	Ionic gelation	Average size: ~140 nm Zeta potential: −19.5 mV	Gentamicin	*L. monocytogenes, S. aureus, P. aeruginosa, E. coli*	[92]
Dextran NPs	Ionic gelation	Average size: 18 nm Zeta potential: −13 mV	SET-M33 peptide	*P. aeruginosa*	[93]

Table 2. *Cont.*

NP Type	Fabrication Method	Physicochemical Properties	Antimicrobial Agent(s)	Targeted Pathogen(s)	Ref.
PLA NPs functionalized with poly-L-lysine	Surfactant-free nanoprecipitation	Average hydrodynamic diameter: 162 ± 2 nm Zeta potential: +40 ± 2 mV	Rifampicin	*S. aureus*	[94]
PLA/PVA NPs	Solvent evaporation method	Average size: 239.9 nm Zeta potential: −29.1 mV	*P. lentiscus* L. var. *chia* essential oil	*E. coli*, *B. subtilis*	[95]
PLA/lecithin	Solvent evaporation method	Average size: 286.1 nm Zeta potential: −34.5 mV	*Pistacia lentiscus* L. var. *chia* essential oil	*E. coli*, *B. subtilis*	[95]
PLGA NPs	Emulsification solvent diffusion method	Average size: 69.2 nm Average hydrodynamic diameter: 224.23 ± 18.87 nm Zeta potential: −32.1 ± 4.53 mV Shape: spherical	Red propolis extract	*S. aureus*, *P. aeruginosa*	[96]
PLGA NPs functionalized with specific aptamers	Oil-in-water emulsification-evaporation method	Average hydrodynamic size: 226.00 ± 5.57 nm Zeta potential: 29.00 ± 1.18 mV	Teicoplanin	*S. aureus*	[97]
PVP-coated silver NPs	Chemical reduction	Average size: 9.23 ± 2.03 nm Shape: spherical	Silver NPs	*S. pneumoniae*, *H. influenzae*	[98]

Abbreviations: CTAB—cetyltrimethylammonium bromide; MRSA—methicillin-resistant *Staphylococcus aureus*; MSSA—methicillin-sensitive *Staphylococcus aureus*; NP—nanoparticle; PLA—polylactic acid; PLGA—poly(lactic-co-glycolic acid); PVA—poly(vinyl alcohol); PVP—polyvinylpyrrolidone.

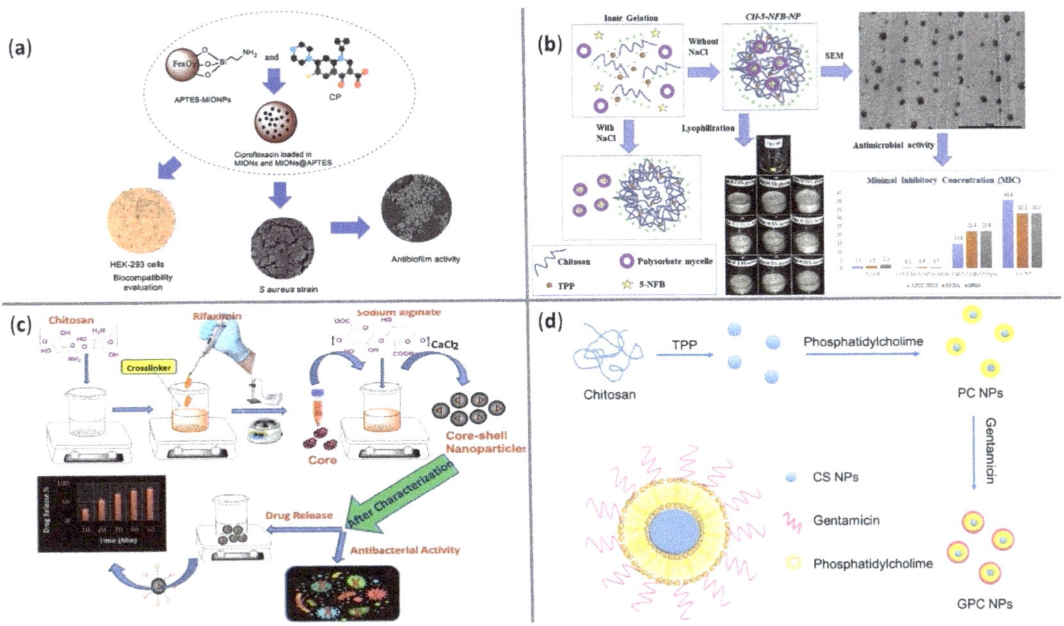

Figure 6. Visual representation of several nanosystems for antibacterial agents' delivery. (**a**) Mesoporous iron oxide NPs loaded with ciprofloxacin. Reprinted with permission from [76], © Elsevier, 2021. (**b**) Polysorbate 20 micelles loaded in chitosan NPs for N′-((5-nitrofuran-2-yl)methylen)-2-benzhydrazide delivery. Reprinted with permission from [86], © Elsevier, 2020. (**c**) Chitosan-alginate core-shell NPs loaded with rifaximin. Reprinted with permission from [89], © Elsevier, 2021. (**d**) Phosphatidylcholine-chitosan liposome NPs for gentamicin delivery. Reprinted with permission from [92], © Elsevier, 2020.

2.3. *Antifungal Nanoformulations*

Fungal infections represent a significant health issue, being associated with high morbidity and mortality. Immunocompromised hosts are particularly susceptible to invasive infections, with the mortality rates in such patients going above 60% in certain situations [3,102,103]. The conventional approach in treating such infections assumes the administration of antifungal agents such as polyenes, azoles, and echinocandins [104]. Nevertheless, these drugs present a series of disadvantages that impede their therapeutic action. Conventional antifungals exhibit non-neglectable toxicity, adverse side effects, acquired resistance, and unclear effects in immunocompromised patients [105–108].

Therefore, recent research tried to solve these issues by orienting to safer strategies, including encapsulation into biocompatible NPs and replacement with natural alternatives. In this regard, scientists have explored a plethora of nanomaterials for designing performant delivery systems for antifungal agents. Promising results have been reported when using metal and metal oxide-based NPs [109,110], natural polymers [90,111,112], biocompatible synthetic polymers [113–118], and lipid-based nanocarriers [119,120]. Regarding the choice of antimicrobial agents, most studies elaborated nanoconstructs for the delivery of synthetic drugs (e.g., nystatin [109], fluconazole [109], amphotericin B [110,117], voriconazole [111], itraconazole [115,116], ketoconazole [118], miconazole nitrate [119], clotrimazole [112]), but several natural antifungal agents (e.g., seedless *Vitis vinifera* (*V. vinifera*) [112], oregano oil [90], *Lippia sidoides* (*L. sidoides*) essential oil [120], pterostilbene [113], farnesol [114]) have also been investigated and led to promising results.

In what concerns the species of interest, most studies have focused on developing antifungal nanoformulations targeting *Candida albicans* (*C. albicans*) [90,109–112,114,115,117–119], as candidiasis is among the most common invasive mycotic diseases, and *C. albicans* is recognized as the leading cause of invasive candidiasis [121,122]. Nonetheless, other fungal pathogens have also been considered, including *Aspergillus brasiliensis* (*A. brasiliens*) [109,113], *Aspergilus niger* (*A. niger*) [112], *Cryptococcus neoformans* (*C. neoformans*) [110], *Histoplasma capsulatum* (*H. capsulatum*) [116], *Trichophyton rubrum* (*T. rubrum*) [118], *Trichophyton mentagrophytes* (*T. mentagrophytes*) [118], *Microsporum gypseum* (*M. gypseum*) [118], *Candida dubliniensis* (*C. dubliniensis*) [118], *Candida krusei* (*C. krusei*) [118], *Candida parapsilosis* (*C. parapsilosis*) [118], *Candida tropicalis* (*C. tropicalis*) [118], and *Candida auris* (*C. auris*) [120].

For clarity, Table 3 correlates NP material, physicochemical properties of the delivery nanosystems, carried antifungal agents, and targeted pathogens, while Figure 7 schematically illustrates a few of these nanostructures.

Table 3. Examples of antifungal nanoformulations.

NP Type	Fabrication Method	Physicochemical Properties	Antimicrobial Agent(s)	Targeted Pathogen(s)	Ref.
Silver NPs	Chemical reduction	Average size: 80 nm Shape: spherical	Nystatin	*C. albicans, A. brasiliensis*	[109]
Silver NPs	Chemical reduction	Average size: 25 nm Shape: spherical	Fluconazole	*C. albicans, A. brasiliensis*	[109]
ZnO-PEGylated NPs	Nanoemulsification	Average size: 662.3 ± 24.7 nm Zeta potential: −14.2 ± 0.94 mV	Amphotericin B	*C. albicans, C. neoformans*	[110]
Chitosan-based NPs	Ionic gelation	Average size: ranging from 167 ± 8.23 to 475 ± 15.30 nm Zeta potential: ranging from 39 ± 2.56 to 45 ± 3.11 mV Shape: spherical	Voriconazole	*C albicans*	[111]
Chitosan NPs	Ionic gelation	Average size: 35.4 nm Zeta potential: +31 mV	Seedless *V. vinifera* and clotrimazole	*C. albicans, A. niger*	[112]

Table 3. *Cont.*

NP Type	Fabrication Method	Physicochemical Properties	Antimicrobial Agent(s)	Targeted Pathogen(s)	Ref.
Chitosan-alginate NPs	Emulsification and consequent electrostatic gelation	Average size: 320 nm Zeta potential: −25 mV	Oregano oil	*C albicans*	[90]
PLGA NPs	n/r	Average size: 50 nm Zeta potential: −25 mV	Coumarin 6 and pterostilbene	*A. brasiliensis*	[113]
PLGA NPs	Emulsion evaporation method	Average size: 140 nm	Farnesol	*C albicans*	[114]
PLGA NPs	Nanoprecipitation and single emulsion solvent evaporation methods	Average size: 176.96 ± 24.32 nm Zeta potential: −24.7 ± 1.04 mV Shape: spherical	Itraconazole	*C albicans*	[115]
PLGA NPs functionalized with anti-F4/80 antibodies	Nanoemulsion	Average size: 226.66 ± 13.05 nm Zeta potential: −27.9 ± 0.26 mV	Itraconazole	*H. capsulatum*	[116]
Aptamer-functionalized PLGA-PEG NPs	Double emulsification method	Average size: 273.9 ± 1.14 nm Zeta potential: −20 mV	Amphotericin B	*C albicans*	[117]
PLA NPs	Nanoprecipitation	Mean size: 188.5 nm Zeta potential: 4.80 mV Shape: spherical	Ketoconazole	*T. rubrum, T. mentagrophytes, M. gypseum, C. albicans, C. dubliniensis, C. krusei, C. parapsilosis, C. tropicalis*	[118]
SLNs	High shear homogenization and ultrasonication	Average size: ranging from 244.2 ± 27.2 to 493.6 ± 35.3 nm Zeta potential: ranging from −21.6 ± 7.05 to −1.4 ± 6.84 mV	Miconazole nitrate	*C. albicans*	[119]
NLC	Hot emulsification method	Average size: ranging from 213.1 ± 1.7 to 445.5 ± 8.7 nm Zeta potential: ranging from −63.8 ± 8.7 to −93.1 ± 2.7 mV	*L. sidoides* essential oil	*C. auris*	[120]

Abbreviations: n/r—not reported; NLC—nanostructured lipid carriers; NP—nanoparticle; PEG—polyethylene glycol; PLA—polylactic acid; PLGA—poly(lactic-co-glycolic acid); SLN—solid lipid nanoparticle.

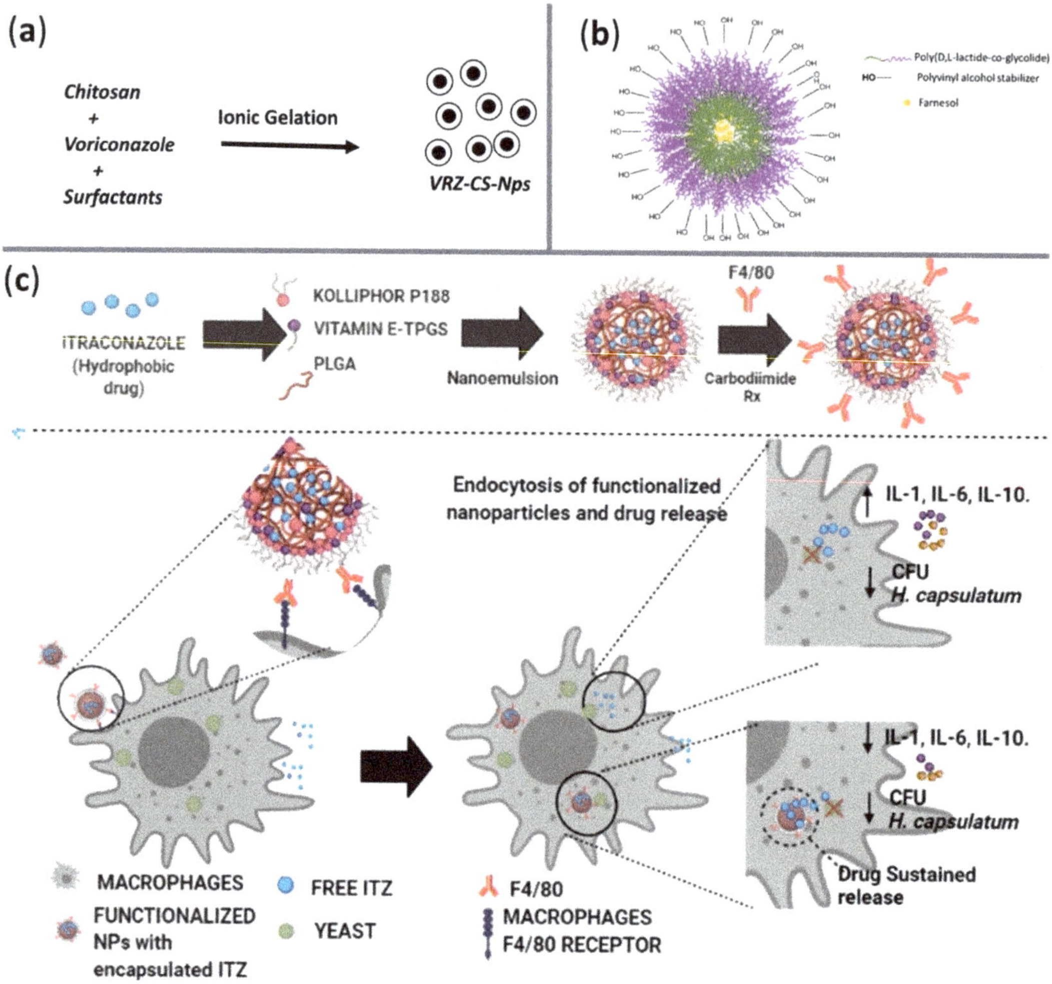

Figure 7. Visual representation of several nanosystems for antifungal agents' delivery. (**a**) Chitosan-based NPs loaded with voriconazole. Adapted from an open-access source [111]. (**b**) PLGA NPs loaded with farnesol. Reprinted from an open-access source [114]. (**c**) PLGA NPs functionalized with anti-F4/80 antibodies loaded for itraconazole delivery. Reprinted from an open-access source [116].

2.4. Antiviral Nanoformulations

Viruses are another class of dangerous pathogens, as they are responsible for around two million deaths per year [123]. Their small size allows viruses to enter the human body through various routes and internalize into living cells. Some of the most pathogenic viruses include human immunodeficiency virus (HIV), human papillomavirus (HPV), herpes simplex virus (HSV), norovirus, hepatitis viruses, and coronaviruses, leading to significant morbidity and mortality. Moreover, the occurrence of viral outbreaks was seen to have devastating effects from economic and social points of view [3,124].

In the fight against infections, NPs offer certain advantages for delivering antivirals to the target sites as they have the ability to surpass biological barriers thanks to their small size and tailored surface characteristics. Through their unique properties, NPs allow antivirals to be released at the infection site, followed by their attachment to viral receptors on the surface of host cells or internalization within the cell resulting in the disruption of the viral replication cycle [124].

Taking into account the benefits of NPs, several research studies explored the antimicrobial potential of a number of antiviral drugs encapsulated in different nanomaterials. Examples of investigated nanocarriers include silver NPs [125], titanium dioxide NPs [126], oligo- and polysaccharide-based NPs [127–131], solid lipid NPs (SLNs) [132,133], and large unilamellar vesicles [134]. The ingenious association with antivirals (e.g., docetaxel [125], flavonoids [126], zidovudine [127,128], dolutegravir sodium [129], efavirenz [130], acyclovir [131,132], ritonavir [133]) and functionalization agents has led to the obtaining of promising anti-infective therapeutic nanoformulations. Specifically, the proposed delivery systems targeted clinically relevant viruses, such as severe acute respiratory syndrome coronavirus 2 (SARS-CoV-2) [125,126], HIV [127–130,133–135], and HSV [131,132].

To better clarify the features of the newly developed antiviral drug delivery systems, Table 4 and Figure 8 summarize several examples of studies in the field.

Table 4. Examples of antiviral nanoformulations.

NP Type	Fabrication Method	Physicochemical Properties	Antimicrobial Agent(s)	Targeted Pathogen(s)	Ref.
NH_2-functionalized silver NPs	UV irradiation and chemical reduction	Average hydrodynamic diameter: 5.0 nm Zeta potential: 22 mV	Docetaxel	SARS-CoV-2	[125]
Amide-functionalized alginate NPs	Emulsion solvent evaporation method	Average size: ranging from 437 ± 2.3 to 473 ± 2.6 nm Zeta potential: ranging from −42.16 ± 3.2 to −34.13 ± 1.61 mV Shape: spherical	Zidovudine	HIV	[127]
Dextran-stearic acid core shell NPs	Double emulsion solvent evaporation method	Average size: ranging from 356 ± 2.06 to 730 ± 1.22 nm Zeta potential: ranging from −31.6 ± 2.12 to −20.9 ± 1.7 mV	Zidovudine	HIV	[128]
HPβCD NPs crosslinked with diphenyl carbonate	Cross-linking	Average size: ranging from 72.47 ± 4.8 to 106.5 ± 5.6 nm Zeta potential: ranging from −30.3 ± 4.1 to −7.77 ± 3.1 mV	Dolutegravir sodium	HIV	[129]
Chitosan-g-HPβCD NPs	Ionic gelation	Average size: ranging from 198 ± 4.4 471.3 ± 5.3 nm Zeta potential: ranging from 3.14 ± 2.6 to 11.5 ± 2.1 mV	Efavirenz	HIV	[130]
Sulfobutyl ether-β-cyclodextrin decorated chitosan nanodroplets	Electrostatic interaction	Average size: 395.4 ± 12.6 nm Zeta potential: 19.98 ± 3.02 mV	Acyclovir	HSV type 2	[131]
SLNs	Emulsification and low-temperature solidification	Average size: 180 ± 1.2 nm Zeta potential: −25 mV	Acyclovir	HSV	[132]
SLNs	Solvent emulsification evaporation and double emulsion methods	Mean size: ranging from 178.7 ± 4.5 to 254.3 ± 16.6 nm Zeta potential: ranging from 39.35 ± 1.2 to 50.80 ± 4.8 mV	Ritonavir	HIV-1	[133]

Abbreviations: HIV—human immunodeficiency virus; HPβCD—2-Hydroxypropyl-beta-cyclodextrin; HSV—herpes simplex virus; LUVs—large unilamellar vesicles; NP—nanoparticle; SARS-CoV-2—severe acute respiratory syndrome coronavirus 2; SLN—solid lipid nanoparticle.

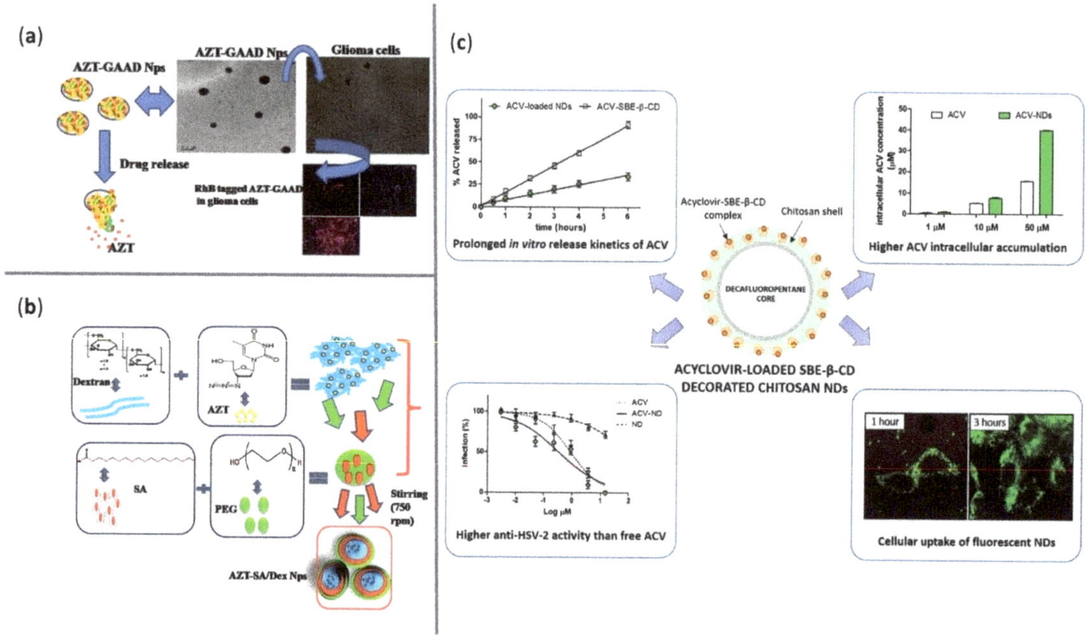

Figure 8. Visual representation of several nanosystems for antiviral agents' delivery. (**a**) Amide-functionalized alginate NPs loaded with zidovudine. Reprinted with permission from [127], © Elsevier, 2018. (**b**) Dextran-stearic acid core shell NPs loaded with zidovudine. Reprinted with permission from [128], © Elsevier, 2018. (**c**) Sulfobutyl ether-β-cyclodextrin decorated chitosan nanodroplets for acyclovir delivery. Reprinted with permission from [131], © Elsevier, 2020.

2.5. Antiparasitic Nanoformulations

Aside from bacteria, fungi, and viruses, several parasites have also been recognized for their infective potential. One example is represented by *Leishmania* spp., which comprises a group of flagellated protozoans responsible for neglected tropical diseases known as leishmaniasis. Characterized by high mortality, disability, and morbidity rates, leishmaniasis represents a major global health concern, being endemic in 102 countries worldwide [136–138]. Consequently, scientific interest arose in finding antiparasitic solutions able to effectively and efficiently fight against *Leishmania*.

For instance, Badirzadeh et al. [137] proposed coating silver NPs with curcumin. In vitro and in vivo tests performed on mouse models demonstrated encouraging results, the nanoformulation significantly reducing the burden of promastigotes and amastigotes of *Leishmania* parasite in a single treatment. Alternatively, Snoussi et al. [136] prepared silver-loaded biochar that exhibited strong antiparasitic activity against the promastigotes stage of *Leishmania donovani*, *Leishmania amazonensis*, and epimastigotes of *Trypanosoma cruzi*. On a different note, Durak and colleagues [139] encapsulated two active ingredients with antibacterial and antiparasitic activities (i.e., caffeic acid phenethyl ester and juglone) into single polymeric NPs. These multifunctional nanoformulations proved synergistic activity, being promising candidates for antiparasitic therapy.

Several studies have also directed their efforts toward creating delivery nanosystems aimed at other pathogens. For example, Kanwal et al. [140] reported the fabrication of silver NPs conjugated with novel bisindole and thiazole derivatives as potential antiamoebic formulations with enhanced activity against *Balamuthia mandrillaris* and *Naegleria fowleri*. In contrast, Real et al. [141] developed a drug delivery system for treating fascioliasis. For this purpose, the researchers loaded triclabendazole into nanocapsules, enhancing its bioavailability and lowering its cytotoxic effects compared to the free drug. Differently, Wei et al. [142] created a nanocarrier for decoquinate, a drug known to have control effects on hematogeneous parasites. The authors encapsulated the antiparasitic agent into disodium glycyrrhizinate NPs with protamine and anionic hyaluronic acid layers (Figure 9), significantly increasing the drug's bioavailability, ensuring a higher concentration in the blood and preferential liver tissue accumulation.

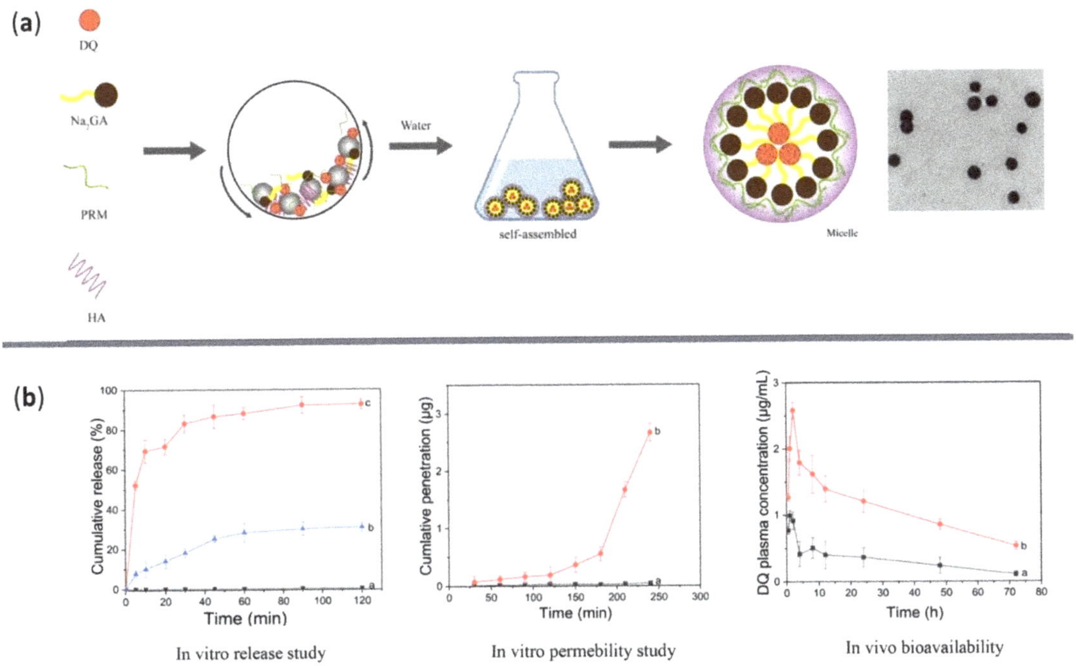

Figure 9. (**a**) Visual representation for the fabrication of decoquinate delivery system and (**b**) systematic testing. Reprinted with permission from [142], © Elsevier, 2022.

3. Discussion

Humans are exposed to numerous pathogens that can trigger burdensome bacterial, fungal, viral, and parasitic infections. Conventional treatment approaches revolve around the systemic administration of synthetic drugs that, due to the emergence of drug-resistant microbial strains, exhibit low efficacy, in addition to the disadvantages of poor solubility, toxicity, and adverse effects. In this context, nanotechnology started being increasingly explored for designing improved antimicrobial agents

NPs of many sorts (Figure 10) have been recently developed as performant carriers of numerous antimicrobial agents, holding promise for improved strategies to combat a wide range of infectious diseases. Nanodimensional materials, such as metallic NPs, metal-oxide NPs, lipid-based NPs, and polymeric NPs, have attracted considerable interest in recent years for fabricating delivery vehicles. Specifically, the variety and versatility of nanomaterials have been extensively explored by researchers for creating innovative therapeutic formulations that can be administrated on different routes, including oral [2,141,142], ocular [82,83,143], intranasal [125,129,130], intratracheal [55,144], intravaginal [145,146], intravesical [147,148], and transdermal [149,150] routes. Compared to free drugs, NP-loaded antimicrobial agents can be administered in so many ways due to their increased safety, reduced systemic adverse effects, enhanced solubility, and improved bioavailability. Moreover, the various natural and synthetic antimicrobial cargos can be released in a targeted manner by adding various functionalization agents onto the nanocarriers' surface. Functionalization agents can also work in synergy with the core delivery system, increasing therapeutic efficacy and reducing drug resistance [11,31].

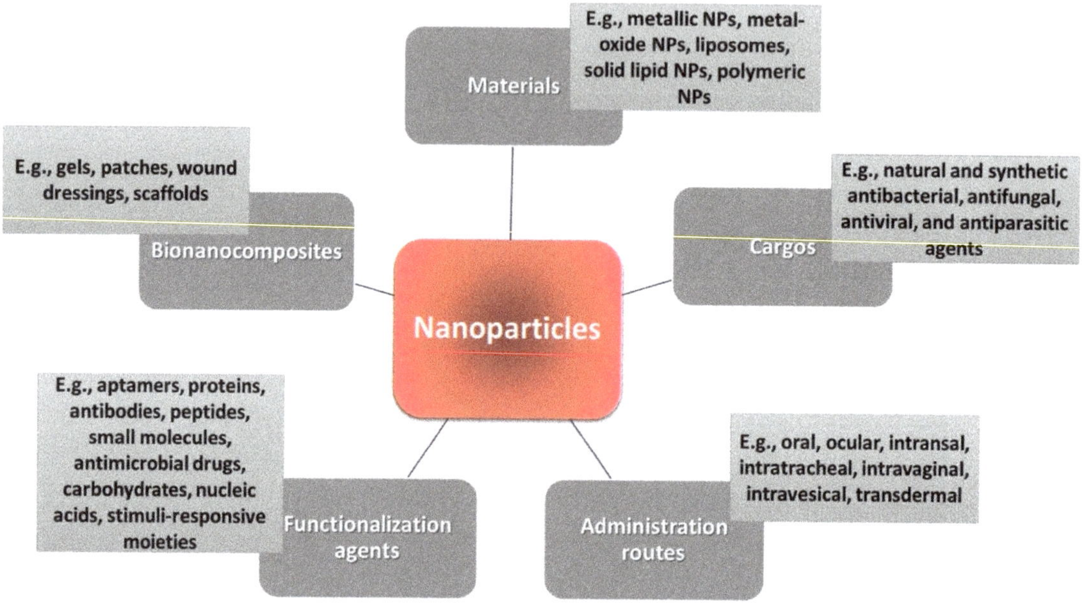

Figure 10. Overview of the possibilities of using NPs as delivery vehicles for antimicrobial applications.

In addition to their stand-alone utility, NPs may further be incorporated in different other materials to create bionanocomposites with enhanced antimicrobial properties. In this respect, researchers propose the use of various nanostructured gels [82,83,98,100,151], patches [149,150,152], wound dressings [153–156], and scaffolds [157–159] as alternative solutions for treating and preventing microbial infections. A particularly exploited application of NPs is the fabrication of coatings for creating surfaces with antimicrobial and antibiofilm properties [12,19,27,29,77,160–164]. Even though aimed mainly at the modification of biomedical devices, such as catheters, implants, and prostheses, applying antimicrobial coatings can also be of high utility in covering other contact surfaces. For instance, they can be used to prevent pathogens from spreading from day-to-day objects, including doorknobs, packaging, and handrails [37].

4. Conclusions and Future Perspectives

To summarize, various NPs have been investigated as drug delivery vehicles to surpass traditional drugs' limitations. Numerous studies have successfully loaded natural and synthetic drugs into inorganic, lipid, and polymeric-based nanosystems, obtaining promising results against a broad range of pathogens, but mostly bacterial strains.

To conclude, there is an increased research interest in developing alternative antimicrobial agents, and current progress demonstrates the great potential of nanostructured materials in preventing and treating infectious diseases. Nonetheless, there is still room for improvement in the field, especially concerning the expansion of antiviral, antifungal, and antiparasitic applications of drug delivery nanosystems. Further studies should also focus on managing complex and mixed biofilms, an understudied and challenging niche of microbial infections. Moreover, being so new, most of the discussed nano-therapeutic options have not yet advanced beyond preclinical testing. Thus, rigorous additional studies are required before they become clinically and commercially available solutions.

Author Contributions: D.-A.M., A.-G.N., and A.M.G. participated in reviewing, writing, and revision. All authors have read and agreed to the published version of the manuscript.

Funding: This research received no external funding.

Institutional Review Board Statement: Not applicable.

Informed Consent Statement: Not applicable.

Data Availability Statement: Not applicable.

Conflicts of Interest: The authors declare no conflict of interest.

References

1. Antabe, R.; Ziegler, B.R. Diseases, emerging and infectious. In *International Encyclopedia of Human Geography*; Springer: Berlin/Heidelberg, Germany, 2020; pp. 389–391.
2. Raza, A.; Sime, F.B.; Cabot, P.J.; Maqbool, F.; Roberts, J.A.; Falconer, J.R. Solid nanoparticles for oral antimicrobial drug delivery: A review. *Drug Discov. Today* **2019**, *24*, 858–866. [CrossRef]
3. Sharmin, S.; Rahaman, M.M.; Sarkar, C.; Atolani, O.; Islam, M.T.; Adeyemi, O.S. Nanoparticles as antimicrobial and antiviral agents: A literature-based perspective study. *Heliyon* **2021**, *7*, e06456. [CrossRef] [PubMed]
4. Gandra, S.; Barter, D.M.; Laxminarayan, R. Economic burden of antibiotic resistance: How much do we really know? *Clin. Microbiol. Infect.* **2014**, *20*, 973–980. [CrossRef] [PubMed]
5. Fadda, A.A.; Mohammed, A.R.; Abdel-Galil, E. Synthesis and antimicrobial evaluation of some 4-quinolinylazo-N-pyrimidinyl benzenesulfonamide derivatives. *Biointerface Res. Appl. Chem.* **2020**, *10*, 4846–4852. [CrossRef]
6. Grozav, A.; Fedoriv, M.; Chornous, V.; Yakovychuk, N.; Kemskyi, S.; Vovk, M. Synthesis and Bioevaluation of 5-Chloro-4-(1,3-Oxazol-5-yl)-1H-Pyrrole-3-Carboxyamides as Antimicrobial Agents. *Biointerface Res. Appl. Chem.* **2021**, *11*, 10595–10606. [CrossRef]
7. Kumar, A.; Kumar, A. Design and synthesis of anti-convulsant and anti-bacterial activity of new hydrazone derivatives. *Biointerface Res. Appl. Chem.* **2020**, *10*, 5229–5236. [CrossRef]
8. Eleraky, N.E.; Allam, A.; Hassan, S.B.; Omar, M.M. Nanomedicine Fight against Antibacterial Resistance: An Overview of the Recent Pharmaceutical Innovations. *Pharmaceutics* **2020**, *12*, 142. [CrossRef]
9. Devrim, B.; Bozkır, A. Chapter 7—Nanocarriers and Their Potential Application as Antimicrobial Drug Delivery. In *Nanostructures for Antimicrobial Therapy*; Ficai, A., Grumezescu, A.M., Eds.; Elsevier: Amsterdam, The Netherlands, 2017; pp. 169–202.
10. Anghel, A.G.; Grumezescu, A.M.; Chirea, M.; Grumezescu, V.; Socol, G.; Iordache, F.; Oprea, A.E.; Anghel, I.; Holban, A.M. MAPLE Fabricated Fe3O4@Cinnamomum verum Antimicrobial Surfaces for Improved Gastrostomy Tubes. *Molecules* **2014**, *19*, 8981. [CrossRef] [PubMed]
11. Canaparo, R.; Foglietta, F.; Giuntini, F.; Della Pepa, C.; Dosio, F.; Serpe, L. Recent Developments in Antibacterial Therapy: Focus on Stimuli-Responsive Drug-Delivery Systems and Therapeutic Nanoparticles. *Molecules* **2019**, *24*, 1991. [CrossRef]
12. Gherasim, O.; Grumezescu, A.M.; Grumezescu, V.; Iordache, F.; Vasile, B.S.; Holban, A.M. Bioactive Surfaces of Polylactide and Silver Nanoparticles for the Prevention of Microbial Contamination. *Materials* **2020**, *13*, 768. [CrossRef] [PubMed]
13. Polívková, M.; Hubáček, T.; Staszek, M.; Švorčík, V.; Siegel, J. Antimicrobial Treatment of Polymeric Medical Devices by Silver Nanomaterials and Related Technology. *Int. J. Mol. Sci.* **2017**, *18*, 419. [CrossRef]
14. Petrachi, T.; Resca, E.; Piccinno, M.S.; Biagi, F.; Strusi, V.; Dominici, M.; Veronesi, E. An Alternative Approach to Investigate Biofilm in Medical Devices: A Feasibility Study. *Int. J. Environ. Res. Public Health* **2017**, *14*, 1587. [CrossRef] [PubMed]
15. French, G.L. The continuing crisis in antibiotic resistance. *Int. J. Antimicrob. Agents* **2010**, *36*, S3–S7. [CrossRef]
16. de la Fuente-Núñez, C.; Reffuveille, F.; Fernández, L.; Hancock, R.E.W. Bacterial biofilm development as a multicellular adaptation: Antibiotic resistance and new therapeutic strategies. *Curr. Opin. Microbiol.* **2013**, *16*, 580–589. [CrossRef] [PubMed]
17. Li, Y.; Li, X.; Hao, Y.; Liu, Y.; Dong, Z.; Li, K. Biological and Physiochemical Methods of Biofilm Adhesion Resistance Control of Medical-Context Surface. *Int. J. Biol. Sci.* **2021**, *17*, 1769–1781. [CrossRef] [PubMed]
18. Vallet-Regí, M.; Lozano, D.; González, B.; Izquierdo-Barba, I. Biomaterials against Bone Infection. *Adv. Healthc. Mater.* **2020**, *9*, 2000310. [CrossRef] [PubMed]
19. Grumezescu, V.; Negut, I.; Grumezescu, A.M.; Ficai, A.; Dorcioman, G.; Socol, G.; Iordache, F.; Truşcă, R.; Vasile, B.S.; Holban, A.M. MAPLE fabricated coatings based on magnetite nanoparticles embedded into biopolymeric spheres resistant to microbial colonization. *Appl. Surf. Sci.* **2018**, *448*, 230–236. [CrossRef]
20. Mihai, A.D.; Chircov, C.; Grumezescu, A.M.; Holban, A.M. Magnetite Nanoparticles and Essential Oils Systems for Advanced Antibacterial Therapies. *Int. J. Mol. Sci.* **2020**, *21*, 7355. [CrossRef] [PubMed]
21. Jamal, M.; Ahmad, W.; Andleeb, S.; Jalil, F.; Imran, M.; Nawaz, M.A.; Hussain, T.; Ali, M.; Rafiq, M.; Kamil, M.A. Bacterial biofilm and associated infections. *J. Chin. Med. Assoc. JCMA* **2018**, *81*, 7–11. [CrossRef] [PubMed]

22. Wu, K.; Yang, Y.; Zhang, Y.; Deng, J.; Lin, C. Antimicrobial activity and cytocompatibility of silver nanoparticles coated catheters via a biomimetic surface functionalization strategy. *Int. J. Nanomed.* **2015**, *10*, 7241–7252. [CrossRef]
23. Koley, S.; Mukherjee, M. Genetic Basis of Biofilm Formation and Spread of Nosocomial Infections. In *Analytical Methodologies for Biofilm Research*; Springer: New York, NY, USA, 2021; pp. 269–298.
24. Jenkins, D.R. Nosocomial infections and infection control. *Medicine* **2017**, *45*, 629–633. [CrossRef]
25. Pircalabioru, G.G.; Chifiriuc, M.-C. Nanoparticulate drug-delivery systems for fighting microbial biofilms: From bench to bedside. *Future Microbiol.* **2020**, *15*, 679–698. [CrossRef] [PubMed]
26. Barros, C.H.N.; Casey, E. A Review of Nanomaterials and Technologies for Enhancing the Antibiofilm Activity of Natural Products and Phytochemicals. *ACS Appl. Nano Mater.* **2020**, *3*, 8537–8556. [CrossRef]
27. Besinis, A.; Hadi, S.D.; Le, H.R.; Tredwin, C.; Handy, R.D. Antibacterial activity and biofilm inhibition by surface modified titanium alloy medical implants following application of silver, titanium dioxide and hydroxyapatite nanocoatings. *Nanotoxicology* **2017**, *11*, 327–338. [CrossRef]
28. Makvandi, P.; Wang, C.-Y.; Zare, E.N.; Borzacchiello, A.; Niu, L.-N.; Tay, F.R. Metal-Based Nanomaterials in Biomedical Applications: Antimicrobial Activity and Cytotoxicity Aspects. *Adv. Funct. Mater.* **2020**, *30*, 1910021. [CrossRef]
29. Gherasim, O.; Popescu, R.C.; Grumezescu, V.; Mogoșanu, G.D.; Mogoantă, L.; Iordache, F.; Holban, A.M.; Vasile, B.Ș.; Bîrcă, A.C.; Oprea, O.-C.; et al. MAPLE Coatings Embedded with Essential Oil-Conjugated Magnetite for Anti-Biofilm Applications. *Materials* **2021**, *14*, 1612. [CrossRef]
30. Ahmadabadi, H.Y.; Yu, K.; Kizhakkedathu, J.N. Surface modification approaches for prevention of implant associated infections. *Colloids Surf. B Biointerfaces* **2020**, *193*, 111116. [CrossRef]
31. Bahrami, A.; Delshadi, R.; Jafari, S.M. Active delivery of antimicrobial nanoparticles into microbial cells through surface functionalization strategies. *Trends Food Sci. Technol.* **2020**, *99*, 217–228. [CrossRef]
32. Xie, S.; Tao, Y.; Pan, Y.; Qu, W.; Cheng, G.; Huang, L.; Chen, D.; Wang, X.; Liu, Z.; Yuan, Z. Biodegradable nanoparticles for intracellular delivery of antimicrobial agents. *J. Control. Release* **2014**, *187*, 101–117. [CrossRef]
33. Liu, Z.; Ye, L.; Xi, J.; Wang, J.; Feng, Z.-g. Cyclodextrin polymers: Structure, synthesis, and use as drug carriers. *Prog. Polym. Sci.* **2021**, *118*, 101408. [CrossRef]
34. Valenti, G.E.; Alfei, S.; Caviglia, D.; Domenicotti, C.; Marengo, B. Antimicrobial Peptides and Cationic Nanoparticles: A Broad-Spectrum Weapon to Fight Multi-Drug Resistance Not Only in Bacteria. *Int. J. Mol. Sci.* **2022**, *23*, 6108. [CrossRef] [PubMed]
35. Tran, H.M.; Tran, H.; Booth, M.A.; Fox, K.E.; Nguyen, T.H.; Tran, N.; Tran, P.A. Nanomaterials for Treating Bacterial Biofilms on Implantable Medical Devices. *Nanomaterials* **2020**, *10*, 2253. [CrossRef] [PubMed]
36. Delfi, M.; Ghomi, M.; Zarrabi, A.; Mohammadinejad, R.; Taraghdari, Z.B.; Ashrafizadeh, M.; Zare, E.N.; Agarwal, T.; Padil, V.V.T.; Mokhtari, B.; et al. Functionalization of Polymers and Nanomaterials for Biomedical Applications: Antimicrobial Platforms and Drug Carriers. *Prosthesis* **2020**, *2*, 117–139. [CrossRef]
37. Lin, N.; Verma, D.; Saini, N.; Arbi, R.; Munir, M.; Jovic, M.; Turak, A. Antiviral nanoparticles for sanitizing surfaces: A roadmap to self-sterilizing against COVID-19. *Nano Today* **2021**, *40*, 101267. [CrossRef] [PubMed]
38. Soares, S.; Sousa, J.; Pais, A.; Vitorino, C. Nanomedicine: Principles, Properties, and Regulatory Issues. *Front. Chem.* **2018**, *6*, 360. [CrossRef] [PubMed]
39. Chapman, J.; Regan, F.; Sullivan, T. *Nanoparticles in Anti-Microbial Materials: Use and Characterisation*; Royal Society of Chemistry: Cambridge, UK, 2012; Volume 23.
40. Nas, F.S.; Ali, M.; Aminu Muhammad, A. Application of Nanomaterials as Antimicrobial Agents: A Review. *Arch. Nano Op. Acc. J.* **2018**, *1*, 59–64.
41. Kim, T.; Hyeon, T. Applications of inorganic nanoparticles as therapeutic agents. *Nanotechnology* **2013**, *25*, 012001. [CrossRef]
42. Gilavand, F.; Saki, R.; Mirzaei, S.Z.; Lashgarian, H.E.; Karkhane, M.; Marzban, A. Green Synthesis of Zinc Nanoparticles Using Aqueous Extract of Magnoliae officinalis and Assessment of its Bioactivity Potentials. *Biointerface Res. Appl. Chem.* **2021**, *11*, 7765–7774. [CrossRef]
43. Samrot, A.V.; Sahithya, C.S.; Sruthi, P.D.; Selvarani, A.J.; Raji, P.; Prakash, P.; Ponnaiah, P.; Petchi, I.; Pattammadath, S.; Purayil, S.K.; et al. Itraconazole Coated Super Paramagnetic Iron Oxide Nanoparticles for Antimicrobial Studies. *Biointerface Res. Appl. Chem.* **2020**, *10*, 6262–6269. [CrossRef]
44. Tyagi, P.K.; Mishra, R.; Khan, F.; Gupta, D.; Gola, D. Antifungal Effects of Silver Nanoparticles Against Various Plant Pathogenic Fungi and its Safety Evaluation on Drosophila melanogaster. *Biointerface Res. Appl. Chem.* **2020**, *10*, 6587–6596. [CrossRef]
45. Masri, A.; Anwar, A.; Khan, N.A.; Siddiqui, R. The Use of Nanomedicine for Targeted Therapy against Bacterial Infections. *Antibiotics* **2019**, *8*, 260. [CrossRef] [PubMed]
46. Niculescu, A.-G.; Chircov, C.; Bîrcă, A.C.; Grumezescu, A.M. Nanomaterials Synthesis through Microfluidic Methods: An Updated Overview. *Nanomaterials* **2021**, *11*. [CrossRef] [PubMed]
47. Arole, V.M.; Munde, S.V. Fabrication of nanomaterials by top-down and bottom-up approaches-an overview. *J. Mater. Sci* **2014**, *1*, 89–93.
48. Shrimal, P.; Jadeja, G.; Patel, S. A review on novel methodologies for drug nanoparticle preparation: Microfluidic approach. *Chem. Eng. Res. Des.* **2020**, *153*, 728–756. [CrossRef]
49. Niculescu, A.-G.; Chircov, C.; Grumezescu, A.M. Magnetite nanoparticles: Synthesis methods—A comparative review. *Methods* **2022**, *199*, 16–27. [CrossRef]

50. Niculescu, A.-G.; Grumezescu, A.M. Polymer-Based Nanosystems—A Versatile Delivery Approach. *Materials* **2021**, *14*. [CrossRef]
51. Zielińska, A.; Carreiró, F.; Oliveira, A.M.; Neves, A.; Pires, B.; Venkatesh, D.N.; Durazzo, A.; Lucarini, M.; Eder, P.; Silva, A.M. Polymeric nanoparticles: Production, characterization, toxicology and ecotoxicology. *Molecules* **2020**, *25*, 3731. [CrossRef]
52. Niculescu, A.-G.; Bîrcă, A.C.; Grumezescu, A.M. New Applications of Lipid and Polymer-Based Nanoparticles for Nucleic Acids Delivery. *Pharmaceutics* **2021**, *13*, 2053. [CrossRef]
53. Lundberg, S.; Karlsson, E.; Dahlberg, H.; Glansk, M.; Larsson, S.; Larsson, S.; Carlsson, K. Exosomes and Lipid Nanoparticles-the Future of Targeted Drug Delivery. Ph.D. Thesis, Uppsala University, Uppsala, Sweden, 2020.
54. Tiplea, R.E.; Lemnaru, G.M.; Trusca, R.D.; Holban, A.; Kaya, M.G.A.; Dragu, L.D.; Ficai, D.; Ficai, A.; Bleotu, C. Antimicrobial Films based on Chitosan, Collagen, and ZnO for Skin Tissue Regeneration. *Biointerface Res. Appl. Chem.* **2021**, *11*, 11985–11995. [CrossRef]
55. Scolari, I.R.; Páez, P.L.; Musri, M.M.; Petiti, J.P.; Torres, A.; Granero, G.E. Rifampicin loaded in alginate/chitosan nanoparticles as a promising pulmonary carrier against Staphylococcus aureus. *Drug Deliv. Transl. Res.* **2020**, *10*, 1403–1417. [CrossRef]
56. Varela, M.F.; Stephen, J.; Lekshmi, M.; Ojha, M.; Wenzel, N.; Sanford, L.M.; Hernandez, A.J.; Parvathi, A.; Kumar, S.H. Bacterial Resistance to Antimicrobial Agents. *Antibiotics* **2021**, *10*, 593. [CrossRef] [PubMed]
57. Mohid, S.A.; Bhunia, A. Combining Antimicrobial Peptides with Nanotechnology: An Emerging Field in Theranostics. *Curr. Protein Pept. Sci.* **2020**, *21*, 413–428. [CrossRef] [PubMed]
58. Nag, M.; Lahiri, D.; Mukherjee, D.; Banerjee, R.; Garai, S.; Sarkar, T.; Ghosh, S.; Dey, A.; Ghosh, S.; Pattnaik, S.; et al. Functionalized Chitosan Nanomaterials: A Jammer for Quorum Sensing. *Polymers* **2021**, *13*, 2533. [CrossRef] [PubMed]
59. Kong, J.; Zhang, S.; Shen, M.; Zhang, J.; Yoganathan, S. Evaluation of copper(I)-doped zinc oxide composite nanoparticles on both gram-negative and gram-positive bacteria. *Colloids Surf. A Physicochem. Eng. Asp.* **2022**, *643*, 128742. [CrossRef]
60. Pietsch, F.; O'Neill, A.J.; Ivask, A.; Jenssen, H.; Inkinen, J.; Kahru, A.; Ahonen, M.; Schreiber, F. Selection of resistance by antimicrobial coatings in the healthcare setting. *J. Hosp. Infect.* **2020**, *106*, 115–125. [CrossRef] [PubMed]
61. Nikaido, H. Multidrug resistance in bacteria. *Annu. Rev. Biochem.* **2009**, *78*, 119–146. [CrossRef]
62. Wang, L.; Hu, C.; Shao, L. The antimicrobial activity of nanoparticles: Present situation and prospects for the future. *Int. J. Nanomed.* **2017**, *12*, 1227–1249. [CrossRef]
63. Fernando, S.S.N.; Gunasekara, C.; Holton, J. Antimicrobial Nanoparticles: Applications and mechanisms of action. *Sri Lankan J. Infect. Dis.* **2018**, *8*, 2. [CrossRef]
64. Varier, K.M.; Gudeppu, M.; Chinnasamy, A.; Thangarajan, S.; Balasubramanian, J.; Li, Y.; Gajendran, B. Nanoparticles: Antimicrobial Applications and Its Prospects. In *Advanced Nanostructured Materials for Environmental Remediation*; Springer: Cham, Switzerland, 2019; pp. 321–355. [CrossRef]
65. Lee, N.-Y.; Ko, W.-C.; Hsueh, P.-R. Nanoparticles in the Treatment of Infections Caused by Multidrug-Resistant Organisms. *Front. Pharmacol.* **2019**, *10*, 1153. [CrossRef]
66. Spirescu, V.A.; Chircov, C.; Grumezescu, A.M.; Vasile, B.Ș.; Andronescu, E. Inorganic Nanoparticles and Composite Films for Antimicrobial Therapies. *Int. J. Mol. Sci.* **2021**, *22*, 4595. [CrossRef]
67. Spirescu, V.A.; Chircov, C.; Grumezescu, A.M.; Andronescu, E. Polymeric Nanoparticles for Antimicrobial Therapies: An up-to-date Overview. *Polymers* **2021**, *13*, 724. [CrossRef] [PubMed]
68. Gupta, K.; Chundawat, T.S.; Malek, N. Antibacterial, Antifungal, Photocatalytic Activities and Seed Germination Effect of Mycosynthesized Silver Nanoparticles using Fusarium oxysporum. *Biointerface Res. Appl. Chem.* **2021**, *11*, 12082–12091. [CrossRef]
69. Khan, K.; Javed, S. Functionalization of Inorganic Nanoparticles to Augment Antimicrobial Efficiency: A Critical Analysis. *Curr. Pharm. Biotechnol.* **2018**, *19*, 523–536. [CrossRef] [PubMed]
70. Malaekeh-Nikouei, B.; Fazly Bazzaz, B.S.; Mirhadi, E.; Tajani, A.S.; Khameneh, B. The role of nanotechnology in combating biofilm-based antibiotic resistance. *J. Drug Deliv. Sci. Technol.* **2020**, *60*, 101880. [CrossRef]
71. Karuppannan, S.K.; Ramalingam, R.; Mohamed Khalith, S.B.; Dowlath, M.J.H.; Darul Raiyaan, G.I.; Arunachalam, K.D. Characterization, antibacterial and photocatalytic evaluation of green synthesized copper oxide nanoparticles. *Biocatal. Agric. Biotechnol.* **2021**, *31*, 101904. [CrossRef]
72. Caciandone, M.; Niculescu, A.-G.; Grumezescu, V.; Bîrcă, A.C.; Ghica, I.C.; Vasile, B.Ș.; Oprea, O.; Nica, I.C.; Stan, M.S.; Holban, A.M.; et al. Magnetite Nanoparticles Functionalized with Therapeutic Agents for Enhanced ENT Antimicrobial Properties. *Antibiotics* **2022**, *11*, 623. [CrossRef]
73. Abduraimova, A.; Molkenova, A.; Duisembekova, A.; Mulikova, T.; Kanayeva, D.; Atabaev, T.S. Cetyltrimethylammonium Bromide (CTAB)-Loaded SiO2–Ag Mesoporous Nanocomposite as an Efficient Antibacterial Agent. *Nanomaterials* **2021**, *11*, 477. [CrossRef]
74. Rashid, M.; Rabbi, M.A.; Ara, T.; Hossain, M.M.; Islam, M.S.; Elaissari, A.; Ahmad, H.; Rahman, M.M. Vancomycin conjugated iron oxide nanoparticles for magnetic targeting and efficient capture of Gram-positive and Gram-negative bacteria. *RSC Adv.* **2021**, *11*, 36319–36328. [CrossRef]
75. Rayegan, A.; Allafchian, A.; Abdolhosseini Sarsari, I.; Kameli, P. Synthesis and characterization of basil seed mucilage coated Fe3O4 magnetic nanoparticles as a drug carrier for the controlled delivery of cephalexin. *Int. J. Biol. Macromol.* **2018**, *113*, 317–328. [CrossRef]

76. Lage, W.C.; Sachs, D.; Nunes Ribeiro, T.A.; Tebaldi, M.L.; de Moura, Y.d.R.S.; Domingues, S.C.; Ferreira Soares, D.C. Mesoporous iron oxide nanoparticles loaded with ciprofloxacin as a potential biocompatible antibacterial system. *Microporous Mesoporous Mater.* **2021**, *321*, 111127. [CrossRef]
77. Caciandone, M.; Niculescu, A.-G.; Roșu, A.R.; Grumezescu, V.; Negut, I.; Holban, A.M.; Oprea, O.; Vasile, B.Ș.; Bîrcă, A.C.; Grumezescu, A.M.; et al. PEG-Functionalized Magnetite Nanoparticles for Modulation of Microbial Biofilms on Voice Prosthesis. *Antibiotics* **2022**, *11*, 39. [CrossRef] [PubMed]
78. de Lacerda Coriolano, D.; de Souza, J.B.; Bueno, E.V.; Medeiros, S.M.d.F.R.d.S.; Cavalcanti, I.D.L.; Cavalcanti, I.M.F. Antibacterial and antibiofilm potential of silver nanoparticles against antibiotic-sensitive and multidrug-resistant Pseudomonas aeruginosa strains. *Braz. J. Microbiol.* **2020**, *52*, 267–278. [CrossRef] [PubMed]
79. Camacho-Jiménez, L.; Álvarez-Sánchez, A.R.; Mejía-Ruíz, C.H. Silver nanoparticles (AgNPs) as antimicrobials in marine shrimp farming: A review. *Aquac. Rep.* **2020**, *18*, 100512. [CrossRef]
80. Hamad, A.; Khashan, K.S.; Hadi, A. Silver Nanoparticles and Silver Ions as Potential Antibacterial Agents. *J. Inorg. Organomet. Polym. Mater.* **2020**, *30*, 4811–4828. [CrossRef]
81. Balakrishnan, K.; Casimeer, S.C.; Ghidan, A.Y.; Al Antary, T.M.; Singaravelu, A. Exploration of Antioxidant, Antibacterial Activities of Green Synthesized Hesperidin Loaded PLGA Nanoparticles. *Biointerface Res. Appl. Chem.* **2021**, *11*, 14520–14528. [CrossRef]
82. Ameeduzzafar; Imam, S.S.; Abbas Bukhari, S.N.; Ahmad, J.; Ali, A. Formulation and optimization of levofloxacin loaded chitosan nanoparticle for ocular delivery: In-vitro characterization, ocular tolerance and antibacterial activity. *Int. J. Biol. Macromol.* **2018**, *108*, 650–659. [CrossRef] [PubMed]
83. Alruwaili, N.K.; Zafar, A.; Imam, S.S.; Alharbi, K.S.; Alotaibi, N.H.; Alshehri, S.; Alhakamy, N.A.; Alzarea, A.I.; Afzal, M.; Elmowafy, M. Stimulus Responsive Ocular Gentamycin-Ferrying Chitosan Nanoparticles Hydrogel: Formulation Optimization, Ocular Safety and Antibacterial Assessment. *Int. J. Nanomed.* **2020**, *15*, 4717–4737. [CrossRef] [PubMed]
84. Mohammadi, A.; Hosseini, S.M.; Hashemi, M. Emerging chitosan nanoparticles loading-system boosted the antibacterial activity of Cinnamomum zeylanicum essential oil. *Ind. Crops Prod.* **2020**, *155*, 112824. [CrossRef]
85. Bagheri, R.; Ariaii, P.; Motamedzadegan, A. Characterization, antioxidant and antibacterial activities of chitosan nanoparticles loaded with nettle essential oil. *J. Food Meas. Charact.* **2021**, *15*, 1395–1402. [CrossRef]
86. Andrade, L.F.d.; Apolinário, A.C.; Rangel-Yagui, C.O.; Stephano, M.A.; Tavares, L.C. Chitosan nanoparticles for the delivery of a new compound active against multidrug-resistant Staphylococcus aureus. *J. Drug Deliv. Sci. Technol.* **2020**, *55*, 101363. [CrossRef]
87. Kaur, J.; Kour, A.; Panda, J.J.; Harjai, K.; Chhibber, S. Exploring Endolysin-Loaded Alginate-Chitosan Nanoparticles as Future Remedy for Staphylococcal Infections. *AAPS PharmSciTech* **2020**, *21*, 233. [CrossRef] [PubMed]
88. Kadhum, W.N.; Zaidan, I.A. The synergistic effects of chitosan-alginate nanoparticles loaded with doxycycline antibiotic against multidrug resistant proteus mirabilis, Escherichia coli and enterococcus faecalis. *Iraqi J. Sci.* **2020**, *61*, 3187–3199.
89. Kumar, D.; Kumar, S.; Kumar, S.; Rohatgi, S.; Kundu, P.P. Synthesis of rifaximin loaded chitosan-alginate core-shell nanoparticles (Rif@CS/Alg-NPs) for antibacterial applications. *Int. J. Biol. Macromol.* **2021**, *183*, 962–971. [CrossRef] [PubMed]
90. Yoncheva, K.; Benbassat, N.; Zaharieva, M.M.; Dimitrova, L.; Kroumov, A.; Spassova, I.; Kovacheva, D.; Najdenski, H.M. Improvement of the Antimicrobial Activity of Oregano Oil by Encapsulation in Chitosan—Alginate Nanoparticles. *Molecules* **2021**, *26*, 7017. [CrossRef] [PubMed]
91. Ciro, Y.; Rojas, J.; Oñate-Garzon, J.; Salamanca, C.H. Synthesis, Characterisation and Biological Evaluation of Ampicillin–Chitosan–Polyanion Nanoparticles Produced by Ionic Gelation and Polyelectrolyte Complexation Assisted by High-Intensity Sonication. *Polymers* **2019**, *11*, 1758. [CrossRef]
92. Qiu, Y.; Xu, D.; Sui, G.; Wang, D.; Wu, M.; Han, L.; Mu, H.; Duan, J. Gentamicin decorated phosphatidylcholine-chitosan nanoparticles against biofilms and intracellular bacteria. *Int. J. Biol. Macromol.* **2020**, *156*, 640–647. [CrossRef]
93. Falciani, C.; Zevolini, F.; Brunetti, J.; Riolo, G.; Gracia, R.; Marradi, M.; Loinaz, I.; Ziemann, C.; Cossío, U.; Llop, J.; et al. Antimicrobial Peptide-Loaded Nanoparticles as Inhalation Therapy for Pseudomonas aeruginosa Infections. *Int. J. Nanomed.* **2020**, *15*, 1117–1128. [CrossRef]
94. Da Costa, D.; Exbrayat-Héritier, C.; Rambaud, B.; Megy, S.; Terreux, R.; Verrier, B.; Primard, C. Surface charge modulation of rifampicin-loaded PLA nanoparticles to improve antibiotic delivery in Staphylococcus aureus biofilms. *J. Nanobiotechnology* **2021**, *19*, 12. [CrossRef]
95. Vrouvaki, I.; Koutra, E.; Kornaros, M.; Avgoustakis, K.; Lamari, F.N.; Hatziantoniou, S. Polymeric Nanoparticles of Pistacia lentiscus var. chia Essential Oil for Cutaneous Applications. *Pharmaceutics* **2020**, *12*, 353. [CrossRef]
96. De Mélo Silva, I.S.; do Amorim Costa Gaspar, L.M.; Rocha, A.M.O.; da Costa, L.P.; Tada, D.B.; Franceschi, E.; Padilha, F.F. Encapsulation of Red Propolis in Polymer Nanoparticles for the Destruction of Pathogenic Biofilms. *AAPS PharmSciTech* **2020**, *21*, 49. [CrossRef]
97. Ucak, S.; Sudagidan, M.; Borsa, B.A.; Mansuroglu, B.; Ozalp, V.C. Inhibitory effects of aptamer targeted teicoplanin encapsulated PLGA nanoparticles for Staphylococcus aureus strains. *World J. Microbiol. Biotechnol.* **2020**, *36*, 69. [CrossRef] [PubMed]
98. Ma, X.; Lang, J.; Chen, P.; Yang, R. Silver nanoparticles as an effective antimicrobial against otitis media pathogens. *AIChE J.* **2021**, *67*, e17468. [CrossRef] [PubMed]
99. Al-Gethami, W.; Al-Qasmi, N. Antimicrobial Activity of Ca-Alginate/Chitosan Nanocomposite Loaded with Camptothecin. *Polymers* **2021**, *13*, 3559. [CrossRef] [PubMed]

100. Rezaei, F.; Damoogh, S.; Reis, R.L.; Kundu, S.C.; Mottaghitalab, F.; Farokhi, M. Dual drug delivery system based on pH-sensitive silk fibroin/alginate nanoparticles entrapped in PNIPAM hydrogel for treating severe infected burn wound. *Biofabrication* **2020**, *13*, 015005. [CrossRef] [PubMed]
101. Hassan, Y.A.; Khedr, A.I.M.; Alkabli, J.; Elshaarawy, R.F.M.; Nasr, A.M. Co-delivery of imidazolium Zn(II)salen and Origanum Syriacum essential oil by shrimp chitosan nanoparticles for antimicrobial applications. *Carbohydr. Polym.* **2021**, *260*, 117834. [CrossRef]
102. Balabathula, P.; Whaley, S.G.; Janagam, D.R.; Mittal, N.K.; Mandal, B.; Thoma, L.A.; Rogers, P.D.; Wood, G.C. Lyophilized Iron Oxide Nanoparticles Encapsulated in Amphotericin B: A Novel Targeted Nano Drug Delivery System for the Treatment of Systemic Fungal Infections. *Pharmaceutics* **2020**, *12*, 247. [CrossRef] [PubMed]
103. Luis Enrique Jerez, P. Chapter 6 – Fungal Infections in Immunosuppressed Patients. In *Immunodeficiency*; Krassimir, M., Ed.; IntechOpen: Rijeka, Croatia, 2012.
104. Patterson, T.F. The role of echinocandins, extended-spectrum azoles, and polyenes to treat opportunistic moulds and candida. *Curr. Infect. Dis. Rep.* **2006**, *8*, 442–448. [CrossRef]
105. Quindós, G.; Gil-Alonso, S.; Marcos-Arias, C.; Sevillano, E.; Mateo, E.; Jauregizar, N.; Eraso, E. Therapeutic tools for oral candidiasis: Current and new antifungal drugs. *Med. Oral Patol. Oral Cir. Bucal* **2019**, *24*, e172–e180. [CrossRef] [PubMed]
106. Araujo, V.H.S.; Duarte, J.L.; Carvalho, G.C.; Silvestre, A.L.P.; Fonseca-Santos, B.; Marena, G.D.; Ribeiro, T.d.C.; dos Santos Ramos, M.A.; Bauab, T.M.; Chorilli, M. Nanosystems against candidiasis: A review of studies performed over the last two decades. *Crit. Rev. Microbiol.* **2020**, *46*, 508–547. [CrossRef]
107. Muñoz, J.E.; Rossi, D.C.P.; Jabes, D.L.; Barbosa, D.A.; Cunha, F.F.M.; Nunes, L.R.; Arruda, D.C.; Pelleschi Taborda, C. In Vitro and In Vivo Inhibitory Activity of Limonene against Different Isolates of Candida spp. *J. Fungi* **2020**, *6*, 183. [CrossRef]
108. Biswas, K.D.; Choudhary, A.; Ghosh, S.K.; Biswas, S. Primary Laryngeal Aspergillosis in an Immunocompetent Host. *Bengal J. Otolaryngol. Head Neck Surg.* **2018**, *26*, 131–133. [CrossRef]
109. Hussain, M.A.; Ahmed, D.; Anwar, A.; Perveen, S.; Ahmed, S.; Anis, I.; Shah, M.R.; Khan, N.A. Combination Therapy of Clinically Approved Antifungal Drugs Is Enhanced by Conjugation with Silver Nanoparticles. *Int. Microbiol.* **2019**, *22*, 239–246. [CrossRef] [PubMed]
110. Alshahrani, S.M.; Khafagy, E.-S.; Riadi, Y.; Al Saqr, A.; Alfadhel, M.M.; Hegazy, W.A.H. Amphotericin B-PEG Conjugates of ZnO Nanoparticles: Enhancement Antifungal Activity with Minimal Toxicity. *Pharmaceutics* **2022**, *14*, 1646. [CrossRef] [PubMed]
111. Shah, M.K.; Azad, A.K.; Nawaz, A.; Ullah, S.; Latif, M.S.; Rahman, H.; Alsharif, K.F.; Alzahrani, K.J.; El-Kott, A.F.; Albrakati, A.; et al. Formulation Development, Characterization and Antifungal Evaluation of Chitosan NPs for Topical Delivery of Voriconazole In Vitro and Ex Vivo. *Polymers* **2022**, *14*, 135. [CrossRef] [PubMed]
112. Elshaer, E.E.; Elwakil, B.H.; Eskandrani, A.; Elshewemi, S.S.; Olama, Z.A. Novel Clotrimazole and Vitis vinifera loaded chitosan nanoparticles: Antifungal and wound healing efficiencies. *Saudi J. Biol. Sci.* **2022**, *29*, 1832–1841. [CrossRef]
113. Orekhova, A.; Palocci, C.; Chronopoulou, L.; De Angelis, G.; Badiali, C.; Petruccelli, V.; D'Angeli, S.; Pasqua, G.; Simonetti, G. Poly-(lactic-co-glycolic) Acid Nanoparticles Entrapping Pterostilbene for Targeting Aspergillus Section Nigri. *Molecules* **2022**, *27*, 5424. [CrossRef]
114. Yenice Gürsu, B. Potential antibiofilm activity of farnesol-loaded poly(DL-lactide-co-glycolide) (PLGA) nanoparticles against Candida albicans. *J. Anal. Sci. Technol.* **2020**, *11*, 43. [CrossRef]
115. Alhowyan, A.A.; Altamimi, M.A.; Kalam, M.A.; Khan, A.A.; Badran, M.; Binkhathlan, Z.; Alkholief, M.; Alshamsan, A. Antifungal efficacy of Itraconazole loaded PLGA-nanoparticles stabilized by vitamin-E TPGS: In vitro and ex vivo studies. *J. Microbiol. Methods* **2019**, *161*, 87–95. [CrossRef]
116. Mejía, S.P.; López, D.; Cano, L.E.; Naranjo, T.W.; Orozco, J. Antifungal Encapsulated into Ligand-Functionalized Nanoparticles with High Specificity for Macrophages. *Pharmaceutics* **2022**, *14*, 1932. [CrossRef]
117. Hou, Y.; Yang, M.; Li, J.; Bi, X.; Li, G.; Xu, J.; Xie, S.; Dong, Y.; Li, D.; Du, Y. The enhancing antifungal effect of AD1 aptamer-functionalized amphotericin B-loaded PLGA-PEG nanoparticles with a low-frequency and low-intensity ultrasound exposure on C.albicans biofilm through targeted effect. *NanoImpact* **2021**, *21*, 100275. [CrossRef]
118. Endo, E.H.; Makimori, R.Y.; Companhoni, M.V.P.; Ueda-Nakamura, T.; Nakamura, C.V.; Dias Filho, B.P. Ketoconazole-loaded poly-(lactic acid) nanoparticles: Characterization and improvement of antifungal efficacy in vitro against Candida and dermatophytes. *J. Mycol. Médicale* **2020**, *30*, 101003. [CrossRef] [PubMed]
119. Al-Maghrabi, P.M.; Khafagy, E.-S.; Ghorab, M.M.; Gad, S. Influence of formulation variables on miconazole nitrate–loaded lipid based nanocarrier for topical delivery. *Colloids Surf. B Biointerfaces* **2020**, *193*, 111046. [CrossRef] [PubMed]
120. Baldim, I.; Paziani, M.H.; Grizante Barião, P.H.; Kress, M.R.v.Z.; Oliveira, W.P. Nanostructured Lipid Carriers Loaded with Lippia sidoides Essential Oil as a Strategy to Combat the Multidrug-Resistant Candida auris. *Pharmaceutics* **2022**, *14*, 180. [CrossRef]
121. Lee, Y.; Puumala, E.; Robbins, N.; Cowen, L.E. Antifungal Drug Resistance: Molecular Mechanisms in Candida albicans and Beyond. *Chem. Rev.* **2021**, *121*, 3390–3411. [CrossRef]
122. Lim, C.S.Y.; Rosli, R.; Seow, H.F.; Chong, P.P. Candida and invasive candidiasis: Back to basics. *Eur. J. Clin. Microbiol. Infect. Dis.* **2012**, *31*, 21–31. [CrossRef] [PubMed]
123. Jagaran, K.; Singh, M. Nanomedicine for COVID-19: Potential of Copper Nanoparticles. *Biointerface Res. Appl. Chem.* **2021**, *11*, 10716–10728. [CrossRef]

124. Delshadi, R.; Bahrami, A.; McClements, D.J.; Moore, M.D.; Williams, L. Development of nanoparticle-delivery systems for antiviral agents: A review. *J. Control. Release* **2021**, *331*, 30–44. [CrossRef] [PubMed]
125. Pokhrel, L.R.; Williams, F.; Cook, P.P.; O'Rourke, D.; Murray, G.; Akula, S.M. Preclinical efficacy and safety of novel SNAT against SARS-CoV-2 using a hamster model. *Drug Deliv. Transl. Res.* **2022**, *12*, 3007–3016. [CrossRef] [PubMed]
126. León-Gutiérrez, G.; Elste, J.E.; Cabello-Gutiérrez, C.; Millán-Pacheco, C.; Martínez-Gómez, M.H.; Mejía-Alvarez, R.; Tiwari, V.; Mejía, A. A potent virucidal activity of functionalized TiO2 nanoparticles adsorbed with flavonoids against SARS-CoV-2. *Appl. Microbiol. Biotechnol.* **2022**, *106*, 5987–6002. [CrossRef]
127. Joshy, K.S.; Susan, M.A.; Snigdha, S.; Nandakumar, K.; Laly, A.P.; Sabu, T. Encapsulation of zidovudine in PF-68 coated alginate conjugate nanoparticles for anti-HIV drug delivery. *Int. J. Biol. Macromol.* **2018**, *107*, 929–937. [CrossRef]
128. Joshy, K.S.; George, A.; Snigdha, S.; Joseph, B.; Kalarikkal, N.; Pothen, L.A.; Thomas, S. Novel core-shell dextran hybrid nanosystem for anti-viral drug delivery. *Mater. Sci. Eng. C* **2018**, *93*, 864–872. [CrossRef] [PubMed]
129. Belgamwar, A.V.; Khan, S.A.; Yeole, P.G. Intranasal dolutegravir sodium loaded nanoparticles of hydroxypropyl-beta-cyclodextrin for brain delivery in Neuro-AIDS. *J. Drug Deliv. Sci. Technol.* **2019**, *52*, 1008–1020. [CrossRef]
130. Belgamwar, A.; Khan, S.; Yeole, P. Intranasal chitosan-g-HPβCD nanoparticles of efavirenz for the CNS targeting. *Artif. Cells Nanomed. Biotechnol.* **2018**, *46*, 374–386. [CrossRef] [PubMed]
131. Donalisio, M.; Argenziano, M.; Rittà, M.; Bastiancich, C.; Civra, A.; Lembo, D.; Cavalli, R. Acyclovir-loaded sulfobutyl ether-β-cyclodextrin decorated chitosan nanodroplets for the local treatment of HSV-2 infections. *Int. J. Pharm.* **2020**, *587*, 119676. [CrossRef]
132. Parthiban, R.; Sathishkumar, S.; Ramakrishnan, P. Design and evaluation of acyclovir-loaded solid lipid nanoparticles for sustained release. *Drug Invent. Today* **2020**, *14*, 108–111.
133. Javan, F.; Vatanara, A.; Azadmanesh, K.; Nabi-Meibodi, M.; Shakouri, M. Encapsulation of ritonavir in solid lipid nanoparticles: In-vitro anti-HIV-1 activity using lentiviral particles. *J. Pharm. Pharmacol.* **2017**, *69*, 1002–1009. [CrossRef]
134. Figueira, T.N.; Domingues, M.M.; Illien, F.; Cadima-Couto, I.; Todorovski, T.; Andreu, D.; Sagan, S.; Castanho, M.A.R.B.; Walrant, A.; Veiga, A.S. Enfuvirtide-Protoporphyrin IX Dual-Loaded Liposomes: In Vitro Evidence of Synergy against HIV-1 Entry into Cells. *ACS Infect. Dis.* **2020**, *6*, 224–236. [CrossRef]
135. Cao, S.; Slack, S.D.; Levy, C.N.; Hughes, S.M.; Jiang, Y.; Yogodzinski, C.; Roychoudhury, P.; Jerome, K.R.; Schiffer, J.T.; Hladik, F.; et al. Hybrid nanocarriers incorporating mechanistically distinct drugs for lymphatic CD4+ T cell activation and HIV-1 latency reversal. *Sci. Adv.* **2019**, *5*, eaav6322. [CrossRef]
136. Snoussi, Y.; Sifaoui, I.; Khalil, A.M.; Bhakta, A.K.; Semyonov, O.; Postnikov, P.S.; Michely, L.; Pires, R.; Bastide, S.; Barroso, J.E.-P.; et al. Facile synthesis of silver decorated biochar as a novel and highly active biosourced anti-kinetoplastid agent. *Mater. Today Commun.* **2022**, *32*, 104126. [CrossRef]
137. Badirzadeh, A.; Alipour, M.; Najm, M.; Vosoogh, A.; Vosoogh, M.; Samadian, H.; Hashemi, A.S.; Farsangi, Z.J.; Amini, S.M. Potential therapeutic effects of curcumin coated silver nanoparticle in the treatment of cutaneous leishmaniasis due to Leishmania major in-vitro and in a murine model. *J. Drug Deliv. Sci. Technol.* **2022**, *74*, 103576. [CrossRef]
138. Antinori, S.; Schifanella, L.; Corbellino, M. Leishmaniasis: New insights from an old and neglected disease. *Eur. J. Clin. Microbiol. Infect. Dis.* **2012**, *31*, 109–118. [CrossRef] [PubMed]
139. Durak, S.; Arasoglu, T.; Ates, S.C.; Derman, S. Enhanced antibacterial and antiparasitic activity of multifunctional polymeric nanoparticles. *Nanotechnology* **2020**, *31*, 175705. [CrossRef] [PubMed]
140. Kanwal; Mungroo, M.R.; Anwar, A.; Ali, F.; Khan, S.; Abdullah, M.A.; Siddiqui, R.; Khan, K.M.; Khan, N.A. Synthetic nanoparticle-conjugated bisindoles and hydrazinyl arylthiazole as novel antiamoebic agents against brain-eating amoebae. *Exp. Parasitol.* **2020**, *218*, 107979. [CrossRef] [PubMed]
141. Real, D.; Hoffmann, S.; Leonardi, D.; Salomon, C.; Goycoolea, F.M. Chitosan-based nanodelivery systems applied to the development of novel triclabendazole formulations. *PLoS ONE* **2018**, *13*, e0207625. [CrossRef]
142. Wei, W.; Lu, M.; Xu, W.; Polyakov, N.E.; Dushkin, A.V.; Su, W.-k. Preparation of protamine-hyaluronic acid coated core-shell nanoparticles for enhanced solubility, permeability, and oral bioavailability of decoquinate. *Int. J. Biol. Macromol.* **2022**, *218*, 346–355. [CrossRef]
143. Taghe, S.; Mirzaeei, S. Preparation and characterization of novel, mucoadhesive ofloxacin nanoparticles for ocular drug delivery. *Braz. J. Pharm. Sci.* **2019**, *55*, e17105. [CrossRef]
144. Sanchez-Guzman, D.; Le Guen, P.; Villeret, B.; Sola, N.; Le Borgne, R.; Guyard, A.; Kemmel, A.; Crestani, B.; Sallenave, J.-M.; Garcia-Verdugo, I. Silver nanoparticle-adjuvanted vaccine protects against lethal influenza infection through inducing BALT and IgA-mediated mucosal immunity. *Biomaterials* **2019**, *217*, 119308. [CrossRef]
145. Lina, T.T.; Johnson, S.J.; Wagner, R.D. Intravaginal poly-(D, L-lactic-co-glycolic acid)-(polyethylene glycol) drug-delivery nanoparticles induce pro-inflammatory responses with Candida albicans infection in a mouse model. *PLoS ONE* **2020**, *15*, e0240789. [CrossRef]
146. Atinderpal, K.; Kapoor, N.; Gupta, S.; Tyag, A.; Sharma, R.K.; Ali, J.; Gabrani, R.; Dang, S. Development and characterization of green tea catechins and ciprofloxacin-loaded nanoemulsion for intravaginal delivery to treat urinary tract infection. *Indian J. Pharm. Sci.* **2018**, *80*, 442–452. [CrossRef]

147. Brauner, B.; Semmler, J.; Rauch, D.; Nokaj, M.; Haiss, P.; Schwarz, P.; Wirth, M.; Gabor, F. Trimethoprim-Loaded PLGA Nanoparticles Grafted with WGA as Potential Intravesical Therapy of Urinary Tract Infections—Studies on Adhesion to SV-HUCs Under Varying Time, pH, and Drug-Loading Conditions. *ACS Omega* **2020**, *5*, 17377–17384. [CrossRef]
148. Sarfraz, M.; Qamar, S.; Rehman, M.U.; Tahir, M.A.; Ijaz, M.; Ahsan, A.; Asim, M.H.; Nazir, I. Nano-Formulation Based Intravesical Drug Delivery Systems: An Overview of Versatile Approaches to Improve Urinary Bladder Diseases. *Pharmaceutics* **2022**, *14*, 1909. [CrossRef] [PubMed]
149. Kothawade, S.; Bagul, U.; Kokare, C.; Giikwad, S.; Wakure, R.; Biyani, S.; Harne, C. Formulation development of antimicrobial zinc oxide nanoparticle loaded trans dermal patch by using 2 3 factorial design. *Indo Am. J. Pharm. Res.* **2019**, *9*, 483–493.
150. Nasrollahzadeh, M.; Ganji, F.; Taghizadeh, S.M.; Vasheghani-Farahani, E.; Mohiti-Asli, M. Drug in adhesive transdermal patch containing antibiotic-loaded solid lipid nanoparticles. *J. Biosci. Bioeng.* 2022, in press. [CrossRef] [PubMed]
151. Elfaky, M.A.; Sirwi, A.; Tolba, H.H.; Shaik, R.A.; Selmi, N.M.; Alattas, A.H.; Albreki, R.S.; Alshreef, N.M.; Gad, H.A. Development, Optimization, and Antifungal Assessment of Ocular Gel Loaded With Ketoconazole Cubic Liquid Crystalline Nanoparticles. *J. Pharm. Sci.* **2021**, *110*, 2210–2220. [CrossRef]
152. Juknius, T.; Juknienė, I.; Tamulevičius, T.; Ružauskas, M.; Pamparienė, I.; Oberauskas, V.; Jurkevičiūtė, A.; Vasiliauskas, A.; Tamulevičius, S. Preclinical Study of a Multi-Layered Antimicrobial Patch Based on Thin Nanocomposite Amorphous Diamond Like Carbon Films with Embedded Silver Nanoparticles. *Materials* **2020**, *13*, 3180. [CrossRef] [PubMed]
153. Rayyif, S.M.I.; Mohammed, H.B.; Curuțiu, C.; Bîrcă, A.C.; Grumezescu, A.M.; Vasile, B.Ș.; Dițu, L.M.; Lazăr, V.; Chifiriuc, M.C.; Mihăescu, G.; et al. ZnO Nanoparticles-Modified Dressings to Inhibit Wound Pathogens. *Materials* **2021**, *14*, 3084. [CrossRef]
154. Ambrogi, V.; Pietrella, D.; Donnadio, A.; Latterini, L.; Di Michele, A.; Luffarelli, I.; Ricci, M. Biocompatible alginate silica supported silver nanoparticles composite films for wound dressing with antibiofilm activity. *Mater. Sci. Eng. C* **2020**, *112*, 110863. [CrossRef]
155. Namviriyachote, N.; Muangman, P.; Chinaroonchai, K.; Chuntrasakul, C.; Ritthidej, G.C. Polyurethane-biomacromolecule combined foam dressing containing asiaticoside: Fabrication, characterization and clinical efficacy for traumatic dermal wound treatment. *Int. J. Biol. Macromol.* **2020**, *143*, 510–520. [CrossRef]
156. Başaran, D.D.A.; Gündüz, U.; Tezcaner, A.; Keskin, D. Topical delivery of heparin from PLGA nanoparticles entrapped in nanofibers of sericin/gelatin scaffolds for wound healing. *Int. J. Pharm.* **2021**, *597*, 120207. [CrossRef]
157. Ahmed, M.K.; Zayed, M.A.; El-Dek, S.I.; Hady, M.A.; El Sherbiny, D.H.; Uskoković, V. Nanofibrous ε-polycaprolactone scaffolds containing Ag-doped magnetite nanoparticles: Physicochemical characterization and biological testing for wound dressing applications in vitro and in vivo. *Bioact. Mater.* **2021**, *6*, 2070–2088. [CrossRef]
158. Ahmed, M.K.; Menazea, A.A.; Abdelghany, A.M. Blend biopolymeric nanofibrous scaffolds of cellulose acetate/ε-polycaprolactone containing metallic nanoparticles prepared by laser ablation for wound disinfection applications. *Int. J. Biol. Macromol.* **2020**, *155*, 636–644. [CrossRef] [PubMed]
159. Saravanan, S.; Nethala, S.; Pattnaik, S.; Tripathi, A.; Moorthi, A.; Selvamurugan, N. Preparation, characterization and antimicrobial activity of a bio-composite scaffold containing chitosan/nano-hydroxyapatite/nano-silver for bone tissue engineering. *Int. J. Biol. Macromol.* **2011**, *49*, 188–193. [CrossRef] [PubMed]
160. Spirescu, V.A.; Niculescu, A.-G.; Slave, Ș.; Bîrcă, A.C.; Dorcioman, G.; Grumezescu, V.; Holban, A.M.; Oprea, O.-C.; Vasile, B.Ș.; Grumezescu, A.M.; et al. Anti-Biofilm Coatings Based on Chitosan and Lysozyme Functionalized Magnetite Nanoparticles. *Antibiotics* **2021**, *10*, 1269. [CrossRef] [PubMed]
161. Spirescu, V.A.; Șuhan, R.; Niculescu, A.-G.; Grumezescu, V.; Negut, I.; Holban, A.M.; Oprea, O.-C.; Bîrcă, A.C.; Vasile, B.Ș.; Grumezescu, A.M.; et al. Biofilm-Resistant Nanocoatings Based on ZnO Nanoparticles and Linalool. *Nanomaterials* **2021**, *11*, 2564. [CrossRef] [PubMed]
162. Prodana, M.; Stoian, A.B.; Burnei, C.; Ionita, D. Innovative Coatings of Metallic Alloys Used as Bioactive Surfaces in Implantology: A Review. *Coatings* **2021**, *11*, 649. [CrossRef]
163. Florea, D.A.; Grumezescu, V.; Bîrcă, A.C.; Vasile, B.Ș.; Mușat, M.; Chircov, C.; Stan, M.S.; Grumezescu, A.M.; Andronescu, E.; Chifiriuc, M.C. Design, Characterization, and Antibacterial Performance of MAPLE-Deposited Coatings of Magnesium Phosphate-Containing Silver Nanoparticles in Biocompatible Concentrations. *Int. J. Mol. Sci.* **2022**, *23*, 7910. [CrossRef]
164. Olar, R.; Badea, M.; Maxim, C.; Grumezescu, A.M.; Bleotu, C.; Măruţescu, L.; Chifiriuc, M.C. Anti-biofilm Fe3O4@C18-[1,3,4]thiadiazolo[3,2-a]pyrimidin-4-ium-2-thiolate Derivative Core-shell Nanocoatings. *Materials* **2020**, *13*, 4640. [CrossRef]

International Journal of *Molecular Sciences*

MDPI

Article

Antibacterial Soy Protein Isolate Prepared by Quaternization

Qi Dong [1], Jingwen Lei [1], Hanjian Wang [1], Meifang Ke [1], Xiao Liang [1], Xindi Yang [1], Hui Liang [2], Céline Huselstein [3], Zan Tong [1,*] and Yun Chen [1,*]

[1] Department of Biomedical Engineering and Hubei Province Key Laboratory of Allergy and Immune Related Disease, TaiKang Medical School (School of Basic Medicine Sciences), Wuhan University, Wuhan 430071, China

[2] Research Center for Medicine and Structural Biology, TaiKang Medical School (School of Basic Medicine Sciences), Wuhan University, Wuhan 430071, China

[3] UMR 7365 CNRS, Medical School, University of Lorraine, 54505 Nancy, France

* Correspondence: ztong@whu.edu.cn (Z.T.); yunchen@whu.edu.cn (Y.C.)

Abstract: Soy protein isolate (SPI) is green, high-yield natural plant protein, which is widely applied in industry (packing material and adhesives) and tissue engineering. It is meaningful to improve the antibacterial property of soy protein isolate to fabricate versatile safe products to meet people's requirements. In this study, quaternized soy protein isolate (QSPI) was synthesized by the reaction between 2,3-epoxypropyltrimethylammonium chloride (EPTMAC) and SPI. The positive charged (17.8 ± 0.23 mV) quaternary ammonium groups endow the QSPI with superior antibacterial properties against multiple bacteria in vitro and in vivo. Notably, QSPI maintains its good biocompatibility and promotes bacterial-infected wound healing in rat models. Furthermore, QSPI possesses superior water solubility in a broad pH range than raw SPI. Altogether, this soy protein isolate derivative with antibacterial property and superior water solubility may extend the application of SPI in industry and tissue engineering.

Keywords: soy protein isolate; quaternization; antibacterial property; wound healing

Citation: Dong, Q.; Lei, J.; Wang, H.; Ke, M.; Liang, X.; Yang, X.; Liang, H.; Huselstein, C.; Tong, Z.; Chen, Y. Antibacterial Soy Protein Isolate Prepared by Quaternization. *Int. J. Mol. Sci.* **2022**, *23*, 9110. https://doi.org/10.3390/ijms23169110

Academic Editor: Helena Felgueiras

Received: 27 July 2022
Accepted: 12 August 2022
Published: 14 August 2022

1. Introduction

Soy protein isolate (SPI) is a natural and nutritious plant-based protein with a protein content higher than 90% [1]. Because of its ready availability, low price, and renewable and degradable properties, SPI has been extensively applied as packaging films and adhesives in the industry to replace petroleum feedstocks [2,3]. Moreover, SPI possesses biocompatibility, non-toxicity of degradation products, and contains abundant functional amino acids that are beneficial for tissue regeneration [4]. Thus, SPI is also widely used as a biomaterials for tissue engineering. As people's requirements for the safety of the living environment increase, the new packaging films should have a good antibacterial property to extend the shelf life of food [5,6]; the new adhesives should possess antibacterial properties to improve the mildew-resistant property and durability [7]. However, the polysaccharides and proteins in SPI are susceptible to corrosion by bacteria and fungi. Thus, many antibacterial agents such as natural compounds, curcumin, tannins, carvacrol; nanoparticles, zinc oxide, silver nanoparticles; and quaternized polymers have been incorporated into SPI-based materials to fabricate antibacterial, environmentally friendly industrial materials [8–11]. For SPI-based biomaterials, a good antibacterial property is especially important. Therefore, some antibacterial agents such as antibiotics, nanosilver, and quaternized chitosan were introduced into SPI-based biomaterials to improve the antibacterial property of biomaterials, especially for the repair of some infected wounds [12]. If endowed with the antibacterial property itself, SPI could be used to fabricate even more versatile products for practical application.

To fabricate SPI-based materials, some harsh dissolving conditions (pH > 10; enzyme treatment; polar solvents) have been employed to dissolve SPI due to the poor solubility of SPI, which limits its further application in cooperation with other materials for the fabrication of composite materials [13–15]. For example, Chen et al. prepared a variety of SPI-based composite sponge conduits by dissolving SPI in a strong alkaline solution [16]. Qi et al. used papain partial hydrolysis of SPI to improve its water solubility [17]. Therefore, a simple modification to improve the antibacterial property and water solubility of SPI will open a new avenue for the application of SPI in industry and tissue engineering.

Quaternization modification not only endows antibacterial properties to the substrate but also improves hydrophilicity by introducing positive charged quaternary ammonium groups. For example, quaternized chitin displays an excellent antibacterial ability against *Escherichia coli* and *Staphylococcus aureus* [18]. Meanwhile, the positive charged quaternized ammonium groups reconstruct the inter- and intra-molecular hydrogen bond of the chitin chains, decreasing the degree of crystallinity and increasing the solubility of the chitin. However, quaternized soy protein isolate (QSPI) with an antibacterial property has not been reported.

In this work, QSPI was synthesized by the reaction between 2,3-epoxypropyltrimethylammonium chloride (EPTMAC) and SPI (Figure 1a). The antibacterial property and water solubility of QSPI were evaluated (Figure 1b,c). The potential application as an antibacterial reagent for infected wound repair was investigated (Figure 1d).

2. Results and Discussion

2.1. Fabrication and Characterization of QSPI

The quaternized soy protein isolate (QSPI) was synthesized in a two-step reaction between SPI and EPTMAC, as shown in Figure 1a. Firstly, SPI was dissolved in NaOH solution (0.125 M) and stirred for 30 min at 25 °C to disrupt the inter- and intra-molecular bonds between SPI chains. Secondly, EPTMAC was added, and the mixture was stirred for 24 h at 25 °C to produce the QSPI. The w/w ratio of EPTMAC to SPI in the reaction was the primary factor influencing the degree of quaternization of QSPI. Therefore, a series of w/w ratios of EPTMAC to SPI (0.5, 1.0, 1.5, 2.0, 2.5) was set. The degree of quaternization was determined by the OPA reagent (Figure 2a). The $-NH_2$ in SPI or QSPI without quaternization can react with the OPA reagent and exhibits a characteristic absorption peak at 336 nm in the UV–vis absorption spectra (Figure 2b). Thus, a higher absorption at 336 nm means that more $-NH_2$ in SPI was not quaternized. A high absorption at a wavelength of 336 nm was observed in SPI, while the absorption of QSPI-n at 336 nm reduced from 2.93 in SPI to 1.83 in QSPI-0.5 and 0.67 in QSPI-2.5 due to more moles of $-NH_2$ of SPI reacting with EPTMAC. The degree of quaternization was calculated and is shown in Figure 2c. As the ratios (w/w) of EPTMA to SPI increased from 0.5 to 2.5, the degree of quaternization increased from 37.69% of QSPI-0.5 to 77.15% of QSPI-2.5. When the ratio exceeds 2.0 (75.29%), the degree of quaternization changes little because the reaction tends to saturate. Therefore, the ratio of 2.0 was selected for the subsequent experiments. To further confirm the formation of the QSPI, the molecular structure of the QSPI was analyzed by Fourier transform infrared (FTIR) and ^{1}HNMR. In the FTIR spectrum of SPI (Figure 2d), the characteristic absorption peaks of O–H stretching, amide I band (C=O stretching), amide II band (N–H bending), and amide III band (C–N stretching) presented at 3274 cm^{-1}, 1640 cm^{-1}, 1532 cm^{-1}, and 1236 cm^{-1} in the spectrum of SPI and QSPI [19]. After quaternization, two new peaks appeared at 969 cm^{-1} and 879 cm^{-1} in the spectrum of QSPI. The peak at 969 cm^{-1} is for the C–N stretching vibration from $-N^+(CH_3)_3$, and 879 cm^{-1} is for the vibration of N–H. These new characteristic absorption peaks indicated that the quaternary ammonium groups were successfully grafted onto the side chain of SPI. The structure of SPI in NaOH/D_2O solution and QSPI-2.0 in D_2O were further analyzed by ^{1}HNMR (Figure 2e). The characteristic peaks of QSPI at chemical shift 3.14 ppm, 2.67 ppm, and 3.52 ppm were attributed to H in the $-N^+(CH_3)_3$ group, methylene and methine, respectively [20]. These results revealed that QSPI was successfully synthesized.

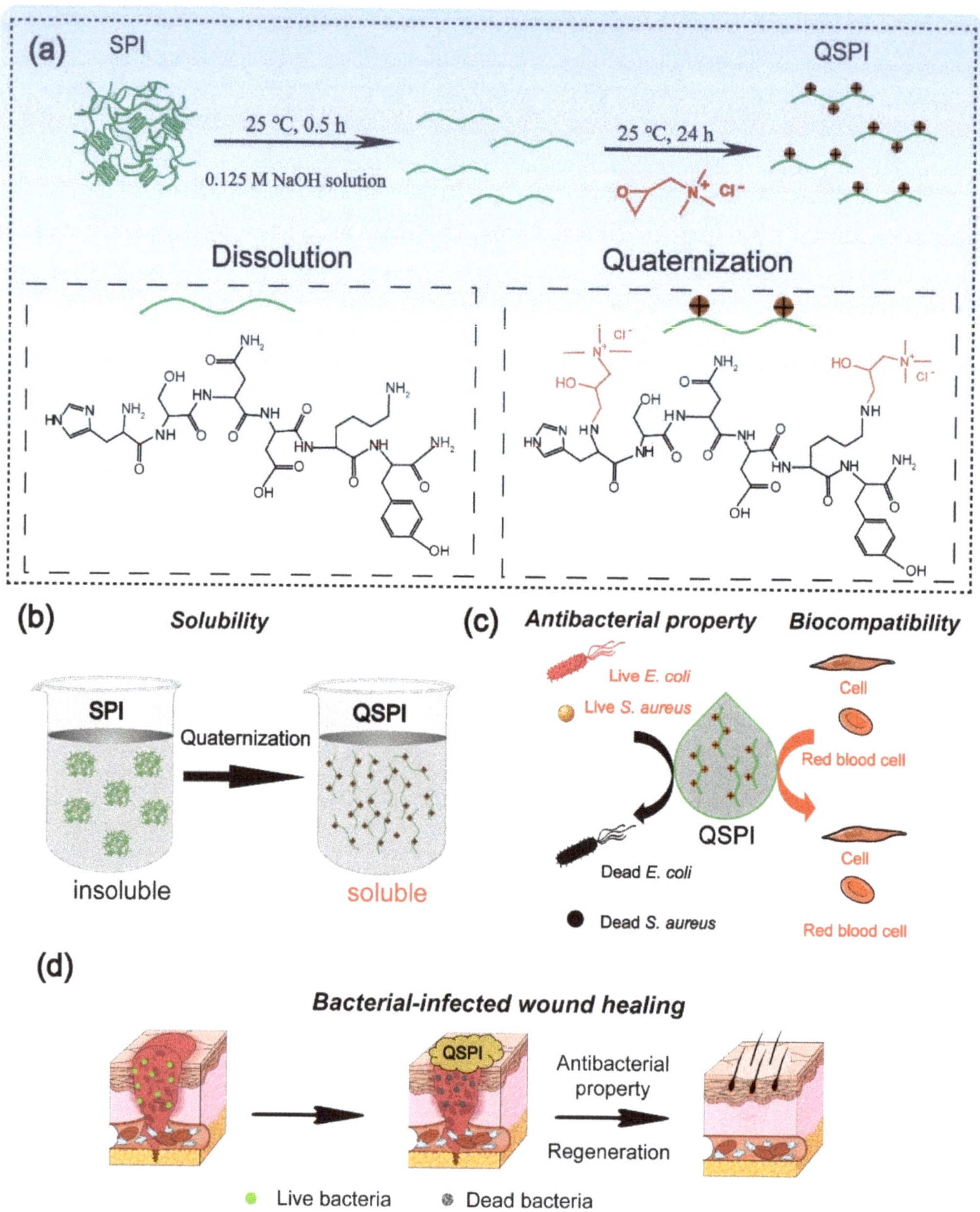

Figure 1. Schematic illustration of the fabrication and application of quaternized soy protein isolate. (**a**) The synthetic method of QSPI. (**b**) The water solubility of QSPI. (**c**) The antibacterial property and biocompatibility of QSPI. (**d**) The therapeutic process of bacterial-infected wound healing by QSPI.

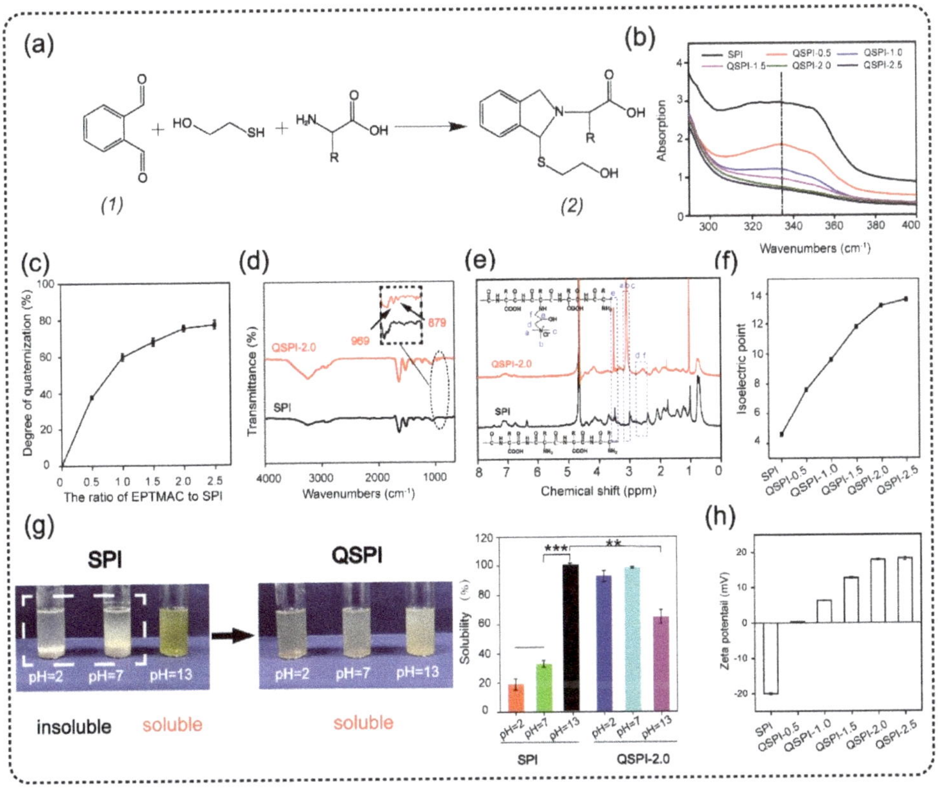

Figure 2. Synthesis and characterization of quaternized soy protein isolate (QSPI). (**a**) The reaction principle of OPA reagent and compound (2) has a marked absorption peak at 336 nm in the UV–vis absorption spectra (**b**). (**c**) The degree of quaternization of the QSPI-n. (**d**) The FTIR spectra of SPI and QSPI-2.0. (**e**) The ^{1}HNMR spectra of SPI and QSPI-2.0. (**f**) The isoelectric point of SPI and QSPI-n. (**g**) The water solubility of SPI and QSPI-n in aqueous solution with pH = 2, 7, and 13. The water solubility of SPI in solution with pH = 13 as control. (**h**) The zeta potential of QSPI-n at pH = 7. ** $p < 0.01$ and *** $p < 0.001$.

The isoelectric point is a remarkable physical characteristic of protein. The isoelectric point of QSPI-n is shown in Figure 2f. The isoelectric point increased from 4.6 of SPI to 13.2 of QSPI-2.0 and 13.6 of QSPI-2.5. As the w/w ratios of EPTMAC to SPI increased, more positive charged quaternary ammonium groups were grafted onto the SPI, resulting in the increase in the isoelectric point. Proteins have the lowest solubility at the isoelectric point and have good water solubility in solutions with a pH far from the isoelectric point [7]. The isoelectric points of QSPI-2.0 and QSPI-2.5 are closer to the pH extreme than other groups, so they may have better water solubility in a broader range of pH.

Water solubility plays an important role in the application of protein biomaterials. The photograph of SPI and QSPI-2.0 in the solution with pH = 2, 7, and 13 is shown in Figure 2g. SPI exhibited aggregation and sedimentation at pH = 2 and 7, and a homogeneous transparent yellow at pH = 13. In contrast, QSPI-2.0 displayed homogeneous transparent at pH = 2, 7, and 13. The quantitative analysis of water solubility showed that SPI displayed good water solubility at pH = 13 (100%), and poor water solubility at pH = 2 (19.11 ± 3.93%) and pH = 7 (33.02 ± 2.45%) (Figure 2g). Interestingly, QSPI-2.0 was almost 100 % soluble at pH = 2, 7 and partially dissolved at pH = 13 (64.90 ± 4.84%), close to the isoelectric point (13.2) of QSPI-2.0. QSPI-2.0 had a better solubility than SPI in a broader range of pH.

The zeta potentials of the SPI and QSPI-2.0 are shown in Figure 2h. With the increase in positive charged quaternary ammonium groups, the zeta potentials increased from −20 ± 0.33 mV of SPI to 17.8 ± 0.23 mV of QSPI-2.0 and 18.1 ± 0.46 mV of QSPI-2.5 at pH = 7. Quaternarization may change the zeta potential of the target protein. For example, Zhang et al. [18] synthesized a quaternized chitin by the reaction between chitin and EPTMAC. After quaternization, the zeta potential of quaternized chitin increased from 0.28 ± 0.03 mV of chitin to 30.57 ± 0.99 mV of quaternized chitin. In summary, QSPI was successfully synthesized, and QSPI-2.0 has good water solubility in a broad range of pH.

2.2. Antibacterial Property Evaluation of QSPI-2.0

Quaternary ammonium compounds are widely used to kill bacteria. The strong positive charges of quaternary ammonium groups may endow QSPI with antibacterial properties. Therefore, three common bacteria, *E. coli* (Gram-negative), *S. aureus* (Gram-positive), and MRSA were used to evaluate the antibacterial property of the raw SPI and QSPI-2.0. In the colony-forming units (CFU) testing, SPI exhibited no antibacterial property (Figure 3a). Meanwhile, the effects of promoting bacterial growth became more obvious as the concentration of SPI increased. The capacity of promoting the growth of bacteria limits the application of SPI. In contrast, QSPI-2.0 exhibited significant antibacterial properties. The statistical analysis results showed that the inhibition rate increased with the increase in QSPI-2.0 concentration, from 20.86 ± 3.16% (0.5 mg/mL) to 89.17 ± 1.15% (2.5 mg/mL) for *E. coli* (Figure 3b), from 40.86 ± 9.31% (0.5 mg/mL) to 94.17 ± 1.15% (2.5 mg/mL) for *S. aureus* (Figure 3c), and from 80.24 ± 5.80% (0.5 mg/mL) to 99.79 ± 9.26% (2.5 mg/mL) for MRSA (Figure 3d). The same tendency was also observed in the results of the proliferation assay. As shown in Figure 3e–g, the proliferation rates of the three bacteria remarkably declined with the increase in QSPI-2.0 concentration. The antibacterial property of QSPI-2.0 is probably due to the positive charges of quaternary ammonium groups. The positive charged groups interact with the negative charged cytomembrane of bacteria, damaging the cytomembrane and killing bacteria [21–23]. Thus, these results demonstrated that the QSPI-2.0 had a good antibacterial property. QSPI can be used as an antimicrobial agent for the fabrication of antibacterial, environmentally friendly SPI-based packaging films and adhesives.

2.3. Biocompatibility Evaluation of QSPI-2.0

The biocompatibility of QSPI was evaluated by cytocompatibility and hemocompatibility in vitro. Firstly, the cytocompatibility of the QSPI-2.0 was accessed by live/dead staining and the (3- (4,5-dimethyl-2-thiazoLyl) -2,5-diphenyl tetrazolium bromide (MTT) assay. L929 cells treated with different concentrations of QSPI-2.0 were stained with calcein-AM (green) for live cells and propidium iodide (PI) for dead cells (Figure 4a). Almost all cells exhibited normal shapes and stained green demonstrating no significant cytotoxicity of the QSPI-2.0. In the MTT assay, the cell viability exhibited no significant difference when the concentrations of QSPI-2.0 increased from 0.5 to 2.0 mg/mL (Figure 4b). However, the cell viability slightly decreased after treatment with a high concentration (2.5 mg/mL) of QSPI-2.0. These data revealed that QSPI had excellent cell biocompatibility.

Furthermore, the hemocompatibility of QSPI-2.0 was tested by a hemolysis experiment (Figure 4c). After being incubated with QSPI-2.0, blood displayed light yellow after centrifugation, which is similar to the blood treated with the negative control (PBS). In contrast, Triton-treated blood as the positive control was bright red. Quantitative evaluation results indicated that when the concentrations of QSPI-2.0 increased from 0.5 to 2.5 mg/mL, the hemolysis ratios were slightly increased from 0.233 ± 0.096% to 2.40 ± 0.49%, which were less than the permissible limit (5%). These results revealed that QSPI-2.0 had good cytocompatibility and hemocompatibility, which may have potential as biomaterials [24].

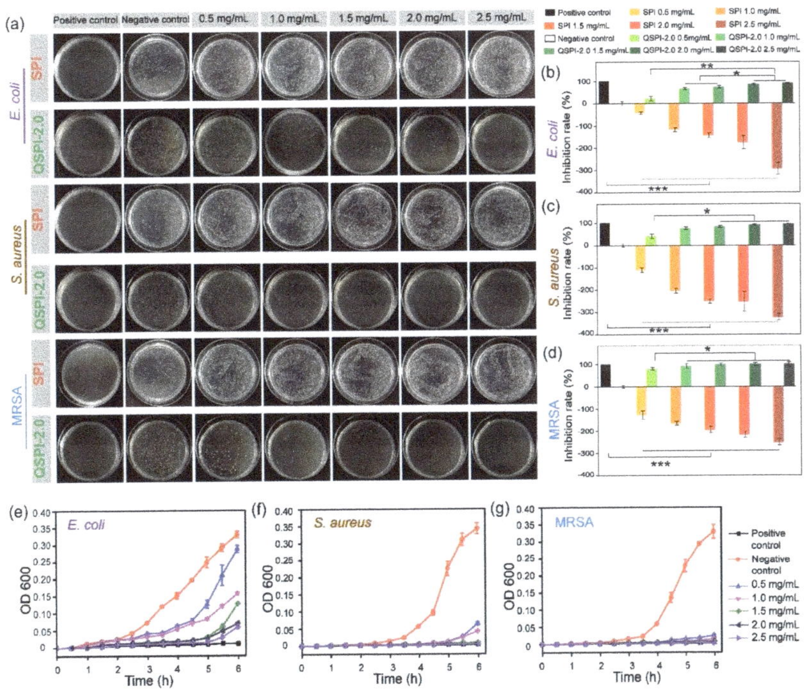

Figure 3. The antibacterial property of SPI and QSPI-2.0. (**a**) Agar plate images of CFU test for antibacterial properties of SPI and QSPI-2.0 against *E. coli*, *S. aureus* and MRSA. The statistical analysis results of the CFU test of *E. coli* (**b**); *S. aureus* (**c**); and MRSA (**d**). The results of proliferation assay of QSPI-2.0 with different concentrations against *E. coli* (**e**); *S. aureus* (**f**); and MRSA (**g**). * $p < 0.05$; ** $p < 0.01$ and *** $p < 0.001$.

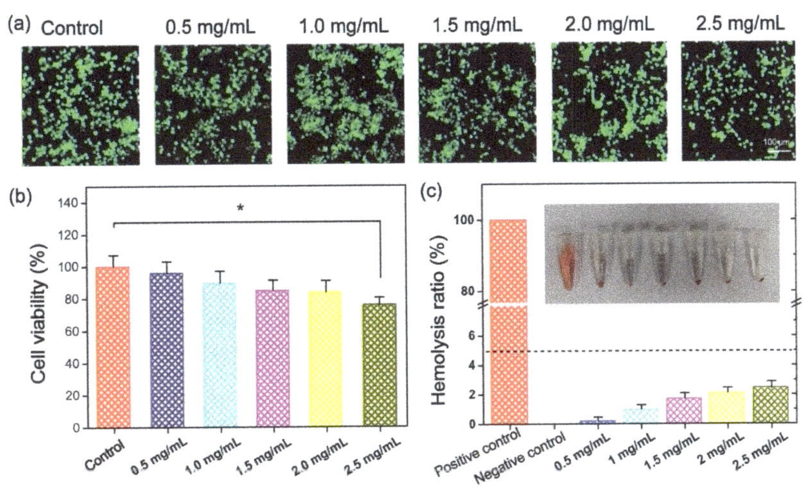

Figure 4. Biocompatibility evaluation of QSPI-2.0. (**a**) Live/dead cell staining of L929 cells after treatment with 0.5–2.5 mg/mL QSPI-2.0 for 48 h. (**b**) Cell viability of L929 cells after treatment with 0.5–2.5 mg/mL QSPI-2.0 for 48 h. (**c**) Hemolysis assay of QSPI-2.0. * $p < 0.05$.

2.4. In Vivo Antibacterial Property and Wound Healing Evaluation of QSPI-2.0

To further examine the antibacterial ability and promotion of infected wound healing of QSPI-2.0 in vivo, an MRSA-infected skin wound model was established in rats. QSPI, Alginate Ag, and Tegaderm^{TM} film were used as wound dressings (Figure 5a). Representative photos of the wound sites in different groups at scheduled time intervals are shown in Figure 5b. The QSPI-2.0 and Alginate Ag showed a significantly higher wound healing rate than the Tegaderm^{TM} film. More specifically, the QSPI-2.0 group and the Alginate Ag group led to 76.03 ± 5.76% and 73.35 ± 2.45% wound healing rates on Day 10, which were significantly higher than the Tegaderm^{TM} film (57.65 ± 3.45%) (Figure 5c). On Day 20, the wound treated with QSPI-2.0 was almost healed with the wound healing rate closed to 100%, while the wound healing rate in the Alginate Ag group and Tegaderm^{TM} film were approximately 90.38 ± 7.25% and 83.69 ± 2.68%. These results indicate that the QSPI-2.0 could promote infected wound healing.

To further evaluate the antibacterial property of QSPI-2.0 in vivo, the amounts of surviving bacteria at the wound site of different groups were evaluated on Day 3 after treatment. As shown in Figure 5d, a few colony units were observed in QSPI-2.0 and Alginate Ag groups, while the Tegaderm^{TM} film group exhibited poor bacterial property with lots of colony units. The QSPI-2.0 group and the Alginate Ag group could efficiently eliminate bacteria in vivo and displayed only 3.42 ± 0.44% and 2.64 ± 0.55% bacterial viabilities compared with Tegaderm^{TM} film (Figure 5e). These results indicated that the bacteria infection in the wound tissue could be effectively inhibited by QSPI-2.0 due to its excellent antibacterial property.

The wound healing effects were further analyzed by histological analyses. The H&E and Masson's trichrome staining of the wound tissue on Day 10 and Day 20 were performed. As shown in Figure 5f, less inflammatory cells in the QSPI-2.0 group and Alginate Ag group were observed compared with the Tegaderm^{TM} film group on Day 10, which was probably due to QSPI-2.0 and Alginate Ag inhibiting bacterial infection-induced inflammation [25]. Granulation tissue formation is a critical stage during the wound healing process, where the injury is filled with a matrix of fibrous connective tissue and blood vessels. Thicker granulation tissue may create a better framework for the other cells to grow, fill in the wound and restore skin function. Thus, the granulation tissue was observed by H&E staining, and its thickness was calculated on Day 10 (Figure 5h). The thickness of the granulation tissue of the QSPI-2.0 group was significantly higher than those of the other groups, suggesting a better wound regeneration environment of the QSPI-2.0 group. The epidermal thickness is a key index to evaluate the wound healing effects at the end stage of wound repairing [26]. Thus, the epidermal thickness was observed by H&E staining on Day 20 after treatment (Figure 5g). The thickest epithelium (66.49 ± 4.26 μm) of the QSPI-2.0 group among the three groups, indicating the best re-epithelialization effect (Figure 5i). In addition, the regenerated tissue of the QSPI-2.0 group exhibited better organized structures including more hair follicles and blood vessels than those in the Alginate Ag group and Tegaderm^{TM} film group.

Moreover, collagen deposition emerges during the ECM remodeling phase of the wound healing process, which is crucial for improving ECM formation [27]. Masson trichrome staining was used to evaluate the collagen levels in the newly formed connective tissue at wound sites. The images of Masson trichrome staining (blue area representing collagen fiber) on Day 20 were shown in Figure 5j. More dense and oriented collagen fibers were observed in the QSPI-2.0 group compared with the Tegaderm^{TM} film group. Quantitative analysis demonstrated that the QSPI-2.0 group had the highest collagen index (relative intensity of collagen deposition) among the three groups, indicating that the deposition of ECM was remarkably increased in QSPI-2.0 treated wounds (Figure 5k). All the data demonstrated that QSPI-2.0 had a superior capacity of promoting infected wound healing.

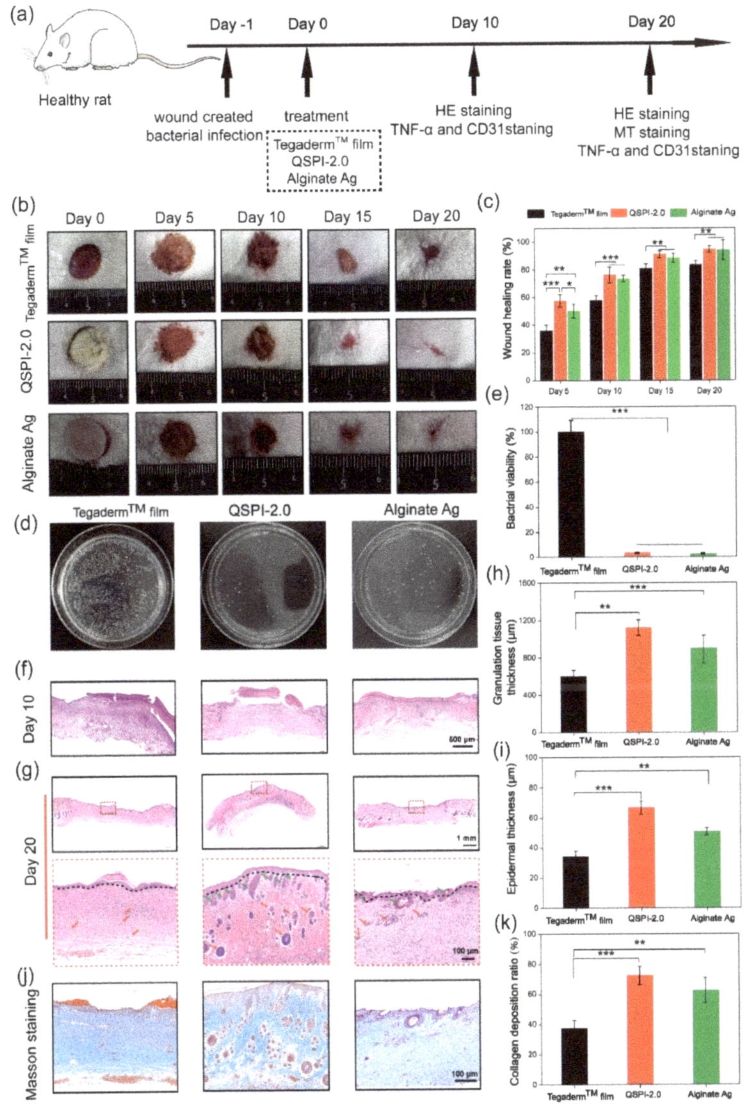

Figure 5. Effects of the QSPI-2.0 in a rat infected wound healing model. (**a**) The schematic diagram illustrates the establishment and treatment of the bacteria-infected wound. (**b**) Representative images of the bacteria-infected wound healing. (**c**) Quantification of wound healing rate. (**d**) Photographs of CFU test of bacteria derived from wound sites on Day 3 after different treatments. (**e**) Bacterial viability was calculated according to the CFU test. Images of histological analysis for wound regeneration on Day 10 (**f**) and Day 20 (**g**). Black lines point to the boundary of epidermis and dermis; green arrows point to the hair follicles and red arrows point to blood vessels. Statistical results of (**h**) granulation tissue thickness of the regenerating skin on Day 10 and (**i**) epidermal thickness of the regenerated skin tissue on Day 20. (**j**) Masson trichrome staining images of the regenerated skin tissue. (**k**) Quantification of the collagen deposition ratio in each group on Day 20. * $p < 0.05$; ** $p < 0.01$ and *** $p < 0.001$.

The continuous bacterial infection-induced prolonged inflammatory phase is the greatest obstacle to wound healing in the early healing stage [28]. Therefore, the expression of the typical inflammatory factor tumor necrosis factor (TNF-α) in the wound site was assayed by immunohistochemistry on Day 10 and 20 (Figure 6a,b) and quantified in Figure 6c. Although the ratios of TNF-α positive cells decreased over time in all three groups, the Tegaderm^{TM} film group showed a severe inflammatory response with the highest expression levels of TNF-α. In contrast, the QSPI-2.0 and Alginate Ag groups exhibited lower levels of TNF-α expressions on Day 10 and Day 20, suggesting less inflammation or infection in the wound sites. These data are consistent with the results that QSPI-2.0 had antibacterial properties, inhibited bacterial infection in the wound sites, and less inflammation was observed in the H&E staining of regenerating skin tissues.

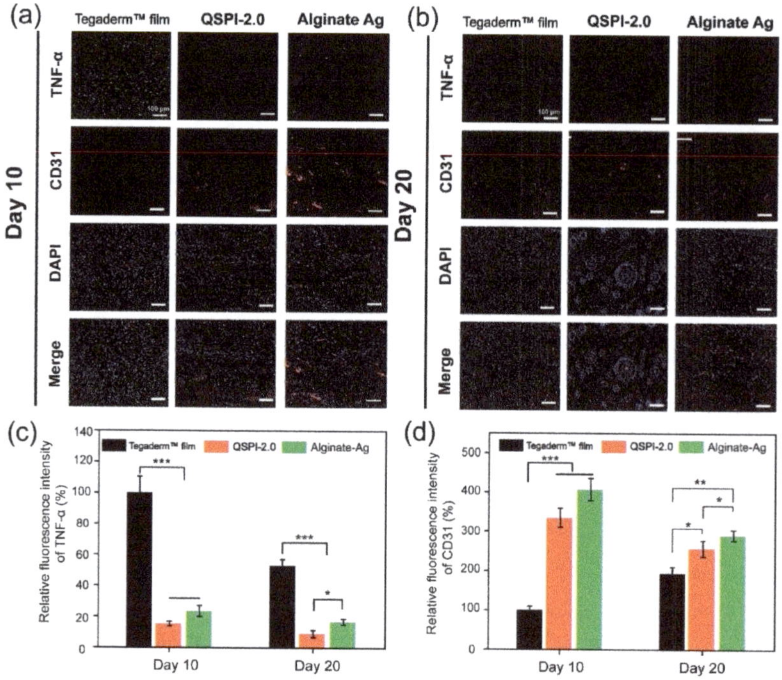

Figure 6. Immunofluorescence images of the regenerated wound tissue. Immunofluorescence staining against TNF-α (green) and CD31 (red) on Day 10 (**a**) and Day 20 (**b**); Statistical analysis of TNF-α (**c**) and CD31 (**d**) relative fluorescence intensity. * $p < 0.05$; ** $p < 0.01$ and *** $p < 0.001$.

On the other hand, neovascularization supports cell migration and proliferation in the newly forming tissue, which is essential for re-epithelialization and chronic wound healing [29–31]. Therefore, the formation of blood vessels in the wound site was detected using the immunofluorescent staining of endothelial cell specific protein marker, CD31. Red fluorescence was hardly detectable in the Tegaderm^{TM} film group. However, QSPI-2.0 and Alginate Ag groups exhibited high levels of CD31 expression. The neovascularization in the QSPI-2.0 group appeared to have grossly more CD31 expression than the Tegaderm^{TM} film on day 10 (Figure 6d). Besides, the CD31 expression level was lowest in the Tegaderm^{TM} film group and there was a significant difference between the QSPI-2.0 and Alginate Ag groups on Day 20. Additionally, more mature round capillaries were found in the QSPI-2.0 group, indicating the promoting angiogenesis effects of QSPI-2.0. In summary, QSPI-2.0 effectively accelerates wound healing by simultaneously reducing bacterial infection-induced inflammation and promoting angiogenesis.

3. Materials and Methods

3.1. Materials

Soy protein isolate (Mw: 205 KDa, protein content: 95%) was purchased from DuPont Protein Technology (Luohe, China), and vacuum-dried for 24 h before use. 2,3-epoxypropyl-trimethylammonium chloride (EPTMAC), sodium hydroxide, acetic acid, and ampicillin (Amp) were supplied by Aladdin (Shanghai, China). Ethanol and o-phthalaldehyde were obtained from Guoyao Chemical Company (Shanghai, China).

3.2. Synthesis of Quaternized Soy Protein Isolate

The synthesis method of quaternized soy protein isolate (QSPI) is as follows: 6 g SPI was suspended in 100 mL 0.125 M NaOH aqueous solution, followed by stirring for approximately 30 min at 25 °C to obtain a homogeneous solution. A specified amount of EPTMAC (3, 6, 9, 12, 15 g) dissolved in 10 mL deionized water was added slowly into the SPI solution, and the mixture reacted at 25 °C for 24 h. Next, the supernatant was collected by centrifugation at 5000 rpm for 20 min, neutralized to neutral with acetic acid, precipitated in 500 mL ethanol, and then filtered. The filter cake was dried at 50 °C for 24 h to obtain quaternized soy protein isolate (QSPI). According to the amount of EPTMAC participated in the reaction, a series of quaternized soy protein isolates were synthesized and coded as QSPI-n (n = 0.5, 1.0, 1.5, 2.0, 2.5), where "n" represents the ratio (*w*/*w*) of EPTMAC to SPI.

3.3. Chemical and Physical Characterization of QSPI

The degree of quaternization was determined using the OPA (o-phthalaldehyde) method [32]. The OPA reagent was freshly prepared by adding OPA (40 mg), methanol (1 mL), β-mercaptoethanol (100 μL), and 20% wt of SDS (2.5 mL) into 0.1 M sodium borate buffer (25 mL, pH 9.75). The mixture was diluted to 50 mL. Subsequently, 4 mL OPA reagent was reacted with 200 μL 0.4% QSPI solution at 37 °C for 2 min. The absorption at 336 nm was measured by a UV spectrophotometer (Shimadzu, Kyoto, Japan).

The degree of quaternization was calculated according to the equation:

$$\text{Degree of quaternization (\%)} = [(A_0 - A_s)/A_0] \times 100\% \tag{1}$$

where: A_0 refers to the absorbance of the SPI, As refers to the absorbance of the QSPI.

The ^{1}HNMR spectra of QSPI and SPI were recorded using a Bruker Avance 500 MHz. The Fourier transform infrared (FTIR) spectra were analyzed from 4000 to 500 cm^{-1} (PerkinElmer System 2000 spectrometer, PerkinElmer, Waltham, MA, USA).

The isoelectric point was measured by Nano-ZS ZEN3600 (Malvern Instruments, Malvern, UK). After the QSPI dissolved in deionized water at a 1 mg/mL concentration, the pH of the solutions was adjusted from 1, 2, 3, to 14, and the zeta potential was measured by Nano-ZS ZEN3600. When the zeta potential is 0, the corresponding pH is the isoelectric point.

After the QSPI was dissolved in deionized water at pH = 2, 7, and 13 with a concentration of 10 mg/mL, the water solubility of the QSPI was measured by the UV spectrophotometer at the absorbance of 280 nm [33]. The SPI could be dissolved thoroughly when the pH ≥ 11 and the absorption of QSPI in deionized water with pH = 13 was used as control [34].

The water solubility was calculated according to the equation:

$$\text{Water solubility (\%)} = (I_s/I_{13}) \times 100\% \tag{2}$$

where: I_{13} refers to the absorbance of the SPI in the solution of pH = 13, Is refers to the absorbance of the QSPI or SPI.

3.4. Antibacterial Property Evaluation of QSPI

The antibacterial property of QSPI against Gram-negative bacteria Escherichia coli (*E. coli*), Gram-positive bacteria Staphylococcus aureus (*S. aureus*) and Methicillin-resistant Staphylococcus aureus (MRSA) was assessed by the proliferation assay and colony forming units (CFU) testing [35,36]. For the proliferation assay, 10 mL of sterilized QSPI solution (PBS as solvent) with increasing concentrations (0, 0.5, 1.0, 1.5, 2.0, 2.5 mg/mL) was added to centrifugation tubes, followed by the addition of 100 μL of bacterial suspension (10^6 CFU/mL). Bacterial suspension with antibiotic (ampicillin) was used as the positive control, and bacterial suspension with 0 mg/mL of QSPI was used as the negative control. Then the centrifugation tubes were incubated at 37 °C for 12 h. 100 μL mixture was added into a 5 mL LB medium and incubated at 37 °C. Then, 200 μL LB medium was taken from each group at each time interval, and its OD value was measured at the wavelength of 600 nm via a microplate reader (Spectra Max M2). The suspension turbidity at 600 nm is proportional to the number of bacteria. For the CFU test, after the bacteria co-cultured with QSPI, the mixture was step wisely diluted with PBS, and then cultured on agar plates. After incubation at 37 °C for 12 h, the bacterial colonies were photographed and counted. Each experiment was repeated three times. The inhibition rate was calculated with the following equation:

$$\text{Inhibition rate } (\%) = (C_N - C_S)/(C_N - C_p) \times 100\% \quad (3)$$

where C_N, C_p, and C_S stand for the average colony counts of the negative control, positive control, and QSPI groups, respectively.

3.5. Cytocompatibility and Hemocompatibility of the QSPI

Fibroblast cell line (L929) was used to evaluate the cytocompatibility of QSPI by 3- (4,5-dimethyl-2-thiazolyl) -2,5-diphenyl-2-H-tetrazolium bromide (MTT) assay and live/dead assay. L929 cells were seeded into a 96-well plate with 2×10^3 cells/well for 24 h at 37 °C. Then, the cell culture medium was updated with 200 μL serial dilutions of QSPI (0.5, 1.0, 1.5, 2.0, 2.5 mg/mL) in the culture medium for another 48 h and without QSPI as control. Then, the culture medium was replaced by 20 μL MTT solution and incubated for 4 h. Finally, the MTT solution was abandoned and 200 μL of DMSO was added to each well to dissolve the formazan crystals. The absorbance at 490 nm was measured via a microplate reader (SpectraMax M2, Molecular Devices, San Jose, CA, USA). The cell viability of L929 cells was analyzed using the following equation:

$$\text{Cell viability } (\%) = (A_S - A_B)/(A_N - A_B) \times 100\% \quad (4)$$

where A_S, A_N, and A_B represent the absorbance of the sample, negative control, and blank control at a wavelength of 490 nm, respectively.

Live/dead staining was conducted with L929 cells seeded in 24-well plate with 2×10^4 per well. After cocultured with 1 mL serial dilutions of QSPI solution (0.5, 1.0, 1.5, 2.0, 2.5 mg/mL) at 37 °C for 48 h, the L929 cells were stained with 500 μL of calcein-AM/propidium iodide dye at room temperature for 30 min. The cells were observed via a fluorescent microscope for the green (492 nm) and red (545 nm) fluorescence.

The hemolysis of QSPI was analyzed by incubating QSPI with rabbit's red blood cells. The red blood cells were obtained via centrifugation of whole blood at 3500 rpm for 10 min and diluted with PBS to 5% (*v*/*v*). The 0.5 mL red blood cells suspension was added into 0.5 mL of gradient concentrations of QSPI solution (0.5, 1.0, 1.5, 2.0 mg/mL) and incubated at 37 °C for 1 h. The red blood cells suspension incubated with PBS was used as the negative control, and 0.1% Triton X-100 solution was used as the positive control. The mixture was further centrifuged at 3500 rpm for 5 min, and 200 μL supernatant was transferred to a

96-well plate. The absorbance was measured at 540 nm via a UV spectrometer (SpectraMax M2). The hemolysis rate of QSPI is estimated by the following equation:

$$\text{Hemolysis ratio (\%)} = (A_S - A_N)/(A_P - A_N) \times 100\% \quad (5)$$

where A_S, A_N, and A_P represent the absorbance of QSPI and the negative and positive controls, respectively.

3.6. In Vivo Bacterial-Infected Wound Healing Assay

All the rats used in the experiments were supplied by ABSL-III Center for Animal Experiment at Wuhan University and handled following Wuhan University Animal Care Committee. The bacterial-infected wound healing assay of QSPI-2.0 was analyzed via a rat MRSA-infected full-thickness skin defect model. Thirty SD rats (280–300 g, 7–8 weeks) were anesthetized by an intraperitoneal injection of 1% sodium pentobarbital and shaved on the dorsum. Then, two round full thickness wounds (diameter: 15 mm) were inflicted on the dorsum of each rat, and the 100 μL suspension containing MRSA (1×10^7 CFU mL^{-1}) was immediately dropped onto the wounds for 24 h to establish the MRSA-infected full thickness wound model. All these rats were randomly divided into three groups with ten rats in each group and treated with Tegaderm^{TM} film (a commercial film dressing) as the negative control, commercial antibacterial dressing Biatain Alginate Ag sponge (coded as "Alginate Ag" in this study) as the positive control, and QSPI-2.0, respectively. The wound sites were photographed on Day 0, 5, 10, 15, and 20. The wound area was analyzed via Image-J. The wound healing rate was calculated by the following formula:

$$\text{Wound healing rate (\%)} = (S_0 - S_t)/S_0 \times 100\% \quad (6)$$

where S_0 is the initial wound area, and S_t is the wound area at a specific time point.

To evaluate the antibacterial property in vivo, the infected wound tissues were collected after 3 days of treatment and homogenized with 5 mL PBS to separate the bacteria. Then, 200 μL of PBS containing bacteria was plated on the LB agar for 12 h. The viable bacteria were photographed and measured by counting the colony-forming units.

For histological analysis, the wound tissues were harvested on Day 10 and 20 after treatments, and fixed in 4% formaldehyde, embedded in paraffin, and sectioned into 3–4 μm thickness sections. Haematoxylin-Eosin (H&E) and Masson's staining were performed for histological analysis. For quantitative analysis of the histopathological changes, the thickness of granulation tissue at day 10, the thickness of the epidermis, and the ratio of the collagen-occupied area at day 20 were calculated based on the staining images by Image-J.

Inflammation on Day 10 and angiogenesis on Day 20 after wound treatment were, respectively, determined by TNF-α and CD31 immunofluorescence staining. The results were photographed by inverted fluorescence microscopy (DMi8, Leica, Germany), and the fluorescence intensity was determined with Image-J.

3.7. Statistical Analyses

All data were displayed as mean ± standard deviation. Statistical significance was determined by one-way ANOVA and Student's *t*-test. $p < 0.05$ was statistically significant (* $p < 0.05$; ** $p < 0.01$; *** $p < 0.001$).

4. Conclusions

In this work, quaternized soy protein isolate (QSPI) was successfully synthesized via quaternization of soy protein isolate (SPI). The QSPI not only has more improved water solubility than SPI, but also shows greatly enhanced antibacterial properties and potential ability for bacteria-infected wound healing. This soy protein isolate-based derivative may open a new avenue for the applications of soy protein in industry and tissue engineering. In the future, it is very necessary to balance the water solubility, antibacterial property, and biocompatibility of QSPI by adjusting the quaternization degree or by introducing other functional groups in the SPI chains.

Author Contributions: Q.D.: Conceptualization, Data curation, Writing—original draft. J.L.: Data curation. H.W.: Investigation, Methodology. M.K.: Visualization, Investigation. X.L.: Methodology. X.Y.: Methodology, Formal analysis. H.L.: Methodology, Conceptualization. C.H.: Supervision. Z.T.: Writing—reviewing and editing. Y.C.: Methodology, Conceptualization, Supervision, writing—reviewing and editing, Funding acquisition. All authors have read and agreed to the published version of the manuscript.

Funding: This work was funded by the Medical Science Advancement Program (Clinical Medicine) of Wuhan University (Grant No. TFLC2018003) and the Seed Fund Program for Sino-Foreign Joint Scientific Research Platform of Wuhan University (Grant No. WHUZZJJ202226).

Institutional Review Board Statement: All the animals were purchased from ABSL-III Center for Animal Experiment at Wuhan University and handled following Wuhan University Animal Care Committee. The project identification approval code was "SO1319011C".

Informed Consent Statement: Not applicable.

Data Availability Statement: Not applicable.

Acknowledgments: We thank the support of Research Center for Medicine and Structural Biology, TaiKang Medical School (School of Basic Medicine Sciences) for excellent technical assistance.

Conflicts of Interest: The authors declare no conflict of interest.

References

1. Percival, N.J. Classification of Wounds and their Management. *Surgery* **2002**, *20*, 114–117. [CrossRef]
2. Rani, S.; Kumar, R. A Review on Material and Antimicrobial Properties of Soy Protein Isolate Film. *J. Polym. Environ.* **2019**, *27*, 1613–1628. [CrossRef]
3. Cai, L.; Li, Y.; Lin, X.; Chen, H.; Gao, Q.; Li, J. High-performance adhesives formulated from soy protein isolate and bio-based material hybrid for plywood production. *J. Clean. Prod.* **2022**, *353*, 131587. [CrossRef]
4. Luo, L.; Gan, L.; Liu, Y.; Tian, W.; Tong, Z.; Wang, X.; Huselstein, C.; Chen, Y. Construction of nerve guide conduits from cellulose/soy protein composite membranes combined with Schwann cells and pyrroloquinoline quinone for the repair of peripheral nerve defect. *Biochem. Biophys. Res. Commun.* **2015**, *457*, 507–513. [CrossRef] [PubMed]
5. Saravanakumar, K.; Sathiyaseelan, A.; Mariadoss, A.V.A.; Hu, X.W.; Wang, M.H. Physical and bioactivities of biopolymeric films incorporated with cellulose, sodium alginate and copper oxide nanoparticles for food packaging application. *Int. J. Biol. Macromol.* **2020**, *153*, 207–214. [CrossRef] [PubMed]
6. Sarengaowa; Hu, W.Z.; Feng, K.; Jiang, A.L.; Xiu, Z.L.; Lao, Y.; Li, Y.Z.; Long, Y. An in situ-Synthesized Gene Chip for the Detection of Food-Borne Pathogens on Fresh-Cut Cantaloupe and Lettuce. *Front. Microbiol.* **2020**, *10*, 3089. [CrossRef] [PubMed]
7. Ma, C.; Pang, H.W.; Shen, Y.L.; Liang, Z.X.; Li, J.Z.; Zhang, S.F.; Shi, J.Y. Plant Polyphenol-Inspired Crosslinking Strategy toward High Bonding Strength and Mildew Resistance for Soy Protein Adhesives. *Macromol. Mater. Eng.* **2021**, *306*, 2100543. [CrossRef]
8. Tao, R.; Sedman, J.; Ismail, A. Characterization and in vitro antimicrobial study of soy protein isolate films incorporating carvacrol. *Food Hydrocoll.* **2022**, *122*, 107091. [CrossRef]
9. Koshy, R.R.; Reghunadhan, A.; Mary, S.K.; Pothen, L.A.; Thomas, S. MgONP/Chitin nanowhisker-based hybrid filler bound soy protein thin films for bioactive packaging applications. *Int. J. Food Sci Technol.* **2022**, *10*, 15805. [CrossRef]
10. Xu, Y.T.; Han, Y.F.; Li, Y.; Li, J.C.; Li, J.Z.; Gao, Q. Preparation of a strong, mildew-resistant, and flame-retardant biomimetic multifunctional soy protein adhesive via the construction of an organic-inorganic hybrid multiple-bonding structure. *Chem. Eng. J.* **2022**, *437*, 135437. [CrossRef]
11. Tansaz, S.; Boccaccini, A.R. Biomedical applications of soy protein: A brief overview. *J. Biomed. Mater. Res. A* **2016**, *104*, 553–569. [CrossRef] [PubMed]
12. Tian, H.; Guo, G.; Fu, X.; Yao, Y.; Yuan, L.; Xiang, A. Fabrication, properties and applications of soy-protein-based materials: A review. *Int. J. Biol. Macromol.* **2018**, *120*, 475–490. [CrossRef] [PubMed]
13. Chen, S.; Wang, X.; Xu, Y.; Zhang, X.; Wang, X.; Jiang, L. Effect of High Pressure Treatment on Interfacial Properties, Structure and Oxidative Stability of Soy Protein Isolate-Stabilized Emulsions. *J. Oleo Sci.* **2019**, *68*, 409–418. [CrossRef]
14. Gharibzahedi, S.M.T.; Smith, B. The functional modification of legume proteins by ultrasonication: A review. *Trends Food Sci. Technol.* **2020**, *98*, 107–116. [CrossRef]
15. Mao, Y.; McClements, D.J. Modification of emulsion properties by heteroaggregation of oppositely charged starch-coated and protein-coated fat droplets. *Food Hydrocoll.* **2013**, *33*, 320–326. [CrossRef]

16. Pirestani, S.; Nasirpour, A.; Keramat, J.; Desobry, S. Preparation of chemically modified canola protein isolate with gum Arabic by means of Maillard reaction under *wet-heating* conditions. *Carbohyd. Polym.* **2017**, *155*, 201–207. [CrossRef] [PubMed]
17. Liu, Y.; Zhang, L.; Li, Y.; Yang, Y.; Yang, F.; Wang, S. The functional properties and structural characteristics of deamidated and succinylated wheat gluten. *Int. J. Biol. Macromol.* **2018**, *109*, 417–423. [CrossRef] [PubMed]
18. Xu, H.; Fang, Z.; Tian, W.; Wang, Y.; Ye, Q.; Zhang, L.; Cai, J. Green Fabrication of Amphiphilic Quaternized β-Chitin Derivatives with Excellent Biocompatibility and Antibacterial Activities for Wound Healing. *Adv. Mater.* **2018**, *30*, 1801100. [CrossRef] [PubMed]
19. Kong, J.; Yu, S. Fourier transform infrared spectroscopic analysis of protein secondary structures. *Acta Biochim. Biophys. Sin.* **2007**, *39*, 549–559. [CrossRef] [PubMed]
20. Ding, F.; Shi, X.; Li, X.; Cai, J.; Duan, B.; Du, Y. Homogeneous synthesis and characterization of quaternized chitin in NaOH/urea aqueous solution. *Carbohydr. Polym.* **2012**, *87*, 422–426. [CrossRef] [PubMed]
21. Jarmila, V.; Vavríková, E. Chitosan derivatives with antimicrobial, antitumour and antioxidant activities—A review. *Curr. Pharm. Des.* **2011**, *17*, 3596–3607. [CrossRef] [PubMed]
22. Xue, Y.; Xiao, H.; Zhang, Y. Antimicrobial polymeric materials with quaternary ammonium and phosphonium salts. *Int. J. Mol. Sci.* **2015**, *16*, 3626–3655. [CrossRef] [PubMed]
23. Joseph, R.; Naugolny, A.; Feldman, M.; Herzog, I.M.; Fridman, M.; Cohen, Y. Cationic Pillararenes Potently Inhibit Biofilm Formation without Affecting Bacterial Growth and Viability. *J. Am. Chem. Soc.* **2016**, *138*, 754–757. [CrossRef] [PubMed]
24. Liu, P.; Wang, Y.; An, L.; Tian, Q.; Lin, J.; Yang, S. Ultrasmall WO(3−x)@γ-poly-l-glutamic Acid Nanoparticles as a Photoacoustic Imaging and Effective Photothermal-Enhanced Chemodynamic Therapy Agent for Cancer. *ACS Appl. Mater. Interfaces* **2018**, *10*, 38833–38844. [CrossRef] [PubMed]
25. Min, J.G.; Sanchez Rangel, U.J.; Franklin, A.; Oda, H.; Wang, Z.; Chang, J.; Fox, P.M. Topical Antibiotic Elution in a Collagen-Rich Hydrogel Successfully Inhibits Bacterial Growth and Biofilm Formation In Vitro. *Antimicrob. Agents Chemother.* **2020**, *64*, e00136-20. [CrossRef]
26. Qu, J.; Zhao, X.; Liang, Y.; Xu, Y.; Ma, P.X.; Guo, B. Degradable conductive injectable hydrogels as novel antibacterial, anti-oxidant wound dressings for wound healing. *Chem. Eng. J.* **2019**, *362*, 548–560. [CrossRef]
27. Patel, S.; Srivastava, S.; Singh, M.R.; Singh, D. Mechanistic insight into diabetic wounds: Pathogenesis, molecular targets and treatment strategies to pace wound healing. *Biomed. Pharm.* **2019**, *112*, 108615. [CrossRef] [PubMed]
28. Liu, J.; Chen, Z.; Wang, J.; Li, R.; Li, T.; Chang, M.; Yan, F.; Wang, Y. Encapsulation of Curcumin Nanoparticles with MMP9-Responsive and Thermos-Sensitive Hydrogel Improves Diabetic Wound Healing. *ACS Appl. Mater. Interfaces* **2018**, *10*, 16315–16326. [CrossRef]
29. Huang, L.; Zhu, Z.; Wu, D.; Gan, W.; Zhu, S.; Li, W.; Tian, J.; Li, L.; Zhou, C.; Lu, L. Antibacterial poly(ethylene glycol) diacrylate/chitosan hydrogels enhance mechanical adhesiveness and promote skin regeneration. *Carbohydr. Polym.* **2019**, *225*, 115110. [CrossRef]
30. Song, A.; Rane, A.A.; Christman, K.L. Antibacterial and cell-adhesive polypeptide and poly(ethylene glycol) hydrogel as a potential scaffold for wound healing. *Acta Biomater.* **2012**, *8*, 41–50. [CrossRef]
31. Yin, G.; Wang, Z.; Wang, Z.; Wang, X. Topical application of quercetin improves wound healing in pressure ulcer lesions. *Exp. Derm.* **2018**, *27*, 779–786. [CrossRef]
32. Li, R.; Wang, X.; Liu, J.; Cui, Q.; Wang, X.; Chen, S.; Jiang, L. Relationship between Molecular Flexibility and Emulsifying Properties of Soy Protein Isolate-Glucose Conjugates. *J. Agric. Food Chem.* **2019**, *67*, 4089–4097. [CrossRef]
33. Chevalier, F.; Chobert, J.-M.; Popineau, Y.; Nicolas, M.; Haertlé, T. Improvement of functional properties of β-Lg glycated through the Maillard-reaction is related to the nature of the sugar. *Int. Dairy J.* **2001**, *11*, 145–152. [CrossRef]
34. He, W.; Xiao, N.; Zhao, Y.; Yao, Y.; Xu, M.; Du, H.; Wu, N.; Tu, Y. Effect of polysaccharides on the functional properties of egg white protein: A review. *J. Food Sci.* **2021**, *86*, 656–666. [CrossRef]
35. Singh, S.B.; Young, K.; Silver, L.L. What is an "ideal" antibiotic? Discovery challenges and path forward. *Biochem. Pharm.* **2017**, *133*, 63–73. [CrossRef] [PubMed]
36. Hu, W.; Wang, Z.; Zha, Y.; Gu, X.; You, W.; Xiao, Y.; Wang, X.; Zhang, S.; Wang, J. High Flexible and Broad Antibacterial Nanodressing Induces Complete Skin Repair with Angiogenic and Follicle Regeneration. *Adv. Healthc. Mater.* **2020**, *9*, 2000035. [CrossRef] [PubMed]

MDPI AG
Grosspeteranlage 5
4052 Basel
Switzerland
Tel.: +41 61 683 77 34

International Journal of Molecular Sciences Editorial Office
E-mail: ijms@mdpi.com
www.mdpi.com/journal/ijms